Applied Economics and Policy Studies

The Applied Economics and Policy Studies present latest theoretical and methodological discussions to bear on the scholarly works covering economic theories, econometric analyses, as well as multifaceted issues arising out of emerging concerns from different industries and debates surrounding latest policies. Situated at the forefront of the interdisciplinary fields of applied economics and policy studies, this book series seeks to bring together the scholarly insights centering on economic development, infrastructure development, macroeconomic policy, governance of welfare policy, policies and governance of emerging markets, and relevant subfields that trace to the discipline of applied economics, public policy, policy studies, and combined fields of the aforementioned. The book series of Applied Economics and Policy Studies is dedicated to the gathering of intellectual views by scholars and poli-cymakers. The publications included are relevant for scholars, policymakers, and students of economics, policy studies, and otherwise interdisciplinary programs.

More information about this series at http://www.springer.com/series/16776

Chunhui Yuan · Xiaolong Li ·
John Kent
Editors

Proceedings of the 4th International Conference on Economic Management and Green Development

Springer

Editors
Chunhui Yuan
School of Economics and Management
Beijing University of Posts
and Telecommunications
Beijing, China

Xiaolong Li
Department of Postal Management
Beijing University of Posts
and Telecommunications
Beijing, China

John Kent
Supply Chain Management
University of Arkansas
Fayetteville, AR, USA

ISSN 2731-4006 ISSN 2731-4014 (electronic)
Applied Economics and Policy Studies
ISBN 978-981-16-5358-2 ISBN 978-981-16-5359-9 (eBook)
https://doi.org/10.1007/978-981-16-5359-9

This Springer imprint is published by the registered company Springer Nature Singapore Pte Ltd.
The registered company address is: 152 Beach Road, #21-01/04 Gateway East, Singapore 189721, Singapore

Contents

Dissecting Corporate Innovation in China

Yixuan Zhang(✉)

Henley Business School, University of Reading, Whiteknights, Reading, England

Abstract. This research aimed to utilize Lasso method to dissect influence factors on corporate innovation based on data collected from Chinese listed companies. Four aspects of factors were considered which are board, firm, industry, and CEO characteristics. Test results showed that, compared with industry and board feature, firm and CEO characteristics were more likely to impact the innovation activities in companies.

Keywords: Corporate innovation · Variable selection · Machine learning · Lasso

1 Introduction

Corporate innovation was an important measurement for corporate profitability or other capacities in companies, especially for technological companies. In addition, corporate innovation contributed to improvement of national economy. Substantial academic interests have focused on technological advances from corporate innovation [1].

There were many factors which had impacts on corporate innovation. These factors could be categorized into characteristics at firm level, at management level, at corporate governance level and at industry level. It was known that corporate governance was impacted by these factors, but which one played the most significant role in corporate governance was a question which needed to be explored. When knowing this, management in corporates could make efficient decisions on improving corporate governance, and market practitioners could make better projections on developments of companies so that better investment decisions could be made. Many researches have investigated different factors impacting on corporate innovation, but rare studies investigated the method to compare these factors as which one makes a bigger difference on the improvement of corporate innovation. Although existing literatures have investigated the priorities of factors impacting corporate governance based on data set in western countries, the related researches in Chinese background is still a deficiency. Different circumstances led to different results for this research topic because of different cultural and regulatory backgrounds. Huse et al. [2] pointed out that the models applied in American corporates may not achieve similar results in corporates from other countries. Therefore, it was worthwhile to explore the corporate innovation in Chinese circumstance separately.

Reeb and Zhao [3] investigated the dissection of corporate innovation in American circumstances. However, corporate innovation in China has not been investigated yet, although substantial literatures have studied many factors which have impacts on corporate innovation. For instance, Xu et al. [4] proved the impact of state shares on corporate

C. Yuan et al. (Eds.): *Proceedings of the 4th International Conference on Economic Management and Green Development*, AEPS, pp. 1–8, 2021.
https://doi.org/10.1007/978-981-16-5359-9_1

innovation performance and strategy. Lin et al. [5] investigated the impact of managerial incentives and CEO characteristics on corporate innovation in China's private sector. Li et al. [6] studied the relationship between internal control and corporate innovation based on evidence from China. Therefore, it was necessary to research and compare these factors based on their impact on corporate innovation. Which one influenced the corporate innovation most significantly based on financial market in China?

The organization of this paper is as follows. Section 2 briefly reviews significant previous literatures. Section 3 introduces the methodologies applied in this study and Sect. 4 discusses the results. The conclusions will be achieved in Sect. 5.

2 Literature Review

There have been many literatures investigating substantial factors of corporate innovation. Hoskisson et al. [7] found that differences existed among governance constituencies' preferences for corporate innovation strategies. The managers of public pension funds preferred internal innovation, but professional investment funds' managers preferred acquiring external innovation. Amore et al. [8] investigated that interstate banking deregulation had significant beneficial effects on the quantity and quality of innovation activities, especially for firms highly dependent on external capital and located closer to entering banks. Chang et al. [9] verified the positive relationship between non-executive employee stock options and corporate innovation. The positive impact was strengthened when employees were more important for innovation, when free-riding among employees was weaker, when options are granted broadly to most employees, when the average expiration period of options was longer, and when employee stock ownership was lower. Huse et al. [2] found that environment and internationalization had positive relations with corporate innovation. However, models developed using U.S. firms may not be generalizable to firms from other countries. Although these literatures have uncovered underlying influence factors on corporate innovation, but they always focused on one or two factors. The main aim of these studies is to prove that there was indeed a connection between one factor and corporate innovation. In my study, a further test is applied to classify these influence factors according to their ability to have an impact on corporate innovation in Chinese corporates.

The sources of dependent variables and independent variables were mainly from Reeb and Zhao [3] and Feng and Wen [10]. Reeb and Zhao [3] compared factors impacting on corporate innovation in America. Feng and Wen [10] analysed the relationship between corporate governance and corporate innovation in listed Chinese companies. I collected variables which influenced or measured corporate innovation, and could be adopted in China to dissect innovation in Chinese companies.

3 Research Method and Data Description

The conventional method to select variables was based on stepwise regression. Compared to other subset regressions, this method was computationally tractable. The best-fitting model was selected among all possible combinations of variables based on certain statistical standard of model fitness. However, some drawbacks existed when stepwise regression was applied. Only a subset of the potential models was focused in conventional method among the possible combinations of factors as the result was contingent which depended on the sequence of the variables inputted in the regression. The sequence of the variables was particularly important when a variable was either dropped or added in a stepwise process. There was no clear and efficient solution to treat this problem. Additionally, all subset or stepwise penalized the model directly to include coefficients. Therefore, stepwise processes lacked validation of the results.

To mitigate the shortcoming of stepwise regression to some extent, machine leaning method was employed in this research. Machine learning methods have been applied by some researchers when researching corporate-related topics. For instance, Zhu et al. [11] predicted Chinese SME credit risk in supply chain finance utilizing machine learning methods. Heaton et al. [12] investigated portfolios with deep learning as the tool. Machine learning consisted of a variety of techniques so that patterns and relationships in the data could be identified. This method was commonly applied in forecasting and simplifying selection processes. In this research, one of machine learning methods, Lasso (Least Absolute Shrinkage Selection Operator), is applied to obtain the regression results. Machine learning included two processes which are training and validation subsamples. In cross-validation process, Lasso clearly took the 'predictive' power of selecting outcomes into account to have a robust and reliable inference. In addition, Lasso did not give a common penalty factor to all coefficients, but it provided a differential weight for penalizing different coefficients. Therefore, I used Lasso to obtain the improved coefficients of independent variables so that the weight factors could be analyzed for conclusions.

The lasso estimation is formulated as follows:

Lasso regression:

$$\frac{\min}{\beta_0, \beta_j} \sum_{i=1}^{n} \left(b_i - \beta_0 - \sum_{1}^{p} \beta_j a_{ij}\right)^2 + \lambda||\beta||_1 \quad (1)$$

Where b was the vector of observations of the dependent variable, a denoted the independent variables, β was the corresponding coefficient, $||_1$ 3was the L1 norm.

To capture the twenty-one identified determinants of corporate innovation, I obtained data mainly from CSMAR database. All A-share companies were considered and if they had all available variables needed by this research, they were included in the data set.

3.1 Dependent Variable

I used one commonly utilized measurement for corporate innovation output which was the number of patents. The patents were on the basis of the patent application and thus the application period of the patents was utilized instead of the year when the patents are granted. This was because the application year was closer to the time of actual innovation [13]. To be specific, I used the log of patents as merits of corporate innovation in this study.

3.2 Independent Variable

I included twenty potential determinants of corporate innovation from earlier literatures that I classified them into four types. In the first group, I included three variables of management characteristics: CEO age, CEO gender, CEO total compensation. The second category considered corporate characteristic: firm size, R&D expenditure, Tobin's q, stock liquidity, tangibility, return on asset, sales growth, organizational capital, manufacturing and capital structure. In the third group, I included corporate governance variables, including institutional ownership, block holder and analyst following. Finally, I included industry characteristics in the fourth group, namely the industry patent intensity, industry average R&D, industry competition, and industry size. I provided specific definitions of these variables in the table.

Table 1. Definition of variables.

CEO age	The log of CEO age
CEO gender	A dummy variable that equals 1 when CEO is male and equals 0 when CEO is female
CEO total compensation	The log of CEO annual total compensation
Firm size	The log of total assets
R&D expenditure	The cumulated R&D expenditure of three years with a 15% depreciation
Tobin's q	(The market value of equity + the book value of debt)/the book value of total assets
Stock liquidity	The log of stock-trade volume during the year
Tangibility	Net PPE/total assets
Return on assets	Income before extraordinary items/total assets
Sales growth	The average of three-year sales growth
Manufacturing	The dummy variable that equals 1 when the firm is in manufacturing industry, otherwise, equals 0
Organizational capital	The sum of selling, general and administrative expenses
Capital structure	Log-term assets/total assets
Analyst Following	The number of financial analysts in the corporate within the year
Institutional ownership	Institutional ownership of a common equity
Industry patent intensity	Total industry patents/total industry assets
Industry average R&D	The average R&D for each industry
Industry competition	The sum of square of the percentage of corporate income to industry income
Industry size	the log of total assets of firms in an industry

Table 2 showed the summary statistics of the sample firms with firm-year observations from 2002–2017. All variables were defined in Table 1.

Table 2. Summary statistic of variables.

	Mean	Median	Std. Deviation	Percentiles	
				25	75
If the firm is in manufacturing	0.28	0	0.58	0	0
Industry patent intensity	0.57	1	0.49	0	1
Industry average R&D	1.34E-09	0	3.88E-09	0	0
Industry competition	45308465.3	3477742.86	148995211	0	62347344.5
Industry size	7.13E + 16	1.15E + 14	8.83E + 17	0	5.51E + 15
Patents	10.84	10.94	0.67	10.44	11.32
CEO gender	0.83	1	0.38	1	1
CEO age	1.680	1.681	0.081	1.623	1.732
CEO compensation	4.018	4.875	2.186	4.301	5.465
Firm size	9.512	9.441	0.547	9.148	9.792
R&D expenditure	122510513	27665796.7	583824598	0	96612986.9
Tobin's q	0.65	0.42	12.63	0.25	0.60
Stock liquidity	9.09	9.12	0.51	8.76	9.42
Tangibility	0.20	0.17	0.16	0.08	0.29
Return on asset	0.35	0.25	0.65	0.08	0.46
Sales growth	36.97	0.17	1261.17	0.04	0.34
Organizational capital	2511730472	569655455	1.98E + 10	203148913	1523936238
Capital structure	0.10	0.03	1.11	0.01	0.11
Institutional ownership	6.46	4	7.11	1.07	9.48
Block holder	0.33	0	0.47	0	1
Analyst following	7.07	4	8.82	0	11

4 Result Description

According to the results from Lasso test, it could be seen that the most significant influence factors were manufacturing, industry patent intensity, industry average R&D, CEO gender, CEO age, CEO compensation, stock liquidity, return on asset and organizational capital. Other factors have weak or negative impact on corporate innovation. It can be seen that most of influence factors focused on characteristics at CEO level and at firm level. The patent was less influenced by industry features and was no relation with board characteristics. CEO made key strategies in companies and set the direction of the whole corporate, so CEO's characteristics were very likely to make a difference on a variety operating activities, including research and development activities. In addition, capital-related indicators, including stock liquidity, return on asset and organizational capital could also imply the ability of a corporate to innovate to some degree. Innovation activities needed capital to provide resources in exchange of raw materials and elites in science and technology field. Therefore, the stable provision of capital enhanced the achievement of corporate innovation. Finally, industry environment also influenced the capability of corporate innovation. Corporates always tended to compare with other companies in the same industry. At the same time, their goals could at least achieve the average level of the industry that they lay in. Therefore, the situation of this industry also had impact on the efforts that a corporate was willing to make, including investing on innovation activities. In particular, whether the company was a manufacturing company was an important indicator to project innovation capacity, because manufacturing company had more demands to produce patents. Their products and sales were based on these patents. However, the board characteristics had litter impact on corporate innovation in companies which were located and listed in China. Board of directors, especially the non-executive board of directors, were not involved in operating and investing decisions directly in most of cases. This explained their tiny influence on corporate innovation (Table 3).

Table 3. Coefficients of independent variables from Lasso method.

Independent variable	Coefficient
(Intercept)	−4.09E−01
If the firm is in manufacturing	7.00E−02
Industry patent intensity	4.20E+07
Industry average R&D	1.63E−09
Industry competition	−2.33E−19
Industry size	−9.61E−03
CEO gender	9.45E−03
CEO age	1.81E−01
CEO compensation	1.90E−04
Firm size	–

(*continued*)

Table 3. (*continued*)

Independent variable	Coefficient
R&D expenditure	–
Tobin's q	–
Stock liquidity	3.26E−02
Tangibility	–
Return on asset	9.37E−03
Sales growth	−2.75E−06
Organizational capital	1.86E−12
Capital structure	−3.18E−03
Institutional ownership	–
Block holder	–
Analyst following	−6.69E−04

5 Conclusion

This research was to compare the influence factors of corporate innovation based on data from Chinese listed companies. The data were mainly from CSMAR database from 2002–2017. The dependent variable was patent. The independent variables were divided into four groups which were at firm level, at board level, at industry level and at CEO level. The Lasso results illustrated that most of influence factors focused on characteristics at CEO level and at firm level. The patent was less influenced by industry features and was no relation with board characteristics.

In conclusion, market practitioners could use the results from this research to project the innovation performance of Chinese companies better, for instance, paying more attention on CEO and firm characteristics.

References

1. Kogan, L., Papanikolaou, D., Seru, A., Stoffman, N.: Technological innovation, resource allocation, and growth. Quart. J. Econ. **132**, 665–712 (2017)
2. Huse, M., Neubaum, D.O., Gabrielsson, J.: Corporate innovation and competitive environment. Int. Entrepreneurship Manage. J. **1**, 313–333 (2005)
3. Reeb, D.M., Zhao, W.: Dissecting innovation. Working paper. Available at SSRN No. 3175055 (2018)
4. Xu, E., Zhang, H.: The impact of state share on corporate innovation strategy and performance in China. Asia Pacific J. Manage. Springer **25**, 473–487 (2008)
5. Lin, C., Lin, P., Song, F., Li, C.: Managerial incentives, CEO characteristics and corporate innovation in China's private sector. J. Comp. Econ. **39**, 176–190 (2011)
6. Li, P., Shu, W., Tang, Q., Zheng, Y.: Internal control and corporate innovation: evidence from China. Asia-Pacific J. Account. Econ. **26**, 622–642 (2017)

7. Hoskisson, R.E., Hitt, M.A., Johnson, R.A., Grossman, W.: Acad. Manag. J. **45**, 697–716 (2017)
8. Amore, M., Schneider, C., Zaldokas, A.: Credit supply and corporate innovation. J. Financ. Econ. **109**, 835–855 (2013)
9. Chang, X., Fu, K., Zhang, W.: Non-executive employee stock options and corporate innovation. J. Financ. Econ. **115**, 168–188 (2015)
10. Feng, F., Wen, J.: An empirical analysis of the relationship between governance of listed companies and technological innovation in China. China Ind. Econ. **7**, 91–101 (2008)
11. Zhu, Y., Xie, C., Wang, G.-J., Yan, X.-G.: Comparison of individual, ensemble and integrated ensemble machine learning methods to predict China's SME credit risk in supply chain finance. Neural Comput. Appl. **28**(1), 41–50 (2016). https://doi.org/10.1007/s00521-016-2304-x
12. Heaton, J.B., Polson, N.G., Witte, J.H.: Appl. Stoch. Model. Bus. Ind. **33**, 3–12 (2017)
13. Griliches, Z., Pakes, A., Hall, B.H.: The value of patents as indicator of inventive activity. NBER Working Paper No. 2083 (1988)

The Review and Theoretical Analysis of ERM Models: A Dynamic Capabilities Perspective

Wenli Li(✉)

School of Professional Study, Columbia University, 2960 Broadway, New York, U.S.
wl2740@columbia.edu

Abstract. Nowadays, enterprise risk management (ERM) has recently become an increasingly concerned topic in academic circles. This study intends to firstly review three ERM models: the resource-based view ERM, the "spring" model and the value-based ERM framework. These three models all incorporate dynamic capability view into enterprise risk management. Then the characteristics of these models are theoretically analyzed and compared, and the joint use of the "spring" model and the value-based ERM framework is proposed. Finally, this paper points out the potential future development of ERM.

Keywords: Enterprise risk management · Dynamic capability · Review

1 Introduction

As defined by the Treadway Commission in 2004, enterprise risk management (ERM) is a procedural suite used by an entity's board of directors, management and other correlate personnel in the development of corporate strategy and business tactics. It aims to analyze potential incidents that may influence the entity and to control risks to a tolerable level for the company, thereby ensuring that the entity can achieve its stated objectives [1]. ERM takes a holistic view of the entire enterprise, and risk management has become an essential part of corporate governance [2]. However, when looking at specific cases of implemented ERM, ERM is often limited to the area of high-level governance [3], ERM is frequently seen as a compliance exercise [4, 5] or a "after-the-fact inspection" activity [6].

The resource-based view is one of the best-known theories in the field of strategic management [7, 8]. The theory argues that there is heterogeneity in resource endowments among firms and the firm's (sustained) competitive advantage stems from the unique resources it possessed. The resource-based view regards that a company can only attain sustainable competitive advantage if it has valuable (V), rare (R), irreplaceable (I) and non-substitutable (N) resources (known as VRIN criteria). These criteria are the fundamental components of holistic ERM. ERM's responsibility instead of managing risks independently by itself, manage all risks under control of a coordinated strategic framework [9]. Transferring the concept of strategic management (e.g. a resource-based

C. Yuan et al. (Eds.): *Proceedings of the 4th International Conference on Economic Management and Green Development*, AEPS, pp. 9–17, 2021.
https://doi.org/10.1007/978-981-16-5359-9_2

view to the field of risk management) may lead academics to build a coherent overarching theory applicable to different areas of the enterprise (strategic) risk management under different (static or dynamic) conditions.

The essential limitation of the resource-based view is its inability to explain well the competitive advantage of firms in a dynamic environment, and to address this shortcoming, it led to the emergence of the dynamic capability theory. The dynamic capability theory is a further development and refinement of the resource-based view [10] and the purpose is to explain how corporations accommodate rapid change in its external environment [11]. Dynamic capability is the ability of an organization to purposefully create, extend, or modify its resource base [12], which can be understood as the ability to change capabilities.

Enterprise risk management is a dynamic process. New emerging risks could from variable aspects, a technological change, newcomer on the market, as well as a political or economic shift. Logically, enterprise-level risk management capabilities are part of dynamic capability. There are designs of ERM models that tend to address company dynamic capabilities, support dynamic strategic planning, and ultimately achieve sustainable competitive advantages. Some of them are pure qualitative methods while some are more mature and provide a set of guidelines.

Based on this, this paper reviews three ERM models that incorporate the idea of dynamic capabilities, including the resource-based view ERM, the "spring" model and the value-based ERM framework, then analyzes the characteristics of each of these models, and finally suggests potential future directions for enterprise risk management.

2 The Review of Three ERM Models

2.1 Resource-Based View ERM

The resource-based view ERM believes that the concept of risk management capability is described as part of the concept of dynamic capability, which is broader in scope. The risk management capabilities mainly relate to the risk management process at the strategic level (assessing the risks associated with VRIN) and at the operational level (the risk management process). Risk resilience at the strategic level is about developing appropriate routines and competencies, while the operational level is about providing simple rules and heuristics.

Firstly, some non-systematic force majeure incidents do not permit responses regularly. Secondly, capabilities and routines remain limited by "blind spots" and fail to recognize all systemic risks [13]. Moreover, a no-routinized operational level is the characteristic of the dynamic capability [14]. Winter (2003) points out that there are various ways to change and puts forward the "fire-fighting model" or "ad-hoc problem-solving" [15]. In this mode, companies need to react without a specific mode of action. Therefore, there is a need to find a non-routinized solution to operational risk management.

For strategic risks, an enterprise should develop a set of routines and capabilities that are prioritized due to VRIN-criteria. Conversely, at the operational level, an enterprise requires developing a simple set of rules and a special way of problem-solving process (Fig. 1).

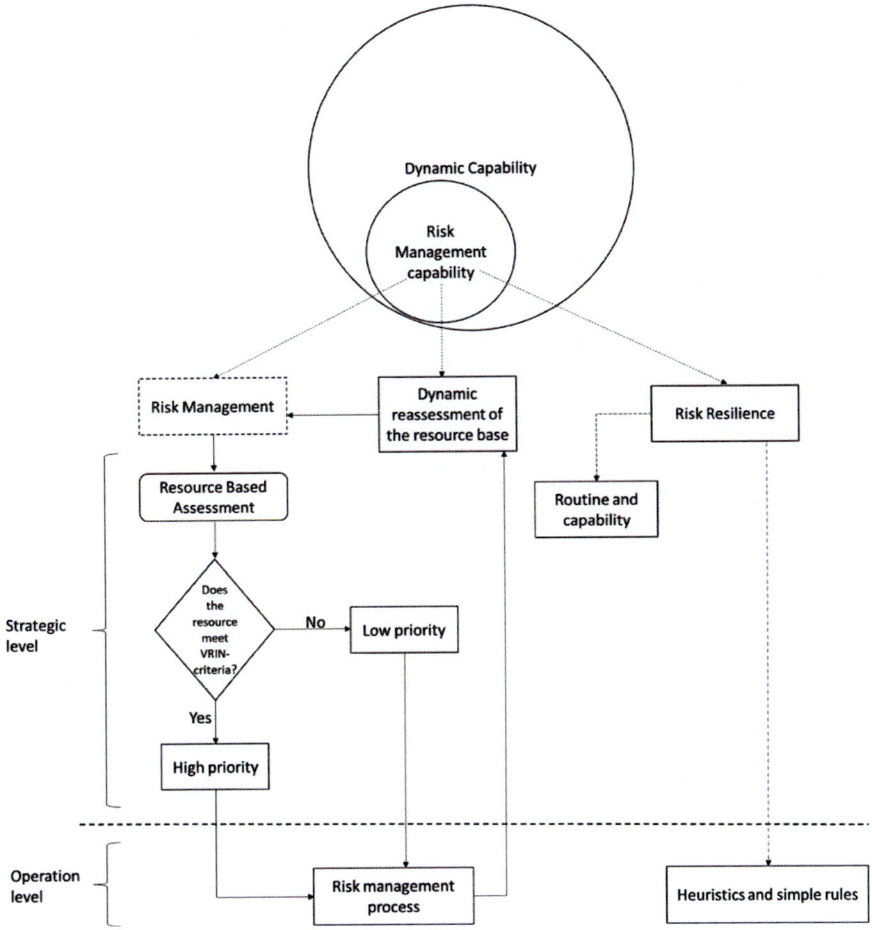

Fig. 1. Outlined framework of the resource-based ERM [16].

Carbone and Tippett (2004) indicated that company use four risk management techniques (avoid, mitigate, transfer, and accept) to handle risk at an operational level according to the priority [17]. The company should handle the high priority first then the low one. Avoidance is the first choice for high priority risk and if that is not applicable, use mitigate or transfer. Acceptance is the last option. For low priority risk, all four techniques could be chosen without any preference.

2.2 The "Spring" Model

The model is so named because capabilities effectively allow organizations to achieve their goals and ensure their resilience while withstanding internal and external departmental pressures [18].

As illustrated in Fig. 2, there are five main elements to the "spring" model: the initial situation, the objectives, the strategy, events, and capabilities. Strategy refers to the combination of the goals that a company strives to achieve and the methods (policies) to achieve them [19]. Events come from internal or external may have impacts on a system, which could bring positive or negative value change. Capabilities can be used to reduce or exploit strategic bias and represent the abilities to leverage organizational resources and organizational and resources to achieve desired results.

When an event causes a deviation from the company's objective, capabilities can be used to control the impact, just as the contraction and release of a spring can be used to visually represent the absorption (threat) or exploitation (opportunity) of an objective under the influence of forces (events) [19].

The "Spring Model" is specifically designed for project-based organizations, where risk management is simultaneously carried out at different organizational levels such as enterprise, portfolio, functional and project [19]. Capabilities are not the same at different levels. The general company's resilience can be divided into structural resilience and organizational resilience. Structural resilience represents the intrinsic ability of each organizational level to deal with external change through a set of capabilities at the current level. Organizational resilience, on the other hand, refers to the ability to take rapid and coordinated action involving the entire resource chain, reflecting the typical characteristics of the organization as a whole, rather than depending on the capabilities available at individual levels.

Implementing the "Spring" model contains two phases. First, identifying the different components of the model and getting a radical picture of all risk management capabilities at different levels within the organization. Second, analyzing risk propagation and resilience mechanisms.

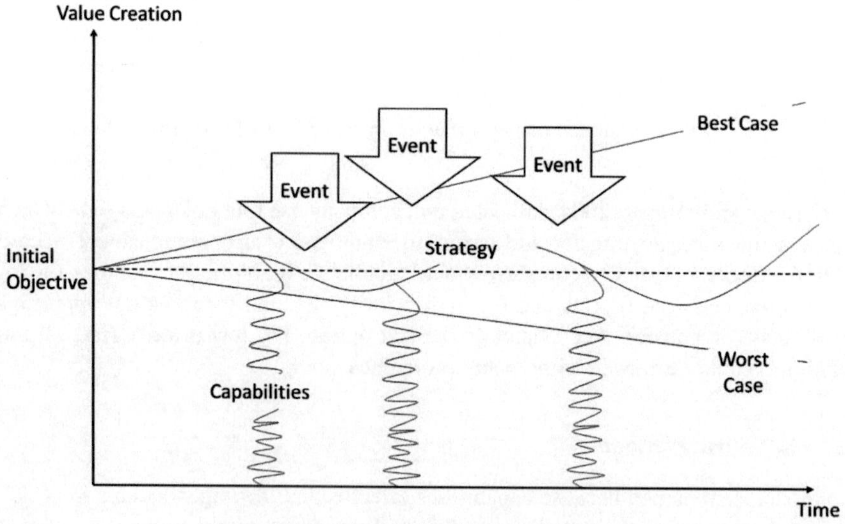

Fig. 2. The "Spring" model for ERM [20].

In the first phase, mapping all relevant internal processes and starting with process analysis, reviewing the company's formal organizational procedures and documents, and interviewing key individuals at each organizational level, thereby identifying currently available capabilities. Based on Teece et al. (1997) [12] and Collis (1994) [20], delivery, integration, and coordination learning and reconfiguration form the four types of capability of the "Spring" model. The identified risk management capability needs to be cross-checked against the planned or implemented risk mitigation measures as documented in the company's risk register and validated by the company managers.

The second phase is to dissect some of the risk mitigation measures and corresponding consequences at different organizational levels through three risk communication/control mechanisms (as shown in Table 1).

Table 1. Three risks propagation/control mechanisms.

Description	(Potential) Impact	Control
Objective variation at project level	Can cause an occurrence affecting another organizational level	Utilization of organizational resilience
Capability unavailability or saturation at portfolio level	Can produce to the other level on an event	Calls for other capabilities (organizational resilience), seeking for the best options in terms of the weigh up between target change and effect
Risk influence on all organizational levels globally	Changes in goals or depletion of several capabilities affect different levels	Based on available resources, activate risk control options between different levels according to compensation mechanisms

A company can respond to the same risk in different ways, proactively or reactively, at different organizational levels. The specific mix of risk management capabilities and the resilience mechanisms activated (supported by a multi-level "Spring" model) will vary depending on approaches and risk mitigation strategies. Appropriate simulation methods can be used to better assess the trade-offs of different risk-control options across management levels.

2.3 Value-Based ERM

The value-based ERM framework is a continuous process that includes risk identification, risk quantification, risk decision making and risk messaging.

In the risk identification process, strategy acts as a filter to eliminate irrelevant risks shown in Fig. 3. The company conducts an internal investigation to rank the risks according to the frequency and severity of their occurrence in order to determine and qualitatively evaluate the major risks of the company. This process narrows down the list of relevant risks to usually 20 to 30 key risks.

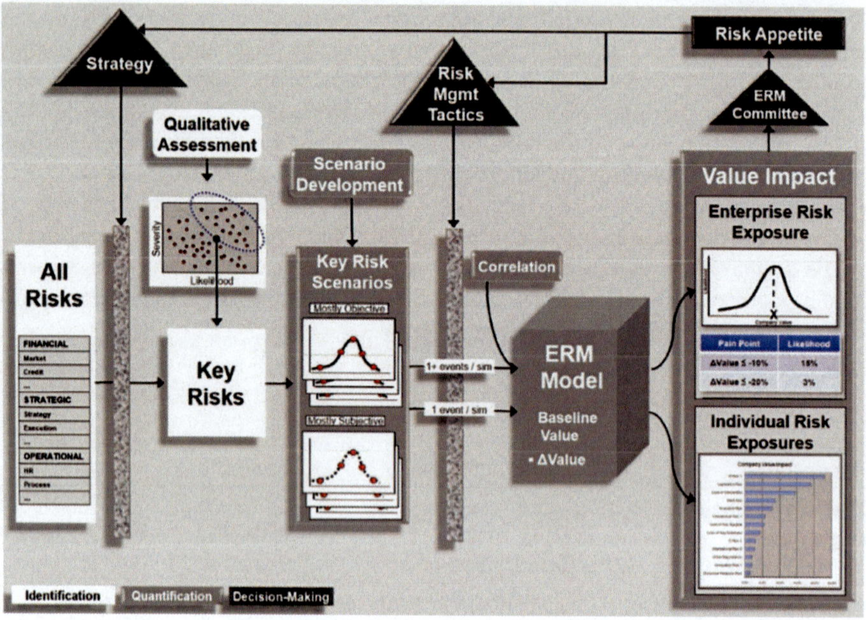

Fig. 3. Value-based ERM framework [21].

In the process of risk quantification, the corresponding deterministic risk plan is formulated for each key risk. Each key risk would have multiple risk scenarios such as very pessimistic, optimistic, extremely pessimistic. Risks could be categorized into three types: financial, strategic, and operational (for insurance companies, insurance risks are added). For financial risk, developing risk scenarios require mostly objective inputs. For operational and strategic risk, developing risk scenarios are mostly subjective, which rely on conducting Failure Mode and Effect Analysis (FMEA) interviews with subject matter experts to define the risk scenario circumstance, designate likelihood and evaluate the quantitative impacts. For all risk categories, FMEA and historical data used combined would be the best. After building the risk scenarios for key risks, the impact of one scenario and multiple risk events (with consideration of risk correlation) on the baseline company value, revenue growth rate should be quantified.

In the value-based ERM model, risk terms as events that cause deviation, either upside or downside, from expected results. The "expected" means that the outcome achieves the baseline of the strategic plan and associated value, i.e., the discounted value of distributable cash flows in compliance with the financial projections of the strategic plan. Thus, in the value-based ERM framework, risks are quantified by using the financial impact containing individual risk scenarios and the combined impact of multiple simultaneous risk scenarios to "shock" the baseline valuation. Company value is a key indicator of the value-based ERM, i.e. management's internal assessment of the company's value to its key stakeholders. These stakeholders are either shareholders (for listed companies) or (for unlisted companies). Quantification of individual exposures produces a ranking of each individual exposure, which gives a holistic view of key risks quantified in terms of their impact on company value. Management could therefore focus on mostly highly ranked risks.

In the process of risk decision-making, the first step is to define risk appetite, which is defined as maximum level of corporate risk reveal that management is satisfied with. By changing strategy or change risk management tactics, managers could manage enterprise risk exposure to within maximum appetite. Value-based ERM gives managers the ability to assess the impact of different risk decisions before truly implement them.

The fourth step in the value-based ERM cycle is risk messaging, which includes internal messaging and external messaging. Internal messaging concerns the incorporation of ERM information into performance measurement and incentive compensation, while external risk messaging pertains to transfer of ERM information to external stakeholders.

3 The Analysis of Three Erm Models

Three models address the dynamic capabilities in different approaches.

The resource-based ERM model mainly draws on the resource-based view and dynamic capabilities. It integrates ERM into a broader strategic management framework, namely, the resource-based view and the dynamic capabilities framework. It suggests that the firms should use the VRIN-criteria to determine the priority of the risks and develop a dynamic capability characterized by a non-routinized operational level. This model does not answer the question of how much effort organizations should put into avoiding, mitigating, or transferring the risks associated with VRINs and it is not helpful for management to select the most suitable risks solution among multiple options. Also, not all firms process VRIN resources, which limits the application of this model.

The resource-based ERM model and the value-based ERM model all focus on key risks but are based on different criteria. The former one uses VRIN-criteria and the latter one uses the impact on company value. But how to determine the priority of multiple risks that all meet the VRIN criteria is missing and misjudgment could happen if determined by the management. While the value-based ERM supplies a clear view of pivotal risks quantified their impact on company's value.

The "Spring" model and the value-based ERM model are addressing the dynamic capabilities on different phases of enterprise risk management. The "Spring" model emphasizes on how to generate corresponding mitigation options for risks by identifying existing capabilities of the company. The value-based ERM model stands from risk experts' perspective, provides a whole routine to identify, quantify risks and generates enough information to support dynamic strategic planning, which is the component of dynamic capabilities [22].

Similarities exist between the "Spring" model and the value-based ERM model. They have a similar definition of risk. The "Spring" model defines risk as "the interaction of internal or external events and the dynamic capabilities of the organization that may influence the degree of achievement of company's objectives". In the value-based ERM model, risk is defined as uncertainty which can cause a deviation, either upside or downside, from expected results. They all define risk as the deviation from expectation which considers upside impacts of risk.

The "Spring" model and the value-based ERM model could be combined to overcome the limitations of the "Spring" model. For the "Spring" model, the analysis was found on qualitative data (interviews and documental analysis). In the place of prospective research should use simulation tools to quantitatively test the divinable latent of the proposed models, so that different management pathways for responding to specific events can be compared [20]. The value-based ERM identifies all the quantifies all the impact of the key risks to support strategic decision making, which provides the comparison of different managerial paths in one metric which is company value. And it can easily reflect changes in the internal and external environment or changes in strategy and tactics, thus the impact of disclosure on projected financial performance [21]. Furthermore, any hypothetical action can also be simply reflected.

4 Conclusion

The resource-based view ERM gives a theoretical way of how to incorporate resource-based view in ERM. But it does not give a specific guidance on how ERM should be conducted in an enterprise. While the other two frameworks are more practical, which provides detailed guidance for managers to follow. From the perspective of enterprise risk management, we could say that dynamic capability is the ability of one company to react to risk in a reasonable amount of time and to control the negative impacts while exploiting the positive impacts. Through the review and discussion of three models, in order to better increase company's dynamic capabilities, the "Spring" Model and value-based ERM could be combined together. By introducing the quantification tools in value-based ERM framework, the "Spring" model could be improved. Future studies could focus on introducing quantification tools into ERM models to improve the quality of outputs and mitigate cognitive bias.

References

1. COSO: Enterprise Risk Management – Integrated Framework. Committee of Sponsoring Organizations, New York, NY (2004)
2. Bromiley, P., McShane, M., Nair, A., Rustambekov, E.: Enterprise risk management: review, critique, and research directions. Long Range Plan. **48**(4), 265–276 (2015)
3. Arena, M., Arnaboldi, M., Azzone, G.: The organizational dynamics of enterprise risk management. Account. Org. Soc. **35**(7), 659–675 (2010)
4. Bruce, R.: Swift message on risk management, Accountancy, 22 April (2005)
5. Collier, P.M., Berry, A.J., Burke, G.T.: Risk and management accounting. CIMA Publishing, London (2007)
6. Bowling, B.M., Rieger, L.: Success factors for implementing enterprise risk management. Bank Account. Financ. **18**(3), 21–26 (2005)
7. Wernerfelt, B.: A resource-based view of the firm. Strat. Manage. J. **5**(2), 171–180 (1984)
8. Barney, J.B.: Resource-based theories of competitive advantage: a ten-year retrospective on the resource-based view. J. Manage. **27**(6), 643–650 (2001)
9. Nocco, B.W., Stulz, R.M.: Enterprise risk management: theory and practice. J. Appl. Corporate Financ. **18**(4), 8–20 (2006)
10. Peteraf, M., Di Stefano, G., Verona, G.: The elephant in the room of dynamic capabilities: bringing two diverging conversations together. Strat. Manage. J. **34**(12), 1389–1410 (2013)
11. Helfat, C.E., et al.: Dynamic Capabilities: Understanding Strategic Change in Organizations, 1st edn. Wiley-Blackwell, Malden (2007)

12. Teece, D.J., Pisano, G., Shuen, A.: Dynamic capabilities and strategic management. Strat. Manage. J. **18**(7), 509–533 (1997)
13. Burisch, R., Wohlgemuth, V.: Blind spots of dynamic capabilities: a systems theoretic perspective. J. Innov. Knowl. **1**(2), 109–116 (2016)
14. Eisenhardt, K.M., Martin, J.A.: Dynamic capabilities: what are they. Strat. Manage. J. **21**(10–11), 1105–1121 (2000)
15. Winter, S.G.: Understanding dynamic capabilities. Strat. Manage. J. **24**(10), 991–995 (2003)
16. Bogodistov, Y., Wohlgemuth, V.: Enterprise risk management: a capability-based perspective. J. Risk Financ. **18**(3), 234–251 (2017)
17. Carbone, T.A., Tippett, D.D.: Project risk management using the project risk FMEA. Eng. Manage. J. **16**(4), 28–35 (2004)
18. Sheffi, Y., Rice, J.B., Jr.: A supply chain view of the resilient enterprise. MIT Sloan Manag. Rev. **47**(1), 41 (2005)
19. Silvestri, A., Arena, M., Cagno, E., Trucco, P., Azzone, G.: Enterprise risk management from theory to practice: the role of dynamic capabilities approach – the "Spring" model. Quantitative Financial Risk Management, pp. 281–307. Springer, Berlin, Heidelberg (2011)
20. Arena, M., Azzone, G., Cagno, E., Silvestri, A., Trucco, P.: A model for operationalizing ERM in project-based operations through dynamic capabilities. Int. J. Energy Sector Manage. **16**(2), 178–197 (2014)
21. Segal, S.: Corporate Value of Enterprise Risk Management: The Next Step in Business Management. Wiley, Hoboken, NJ, USA (2011)
22. Dibrell, C., Down, J., Bull, L.: Dynamic strategic planning: achieving strategic flexibility through formalization. J. Bus. Manage. **13**(1), 21–36 (2007)

Blockchain Technology and Its Applications in Digital Content Copyright Protection

Jiyin Shen(✉)

School of Professional Studies, Columbia University, Broadway, NY, U.S.

Abstract. Nowadays, digital products and related copyright disputes are growing at an accelerating speed, the field of digital copyright protection is facing the problems of numerous infringements, toughness in digital right confirmation and safeguarding. Fortunately, blockchain technology provides a potential technical solution to this dilemma because of its features of decentralization and distributed storage. On the basis of a brief introduction to blockchain technology and its features, this paper analyzes the applicability of blockchain to the problems facing the digital copyright field and discusses the possible challenges and corresponding countermeasures when blockchain is applied in this field.

Keywords: Blockchain · Digital right protection · Decentralization · Distributed storage

1 Introduction

Digital right is the copyright of digital-format products, such as digital text, digital picture, digital video, and digital music. Current digital right protection is facing with a dilemma, such as complexity in digital right confirmation, simplification and unconsciousness in infringement, and toughness in digital right safeguarding, due to the high speed development in Internet dissemination and traditional attitude towards copyright. However, digital products are also booming in recent decades, leading to an increasing number of infringement disputes. Digital right protection needs to be consummated.

The application of blockchain in digital right protection could be a solution to the dilemma. Blockchain is a data structure connected by chain-analogous structure chronologically. It could become an underlying technology with the features of decentralization, timestamp and distributed storage, all of which are key factors contributing to digital right protection since these features are significant in improving efficiency and authentication of digital right verification and digital data itself while satisfying storage requirement for the blooming digital products.

Against this background, this paper first introduces blockchain technology, including its origin, definition, and features. Then, based on the analysis of the current dilemma faced by digital copyright protection, the advantages of applying blockchain technology in this field are presented. Finally, challenges to apply blockchain technology and corresponding solutions are also discussed.

C. Yuan et al. (Eds.): *Proceedings of the 4th International Conference on Economic Management and Green Development*, AEPS, pp. 18–25, 2021.
https://doi.org/10.1007/978-981-16-5359-9_3

2 Blockchain Technology

Blockchain was first perceived as a term in the paper, Bitcoin: A Peer-to-Peer Electronic Cash System, written under the pseudonym of Satoshi Nakamoto in 2008 [1]. As a newly invented term, blockchain is not authorized with officially recognized definition and is currently defined both in a narrow and broad sense.

In terms of narrow definition, blockchain is a specifically assembled data structure connected by chain-analogous structure chronologically, ensuring its decentralized distributed storage for the security of those simple data with contextual connection and verification from system. In terms of broad definition, blockchain is a decentralized distributed algorithm or structure for the operation of data verification, data storage, data updating, and coding, using chain-block structure with encryption protection [2]. In general, blockchain is an Internet contract for data credit using decentralized distributed structure. For detailed explanation, blockchain can be regarded as a digital record of transactions, each of which is validated by multiple computers on the Internet. The systems that configured to monitor specific types of blockchain transactions form a peer-to-peer network (P2P), ensuring each transaction is valid before it is added to the blockchain, which enables new data insertion when processing a data sharing. Even if one does the orientation, it just gives the individual an interface. After the occurrence of blockchain, the participants realize the sharing of data credit [3].

Operation of blockchain contributes to its identical characteristics.

(1) Distributed storage
Distributed storage is a data storage technology that uses the disk space on each machine through the network and forms these scattered storage resources into a virtual storage device. Data is distributed and stored in every corner of the network. Therefore, distributed storage technology does not store complete data on every computer. Instead, it cuts the data and stores it on different computers. It's like storing 100 eggs, not in the same basket, but in separate places, adding up to a total of 100. This contributes to the unlimited capacity of data storage since hard drive capacity would become infinite by adding up all computers in the world, though the capacity of each computer is limited. Even if transaction records is hacked in one specific computer, the whole transaction records for each computer in the world would still stay unmodified [4].

(2) Decentralization
Another feature of blockchain is decentralization, defined as a phenomenon or structure that can only occur in systems with multiple users or numerous nodes, each of which can connect and influence other nodes. In a nutshell, everyone is the center, everyone can connect and influence other nodes. This structure is summarized as flattening, open source and equalization [3].

These two features lead to several advantages [5].

(1) System security
In a traditional centralized network system, a hacker can destroy a whole node by attacking a central node. In a decentralized blockchain network, no central node can attack.
(2) Transaction security
Decentralized trading methods are convenient and simple, because there is no third-party intervention, and do not need to worry about the disclosure of information.
(3) Resource-saving
Since the decentralized processing method is simpler and more convenient than the traditional processing method, the decentralized method saves resources when large data volume transactions are simultaneously performed.
(4) Self-efficiency
The decentralized Blockchain technology eliminates the need for third-party intervention and involves direct peer-to-peer interaction, making high-efficiency, non-centralized agents and large-scale information interactions a reality.

Blockchain has just become a newly prevalent term, however, it is promising in various contemporary popular areas relevant with data credit.

3 Dilemma of Digital Right Protection and Applying Blockchain on Digital Right Protection

3.1 Dilemma of Contemporary Digital Right Protection

The concepts of digital right and digital right management (DRM) are perceived in the 1980s and 1990s when computer program occurred [6]. Digital right acquires various definitions in multiple countries [7], mainly referring to the right of digital-format products, such as digital text, digital picture, digital video, and digital music. Digital product and relative digital right are disseminating into different perspectives in modern routine life at an unignorable high speed with the development of digital technology in recent decades, for example, photos from Internet, electronic book, online music, and short videos, showing significance in digital right protection for fairness and sustainable development.

Digital right management is the current mechanism to deal with digital right infringement [8]. Traditional DRM is a system to protect against the infringement of copyrighted material in a digital format. It contains several phases, including production, digitalization, distribution, identification, ascription of descriptions, the use from a consumer, monitoring, and payment collection [9]. However, it is aimed at copyrighted products, excluding those without registration. Hence traditional DRM needs further improvement.

Current digital right protection is facing with some challenges.

(1) Difficulty in digital right confirmation
Copyright registration is not compulsory since copyright is inherently attributed to its author [2], but it is strong evidence for ownership and legal support for copyright disputes, indicating the importance of copyright registration. However, digital right proving is tough. For example, one declares one's original creation of certain digital texts. But how to prove it? One cannot deny the possibility that the digital text originated from other streams which are not identified, which leads to the second problem. Complexity for operation on digital right registration restricted authors' gaining on authorized protection. Difficulty in confirmation for originality results in requirement for complicated material and toughness on originality check [2]. In most cases, originality check cannot be accomplished in the expected level, resulting in another issue that even approved copyright registration has limited power for legal certification [10]. What worth mentioning is these problems are not only targeting in digital right, but for all copyright.
(2) Simplicity and unawareness in infringement
Digital products are easy to download and disseminate for commercial use via multiple digital streams, such as search engines, websites, and applications [7], without permission of authors, defined as an infringement. Some websites and applications adopt encryption technology to prevent infringement by only permitting visitors to view products on assigned pages, however, it is still easy to crack this encryption using simple software. Besides, weak consciousness on digital right protection leads to unawareness infringement. Individuals have been adapted to use digital products for free, which is effortlessly feasible due to simplicity on circulation of digital products, compared with their being accustomed to entity product bills. Google Library Project in 2014 was a piece of evidence for this point. Google entailed digitizing, indexing, and displaying "snippets" of print books without seeking the permission of copyright holders, asserting its proposed fair uses, which still obviously violated copyright of authors [11]. People pay for printed books, but when transferring into digital ones, even famous enterprise regarded them as free products.
(3) Difficulty in safeguarding digital right
Simplicity on spreading digital products and unconsciousness in digital right lead to a large quantity of infringement cases. However, safeguarding digital right is in predicament due to toughness on digital right confirmation since the arduous process to check originality increases the human cost and technology cost for acceptor, further leading to high time cost and high financial cost for authors to apply for adjudication. Disproportion between cost and revenue diminishes the willingness and possibility of authors for safeguarding [2]. Limited channels for safeguarding is another element because it would increase time cost for safeguarding.

3.2 Advantages of Applying Blockchain on Digital Right Protection

There are three advantages when we apply blockchain technology on digital right protection.

(1) Timestamp: authentication on copyright confirmation
Blockchain operates chronologically, contributing to occurrence and utilization of timestamp, recording immutable time for the appearance of the digital materials. Timestamp could become solid verification for originality of digital products since the one who created the digital product firstly is an undoubted owner for its copyright. Even those preferring recognized as an anonymous author could identify their copyright utilizing timestamp. The following application of digital materials and its respected commercial revenue could be easily attributed to their creators.

(2) Decentralization: improvement on efficiency for operation, cost-saving & security of data
Decentralization, the key feature of blockchain, enables data recording and storage by numerous users and multiple nodes across Internet, instead of relying on traditional centralized node [12]. Thus blockchain could become underlying technology for a public copyright system. An analogy could simplify the concepts: each author could be regarded as the node scattered through Internet in blockchain system while normal copyright agent, such as national copyright center, is the centralized node in traditional system. Each author publishes their product and its information for trade, in which copyright could be confirmed, as mentioned previously, and transmission of the digital product would reach a high level of efficiency, compared with that of traditional system since the process of centralized node manually confirming and publishing, which requires large human resource and time cost, is replaced by product automatically flowing along the chain of the decentralized blockchain system. Diminishing in human resource and time cost and elimination in entity investment contribute to cost-saving both in finance and time perspectives. It also leads to alleviation in difficulty for individuals to safeguard against infringement because both individual and enterprise are treated identically and equally in blockchain, unlike in current system where enterprise with adequate financial support and professional legal advisor has privilege in digital right protection.
Decentralization of blockchain guarantees permanent immutability of data in blockchain since there is no centralized node to be hacked or edited. Original information input by author, such as copyright and price, is kept and flows along the chain with high efficiency. Data credit is authentic in blockchain.
Bilateral trust dilemma, usually appears in commercial culture creation, in which creator of digital product, such as the screenwriter and painter, worries about plagiarism and no guarantee for next-stage income while investor, such as film company, worries about its capital being stolen by creator [10]. Blockchain could mitigate this phenomenon since the copyright could be confirmed from the start and the contracting process would be operated automatically along the chain. For detailed explanation, there is an immutable and complete record of creation of the digital material by its unique author, from raw brainstorm to the final version, preventing plagiarism. Update, editing, distribution, and payment for the digital material are operated under the automatic contract, preventing the breach of contract.

(3) Distributed storage: satisfying demand of digital product
Storage of blockchain is equal to the sum of storage of each node scattered distributed in blockchain, indicating that blockchain storage could be viewed as unlimited. Digital product shares global transmission and the number of digital products has reached a high level, with potential unlimited quantity in the future. Distributed storage of blockchain provides adequate space for digital product storage and convenience to trading internationally.

4 Challenges and Corresponding Solutions

As a new concept with only ten-year history, blockchain application is still doubtful since no adequate experiments, application data and experts to support. It has completed the proof of concept, but several challenges might occur when applying blockchain to digital right protection in reality.

(1) Immaturity of blockchain technology
The immature blockchain technology leads to a gap between truly accomplished storage of blockchain and theoretically reached storage. As mentioned previously, the storage of blockchain is theoretically unlimited because of its distributed storage. However, the actual storage is limited to a small number currently due to immature technology. Unfortunately, a noticeable fact is that storage requirement for data, including metadata and digital content, is at a high level and is increasing at an exponential speed [13]. Take bitcoin as an instance, the size of bitcoin blockchain reached almost 130 GB in September, 2017 and it doubles size every year. In such case, bitcoin blockchain only consists of metadata, excluding digital content. However, current bitcoin block size is only reaching a level at around 1 MB per block, restricted by code. This results in the problem about the storage of metadata and digital content. Digital right protection should be validation internationally, indicating the globally large number of users and work and exponentially increasing transaction amount [13]. Moreover, digital work contains both metadata and digital content, unlike bitcoin, which means a greater demand in digital work storage is required. Blockchain is at the beginning stage while digital product and digital copyright protection have reached its prosperous development. The conflicts between these two stages would lead to problem when applying blockchain into digital right protection.
To solve this problem, one of the most obvious measures is to input greater investment in the development on blockchain, which would lead to the second challenge to be discussed in the following text. Another method is to store digital files with copyrighted works outside of blockchain [13] and only copyright information would be left within blockchain. However, such solution is also facing with issues, such as credibility of data and accessibility to the content.
(2) High development cost
It is true that information transferring along the blockchain is cost-saving, however, infrastructure for the blockchain development is costly, such as customer ICT equipment and software [14]. This challenge diminishes competitiveness of blockchain, especially when current available conventional database is less costly [15].

One potential solution to this problem is to optimize the credibility of data within blockchain to leverage cost and benefits of developing blockchain since conventional database is inferiors in immutability of data records and cannot guarantee authentication of data and information. Reliability of information is incomparable advantages of blockchain and maximizing the superior field could increase the value of high investment in blockchain development.

(3) Legal issues
Blockchain is a decentralized system, but still need supervision and attribution institution when facing with disputes. Disputes are likely to appear for the following two reasons, lack of standardization and global circulation for blockchain. Standardization difference would result in controversy in different categories of work, regions, and countries. Global circulation would result in different laws in different countries, leading to severe legal issues.
Legal issues could be solved by constructing a standardized criteria under participation from multi-field and multi-region representatives. Besides, a global supervision mechanism and an attribution institution also need to be considered since blockchain is distinguished from traditional affairs, thus need targeted regulation and supervision.

(4) Human resource issues
The first human resource issue is lacking experts in blockchain field. There is no major in blockchain at present and relative research is limited, restricting development of this technology. Thus professional scholar is in huge demand. The second issue is the difficulty for individuals to utilize blockchain, and furthermore choose blockchain to verify digital copyright. Blockchain is a new concept and has not been spread into routine life. The former two issues lead to the third one, the difficulty in attracting large number of users to participate in blockchain copyright verification at beginning. This problem would abate value of network effect from blockchain [13]. Network effect is a phenomenon where the value of a product depends on the number of others using it [16]. With few experts leading and minority of right owners following in blockchain application, network effect cannot be optimized and mitigate the value of blockchain application in real world.
Human resource issues are relatively easier to be solved, compared with other issues. Universities could list advanced blockchain courses with scholarships to attract elites to study and prepare to contribute to development of blockchain. Secondary school and high school could provide courses on fundamental knowledge of blockchain to get teenagers familiar with this concept. Colleges and institutions could consider propagation on participating in digital right verification via blockchain system.

References

1. Arner, D.W., Zetzsche, D.A., Buckley, R.P., Barberis, J.N.: The identity challenge in finance: from analogue identity to digitized identification to digital KYC utilities. Eur. Bus. Org. Law Rev. **20**(1), 55–80 (2019)

2. Zhao, F., Zhou, W.: Exploring the problem of protecting digital copyright based on blockchain technology. Technol. Law **2017**(01), 59–70 (2017). (in Chinese)
3. Crosby, M., Pattanayak, P., Verma, S., Kalyanaraman, V.: Blockchain technology: beyond bitcoin. Appl. Innov. **2**(6–10), 71 (2016)
4. Zwitter, A., Hazenberg, J.: Decentralized network governance: blockchain technology and the future of regulation. Front. Blockchain **3**(2624–7852), 12 (2020). https://doi.org/10.3389/fbloc.2020.00012
5. Singh, N.: Benefits of blockchain technology. https://101blockchains.com/benefits-of-blockchain-technology/ (2019). Accessed 3 July 2020
6. Martén, M.: Digital rights management: blockchain and digital music content management (2017)
7. Ma, Z., Liu, H.: Building a digital copyright governance system under the perspective of blockchain technology. Technol. Law **2**, 1–9 (2018). (in Chinese)
8. Tresise, A., Goldenfein, J., Hunter, D.: What blockchain can and can't do for copyright (2018)
9. Paskin, N.: Components of drm systems identification and metadata. In: Becker, E., Buhse, W., Günnewig, D., Rump, N. (eds.) Digital Rights Management. LNCS, vol. 2770, pp. 26–61. Springer, Heidelberg (2003). https://doi.org/10.1007/10941270_4
10. Shi, D.: On the value and risks of blockchain technology for digital copyright governance. Technol. Publish. **6**, 111–120 (2019). (in Chinese)
11. Manuel, K.M.: The Google Library Project: is digitization for purposes of online indexing fair use under copyright law. Library of Congress Washington DC Congressional Research Service (2010)
12. Litman, J.: Digital copyright. Prometheus Books (2001)
13. Savelyev, A.: Copyright in the blockchain era: promises and challenges. Comput. Law Secur. Rev. **34**(3), 550–561 (2018)
14. Andoni, M., et al.: Blockchain technology in the energy sector: a systematic review of challenges and opportunities. Renew. Sustain. Energy Rev. **100**, 143–174 (2019)
15. Hasse, F., Von Perfall, A., Hillebrand, T., Smole, E., Lay, L., Charlet, M.: Blockchain – An Opportunity for Energy Producers and Consumers. PwC Global Power & Utilities, pp. 1–45 (2016)
16. Shapiro, C., Varian, H.R.: Information Rules: A Strategic Guide to the Network Economy. Harvard Business School Press, Boston, Massachusetts (1999)

Research on the Operation of the Takeaway Platform During COVID-19 Based on the Theory of Two-Sided Market

Taking Meituan Takeaway as an Example

Linfeng Li(✉)

Business School, Sun Yat-Sen University, No. 135 West Xingang Road, Guangzhou City, China

Abstract. Under the influence of the COVID-19, Meituan Takeaway, the leading Chinese takeaway O2O (Online to Offline) platform, has been greatly affected. At the same time that consumers' demand and delivery staff are decreasing, the supply of merchants (refer to restaurants) has greatly increased due to all online operations, resulting in a mismatch between supply and demand. This article analyzed the impact of the COVID-19 on three perspectives of Meituan Takeaway, including consumers, platform operations, and merchants, in conjunction with two-sided market theory, and made strategic advice from the strategic level and operational level based on Meituan Takeaway's existing coping strategies.

Keywords: Two-sided market · O2O · COVID-19 · Meituan takeaway

1 Introduction

In recent years, with the rapid development of the Internet technology, the Chinese takeaway O2O (Online to Offline) market has emerged in a very short period of time and has formed a certain scale. At the same time, the COVID-19 that broke out in January 2020 had brought a tremendous impact on the Internet economy, which seriously affected the operation of takeaway O2O. Meituan Takeaway as a leading Chinese O2O food delivery company was severely affected during that period. This article will analyze the impact of the COVID-19 on Meituan's takeaway in conjunction with two-sided market theory, and make strategic advice based on Meituan Takeaway's existing coping strategies.

2 Review of the Two-Sided Market Theory

2.1 Definition of Two-Sided Market

The concept of the two-sided market was first proposed by the "Penny Newspaper" movement initiated by the United States in 1833. After entering the 21st century, its related theories and fields has been continuously improved. Rochet and Tirole (2002,

C. Yuan et al. (Eds.): *Proceedings of the 4th International Conference on Economic Management and Green Development*, AEPS, pp. 26–36, 2021.
https://doi.org/10.1007/978-981-16-5359-9_4

2003) [1, 2] proposed "non-neutral price structure" and defined the two-sided market as: When the total price level P=PB+PS that the platform demands from both parties is unchanged (PB is the price of user B, and PS is the price of user S). If any change in the user's price will have a direct impact on the total demand and transaction volume of the platform, the platform market is called a two-sided market. Considering the indirect network externality, Evan (2003) [3] and Armstrong (2006) [4] and others defined it as: two groups of participants need to conduct transactions through the intermediate platform, and the revenue or the utility of one user will increase as the number of users of the other party increases. The price structure is non-neutral, and cross-network externalities and interdependencies are important characteristics of two-sided markets.

2.2 The Classification and Application of Two-Sided Market

Evan (2003b) divides two-sided markets into three categories from an empirical perspective:

(1) Audience creation: when there are enough users or traffic to attract another party to join the platform, such as magazines, newspapers, and televisions;
(2) Market creation: the increase of users on either side will increase the utility of users on the other side. For example, O2O e-commerce platform, headhunting company, etc.;
(3) Coordinated demand type: to meet the differentiated needs of two-sided users through platform transactions, such as the IT industry and financial institutions.

The pricing methods of the two-sided market can also be roughly divided into three types. The first one is a pricing method that only charges a registration fee. The platform only charges the membership fee to enter the platform for two-sided users; the second one charges a transaction fee, which charges a certain percentage of transaction fees based on the transaction volume of users participating in the platform; the third is the two-step fee collection system charges both registration and transaction fees for two-sided users participating in the platform.

The two-sided market has a large number of application examples in reality, such as e-commerce platforms such as Amazon and Taobao, takeaway platforms such as Uber Eats and Meituan Takeaway, etc. Taxi platforms such as Uber and Didi Taxi also have applications on some travel platforms, such as TripAdvisor, Ma Honeycomb, etc. Some shared rental and hotel reservation platforms are also very extensive, such as Airbnb, Booking, Ctrip, etc.

In summary, the two-sided market is generally connected by platform companies to two different groups, while meeting the existing conditions of non-neutral price structure and extensive cross-network externalities.

3 The Specific Application of the Two-Sided Market in the O2O Takeaway Platform

3.1 What is O2O?

O2O (Online to Offline) refers to combining offline business opportunities with the Internet to make the Internet a front desk for offline transactions. Through the online way, O2O brings convenience to the consumer group, increasing the user's sense of experience, and at the same time, it can facilitate the operation of the physical store and expand the revenue of the merchants by online promotion.

The takeaway O2O refers to the establishment of a website or the display of offline goods or service information on a third-party take out platform by means of the Internet. Having browsed the takeout information online, consumers can book online, generating orders and completing online payment. After receiving the orders, merchants send the food to the designated place [5] According to the its operation mode, the takeaway platform can be divided into three categories: 1. relying on open platform, such as WeChat official account, Baidu map, etc. 2. The form of self-supporting taking out by merchants refers to the way of ordering and distribution by merchants themselves. 3. Third party takeaway platforms, such as Meituan Takeaway, Uber eats, etc. The research subject of this paper will focus on the third one.

3.2 Platform Operation Mode Under the Theory of Two-Sided Market

Two-sided market has existed for a long time in China. One of the typical examples of the two-sided market in the market-creating two-sided market is the takeaway platform. Generally, the two-sided market connected by takeaway platform is composed of three parties: the platform enterprises, consumers and suppliers. Using O2O mode, consumers place orders on the online platform, pay related fees through the third-party payment platform. After receiving the orders, the restaurants prepare the orders, and then distribute them by themselves or by the riders of the platform. After receiving the orders, consumers evaluate the orders and eventually complete the order. The profit or utility of consumers will increase with the number of suppliers, and vice versa. Usually, the platform makes use of indirect network externalities, usually through free pricing strategy to expand the number of consumers. By exploiting positive feedback to promote the franchise of restaurants, the platform company can earn profits using the pricing strategy of combination of registration fee and transaction fee. The operation mode of the platform is shown in Fig. 1.

3.3 Development of O2O Takeaway Platform in China

In recent years, O2O delivery platforms have developed rapidly in China. According to QuestMobile's data, in 2015, the number of Chinese online takeaway market orders exceeded 2 billion. As of 2018, the number of takeaway market orders have reached 11 billion, with an average year-on-year growth of more than 80%. From the perspective of revenue, the takeaway platform revenue was 44.7 billion yuan in 2015. As of 2018,

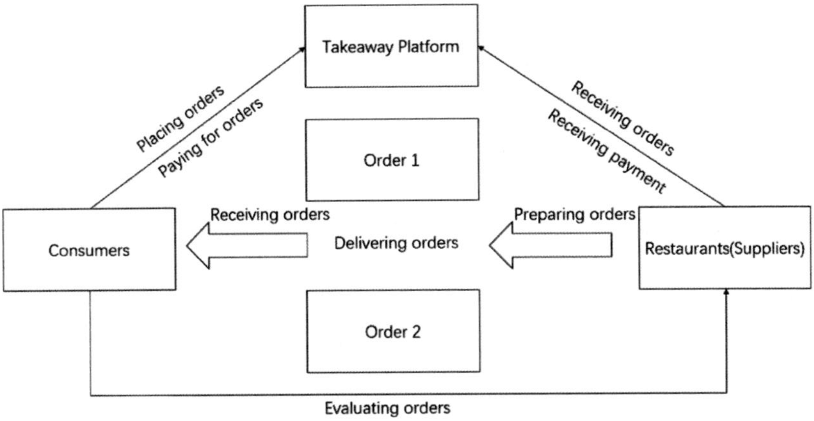

Fig. 1. The operation mode of the takeaway platform.

the takeaway platform revenue has reached 474.6 billion yuan, with a CAGR of approximately 120.63%. China's O2O takeaway platform industry is an oligopolistic market, and Meituan Takeaway platform has always occupied the first place in market share. The following is a brief analysis of Meituan Takeaway platform.

4 Research on the Operation of the Takeaway Platform During the COVID-19 Period—Take Meituan as an Example

4.1 Brief Introduction of Meituan Takeaway Platform

The Development of Meituan Takeaway. Meituan Takeaway is an online ordering platform under Meituan.com, which was officially launched in November 2013 and is headquartered in Beijing. After Meituan Takeaway was launched, it rapidly expanded through its marketing strategy and covered most domestic first- and second-tier cities in a short period of time. In October 2015, it merged with Dianping.com to develop third- and fourth-tier cities. Statistics show that in 2018, Meituan Takeaway has 250 million takeaway users, more than 2 million cooperative merchants, more than 500,000 active delivery riders, covering more than 1,300 cities, and 21 million orders per day.

As one of the oligarchs in the O2O takeaway industry, Meituan Takeaway has a market share of 51.3% in the fourth quarter of 2018, which is the second highest food takeaway platform with a market share of 28.6%. It has become a unicorn enterprise in the Chinese food takeaway industry. Meituan Takeaway is an online ordering platform under Meituan.com, which was officially launched in November 2013 and is headquartered in Beijing. As of 2018, the number of users reached 250 million, the number of cooperative merchants exceeded 2 million, the number of active distribution riders exceeded 500,000, covered more than 1,300 cities, and completed 21 million orders per day.

Business Model

Value proposition. Meituan Takeaway aims to provide consumers with a full range of choices and convenience. Relying on a large number of user groups to attract a large number of merchants to settle in, providing consumers with different levels of choice, improve customer satisfaction, and then cultivate customer loyalty.

Target customers. The target groups of Meituan Takeaway include college students, young white-collar workers and home users. This group of people has a fast pace of life, a pursuit of high-quality life, low price sensitivity, and high user loyalty, so it is highly profitable.

Age of users. In 2017, the age of users is getting younger, among which users aged 20–30 are as high as 63.32%.

Geographical distribution. In 2017, Meituan Takeaway users were mainly distributed in cities with highly developed economies, of which Shanghai accounted for 3.94%. From the perspective of city level, users in first- and second-tier cities are as high as 74.31%.

Core competence. Scale and operational efficiency. Meituan Takeaway attracts a large number of consumer groups through price marketing, thereby attracting a large number of merchants to settle on the platform. The positive feedback effect of the two-sided market occupies most of the takeaway market. Its transaction volume, number of users, and number of merchants are higher than those of other O2O takeaway platform companies, and it has reached a large amount of transactions while meeting customer needs. At the same time, with its advanced algorithm system, Meituan Takeaway quickly matches consumers, restaurants and riders to conclude transactions, and operates the entire system in an efficient and low-cost way.

Profit model. The income source of Meituan.com is mainly through the collection of distribution fees, advertising income and commissions from merchants. The specific method is as follows:

Commission (transaction fee): The most basic profit model of Meituan.com is to extract commissions from the service fees and product promotion fees of the merchants that have settled in.

Advertising: With its large number of user groups and its advanced algorithms, it conducts peer-to-peer marketing for merchants to earn fees.

Delivery cost: Connect the consumer with the store, and after the consumer places an order, the rider will deliver the store's products to the consumer and earn service fees from it.

Fund pool: Every two or three weeks, Meituan Takeaway will pay its merchants for its operating income, and the account period will be used for the investment of the deposited funds.

4.2 Impact of COVID-19 on Meituan Takeaway

The new COVID-19 broke out in various parts of China from January 2020 to May 2020. Due to its high infectivity, the government had to take measures such as travel restrictions and state control. Enterprises switched to online offices and people were isolated at home. Without leaving home, has caused a huge negative influence to the real economy. For Meituan Takeaway O2O takeaway platform, it has caused a high negative impact on its offline part. The following will analyze the impact of the COVID-19 on Meituan Takeaway from the supply side (merchants), consumers and platform perspectives.

Impact from Merchants' Perspective

The shift of the focus of business strategy. The decrease in offline demand and the limitation of in-store personnel flow have caused the main business of the merchant to shift to the online part. Due to national policies and the government's call, the consumers' flow of offline restaurants has a significant reduction. According to iiMedia Research data, 95.0% of the interviewed restaurant businesses experienced a significant decline in store revenue during the COVID-19, and more than half of the surveyed businesses suffered a significant decline in turnover. According to the National Bureau of Statistics, from January to February 2020, the national takeaway industry revenue was only 419.4 billion yuan, a year-on-year decrease of 43.1%. At the same time, because most of rents of restaurants have not been significantly reduced, the offline restaurants are obviously in a state of negative income. In order to pay for its rentals, employee salaries and other administration expenses, the merchants can only choose to give up offline store income and invest the proportion in O2O takeaway platforms, such as Meituan Takeaway. According to data from iiMedia Research, 78.0% of takeaway merchants using online as the main battlefield before and after the 2020 COVID-19 have increased by 63.1 percentage points compared with that before the outbreak. The strategic shift of merchants has greatly strengthened the supply side of Meituan Takeaways.

Merchants' operation problems. Lack of employees. Since Meituan Takeaway is targeted at first-tier and second-tier cities, the proportion of business distribution in first-tier and second-tier cities is higher than that in third-tier and fourth-tier cities. Most of the employees in the first-tier and second-tier cities are migrant workers. Before the Spring Festival, most of the employees have left the city where the business is located. After the outbreak of COVID-19, due to government restrictions on foreign (not local) personnel, some employees were in quarantine and could not return to their original jobs in time. According to the China Hotel Association, in February 2020, 25% of the quarantine personnel of Chinese takeaway companies resumed work, Recovery work only accounts for 35%. During the COVID-19, some employees chose to resign due to their own safety (lack of protective materials, etc.), and insufficient supply of employees caused the merchants to fail to open in time.

Increase in operating costs. The improvement of sanitation standards makes it necessary for businesses to strictly sterilize and control the environment in their restaurants, forcing merchants to purchase additional thermometers, a large number of disinfection appliances, masks and other materials. Some foods, such as meat and meat products, require certificates and tickets to ensure that the origin of the meat can be traced, resulting in merchants needing to spend a higher fee to purchase higher-quality meat to ensure compliance with regulations. On the other hand, merchants' takeaway platform commissions, restaurant rents, and staff management costs have not been significantly reduced. Due to the above and other reasons, the total cost of business operations has increased.

Impact from Consumers' Perspective

Change of the target consumers. The target customers of Meituan Takeaway are mainly aimed at college students and white-collar workers in first-tier and second-tier cities, and the school was closed during the COVID-19, which resulted in the failure of students to return to school and the demand around the campus was greatly reduced. On the other hand, due to the online office solution, some white-collar workers will not go offline for a short period of time, resulting in a decrease in the demand for selling out in major commercial centers.

Increase in price sensitivity. Some white-collar workers in first- and second-tier cities are affected, their salaries only include the basic salary with low bonus, resulting in higher sensitivity to prices. Therefore, the demand for takeaway has also decreased.

Policy and safety. Due to the community blockade caused by the policy, takeaway can only be delivered to the community door, and users need to go to the community door to take food which reduces the convenience of takeaway. On the other hand, due to increased hygiene awareness, consumers have concerns about the safety, and their willingness to buy has decreased, making demand less than before.

Impact from Platform Perspective

Increased cost for distribution personnel. Most of Meituan Takeaway's delivery staff come from second-tier and third-tier cities, and Meituan Takeaway's main delivery area is first-tier and second-tier cities. This has led to a 14-day quarantine when those staff return to work from their homes. As the COVID-19 intensified, some delivery staff quit their jobs out of consideration for their own safety. The reduction in the number of employees has led to the need for Meituan Takeaways to hire new staff through headhunting companies which increases cost. The price of each order delivered by the employees during the COVID-19 will have a huge increase, which also increases the administration cost.

Transformation of services. During the COVID-19, due to problems such as merchants and operating hours adjustments, most businesses only operate for lunch and dinner peak hours, and do not operate at other times. Due to the adjustment of the business of the merchants, some merchants are in the state of being closed all day and cannot provide services to consumers. Consumers' demands for daily necessities such as wine, fresh food, etc. increase. Therefore, in the early stage of the COVID-19, while maintaining the main takeaway business, Meituan Takeaway focuses on the businesses of "supermarket convenience" and "vegetables and fruits". With its content function, the number of merchants (in this case, supermarkets) is expanded.

5 Strategies Respond to COVID-19

In order to cope with the COVID-19, Meituan Takeaway launched various strategies and tactics. The following is a brief analysis of Meituan's strategy for dealing with negative impacts of COVID-19 in two levels, namely, the operational level and the strategic level.

5.1 Meituan Platform Operation Strategy During the COVID-19

Consumer Perspective. Strengthen product diversification. During the COVID-19, consumers' consumption of fresh and snacks such as vegetables and fruits increased. However, due to safety considerations, consumers are resistant to offline purchases. Meituan Takeaway has strengthened the operation of its supermarket convenience, vegetables and fruits, dessert drinks and other functional modules, increased the proportion of distribution personnel, effectively reduced the waiting time of consumers in order to improve the perceived value of consumers. At the same time, strengthened cooperation with existing fresh supermarkets and convenience stores, enriching its product range to meet different levels of consumer demand, and increase consumer satisfaction. For the above services, the platform's advanced algorithms are used to carry out peer-to-peer marketing to customers and recommend them according to their characteristics, so as to increase the popularity of such products and increase the conversion rate of customers.

Price concessions. While establishing contactless delivery services to ensure consumer safety and stimulate consumption, Meituan Takeaway pricing concessions policies were adopted for low-end markets and consumers who are more sensitive to prices. Through full reduction, cashback, member discounts, etc., using its advanced algorithms, it is accurately pushed to consumers to appropriately reduce prices while expanding demand, and ultimately achieve the goal of economies of scale and greater profits.

Platform Perspective. Service area adjustment. Meituan Takeaway adjusted some of the main operating areas. The area is transferred to residential areas where consumers are densely located, mainly maintaining the surrounding merchants to ensure the supply of food during peak periods, and by reasonably arranging the distribution of its delivery staff, strengthen the supply to residential areas, so that the existing demand is met. According to the data of iResearch, the proportion of takeaway users in the second-tier, third-tier, and fourth-tier cities during the Spring Festival in 2020 is significantly higher than that in the same period in 2019. Therefore, the adjustment of the service area of Meituan takeaway during the COVID-19 also includes serving newly-increased users in those cities. In the mid-to-late period of the COVID-19, as companies began to resume production, employees returned to work offline, and the demand for takeaway in the major commercial areas of first- and second-tier cities rebounded during the peak period. Therefore, the main area of service shifted from residential areas to commercial areas. Ensuring the speed of food delivery during peak periods to meet consumer demand, Meituan improves service quality and increases customer loyalty to obtain revenue.

Employee Management. Delivery staff's contactless service. In order to ensure consumer safety, the takeaway platform has changed the delivery service model in a timely manner and has been recognized by many parties. Meituan Takeaway launched the "Contactless Delivery" program in Wuhan on Jan 26, which has been followed and promoted nationwide. That is, consumers can choose contactless delivery in the APP, and the delivery staff puts the food in the designated location or smart dining cabinet to avoid direct contact with the consumer and protect the safety of the user as well as the delivery staff.

Strict hygiene control. In Wuhan and other COVID-19-affected areas, fully equipped with masks for delivery personnel, and set up special person to conduct random sampling inspections; Upgrading site disinfection and temperature measurement policies. The site is equipped with thermometers, masks, disinfectant and alcohol gel. Delivery man's meal boxes are disinfected once in the morning and evening. Measure the delivery staff's body temperature every day, stop them from working immediately if there are any abnormal symptoms.

Employee sharing. Due to the decrease of delivery staff, Meituan Takeaway launched an employee sharing plan and hired employees from temporarily closed restaurants, hotels, and cinema for short-term cooperation.

Merchant Perspective. Increase the commission appropriately. On May 25, Meituan Takeaway disclosed its first quarter 2020 results announcement. The data shows that the company achieved revenue of 16.754 billion yuan in the first quarter, a year-on-year decrease of 12.6%; total operating losses of 1.716 billion yuan, a year-on-year increase of 31.6%; adjusted net loss of 216 million yuan, a 79.4% narrower than the same period last year. Among them, the commission income was 10.8 billion yuan, and the takeaway commission income was 8.564 billion yuan, a year-on-year decrease of 13.68%, accounting for 79.29% of the total commission income. However, considering the decreases of the demand, it could be concluded from the data that Meituan Takeaway increased its commission. Before the COVID-19, Meituan Takeaway commission was about 18%, which was lower than ELEME Takeaway, Meituan Takeaway's largest competing food delivery platform. In the late period of COVID-19, Meituan Takeaway increased the commission in some areas to 20% on the premise of not violating antitrust laws nor causing public problems. On the other hand in order to enhance the reputation and social image of Meituan Takeaway, in the mid-COVID-19 period, in order to help merchants better recover their operations and improve their motivation, Meituan Takeaway launched the "Spring Breeze Action". Meituan Takeaway launched a monthly bonus of 500 million RMB, 400 million RMB merchant subsidies. At the same time, the "Commercial Partner Commission Rebate Program" was launched to refund high-quality food delivery staff nationwide, especially those whose business conditions were greatly affected by the COVID-19, at a rate of not less than 3% to 5%.

"Green channel" registration for merchants. To solve problem of large increases of new merchants, Meituan Takeaway launched "green channels": streamline for new registers. The process can be completed in only 5 steps, and the qualification review can be completed within 6 h at the earliest. It can be opened on the same day to ensure the existing supply for the meal period, ensure the richness of food types, and attract users to purchase.

Strengthen the examination. With the opening of the green channel and accelerating the qualification, the environment and safety of the merchants became major concerns for consumers. Considered that the original audit procedure was relatively weak for the merchant's sanitation inspection, Meituan Takeaway improved the inspection of the store's hygienic environment, storefront, food safety, and employee safety in order to avoid cross-infection, it will have a significant impact on its reputation and social image, thereby affecting user needs. It also responded to the government's relevant measures on wild animal prevention and control, and comprehensively investigated wild animal products on the platform.

5.2 Strategic Thinking at the Platform Level

Impeding the "De-Platform" Trend. The development trend of "de-platform" of takeaway due to the widespread commissions in the takeaway industry (10%–20%) and the long time to return the commission, as well as the problems of higher merchant promotion fees and longer service feedback time, etc. Many merchants found that relying solely on the platform to operate, the profit margin is low. Therefore, some merchants use multiple channels. For example, in May 2018, HEYTEA opened HEYTEA GO takeaway, and accumulated 24 million members in the past 2 years, users can use WeChat applet to order. More typical companies such as KFC's timely delivery, McDonald's McLeod delivery, etc., consumers can order food on their official website, and the merchants deliver independently. This way gets rid of the shackles of the platform and makes the transaction de-platform.

Limitations of "de-platform" of takeaway. Because de-platform requires consumers to accurately retrieve the company name and official purchase channel of the company, de-platform of takeaway is only for those who already have a certain scale of operation, a certain reputation and a certain scale of loyal users of the takeaway companies. For small and medium-sized micro-takeaway companies, due to their low popularity, consumers do not know it, which prevents them from accurately retrieving the name which lowers the sales. On the other hand, traditional marketing methods usually have low conversion rates and high costs, making it impossible for small and medium-sized enterprises to achieve a certain scale of de-platform transaction volume. Due to consumers' preference for convenience when purchasing, unless it is a company with high service quality and good reputation, for ordinary consumers will not spend time and cost to search one by one, thus limiting the platform for small and medium-sized enterprises to change to operating takeaway itself. Although there are some of enterprises have got rid of the high commissions of the platform, they have to face high promotion fees and low transaction volume. Therefore, the online takeaway business of small and medium-sized enterprises cannot achieve completely "de-platform."
Platform strategy to deal with "de-platform":

Price strategy. For merchants, partly lowering the commissions of merchants with higher transaction volume and greater reputation can slow down the process of "de-platform" of such enterprises. Sign contracts with merchants on the conditions of low-price promotion, restricting the sales of other channels of the store which can effectively suppress the de-platform process. For merchants with low transaction volume and average popularity, it is difficult to survive due to its "de-platform", so it can keep the commission unchanged, but it is necessary to speed up the service response and improve the service quality of the platform to the merchants to reduce merchants' dissatisfaction. For consumers with high price sensitivity, price promotion is possible, and the platform capital is used to conduct price wars against de-platform enterprises, which hinders their development.

Diversification strategy. Through the introduction of different functions, such as Meituan Takeaway food, desserts, supermarket convenience, vegetables and fruits and other functional modules, to meet the needs of consumers at all levels, to make it convenient for consumers. At the same time, regular updates should be made within each module, and recommendations for store foods, etc., should be rotated regularly to ensure their diversification, which could improve consumer perception of service quality, and thus improve consumer satisfaction, thereby fostering customer loyalty. Cultivate customer stickiness through diversification of products and services, increase customer switching costs, and achieve effective user retention.

Supply and Demand Adjustment. During the COVID-19, a large number of takeaway companies switched to online operations, resulting in increased supply. At the same time, due to safety and economic factors, consumers' willingness to take out food has declined compared with the past. This leads to the problem that supply exceeds demand. At the same time, there are changes in the demand side, such as delayed school opening, corporate online office and so on. Based on this, the platform first needs to ensure the existing demand, and ensure the user's demand by means of improving consumers' perceived quality and product safety. At the same time, the platform needs to adjust the existing staffing, etc., so that the existing needs are met. After satisfying the existing demand, a diversified strategy, differentiated strategy, price promotion, cross-price subsidies, etc. can be adopted to improve the quality of services and increase the types of products and businesses to develop different needs of consumers. As for the supply, due to the large supply, while ensuring product diversification, for homogeneous products, we can increase profits from the way of increasing commissions.

6 Conclusion

Based on the rapid development stage of the O2O model, Meituan Takeaway takes advantage of the opportunities provided by the Internet and the external characteristics of the cross-network in the two-sided market to flourish, but during the COVID-19, its supply side, demand side and platform operation have been seriously affected. As a leading O2O takeaway company, Meituan Takeaway has effectively responded to the negative impact of the COVID-19 and the trend of "de-platform" through diversification, differentiation and other operational strategies in response to the COVID-19, expanding revenue, and then promoting the optimization of the takeaway O2O model upgrade.

References

1. Rochet, J., Tirole, J.: Cooperation among competitors: some economics of payment card associations. RAND J. Econ. **33**(4), 549–570 (2002)
2. Rochet, J., Tirole, J.: Platform competition in two—sided markets. J. Eur. Econ. Assoc. **4**, 990–1029 (2003)
3. Evans, D.S.: The antitrust economics of multi-sided markets. Yale J. Reg. **20**, 325–382 (2003)
4. Armstrong, M.: Competition in two-sided markets. RAND J. Econ. **37**(3), 668–691 (2006)
5. Cui, X.: Research on Factors Affecting Repeated Purchase Intention of Takeaway O2O Users. Harbin Engineering University (2018)

Prospect Theory and Its Applications in Travel Behavior Research

Nan Zhou(✉)

Leonard Davis School of Gerontology, University of Southern California, McClintock Ave, Los Angeles, USA
nzhou352@usc.edu

Abstract. Prospect theory provides a theoretical basis for the study of decision behavior under uncertainty. In reality, it is difficult for travelers' decisions and behaviors to be completely rational and based on utility maximization, and in recent years, some researchers have attempted to use prospect theory for travel behavior analysis and modeling and have made some research progress. This paper first elaborates the basic framework and main contents of prospect theory, then analyzes the applicability of prospect theory in travel behavior research by reviewing related research. For specific travel behavior research, the application of prospect theory not only needs to consider the nature of the decision problem, but also to see the personality characteristics of the traveler.

Keywords: Prospect theory · Travel behavior · Application

1 Introduction

Dominating the study of decision behavior under uncertainty and risk is expected utility theory, based on the assumption that decision makers are perfectly rational human beings. Expected utility theory holds that decision makers make decisions based on the value of expected utility. Utility is the basic analytical tool for decision making, and rational decision makers usually choose the solution with the greatest utility as the optimal solution. However, in uncertain and risky decisions, predictions and judgments based on utility size using expected utility theory often do not correspond to actual human decision-making behavior in reality. In this context, Kahneman and Tversky (1979) extensively observed and studied the judgment and decision-making behavior of decision makers under uncertainty and proposed a prospect theory by incorporating the findings of psychology into economics [1]. Prospect theory reveals the psychological and behavioral mechanisms of finite rational human beings in the process of risky decision making, which more realistically reflects the characteristics and rules of people's decision making under uncertainty [2].

In recent years, researchers have paid increasing attention to the limited rational decision-making characteristics exhibited by travelers in their daily travel activities, and while questioning the traditional travel behavior research framework, they have

C. Yuan et al. (Eds.): *Proceedings of the 4th International Conference on Economic Management and Green Development*, AEPS, pp. 37–43, 2021.
https://doi.org/10.1007/978-981-16-5359-9_5

attempted to seek a more realistic decision theory. Among them, some researchers have attempted to use prospect theory for travel behavior analysis and modeling and have made some research progress. Since this research involves not only knowledge of transportation science, but also interdisciplinary knowledge such as psychology, behavioral economics, and experimental economics, it places a higher demand on the knowledge structure of the researchers. Many of these researchers view prospect theory as merely an improved model of expected utility theory, lacking a deep understanding of what prospect theory entails.

Wee (2010) points out that only those who truly understand the content of prospect theory understand its importance and value in travel behavior research [3]. Based on this, this paper first elaborates the basic framework and main contents of prospect theory, then discusses the applicability of prospect theory in travel behavior research by reviewing related research.

2 Basic Framework of Prospect Theory

The study of human decision making behavior under uncertainty is a popular area of research in economics, with the most classical model of rational choice being expected utility theory. In recent decades, however, the theory has encountered many problems and it cannot explain the numerous phenomenon. Psychological research has found that people's decision-making processes are not completely rational as assumed in traditional theories. To correct and compensate for the deficiencies of traditional economic theory's assumptions about rationality, the prospect theory proposed by Tversky and Kahneman is based on psychological experiments and directly incorporates factors such as individuals' decision-making behavior and psychological perceptions into the analysis of decision-making behavior [4]. One of the basic components of prospect theory is the framing effect, and prospect theory uses both the weighting function and value function to describe the subjective overall value of an individual.

2.1 Framing Effects

There exist different phases within the prospect theory. There is the first spectrum which is termed as the Framing Effects. It is termed as framing or as editing in the first phase. In this phase, the main concept is the rule of having to emerge the concepts of the possible outcomes that may be linked to a particular set of aspects. In a situation where one is able to extrapolate the possible ideologies that are built around a certain choice, then the whole idea of having to make a choice is simplified. The framing effect can thus be transpired as a way in which the choice can be affected in the way that it is transpired to the one making the actual decision. In most of the cases, the set party may be faced with the problem of not actually knowing what choices they may have. It forces them to have a sit-down and evaluate their actual options.

The main purpose of framing is to be able to reduce and actually simplify the choices that are being presented at a certain stage. We understand that the decisions at times might be rather difficult and numerous with a close relationship or having to follow a set of similar ideologies. Framing simplifies this out. The process that is used for the simplification includes acceptance and segregation. A varied run to evaluate these choices is what forms this phase. In the construction and formation of such ideologies and options, it becomes clear that the order by which the choices have emerged actually determines the complete set of the options to take and those to drop. It all falls on the basic fact of how the options are presented with the options in a cognitive manner. Their situations and the psychological structure are what phases out what remains important to them and that which is rather a brush-off.

2.2 Value Function

There is also the factor of value function as the other phase. Once the choices have been edited, then the decision-maker is presented with the case scenario of having to evaluate the options and make a decision amongst all of them. This is what is called the evaluation aspect or phase. It consists of two main aspects. These are the value function and the weighting function. The value function is what takes on a prior step. The aspects of the value function go on to take an evaluation of the gains and the losses with regard to the reference point. It should be noted that this does not take on the final assumption of data and calculation of wealth. The one that took such a form was based on the spectrum of the utility theory. The main aspects with the prospect theory are that the variables tend to take on a more shifting perspective as compared to a stagnant and rested aspect. The main regard that is depicted with this phase is giving more prominence to the starting point of the process. In the theory, it is what is termed as the reference point [4].

The shape of the value function specified by Kahneman and Tversky is an exponential function:

$$v(x) = \begin{cases} x^{\alpha}, & x \geq 0 \\ -\lambda(-x)^{\beta}, & x < 0 \end{cases} \quad (1)$$

In Eq. (1), parameters α and β, respectively represent the unevenness of the power function of the value in the gain and loss area, and $\alpha, \beta < 1$ indicate diminishing sensitivity. λ is used to express the characteristic that the loss area is steeper than the gain area. $\lambda > 1$ means loss aversion, and its function image is shown in the Fig. 1 below:

There is the asymmetrical point of the value function. The value curve is depicted to be steeper on the side of the losses than it is on the side of the gains. It depicts a situation of relative loss aversion. The main spectrum of having to experience the adverse pleasures of a gain is incomparable to what it seems when coming into a closer battle with the pain of having to face any form of loss. Depicted that a certain status quo is depicted to be quite stable or stagnant then the set individuals are presented with being totally fine with the current position as opposed to having downgrade or fall below what they might term as being quite satisfactory on their side might be quite uncomfortable and bring in an arraigned aspect of problems and issues.

2.3 Weighting Function

There is then the weighting function. The outcomes of the options are given a value that is defined as a decision weight. This weight is not linked in any case with the

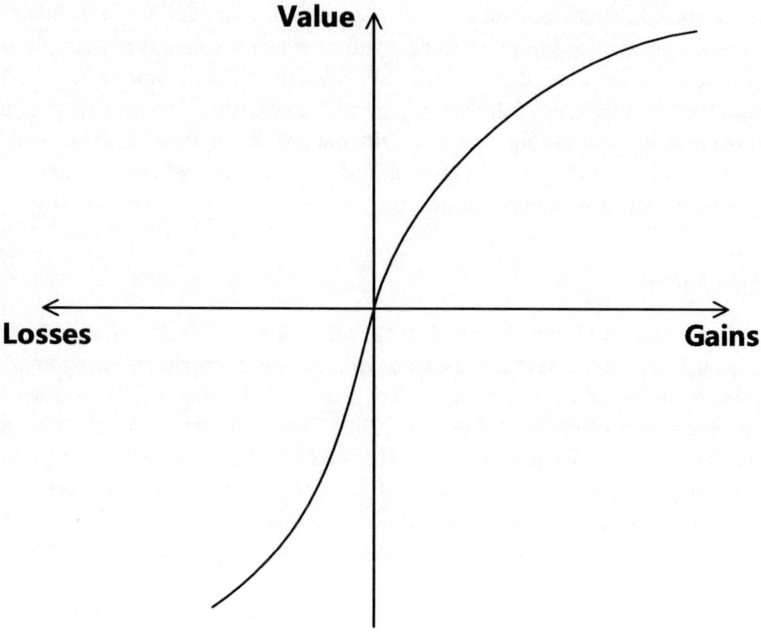

Fig. 1. Value function

probability assumptions. In the complete decision of having to make the decision then the decision-maker will have to multiply the value of each outcome to the complete decision weight that is given to it. The decision weights should be noted not to be used or served as the sole aspect of having to identify with the likelihood of having a specific choice occurring. It is used in accordance to other factors. It actually gives a path and an explanation of how the individuals actually went on to arrive at their final decision. The main domain of the decision weights is seen to be affected by factors such as that one of ambiguity. Therefore, a nonlinear function of the decision weight π and the objective probability can be constructed. $\pi(p)$ is the proportion of the weight of p to the weight of a deterministic event. The following Fig. 2 shows the decision weight function relative to p: $\pi(p)$, the dotted line in the figure is the objective probability p, and the solid line is $\pi(p)$, which embodies the decision weight function.

The main domain of the weighting function is the fact that it does not go on to operate in the far end. Probability is what is termed to give a envision of what the factors are bound to discuss. Some factors that are described as being impossible to some level by certain individuals, it is termed that such ideological sets are termed to be much more depicted with more psychological weights than all other events. The main problematic aspect kicks in when the most obvious and bountiful ideology is bound to take place then it happens that it does not.

The other characteristic of the weighting function is the fact that the low probability events are overweighed while the medium and the high probability events go on to be underweighted to some specific level. The aspect in this case is that the events that prove

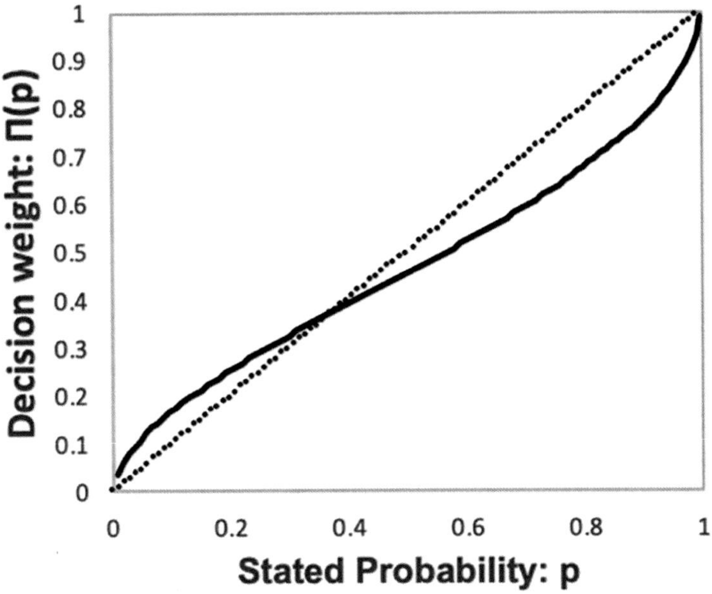

Fig. 2. Weighting function

to be very likely to occur are not given the adequate amount of attention that they deserve. The aspects that go on to show some form of unlikeliness in the events showcasing occur depict a more escalated form of attention-drawing onto this fact. Such is what remains to be the very sense of having to evolve the ideologies of insurance and that of betting.

3 Application of Prospect Theory in Travel Behavior Research

One of the biggest problems with traditional travel decision models is the problem of behavioral assumptions. Due to the complexity of human behavior, many models in transportation science are often somewhere between completely right and completely wrong, and their theories may be too simple to be suspected, or too complex to be discarded. The formation of traffic problems is usually inseparable from the decision-making choices of travelers, and is in fact a travel behavior problem, the study of which should draw more on the findings of behavioral science.

Prospect theory, which is based on gambling experiments and measures gains and losses relative to reference points in monetary terms, has shown good explanatory and predictive power in studies of investor behavior in financial markets and consumer behavior in commodity markets. However, travel decision situations differ from gambling experiments in that travel choices generally do not involve wealth gains and losses, and potential gains and losses cannot be directly measured in monetary terms. When applying prospect theory to travel behavior research, it is first necessary to clarify its boundary conditions or situational constraints and the question of whether the specific subject satisfies the same conditions or situations as such, i.e., the question of validating the applicability of the theory. Katsikopoulos et al. (2002) found experimentally that travelers' risk preferences are reversed during the travel decision process, with travelers exhibiting risk aversion when the path average travel time is less than a particular value, and risk-seeking vice versa, which is consistent with prospect theory [5]. The experimental results of Avineri & Prashker (2005) [6] show that the path selection behavior of travelers has nonlinear probability weights and loss aversion effects. However, travel choice behavior, which includes not only the choice of travel route and departure time, but also the choice of travel mode, destination, and the decision to travel or not, is a complex decision-making process undertaken by travelers. There is a lack of sufficient evidence on the applicability of prospect theory to the study of other travel choice behaviors. In addition, most of the option attributes considered in existing studies are based on time factors, usually measuring gains and losses in terms of travel time, and less so for other attributes. In fact, there are numerous option attributes that influence travel decisions, some of which are uncertain and some of which are deterministic, and the applicability of prospect theory depends crucially on which attributes are primarily considered by decision makers. For example, for the travel mode choice problem, travelers consider influencing factors such as travel time, reliability, comfort, safety, accessibility, and cost of travel, and if travelers are primarily concerned with the cost of travel, and the cost is deterministic, prospect theory no longer applies to this problem. In reality, travelers typically consider multiple attributes when making decisions, so not only does the study involve the problem of translating between different attributes, but also the reference point is often difficult to determine. Prospect theory should have a broader application in the field of transportation research, for example, for problems such as traffic violations, noise and exhaust pollution, congestion pricing, and information response mechanisms, if there is uncertainty in the properties of the choice options, the application of prospect theory to the analysis of the travel decision process can be considered. For example, Han et al. (2005) [7] found that travelers' responses to information depend not only on whether the content of the information is consistent with subjective beliefs, but also on how the information is presented when it is provided, suggesting that travelers' responses to information have a framing effect, which, as a heuristic decision rule, can be explained by prospect theory.

Daily travel is usually not a one-time activity, and more often than not, travelers are faced with a sequential decision-making situation that includes an information feedback mechanism. Travelers may not be fully aware of the current road network conditions before each trip and make decisions based mainly on experience, information and habits, exhibiting an adaptive learning behavior. The traditional view is that adaptability makes people behave in a risk-averse and expected utility maximizing manner [8]. Benartzi and Thaler (1995) found that people's decision-making behavior in sequential decision-making situations that include information feedback can still be described by prospect theory [9]. Most existing prospect theory-based travel behavior research revolves around one-time decisions made by travelers, with little consideration given to the sequential nature of the decision process and the adaptive learning behavior of travelers. As the number of trips increases, the traveler's familiarity with the condition of the road network will increase, the level of subjective uncertainty will decrease accordingly, the reference point will usually be dynamically updated and adjusted, and the decision maker's risk appetite and loss aversion may change. Of course, for specific travel behavior studies, the applicability of prospect theory depends not only on the nature of the decision problem, but also, crucially, on the personal characteristics of the traveler and whether the properties of the options under consideration are uncertain.

4 Conclusions

This paper summarizes the basic elements of prospect theory and analyzes its application to the study of travel behavior. The introduction of prospect theory into the field of transportation science can not only broaden the scope of application of behavioral science theory, but also better take into account the traveler's decision-making process and choice behavior mechanism, which can more realistically depict the interaction between the traveler's microscopic decision-making behavior and the macroscopic operating conditions of the transportation system, and promote the development of transportation demand prediction theory. Prospect theory is also not a fully developed theory, it is still in the process of improvement and development, and it cannot completely replace expected utility theory. In addition to the prospect theory, some other non-expected utility theories are also evolving and are even supported by more experimental evidence. Instead of being boxed into a particular theory before undertaking a study, different theories should be weighed according to their characteristics and in relation to what is to be studied, and ultimately the more appropriate theory should be chosen.

References

1. Tversky, K.A.: Prospect theory: an analysis of decision under risk. Econometrica **47**(2), 263–291 (1979)
2. Barberis, N.C.: Thirty years of prospect theory in economics: a review and assessment. J. Econ. Perspect. **27**(1), 173–196 (2013)
3. Van Wee, B.: Prospect theory and travel behaviour: a personal reflection based on a seminar. EJTIR **10**(4), 385–394 (2010)
4. Hwang, I.D.: Prospect theory and insurance demand. SSRN Electron. J. (2015)
5. Katsikopoulos, K.V., Duse-Anthony, Y., Fisher, D.L., Duffy, S.A.: Risk attitude reversals in drivers' route choice when range of travel time information is provided. Hum. Factors **44**(3), 466–473 (2002)
6. Avineri, E., Prashker, J.N.: Sensitivity to travel time variability: travelers' learning perspective. Transport. Res. Part C Emerg. Technol. **13**(2), 157–183 (2005)
7. Han, Q., Dellaert, B.G., Van Raaij, W.F., Timmermans, H.J.: Integrating prospect theory and stackelberg games to model strategic dyad behavior of information providers and travelers: theory and numerical simulations. Transp. Res. Rec. **1926**(1), 181–188 (2005)
8. Becker, G.S.: Crime and punishment: an economic approach. J. Polit. Econ. **76**(2), 169–217 (1968)
9. Thaler, R.H., Benartzi, S.: Myopic loss aversion and the equity premium puzzle. Quart. J. Econ. **110**(1), 73–92 (1995)

Prediction of Stock Prices Based on Markov Chain

Ke Wu(✉)

Herbert Business School, University of Miami, 1320 S Dixie Hwy, Miami, USA

Abstract. Markov Chain is a random process with Markov property in probability theory and mathematical statistics, which exists in discrete exponential set and state space. The essence of the Markov chain prediction model is "no after effect". No after effect generally refers to the state of things in the future is only related to the state of this stage and has nothing to do with the state in any previous stage [3]. The Markov chain suitable for continuous exponential set is called Markov process. But it is sometimes regarded as a subset of Markov chain, namely Continuous-Time MC, CTMC, and Discrete-Time MC, DTMC correspondingly. So, Markov chain is a relatively broad concept. Based on the Markov chain, this paper makes a prediction on the closing price of Shanghai Stock Exchange Index. The stock market is risky. There are many ways to predict the stock market, which can be summarized into two categories: stock price fluctuation prediction models based on statistical theory [1] and artificial intelligence prediction models [2]. The paper introduces a brief introduction of Markov chain and uses the case of the closing price of the above stock index to measure the accuracy of Markov chain price prediction.

Keywords: Markov chain · Stock price prediction · Shanghai stock exchange index

1 Introduction of Markov Chain

1.1 Definition

Markov chain is a stochastic process. We can also call it Discrete Time Markov Chain (DTMC). Simply put, a stochastic process has the Markov property if its future evolution depends only on its current position, not on the whole history of how it got there.

A stochastic process $\{X_n,\ n = 0, 1, 2, \ldots\}$ is a Markov Chain if it satisfies the Markov property, i.e.,

$$\begin{aligned}\Pr(X_{n+1} = i_{n+1}|X_n = i_n, X_{n-1} = i_{n-1}, \ldots, X_1 - i_1, X_0 = i_0) \\ = \Pr(X_{n+1} = i_{n+1}|X_n = i_n)\end{aligned} \tag{1}$$

for all n and for all $i_0, i_1, \ldots, i_{n+1}$.

C. Yuan et al. (Eds.): *Proceedings of the 4th International Conference on Economic Management and Green Development*, AEPS, pp. 44–50, 2021.
https://doi.org/10.1007/978-981-16-5359-9_6

1.2 Elements

A Markov Chain $\{X_n, \text{n} > 0\}$ is defined by 3 elements.

- State Space S
  State Space S is a finite or countable set of states that the random variables X_n may take on.
- Transition Probability p_{ij}

$$p_{ij} = \Pr(X_{n+1} = \text{j}|X_n = \text{i}|), \text{ for all } i, j \in \text{S}, \text{ for all } n > 0 \quad (2)$$

- Initial Probability Distribution α

$$\alpha_i = \Pr(X_0 = \text{i}), \text{ for all } i, \in \text{S}, \text{ with } \Sigma_{i \in \text{S}} \alpha_i = 1 \quad (3)$$

1.3 Probability Distribution of X_n

$$\pi_n = \alpha P^n, \text{ where } \pi_n \in \mathbb{R}^{1 \times m}, \ \alpha \in \mathbb{R}^{1 \times m}, \ \text{P} \in \mathbb{R}^{m \times m} \quad (4)$$

We use the notation $\pi_n(i) = \Pr(X_n = i)$.

1.4 Stationary Distribution

Let $\{X_n, \text{n} \geq 0\}$ be a Markov Chain with initial probability distribution α, state Space S, and transition probability matrix P. A probability distribution π is called stationary distribution for the Markov chain if

$$\pi = \pi P \text{ or } \pi_j = \Sigma_{i \in \text{S}} \pi_j p_{ij} \text{ for all j} \in \text{S} \quad (5)$$

$$\Sigma_{i \in \text{S}} \pi_i = 1 \quad (6)$$

2 Markov Model Establishment

2.1 Three-State Model

The model in this article is based on three states of stocks: up, flat, and down. Set the stock price of the previous day as the benchmark, the status is flat if the stock of prices' fluctuation is within a certain range of points. Likewise, the status is up if the stock of price goes up to exceed a certain range of points. The status is down if the stock of price goes down exceeds a certain point is a decline. The number of points can be determined based on different data. In the following case, we set that the stock goes up if it increases more than 40 points based on the last day's stock price. The stock goes down if it decreases more than 40 points based on the last day's stock price. Its state space is expressed as i = $\{1, 2, 3\}$, which represents up, flat, and down respectively.

Transition probability matrix P is as following:

$$P = \begin{bmatrix} P_{11} & P_{21} & P_{31} \\ P_{12} & P_{22} & P_{32} \\ P_{13} & P_{23} & P_{33} \end{bmatrix} \tag{7}$$

In the matrix, P_{ij}(i = 1, 2, 3; j = 1, 2, 3) represents the probability from state i to state j. For example, P_{11} represents the probability from state 1 to state 1, which means if today's stock goes up, then the probability of the next day's stock also going up. Let the initial state probability vector be $\pi(0)$, then the probability distribution of the nth period is π(n), which is $\pi(n) = P^n\pi(0)$.

2.2 Mixed Markov Model

The model in this article has more than three intervals. By collecting the daily closing prices of the Shanghai Stock Exchange Index from December 10, 2010, to July 8, 2011, we found the highest and lowest prices, which are used as the range of stock price changes. Taking the change of 40 points as the benchmark, dividing the difference between the highest price and the lowest price by 40, we get the number of intervals, which is n.

Transition probability matrix P is:

$$P = \begin{bmatrix} P_{11} & P_{21} & \ldots & P_{(n-1)1} & P_{n1} \\ P_{12} & P_{22} & & P_{(n-1)2} & P_{n2} \\ \vdots & & \ddots & \vdots & \\ P_{1n} & P_{2n} & \ldots & P_{(n-1)n} & P_{nn} \end{bmatrix} \tag{8}$$

In the matrix, P_{ij}(i = 1, 2,..., n; j = 1, 2, ..., n) represents the probability from state i to state j. The initial probability distribution is π(0). The nth period's probability distribution is π(n), which is $\pi(n) = P^n\pi(0)$.

3 Example

3.1 Data

This paper selects the Shanghai Stock Exchange Index and uses a total of 140 data on the daily closing price of the Shanghai Stock Exchange Index from December 10, 2010, to July 08, 2011, as a sample. The highest value of the closing price of the Shanghai Stock Exchange Index is 3057.33 and the lowest value is 2621.25.

3.2 Price Change

3.3 Transition Probability P

In the model, the three states of up, flat, and down of the Shanghai Stock Exchange Index are marked as $i = \{1, 2, 3\}$. From Table 1, the frequency of the up state is 11. The frequency of the flat state is 114. The frequency of the down states is 16. Because the

Table 1. The ups and downs of the Shanghai stock exchange index.

1	up	21	flat	41	flat	61	down	81	flat	101	flat	121	flat
2	flat	22	flat	42	flat	62	flat	82	flat	102	flat	122	flat
3	flat	23	flat	43	flat	63	flat	83	flat	103	flat	123	flat
4	flat	24	flat	44	flat	64	flat	84	down	104	flat	124	down
5	flat	25	down	45	flat	65	flat	85	flat	105	flat	125	flat
6	down	26	flat	46	down	66	flat	86	flat	106	flat	126	flat
7	up	27	up	47	flat	67	flat	87	flat	107	down	127	flat
8	flat	28	down	48	flat	68	flat	88	down	108	flat	128	flat
9	flat	29	flat	49	flat	69	flat	89	flat	109	flat	129	flat
10	flat	30	flat	50	flat	70	flat	90	flat	110	flat	130	up
11	down	31	flat	51	flat	71	flat	91	flat	111	flat	131	flat
12	down	32	flat	52	flat	72	flat	92	flat	112	flat	132	flat
13	flat	33	up	53	flat	73	flat	93	flat	113	flat	133	flat
14	flat	34	flat	54	flat	74	flat	94	down	114	flat	134	flat
15	up	35	flat	55	up	75	flat	95	flat	115	flat	135	flat
16	up	36	flat	56	flat	76	flat	96	flat	116	flat	136	up
17	flat	37	flat	57	flat	77	flat	97	flat	117	flat	137	flat
18	flat	38	up	58	down	78	flat	98	flat	118	flat	138	flat
19	flat	39	flat	58	flat	79	flat	99	flat	119	down	139	flat
20	down	40	up	60	flat	80	flat	100	flat	120	flat	140	flat

last state is flat, which means there is no subsequent state, so the frequency of the flat state is 113. Then count the transition frequency of each state. Among the 11 up states, 1 continues to maintain the up state, 9 moved to flat state, and 1 moves to down state. So, $P_{11} = 1/11 = 0.0909$, $P_{12} = 9/11 = 0.8182$, and $P_{12} = 1/11 = 0.0909$. Likewise, we can get other transition probability.

So, transition probability matrix is as following:

$$\mathrm{P} = \begin{bmatrix} P_{11}\ P_{21}\ P_{31} \\ P_{12}\ P_{22}\ P_{32} \\ P_{13}\ P_{23}\ P_{33} \end{bmatrix} = \begin{bmatrix} 0.0909\ 0.8182\ 0.0909 \\ 0.0708\ 0.8142\ \ 0.115 \\ 0.0667\ 0.8667\ 0.0666 \end{bmatrix} \tag{9}$$

So, the two-step transition probability matrix is as following:

$$P^2 = \begin{bmatrix} 0.072254\ 0.81934\ 0.10841 \\ 0.071752\ 0.82052\ 0.10773 \\ 0.071868\ 0.81796\ 0.11017 \end{bmatrix} \tag{10}$$

Likewise, the n-step transition probability matrix is:

$$P^n = \begin{bmatrix} P_{11} & P_{21} & P_{31} \\ P_{12} & P_{22} & P_{32} \\ P_{13} & P_{23} & P_{33} \end{bmatrix}^n \tag{11}$$

3.4 Mixed Markov Model

In the dataset, the highest value of the closing price of the Shanghai Stock Exchange Index is 3057.33 and the lowest value is 2621.25. Taking 40 points as the interval difference, the stock price can be divided into 11 intervals. Each interval has a separate state, i = {1, 2, 3..., 11}. The 11 states (2620, 2660), (2660, 2700), (2700, 2740), (2740, 2780), (2780, 2820), (2820, 2860) 7(2860, 2900), (2900, 2940), (2940, 2980), (2980, 3020), (3020, 3060). This is the mixed state of the Markov chain.

By counting the frequency of each interval state and the frequency of state transition, the state transition probability matrix is as following:

$$P = \begin{bmatrix} 0.7500 & 0.2000 & 0.0000 & 0.0000 & 0.0000 & 0.0000 & 0.0000 & 0.0000 & 0.0000 & 0.0000 & 0.0000 \\ 0.2500 & 0.2000 & 0.1250 & 0.0556 & 0.0000 & 0.0000 & 0.0000 & 0.0000 & 0.0000 & 0.0000 & 0.0000 \\ 0.0000 & 0.4000 & 0.5000 & 0.2222 & 0.1667 & 0.0000 & 0.0000 & 0.0000 & 0.0000 & 0.0000 & 0.0000 \\ 0.0000 & 0.2000 & 0.3750 & 0.5000 & 0.4167 & 0.0588 & 0.0000 & 0.0000 & 0.0000 & 0.0000 & 0.0000 \\ 0.0000 & 0.0000 & 0.0000 & 0.2222 & 0.1666 & 0.1765 & 0.0000 & 0.0000 & 0.0000 & 0.0000 & 0.0000 \\ 0.0000 & 0.0000 & 0.0000 & 0.0000 & 0.2500 & 0.4118 & 0.2500 & 0.0455 & 0.0000 & 0.0000 & 0.0000 \\ 0.0000 & 0.0000 & 0.0000 & 0.0000 & 0.0000 & 0.2353 & 0.4500 & 0.3182 & 0.0000 & 0.0000 & 0.0000 \\ 0.0000 & 0.0000 & 0.0000 & 0.0000 & 0.0000 & 0.1176 & 0.3000 & 0.5000 & 0.3333 & 0.0000 & 0.0000 \\ 0.0000 & 0.0000 & 0.0000 & 0.0000 & 0.0000 & 0.2353 & 0.0000 & 0.1363 & 0.3333 & 0.3333 & 0.0000 \\ 0.0000 & 0.0000 & 0.0000 & 0.0000 & 0.0000 & 0.2353 & 0.0000 & 0.0000 & 0.3334 & 0.4445 & 0.2500 \\ 0.0000 & 0.0000 & 0.0000 & 0.0000 & 0.0000 & 0.2353 & 0.0000 & 0.0000 & 0.0000 & 0.2222 & 0.7500 \end{bmatrix}$$

3.5 Reverse Test of Markov Chain

We select the stock market closing price from July 11, 2011, to July 22, 2011, to test the accuracy of the Markov chain.

It is known that the one-step state transition matrix is P in the mixed Markov Chain model. According to the definition, the n-step probability distribution is the initial state matrix multiplied by the P^n matrix. According to dataset, the closing price of the last day is 2797.77, which belongs to the state (2780, 2820), which is state 5. So, the initial state matrix α is:

$$\begin{bmatrix} 0 & 0 & 0 & 0 & 1 & 0 & 0 & 0 & 0 & 0 & 0 \end{bmatrix}$$

And

$$\pi_1 = \alpha P^1$$

$$\pi_2 = \alpha P^2$$

$$\pi_3 = \alpha P^3$$

$$\vdots$$

$$\pi_{10} = \alpha P^{10}$$

$$\pi_{11} = \alpha P^{11}$$

We can get the predicted value from the above equations. Compared with the actual value, the corresponding accuracy rate can be obtained. It can be seen from Table 2 that the average accuracy rate is 99.155% and the average error is only 0.845%. It can be seen that the accuracy of the Markov chain can be recognized to a certain extent.

Table 2. Reverse test of Markov Chain.

Date	Closing price	Predicted value	Accuracy rate	Error
07/11/2011	2,802.69	2779.996	99.190%	0.810%
07/12/2011	2,754.58	2778.2412	99.148%	0.852%
07/13/2011	2,795.48	2777.2552	99.348%	0.652%
07/14/2011	2,810.44	2777.2008	98.817%	1.183%
07/15/2011	2,820.17	2777.5124	98.487%	1.513%
07/18/2011	2,816.69	2778.0492	98.628%	1.372%
07/19/2011	2,796.98	2778.7276	99.347%	0.653%
07/20/2011	2,794.20	2779.4648	99.473%	0.527%
07/21/2011	2,765.89	2780.2628	99.483%	0.517%
07/22/2011	2,770.79	2781.1828	99.626%	0.374%
Average accuracy rate		Average error		
99.155%		0.845%		

4 Conclusion

Markov chain can only predict the price of stocks to a certain extent. In addition to normal changes in stock prices, it will also be subject to human factors such as national policies, company changes, natural and man-made disasters. These factors cannot be predicted by the Markov chain.

References

1. Yan, C.-N.: Application of density evolution method for obtaining up days of stock market. J. Shanghai Univ. **9**(2), 184–187 (2003)
2. Filippo, C.: Forecasting price increments using an artificial neural network. Adv. Comp. Syst. **4**(1), 45–56 (2001)
3. Tai, W.-Z.: Using Markov chain model to forecast stock market's short-term trend. J. Southwest Univ. National. **34**(3), 477–481 (2009)

Business Analysis on Philanthropy Activities of Vulnerable Groups Based on the Theory of Emotional Contagion

Yujie Shan(✉)

North Broward Preparatory School, 7600 Lyons Road, Coconut Creek, FL 33073, US

Abstract. In order to overcome the limitations of traditional philanthropy activities between philanthropy organizations and participants, with the application of the theory of emotional contagion, this paper adopts a business analysis approach combining fashion and charitable activities, focusing on the hearts of the disadvantaged by commercializing their creations and linking charitable organizations and consumers in order to help the disadvantaged. As the form of philanthropy activities will be innovated from the new proposal, assistance to vulnerable groups will also be improved and public participation will be enhanced. The business research from the perspective of economic model, competitors, market research, and implementation, provides a new idea for rescuing the disadvantaged.

Keywords: Business analysis · Emotional contagious theory · Economic model

1 Introduction

All the time, various institutions, organizations and individuals have tried to participate in the activities of helping vulnerable groups in different forms. Contagious Emotion is a company that aims to effectively connect the thoughts and feelings of psychologically disadvantaged groups with others, and to promote the awareness of psychologically vulnerable groups among young people in a daily fashion. This paper is based on the application of the theory of emotional contagion, we believe that fashion can be a great way to portrait the people's inner world. By showing the creative achievements of vulnerable groups to others, it opens a new channel for people to know more about vulnerable groups.

2 Problems with Traditional Philanthropy Activities

There are problems and pain from the both sides, philanthropy organizations and people who would like to contribute. From the philanthropy organization side, these organizations are actively organizing events and activities to improve people's awareness about disadvantage people. However, most of these activities are organized through forms of volunteers and donations. In this case, there are three problems. Firstly, only the small

C. Yuan et al. (Eds.): *Proceedings of the 4th International Conference on Economic Management and Green Development*, AEPS, pp. 51–56, 2021.
https://doi.org/10.1007/978-981-16-5359-9_7

number of volunteers may deeply communicate with the disadvantaged people. People who participate through donations still cannot understand. Secondly, these activities are hold occasionally. People are not consistently involved, thus it's hard for them to become empathetic with disadvantage people, and they can forget easily. Thirdly, these activities require people's relatively large amount of time and efforts, especially volunteering, which people may not able to dedicate. Therefore, philanthropy organizations are looking for solutions that can get people involved more easily, more often, and more effectively.

From people's perspective, they are willing to taking care of others, and want to contribute their parts of efforts. However, they find it hard. Based on the research, there are mainly three reasons. Firstly, as stated above, almost all philanthropic activities require time and efforts, but this requirement, or we can call it demand, can hardly match with what people can contribute, which we can call it the supply. When the demand and supply doesn't match, the effectiveness would be unsatisfactory. Secondly, some people find a lot of the philanthropic events repetitive, boring, and lack of creativity. Therefore, they cannot really connect themselves with those who really need help through just several occasions of the events. Thirdly, and the most importantly, a lot of people find it hard to know all the events and activities going on. For instance, whenever people come up with dining they use Yelp; whenever people think of traveling, they use Expedia or TripAdvisor. However, when people think of making a difference in disadvantaged people's life, they don't know how to start.

In conclusion, there is a mismatch from both the demand and the supply sides, so either the philanthropy organizations or the people who want to contribute are having problems. It is significant to solve the problem and match supply with demand.

3 Theory and Application of Emotional Contagion

3.1 Theory about Contagion and Emotion

The word "Contagion" is derived from the Latin word "contagio", which means "from contact". The definition of contagion can be divided into two categories: medical contagion and psychological contagion. Emotional contagion is a psychological term, first defined by McDougall (1923) as "the direct sensory law of emotion produced by the primitive sympathetic response". Hatfield (1993) enriched and developed the theory of emotion contagion, arguing that emotion contagion is the process of perceiving emotions through direct contact with others, and is an individual's sensitivity to the emotional perceptions of others. [1] Hoffman (2000) defined emotional contagion as an emotional experience that is elicited by an emotional trigger and converges with the emotional state of the recipient. [2] Lundqvist (2006) proposed that emotional contagion includes the readiness or inertia in the emotional system to yield to the emotions expressed by others [3].

The term "emotional contagion" is a psychological term borrowed from the medical terminology of "Contagion", which in medical terms refers to pathological reactions and damage to an organism caused by the invasion and multiplication of pathogenic microorganisms and parasites, or the invasion of pathogens by an infectious agent through a certain route. The other, because emotions, like pathogens, can be transmitted from one individual to another, in the environment, using specific messages, body language, facial expressions as a means of transmission, either unconsciously imitated or consciously exaggerated, thus realizing the "transmission" of emotions.

3.2 Application of Emotional Contagion Theory—Contagious Emotion

After observing people with depression for a long time, we have found that many disadvantaged people enjoy drawing. When they don't want to talk or don't know how to communicate with others, or when they just want to express how they feel, they draw them down. Their drawings may not have the traditional meaning of art, but the drawings demonstrate and introduce their world to others. Through these drawings and a few words, they express themselves.

We believed that clothing could be a great canvas for them. Therefore, we created *Contagious Emotion*, a fashionable streetwear brand that could reintroduce philanthropy to people of all ages. *Contagious Emotion* will be the link between philanthropy and its customers.

Contagious Emotion will make clothing and other fashion products based on drawings from people with psychological illness. On the one hand, we help philanthropy organizations and the disadvantaged people spread the words. Using this method, we are able to help philanthropy organizations integrate events and activities into people's daily lives. On the other hand, we provide an easy way for people who would like to get involved in philanthropy events and activities. At the same time, people can learn more about the psychologically disadvantaged groups, then get better understanding about them.

In this way, *Contagious Emotion* not only improves people's awareness towards psychological disadvantaged people's feel, but also raises the awareness of those around them and protects their loved ones.

3.3 Combination of Emotional Contagion Theory and Philanthropy

The United States has always been a country and a culture that promotes diversity. Although a lot of efforts and actions have been taken, there is still much room for improvement and refinement. We have noticed that diversity is not just about different ideas and behaviors from different cultures, but also all thoughts and feelings from people who may not be able to communicate easily with others, such as those who are depressed or otherwise psychological disadvantaged. In order to build a connection between this number of people and others, we decided to create a social enterprise. We believe that business can be a successful channel to transforming attitudes and behaviors, thus effectively facilitating the exchange of ideas and emotions between different groups of people.

Consequently, it will change people's perceptions of psychologically disadvantaged people and how they see the world. We hope through Contagious Emotion, we can not only create better communication between psychological disadvantaged people and others, but also change the way people see the world and how they feel about each other, thus making the world better.

As mentioned above, many people in our research indicated that they found most of the philanthropic events boring and repetitive. They still participate simply because it is their responsibility. For some young people, participating in these activities is simply a new form of workplace socialization. Participation in philanthropic activities is gradually becoming a burden. However, it shouldn't be.

We believe that philanthropy should not be an extra burden or a boring mission for anyone, but a part of people's lives. We want to create a culture that embrace our belief. We believe that fashion is a great way to portray people's inner world. By wearing *Contagious Emotion*, customers can get a better understanding of the disadvantaged groups, and part of the profits will go towards healing them. At the same time, by wearing the clothes, our customers broadcast our ideas to wider audience.

4 Business Analysis of *Contagious Emotion*

4.1 Business Model

Contagious Emotion will be working intensively with charities, providing supplies and production channels for philanthropy, who will provide drawings and ideas from disadvantaged groups. We will supply the charity with supplies and production channels, who will provide drawings and ideas from disadvantaged groups. *Contagious Emotion* will be responsible for transforming these drawings and ideas into stylish streetwear products that will be sold to customers. 50% of the profits will then be used to help the psychologically disadvantaged.

Our income comes from the sale of products. Our expenses can be divided into two parts. Firstly, our initial costs are renting a factory, working space and salaries for our team members. Second, the operating costs include labor, rent, utilities, distribution, marketing, and legal fees and so on.

4.2 Competitors

While our competitors could be all streetwear fashion brands, due to our distinctive positioning, our main competitors are brands that also focus on social issues through their clothing. Here are the three main competitors in the market.

NOAH, a brand that focuses on packing waste and single-use plastics. The brand frequently uses its platform to support and discuss environmental causes and hosts very open conversations with customers through its blog and social media about packaging waste, factory overrun disposal and the problems with single-use plastic.

CHNGE, a brand focuses on organic cotton. CHNGE was created to " give concerned consumers an avenue to make a positive impact on the world through their purchases". The brand, which donated 50% of its net profits to charities such as the Malala Fund and Charity, their factory produces using 100% organic cotton to embrace ethical.

CATCHING AFISH IN NORWAY focuses on completely organic and fair trade materials. "I've noticed a real lack of ethical or sustainable alternatives for 'streetwear,'" said founder Paul Donati via email. "Everytime I searched for fair trade clothing I always got the same result—a hippie/hemp/middle-aged clothing ranges, but no good alternatives for people our age. I decided to set up my own social enterprise to address the various issues in today's fashion world - mainly the treatment of designers, factory workers and cotton farmers.

While these three brands already have a part of the market, they focus on different perspectives, none of them work very much with philanthropy organizations, and none of them directly use the ideas of the disadvantaged group as their products. Moreover, their products are more focus on ideas and slogans than on design. Our unique design will be one of the key resources for success.

4.3 Marketing Research

PLACE: Contagious Emotion will operate primarily online. The main sales channel will be our official online store website. Depending on marketing needs and timing, we will have pop-up carts and stores.

PRICE: T-shirts and shirts will be priced from $100 to $150. The average market price for hoodies and jackets will be around $150 to $500.

PROMOTION: Our main marketing channel will be social medias. Contagious Emotion will create trends on Instagram, Facebook, and other social media channels. We will collaborate with trending Instagram users and YouTubers who have a large number of followers. When needed, we will choose celebrities by personality to collaborate as well.

In addition, all fashion brands can become our business partners. We welcome more fashion designers to join us to design our products and cocreate with the psychological disadvantaged. The fame of these business partners will better promote our brand and ideas.

PRODUCT: *Contagious Emotion*'s product would be all daily wear clothing and accessories. At the beginning, the product line will focus on T-shirts and shirts. With proper business growth, we will expand the product line into hoodies, jackets, bags and so on.

After reaching a certain number of customers, *Contagious Emotion* will use the intellectual property we have created to host exhibitions, contests, and other events.

4.4 Implementation of Contagious Emotion

First of all, we work with philanthropy organizations across the country to support the disadvantaged people, which indicates our distinct positioning and our difference from other streetwear brands.

Currently, almost all philanthropic T-shirts are very simple. Most of them just have a simple color with the organization's logo. The quality of the clothing is also unsatisfactory. Therefore, people may not wear them twice.

We choose artwork and ideas from disadvantaged people, focusing on their opinions and voices. We have designers who create drawings from the psychological disadvantaged people. The drawings will be turned into street wear and become fashionable. The quality of the clothing will be strictly controlled as well. As a result, while people pursue our brand as a street fashion brand, they are going to wear the clothing for many times. While creating fashion and bring awareness to people, our clothing and ideas are forming a culture.

By creating this brand, we are not only creating a trend, but also creating intellectual properties. We are enabling disadvantaged people to gain recognition or economic benefit from what they create. By striking the right balance between the ideas of disadvantage people and wider public interest, we aim to foster an environment in which communication and creativity can flourish.

Furthermore, intellectual property here is not just about indicate clothing. In certain products, we can extend it to exhibitions, talk shows and so on. The potential for these properties is enormous.

5 Conclusion

With the development of economy and the attention of society, philanthropy is being carried out in a variety of forms. Starting from the theory of emotional contagion, it pays attention to the inner activities of vulnerable groups and adopts novel ways to guide the vulnerable groups to express themselves through creation. Take fashion as the entry point, guide the general public to pay attention to vulnerable groups in their daily life, so as to enhance the social cognition of this group. *Contagious Emotion* as an attempt behavior, the intention is good, may be deviated in the process of implementation, but the intention of focusing on vulnerable groups is worthy of recognition. It is hoped that the practice will continue to be refined and improved.

References

1. Hatfield, E., Cacioppo, J.L., Rapson, R.L.: Emotional contagion. Curr. Dir. Psychol. Sci. **2**, 96–99 (1993)
2. Hoffman, M.L.: Empathy and Moral Development: Implications for Caring and Justice. Cambridge University Press, Cambridge, England (2000)
3. Lundqvist, L.O.: A Swedish adaptation of emotional contagion scale factor and psychometric properties. Scand. J. Psychol. **47**, 263–272 (2006)

Research on the Spatial Differences and Influencing Factors of Regional Economic Development-Taking 21 Cities in Guangdong Province as Examples

Gang Deng(✉)

School of Economics, South-Central University for Nationalities, Wuhan, China

Abstract. Firstly, Moran's *I* in the spatial autocorrelation analysis method is used, and Geoda software is applied to sequentially analyze the spatial distribution of GDP per capita of 21 cities in Guangdong Province in 2018, the global autocorrelation test, the local autocorrelation test, as well as explore its spatial correlation and spatial heterogeneity. It is concluded that there is a significant spatial autocorrelation of GDP per capita of 21 cities in Guangdong Province in 2018. Secondly, the trend surface analysis method of ArcGIS is applied to conduct a three-dimensional trend analysis of the GDP per capita of each city in Guangdong Province in 2018. Finally, a spatial regression model is used to calculate the influencing factors of the spatial differences of GDP per capita of 21 cities in Guangdong Province in 2018, and it is concluded that the three factors of total retail sales of consumer goods, foreign exchange income from international tourism, internal expenditure of R&D funds have a positive relationship with GDP per capita, however the number of employed persons at the end of the year has a negative relationship with GDP per capita in Guangdong province in 2018.

Keywords: Economic development · Spatial differences · Influencing factors

1 Introduction

The Guangdong-Hong Kong-Macao Greater Bay Area is committed to building a world-class Greater Bay Area with good economic development. Guangdong Province has nine cities located in the Greater Bay Area. Therefore, Guangdong Province's geographic location is relatively important and its regional economic development themes are of research significance. The unbalanced economic development has always been a research hotspot in regional economics. Facing the new normal of economic development, all regions should persist in opening to the outside world, deepening exchanges and cooperation, strengthening regional innovation and development, and ultimately narrow the gap in regional economic development. At present, there are not too many studies on the spatial differences of economic development in Guangdong Province. When calculating the influencing factors of economic development, most studies use

C. Yuan et al. (Eds.): *Proceedings of the 4th International Conference on Economic Management and Green Development*, AEPS, pp. 57–64, 2021.
https://doi.org/10.1007/978-981-16-5359-9_8

indicators such as government fiscal expenditure, retail sales of consumer goods, and investment in fixed assets, while ignoring the tourism industry development, technological innovation and other related indicators, but tourism and high-tech industries are both important sources of economic growth in the Guangdong-Hong Kong-Macao Greater Bay Area. Economic growth and tourism development are related [1], and technological innovation also affects economic development [2]. 21 cities in Guangdong Province are taken as the research object, spatial measurement models are constructed to analyze the spatial differences and influencing factors of regional economic development, and adds tourism development, technological innovation and other indicators to analyze various factors affecting economic development.

2 Research Area, Data Source, Research Method

2.1 Research Area, Data Source

Guangdong Province includes 21 cities: Guangzhou, Shenzhen, Zhuhai, Shantou, Foshan, Shaoguan, Heyuan, Meizhou, Huizhou, Shanwei, Dongguan, Zhongshan, Jiangmen, Yangjiang, Zhanjiang, Maoming, Zhaoqing, Qingyuan, Chaozhou, Jieyang, Yunfu, data Sourced from "Guangdong County and City Statistical Database-Annual Data (City Level)" (2018).

2.2 Research Method

Spatial Autocorrelation Analysis Method. The statistics of spatial autocorrelation are diverse, and Moran's I is the most widely used.

$$I = \frac{n}{\Sigma_i \Sigma_j w_{ij}} \frac{\Sigma_i \Sigma_j w_{ij}(x_i - \overline{x})(x_j - \overline{x})}{\Sigma_i (x_i - \overline{x})^2} \tag{1}$$

Where n is the number of samples, w_{ij} is the (i, j) element of the spatial matrix W, x_i and x_j are the observations of the spatial unit i, j. $\overline{x}$ are the average of the observations. A positive value of I reflects a positive correlation, and a negative value of I reflects a negative correlation.

The above is the global Moran's I, and the corresponding is the local Moran's I.

$$I_i = \frac{n(x_i - \overline{x}) \Sigma_{j \neq i} w_{ij}(\chi_i - \overline{x})}{\Sigma_i (\chi_i - \overline{x})^2} \tag{2}$$

Local Moran's I reflects the spatial autocorrelation of a certain spatial unit, that is, the direct correlation between a certain spatial unit and its neighboring spatial units.

Trend Surface Analysis Method. The trend surface is an approximation of the actual surface, which can accurately simulate the spatial distribution of geographic elements.

Spatial Regression Model. The mathematical expression is as follows

$$y = \rho W y + X\beta + \varepsilon \tag{3}$$

3 Empirical Analysis Results

3.1 Spatial Autocorrelation Analysis

Here, 2018 GDP per capita (yuan) of 21 cities in Guangdong Province is applied to measure the spatial difference of economic development in Guangdong Province. Geoda software is applied, the spatial distribution, global autocorrelation test, and local autocorrelation test of the regional GDP per capita (yuan) of 21 cities in Guangdong Province are used to explore their spatial correlation and spatial heterogeneity.

It can be seen from Fig. 1 that the GDP per capita of 21 prefecture-level cities in Guangdong Province is divided into 4 levels. A prefecture level city with a higher GDP per capita, that is, a prefecture level city with better economic development, will have a darker color on the map. From shallow to deep, it can be divided into backward areas, relatively backward areas, more developed areas and developed areas. Figure 1 is the regional level distribution map of the GDP per capita of each city in Guangdong Province in 2018. It can be seen from the figure that there are 3 developed regions, namely Guangzhou City (155,491 yuan) and Shenzhen City (189,568 yuan) as well as Zhuhai (159,428 yuan). There are 4 more developed regions, namely Foshan City (127,691 yuan), Zhongshan City (110,585 yuan), Dongguan City (98,939 yuan), and Huizhou City (85,418 yuan). There are 6 relatively backward areas, namely Maoming City (49,406 yuan), Yangjiang City (52,969 yuan), Jiangmen City (63,328 yuan), Zhaoqing City (53,267 yuan), Shaoguan City (44,971 yuan), Shantou City (44,672 yuan).There are 8 backward areas, namely Zhanjiang City (41,107 yuan), Yunfu City (33,747 yuan),

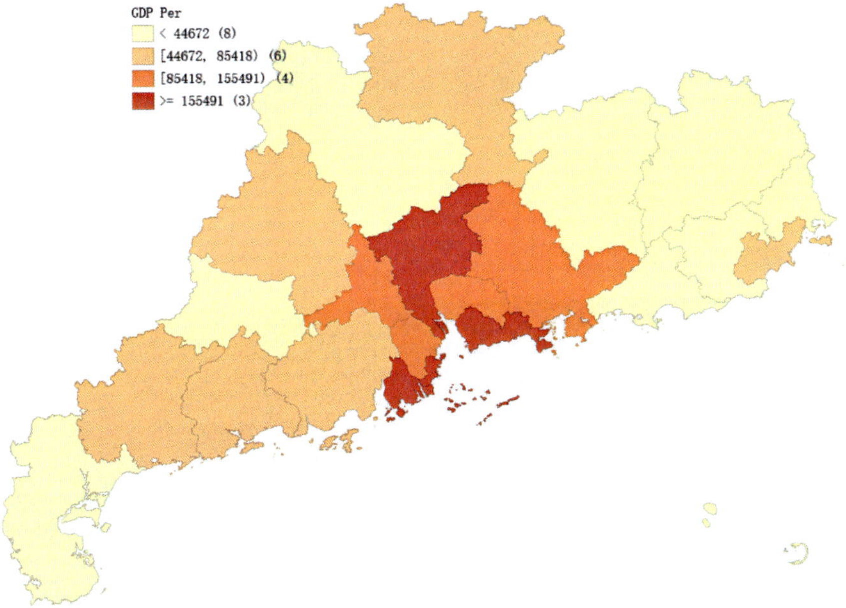

Fig. 1. Distribution map of the GDP per capita of 21 cities in Guangdong Province in 2018.

Qingyuan City (40,476 yuan), Heyuan City (32,530 yuan), Meizhou City (25,367 yuan), Shanwei City (30,825 yuan), Jieyang City (35,358 yuan), Chaozhou City (40,219 yuan).

It can be concluded from Fig. 2 that the GDP per capita of 21 cities in Guangdong Province in 2018 has a positive spatial autocorrelation, which may be due to the better economic development of Guangdong Province in recent years, and the GDP has increased year by year.

Global autocorrelation reflects the level of spatial agglomeration in all regions. In order to study the average correlation degree and significance between the GDP per capita of 21 prefecture level cities in Guangdong Province, the global autocorrelation test was conducted on the GDP per capita data of Guangdong Province in 2018 by Moran's *I* index method. The results are shown in Table 1.

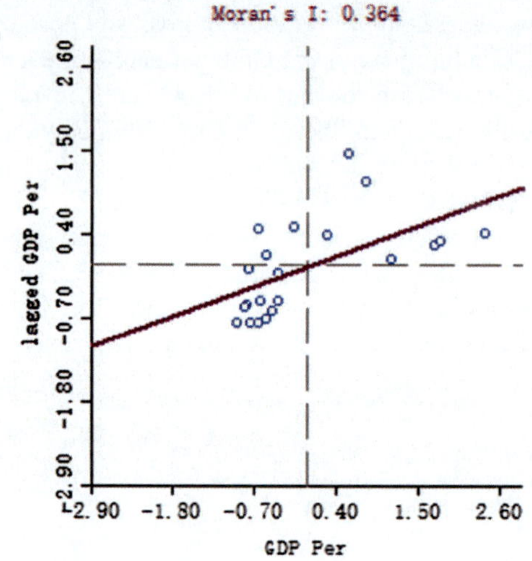

Fig. 2. Scatter chart of global autocorrelation test of GDP per capita of 21 cities in Guangdong Province in 2018.

Table 1. Global autocorrelation test value of GDP per capita of 21 cities in Guangdong Province in 2018

Test category	Moran's *I*	P value
Value	0.364	0.011

The value range of the Global Moran's *I* index is $[-1, 1]$, the index is greater than 0 means positive correlation, the larger the index value, the higher the degree of regional economic attributes clustering due to similarity, the index is less than 0 means negative correlation, the smaller the index value, the higher the degree of regional economic attributes clustering due to differences, the index is equal to 0 means that there is no spatial autocorrelation. According to Table 1, the Moran's *I* index of GDP per capita of

Guangdong Province in 2018 is 0.364, which indicates that there is a significant positive spatial correlation between GDP per capita of Guangdong Province in 2018, and the p value is 0.011, which is less than the significance level of 0.05, that is to say, there is significant spatial autocorrelation in the GDP per capita of Guangdong Province. It shows that the GDP per capita of Guangdong Province in 2018 presents a spatial agglomeration trend, which indicates that the GDP per capita of Guangdong Province in 2018 may have spatial spillover phenomenon.

From the figure on the left of Fig. 3, it can be concluded that the gray area is not significant area, and the bright red area is the area of high-high agglomeration type, which indicates that the value of GDP per capita variable in this region is higher than the average value of the variable. This region includes Zhongshan City and Dongguan City. The dark blue area is low-low cluster type, which indicates that the value of GDP per capita variable in this region is lower than the average value of the variable. This region includes Meizhou City, Jieyang City and Chaozhou City. From the figure on the right of Fig. 3, it can be concluded that the gray area is not significant area, and the dark green area represents the significant area at the level of 0.001, this area includes Meizhou city. The darker green area indicates the significant area at 0.01 level, which includes Dongguan City and Jieyang City. The light green area is significant at 0.05 level, which includes Zhongshan City and Chaozhou City.

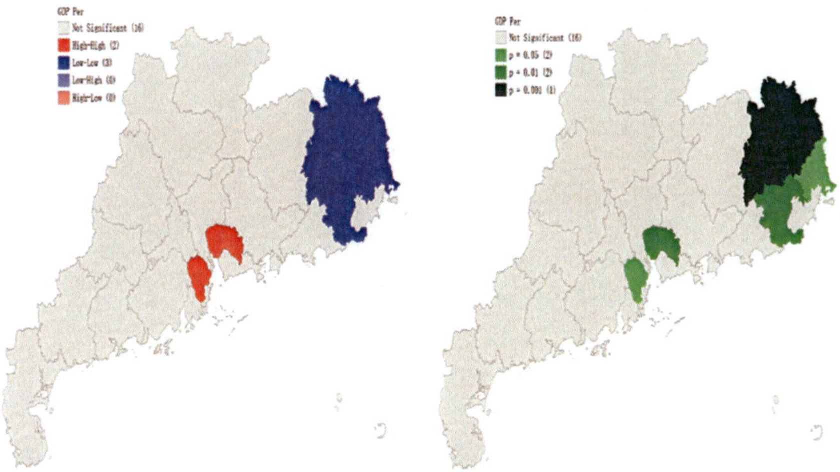

Fig. 3. Lisa clustering and Lisa significance map of GDP per capita of 21 cities in Guangdong Province in 2018.

3.2 Trend Surface Analysis

Each vertical bar in the trend analysis chart represents the value (height) and position of a data point. The point is projected onto an east-west and a north-south orthogonal plane. Best fit lines (blue and green lines) are used to simulate trends in a particular direction. If the line is straight, it indicates that there is no trend, otherwise, there is a global trend. From Fig. 4, it can be concluded that the GDP per capita variable shows an obvious U-shaped trend in both the north-south direction and the east-west direction, indicating that there are spatial trends in both the north-south direction and the east-west direction. Moreover, this variable shows an increasing trend from west to east, and the growth trend of the transition zone is obvious. It shows that the economic growth trend of the transition area from the southwest to the northeast is obvious, and the economic development of eastern cities is gradually accelerating.

3.3 Spatial Regression Model Analysis

Here, the GDP per capita of the region is used as the explained variable. The total retail sales of consumer goods (100 million yuan), the number of employees at the end of the year (10,000 people), internal expenditure of R&D funds (100 million yuan), foreign exchange income from international tourism (100 million US dollars) are used as explanatory variables. In order to explain the variables, a spatial regression model is applied and Geoda software is applied to calculate the factors affecting the spatial difference of economic development in Guangdong Province. The results are as follows.

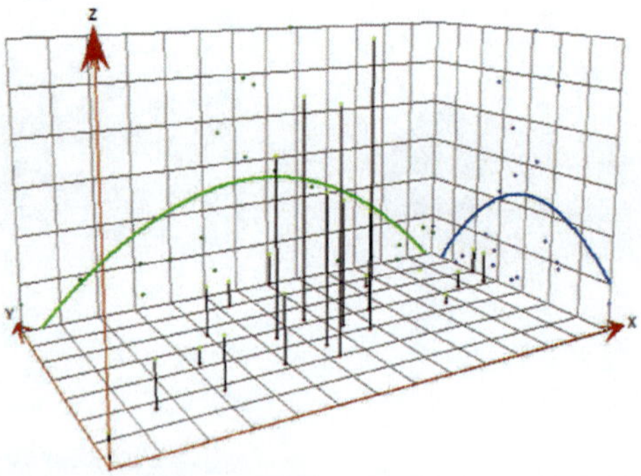

Fig. 4. Trend surface analysis of GDP per capita of 21 cities in Guangdong Province in 2018.

As can be seen from Fig. 5, the value of adjusted R-squared is 0.878168, the independent variable has a strong ability to explain the variation of dependent variable, the fitting effect of the model is good, and the F statistic is significant, indicating that the model is valid on the whole. The elasticity coefficient of GDP per capita to the number of employed people at the end of the year is −0.559916, which indicates that the employment scale cannot be expanded blindly. The elasticity coefficient of GDP per capita to total retail sales of consumer goods is 0.423243, the elasticity coefficient of GDP per capita to foreign exchange income from international tourism is 0.05695, and the elasticity coefficient of GDP per capita to internal expenditure of R&D funds is 0.264854, which is positive correlation. These are consistent with the theoretical expectations.

```
Dependent Variable    :  LN_GDPper  Number of Observations:   21
Mean dependent var    :   10.9982   Number of Variables   :    5
S.D. dependent var    :   0.592527  Degrees of Freedom    :   16

R-squared             :   0.902535  F-statistic           :      37.0403
Adjusted R-squared    :   0.878168  Prob(F-statistic)     :6.69399e-008
Sum squared residual:     0.718597  Log likelihood        :      5.63955
Sigma-square          :  0.0449123  Akaike info criterion :     -1.2791
S.E. of regression    :   0.211925  Schwarz criterion     :      3.94351
Sigma-square ML       :  0.0342189
S.E of regression ML:     0.184984

-----------------------------------------------------------------------------
       Variable      Coefficient       Std.Error     t-Statistic   Probability
-----------------------------------------------------------------------------
          CONSTANT      10.1572        0.718918        14.1284       0.00000
         LN_EMPLOY     -0.559916       0.185221       -3.02296       0.00808
        LN_CONSUME      0.423243       0.189753        2.23049       0.04038
        LN_TOURISM      0.05695        0.0482378       1.18061       0.25501
             LN_RD      0.264854       0.0682988       3.87788       0.00133
-----------------------------------------------------------------------------

REGRESSION DIAGNOSTICS
MULTICOLLINEARITY CONDITION NUMBER   73.991112
TEST ON NORMALITY OF ERRORS
TEST                  DF           VALUE            PROB
Jarque-Bera            2           0.5698           0.75208

DIAGNOSTICS FOR HETEROSKEDASTICITY
RANDOM COEFFICIENTS
TEST                  DF           VALUE            PROB
Breusch-Pagan test     4           3.0809           0.54437
Koenker-Bassett test   4           4.0318           0.40171
------------------------------ END OF REPORT ------------------------------
```

Fig. 5. Analysis of spatial regression model results.

4 Conclusions

To sum up, the economic development of 21 cities in Guangdong Province has significant spatial differences, and with positive spatial correlation. The number of employees at the end of the year, total retail sales of consumer goods, foreign exchange income from international tourism and internal expenditure of R&D funds have significant impact on GDP per capita. Most of the relevant elastic coefficients are positive, which proves the total retail sales of social consumer goods, foreign exchange income from international tourism and internal expenditure of R&D funds have a positive impact on GDP per capita. Therefore, to improve the efficiency and quality of economic development in Guangdong Province, measures should be taken to promote social consumption, advancing the development of tourism and high-tech industries. In order to improve the quality of tourism and technological innovation, the government ought to give more incentives to important scientific achievements [3], so as to promote the coordinated development of regional economy.

References

1. Brida, J.G., Gómez, D.M., Segarra, V.: On the empirical relationship between tourism and economic growth. Tour. Manag. **81**
2. Mansfield, E.: Innovation, Technology, and the Economy
3. Hua, Y.-H., Zhu, K.-L.: Co-integration analysis between technology innovation and economy development in Shandong Province. J. Shandong Inst. Commer. Technol. (2010)

Economic Growth and Environmental Degradation in China: An Institutional Perspective

Fang Yi(✉)

School of Business, George Mason University, 4400 University Drive, Fairfax, VA 22030, USA
fyi2@gmu.edu

Abstract. As one of the significant carbon emissions countries, China faces enormous pressure to protect the environment. This paper discusses the Environmental Kuznets Curve (EKC) movements influenced by China's national and regional macro policies. This comparative study analyzes the economic growth and corresponding environmental policy changes of individual economic regions in the east, middle, and west of China. Under China's macro policies and critical policy orientations, the analysis found that the EKC of three economic regions shows different movements trends. The eastern region will first reach its peak environmental degradation level and initiate environmental improvement. The central region will second meet its turning point with further regulations and policies. Due to its backward economy, the western region has a slow pace to get to the turning point at last. This paper also predicts the future development in the environmental policies of the three regions. Using the EKC methodology, policymakers can have a long-term perspective on ameliorating environmental degradation by formulating macro policies that are more in line with national conditions, and conducive to environmental improvement. China will expect to usher in a significant turning point in the EKC in the next decade.

Keywords: Environmental Kuznets Curve (EKC) · Carbon emissions · Environmental policy · Economic development · Regional disparity · Ecological environment

1 Introduction

Carbon emission is the primary gas causing the greenhouse effect and climate warming, while human activities make it incredibly difficult for the environment to repair itself. The combination of a vast population and rapid GDP growth has made China the world's largest carbon emitter. In the past two decades, China has prioritized environmental development, reducing the annual emission growth rate from 9.3% to 0.6% in consecutive years from 2002–2011 and 2012–2016, respectively [1]. China is committed to improving the environment and climate by reducing carbon emissions, supporting the development of energy-saving and low-carbon industries, and reversing the ecological environment's deterioration. The Chinese government has continuously issued

C. Yuan et al. (Eds.): *Proceedings of the 4th International Conference on Economic Management and Green Development*, AEPS, pp. 65–74, 2021.
https://doi.org/10.1007/978-981-16-5359-9_9

policies related to environmental governance; starting from the 11th Five-Year Plan, it has emphasized the construction of ecological civilization and sustainable development. Furthermore, China also signed the Paris Agreement in 2016 and committed to a global response to climate change after 2020.

In terms of global climate and environmental governance, China has attracted much attention to carbon emission governance effectiveness. The Environmental Kuznets Curve (EKC) can predict future environmental and economic development and assess current environmental degradation. Many scholars have used the EKC model to study China's economic and environmental development after the reform and opening-up policy. Previous research is mainly focused on economic growth and environmental pollution through the time axis, the impact of pollutants on changes in EKC, and analyzing the spillover of environmental pollution and economic growth space based on spatial measurement effect [2]. While there are very few studies on EKC policies' impact, research neglected policies as a significant factor in environmental governance reform.

This paper explores new ground in that it continues the previous research, takes policy as the research object of EKC changes, and uses carbon emissions as the leading indicator to assess China's degree of environmental pollution. At the same time, it uses the Environmental Kuznets Curve to study the characteristics of China's ecological changes and analyzes the relationship between Chinese policies and its EKC.

2 Theory Background of the Environmental Kuznets Curve

The Kuznets Curve is a hypothesis used to analyze the per capita income and the fairness of distribution. Grossman and Krueger proposed the Environmental Kuznets Curve to describe the mutual change of economic level and environmental degradation, expressed as an inverted U-shaped or bell-shaped curve: when the level of economic development is low, the degree of environmental degradation is relatively mitigatory; with the increase in per capita income, the environmental degradation will intensify with economic growth; and when the economy develops to a certain high level, there will be an inflection point, then the degree of environmental degradation gradually slows down as income further increases [3].

Researchers explored EKC - one of the most popular tools to explore the relationship between economic development and environmental sustainability, existing literature mostly focused on two aspects: First, the influence of variant pollution indicators and other variables (such as technology, industrial structure, regulation). For example, Wang mentioned that more efficient and less polluting technologies would replace traditional technologies [4]. Gao found that the curve of economic growth, exhaust gas, and sulfur dioxide follows the inverted U-shaped [5]. Second, the prediction of whether the inflection point and long-term EKC trend conform to the inverted U-shaped curve. Galeotti and Lanza empirically tested the inverted U-shaped relationship between carbon emissions and per capita income [6]. However, some researchers found an N-shaped relationship [7, 8]. Besides, Lantz and Feng found no significant relationship between GDP per capita and carbon emissions [9]. The imparity between the literature mentioned above is mainly due to the significant difference between the research area and the choice of empirical methods.

In China, policymakers put high pressure on the issue of energy conservation and emission reduction. Wang found most studies on whether China's environmental policies can promote carbon emission reduction are discussing issues such as population size, urbanization, industrial structure, and energy structure [10]. Some studies, such as Zhang and Wei's research shows that China currently conforms to the inverted U curve, and environmental policies can effectively curb carbon emissions [11]. Zhou believes that a separate environmental policy could reduce carbon emissions, but decentralization would enable environmental policies to promote carbon emissions significantly [12]. While there are very few studies on the impact of EKC policies, research neglected policies as a significant factor in environmental governance reform.

This paper is a supplement to the existing literature in conjunction with EKC policies. It aims to find the dynamic development trends of China's environmental policies on the environment and economy and promote carbon emission governance calls.

3 Relationship Between China's Macro Economy and Environmental Degradation

The prediction EKC model reveals a schematic diagram of China's overall economic environment prediction in Fig. 1 [13]. The graph shows that the slope represents environmental degradation speed as the economic level increases relying on the EKC theory. Therefore, with the improvement of economic level, environmental degradation is a revert from acceleration to deceleration, and environmental quality gradually shifts from deterioration to recovery. Between 2020 and 2035, China may realize socialist modernization, and the environment may also reach a turning point at this stage [14]. As shown in Fig. 1, China is currently in a stage where EKC is predicted to rise slowly and is expected in the next ten to fifteen years to reach a turning point in the deterioration of the ecological environment. Since then, as the economic level increases, environmental degradation will slow down, and the environment will be significantly improved.

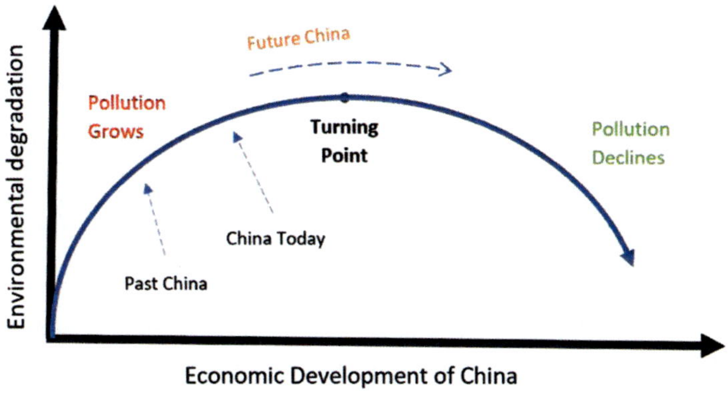

Fig. 1. Predicted EKC of China.

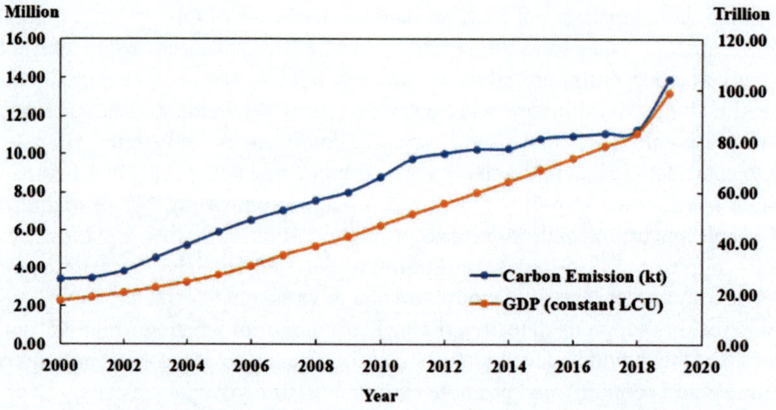

Fig. 2. China's GDP and CO2 emissions change since 2000.

Figures 2 and 3 describe China's GDP and CO2 emissions changes and China's CO2 and GDP growth rates since 2000, respectively, based on the World Bank data, the National Bureau of Statistics of China, and the Rhodium Group [15–17]. Figure 2 shows that carbon emissions increase with the GDP per capita, and the growth rate is the same. It shows that economic conditions still influence carbon emissions. Under the two forces of economic development and environmental pollution, carbon emission is still the top priority as an essential indicator that can represent environmental pollution.

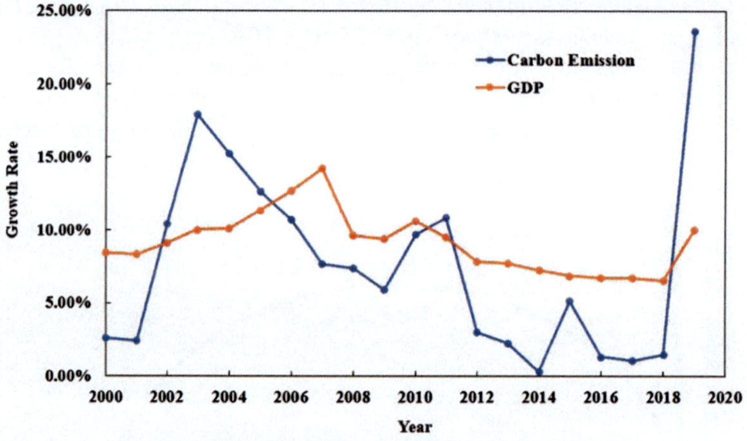

Fig. 3. China's CO2 and GDP growth rate.

Figure 3 shows that the economic growth rate tends to stabilize, while the growth rate of carbon dioxide fluctuates with high speed and apparent changes, and both show positive trends. It shows that economic conditions still influence carbon emissions. Under the two forces of economic development and environmental pollution, carbon emission, as an important indicator that can represent environmental pollution, is still the top priority.

Studying the relationship between GDP and carbon emissions where combining EKC models and influencing factors can predict future environmental degradation and assess China's current environmental degradation. Based on China's economic development laws and characteristics, policymakers can formulate appropriate macroeconomic policies and set long-term development goals.

4 Institutional Impact on Environmental Degradation

4.1 National Policy of Environmental Protection in China

Through the analysis of the "Five-Year Plan for China's Economic Environment," significant macroeconomic policies display China's environmental governance effectiveness.

In the 11th Five-Year Plan, China reduced the total discharge of major pollutants by 10%. To achieve the specified targets, China directly shut down factories with outdated production technology and low production capacity during 2005–2010 [18]. Although crude and straightforward, this is equivalent to saving 750 million tons of coal and reducing carbon emissions by about 1.5 billion tons [19]. Environmental governance has achieved initial results.

Compared with the 11th Five-Year Plan, the 12th Five-Year Plan is more detailed and standardized. It expanded restrictive indicators and strengthened environmental protection standards. It solves major environmental problems related to residents' lives, such as putting forward PM 2.5 governance requirements and further solves the problems of national economic development and people's livelihood [20]. Furthermore, the revised and implemented Environmental Protection Law clearly stated that the purpose of the state's environmental protection policy is to coordinate economic development and environmental protection, which has brought China to a higher level [21].

The 13th Five-Year Plan is more extensive with the continuous deepening of policies. It emphasized the stability of the ecosystem and the ecological environment's safety, with up to nine binding indicators [22]. Climate change will be a vital part of ecological and environmental protection in the 14th Five-Year Plan. Besides, the new plan is currently evaluating, optimizing, and adjusting the protection of the ocean, water, air, soil, and other resources. For instance, the number of air quality monitoring points will increase from the current 1,436 to nearly 1,800, and China's environmental monitoring ability to detect atmospheric particulates will be strengthened [23]. China is now striving to maintain the ecological protection results and have more concrete plans for the critical water resources and air quality requirements in people's livelihoods.

In addition to the five-year plan, China's commitment in the 2015 Paris Agreement to global climate change that carbon emissions will peak as soon as possible around 2030; and compared with 2005, CO2 emissions per unit of GDP will be reduced by 60% to 65% from 2021–2030 [24]. In the same year, China put forward the "The Belt and Road" initiative to provide an impetus for regional economic development, to make the domestic economy develop more effectively, and promote regional economic integration [25]. Energy conservation and emission reduction through national macro-control measures and national economic mid- and long-term plans are China's current main measures to control carbon emissions.

According to the aforementioned national policy analysis, China's overall macro-environmental policy orientation has changed from a mere control of pollutant emissions and energy intensity to the current ecological considerations, civilization construction, and environmentally sustainable development; from the improvement of one or two indicators to the standardization of the overall indicators. While China's overall development level continues to improve, the depth and breadth of environmental protection policies are also increasing. National macro policies will inevitably promote environmental supervision and policy reforms at various regional levels. However, due to the regional disparity of economic development levels in China and the unique industrial structure, each region's environmental and economic development presents different trends and characteristics.

4.2 Analysis of Regional EKC Movement from an Institutional Perspective

Regional Industrial Structure Characteristics. China's macroeconomic policies require a high level of regional cooperation between provinces. Based on China's industrial structure characteristics, the 31 provincial-level administrative units can be categorized into three regions, as shown in Fig. 4 below (excluding Hong Kong, Macau, and Taiwan).

Fig. 4. The regional map of China [26].

In the eastern region, the Beijing-Tianjin-Hebei, the Guangdong-Hong Kong-Macao Greater Bay Area, and the Yangtze River Delta are valuable economic development and technological innovation zones. They are technology-intensive regions, mainly developing tertiary industries, taking the most critical tasks of leading China's economic development. The eastern region is changing priority from economic to environmental development by promoting the use of new energy, building a national-level economic and technological development zone, developing a new industrial trade zone dominated by high-tech industries, and continuing to optimize the industrial structure. In the first half of 2019 GDP statistics, nearly 60% of the top hundred cities come from the east [27]. Compared to others, the eastern region will tighten economic policies and strengthen environmental protection. Furthermore, the eastern region will develop emissions trading and expropriation, execute higher pollution fees, resource taxes, and environmental taxes to balance the ecological environment and economic development.

In the central region, agriculture and light industries are the top leading industry. It's per capita GDP is catching up with the eastern region, but the level of industrialization and economic development is lagging. Due to the uneven distribution of resources, the central region is still developing the secondary and tertiary industries. The ecological environment issue is severe, economic growth and environmental pollution continue the law of "pollution first, manage later" [28]. Coordinated development across regions will be a trend, and the eastern development will effectively drive its economic development. For instance, due to the vigorous development of coal, chemical, metallurgical and other industries developed in Shanxi Province, industrial pollutant emissions have seriously exceeded standards. The ecological environment has degraded. The central region's future development will optimize and adjust the industrial structure, formulate strict environmental standards and environmental supervision, force the upgrading of traditional industries, and coordinate economic development with the eastern region.

The level of industrialization and technology in the western region is relatively backward. Due to its location and unique social and economic development, the western region of China's environmental policies and governance is weak. Commercial construction under the economic development framework of the Belt and Road is primary. The current economic policy trend is to strengthen infrastructure construction, develop a characteristic industrial economy, optimize and upgrade the industrial structure. The western region should strengthen its environmental protection standards, prevent excessive concentration of high-polluting industries. In the future, the western region will still mainly carry out infrastructure construction, strive to realize the rational allocation and use of production capacity resources and prevent industrial structure degradation.

Moreover, in advancing the construction of the "The Belt and Road," China will give full play to the comparative advantages of various domestic regions, implement a more proactive opening strategy, and strengthen interactive cooperation between the east, middle, and the west. China is still in a period of uneven economic development.

Regional EKC Practice. Since the 12th Five-Year Plan, while the central and western regions have relatively looser energy-saving and emission-reduction indicators, the eastern region has stricter energy-saving and emission-reduction targets. China's overall economy is in a trend of high east and low west. Some areas with high total carbon emissions are energy export provinces and large industry-intensive areas, resulting in higher overall regional carbon emissions [13]. Figure 5 is a forecast of the EKC trend in the eastern, central, and western regions by comparing among three regions in environmental and economic development. The standard line is a reference, representing the state of a customarily defined EKC curve without being affected by policies or other influences.

The eastern region is densely populated and has the highest environmental quality requirements. It is feasible for the eastern region to accelerate the tertiary industry's economic development and promote regional environmental policies. The future slope of EKC in the east will be lower than the current curve, and the turning point will appear earlier than the other two regions. The central region needs to accelerate the tertiary industry's economic development and promote the establishment and improvement of regional environmental policies to reach an advanced level of industrialization as soon as possible. After observing the central and eastern regions' development experience, the western region must strengthen environmental governance while developing the economy. Therefore, assume that the western region's economic growth implements sustainable development strategies and appropriate government environmental and economic policies. In this case, the western region's inverted U-shaped curve slope will slow down, and the inflection point of the EKC will appear relatively early. In Fig. 5, the eastern region will be the first to reach its turning point. Due to rapid economic development, its environmental governance will be very fast. Policy intervention will speed up the EKC process in the central region so that the central region's EKC slope will be steeper than the standard line. However, because of the level of economic development, the central's turning point will appear later than the eastern region.

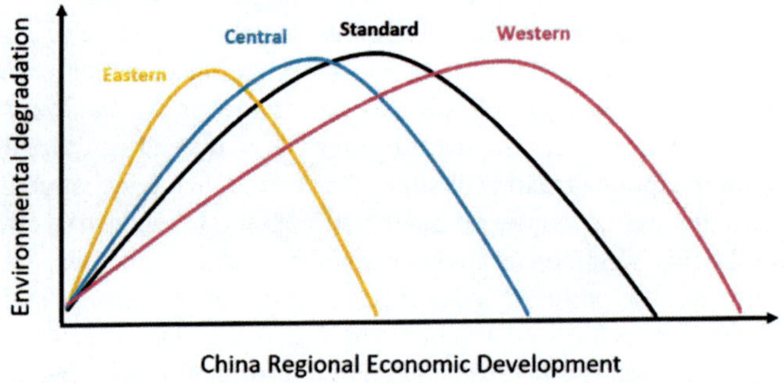

Fig. 5. Predicted regional EKC trend of China.

5 Conclusion

This paper discussed the status quo of China's macro-economic policy implementation to regional development and industrial level. China's economic structure is developing towards advanced development and easing environmental pressures. Analysis cannot thoroughly reflect the relationship between environmental sustainability from an institutional perspective based on a single environmental pollutant indicator and per capita GDP. However, this paper indicates that China's overall EKC slope will slow down in the next five to ten years, and it may reach an inflection point earlier. According to the changes in turning points under different systems in various regions, vigorously promoting green and low-carbon development is an important policy to improve the pollution of China's ecological environment. All provinces should step up cooperation such as improving taxation policies, increasing environmental investment, implementing green and low-carbon new urbanization roads, paying attention to ecology, advanced energy-saving technologies, improving resource utilization efficiency, and reducing the environmental pollution. Besides, future researchers need to look for variables that are beneficial to alleviating environmental degradation from other aspects to supplement current research gaps.

References

1. The Economist Group Limited. China is surprisingly carbon-efficient—but still the world's biggest emitter. The Economist (2019). https://www.economist.com/graphic-detail/2019/05/25/china-is-surprisingly-carbon-efficient-but-still-the-worlds-biggest-emitter
2. Dan, F.: Spatial metrological analysis of China's carbon dioxide EKC curve expansion model. Macroeconomics (05), 83–91 (2014)
3. Grossman, G.M., Krueger, A.B.: Environmental impacts of a north American free trade agreement. NBER Working Paper, No. 3914 (1991)
4. Wang, M., Li, Y., Guan, K.: Impact of market regulations on decision-making and performance of enterprise' green technological innovation. Syst. Eng.-Theory Pract. **40**(05), 1158–1177 (2020)
5. Gao, H., Yang, L., Fu, H.: Research and forecast on the relationship between economic growth and environmental pollution in china's provinces—based on the empirical analysis of environmental Kuznets curve. Econ. Dyn. (1), 52–57 (2012)
6. Galeotti, M., Lanza, A.: Desperately seeking environmental Kuznets. Environ. Model. Softw. **20**(11), 1379–1388 (2005)
7. Friedl, B., Getzner, M.: Determinants of CO2 emissions in a small open economy. Ecol. Econ. **45**(1), 133–148 (2003)
8. Martínez-Zarzoso, I., Bengochea-Morancho, A.: Pooled mean group estimation of an environmental Kuznets curve for CO2. Econ. Lett. **82**(1), 121–126 (2004)
9. Lantz, V., Feng, Q.: Assessing income, population, and technology impacts on CO2 emissions in Canada: where's the EKC? Ecol. Econ. **57**(2), 229–238 (2006)
10. Wang, X., Ren, S., Li, X.: The effects of different types of environmental policies on regional carbon emissions in China. J. Dali. Univ. Technol. (Soc. Sci.) (02), 55–64 (2018)
11. Zhang, H., Wei, X.: Green paradox or forced emission-reduction: dual effect of environmental regulation on carbon emissions. China Popul. Resour. Environ. **24**(9), 21–29 (2014)
12. Zhou, Y., Li, Y.P., Huang, G.H.: Planning sustainable electric-power system with carbon emission abatement through CDM under uncertainty. Appl. Energy (140), 350–364 (2015)
13. Liu, Q.: Can China hit a 2025 'turning point' in its pollution fight? China dialogue (2015). https://chinadialogue.net/en/business/7959-can-china-hit-a-2-25-turning-point-in-its-pollution-fight/
14. Wang, J., Wan, J., Wang, Q., Su, J., Yang, L., Xiao, Y.: The development of China's ecological and environmental planning in forty years of reform and opening-up. China Environ. Manag. **10**(06), 5–18 (2018)
15. The World Bank Group. CO2 emissions(kt) – China (2020). https://data.worldbank.org/indicator/EN.ATM.CO2E.KT?end=2014&locations=CN&start=1960&view=chart
16. National Bureau of Statistics. Statistical Communique of the People's Republic of China on the 2019 National Economic and Social Development (2020). http://www.stats.gov.cn/tjsj/zxfb/202002/t202002 28_1728913.html

17. Grant, M., Larsen, K.: The Rhodium Group (2019). https://rhg.com/research/preliminary-china-emissions-2019/#:~:text=China's%20GHG%20emissions%20grew%202.6%25%20in%202019&text=To%20fill%20this%20information%20gap,estimates%20for%202019%20by%20gas
18. The Central People's Government of the People's Republic of China. Notice of the State Council on Issuing the National Environmental Protection "the Eleventh Five-Year Plan" (2007). http://www.gov.cn/zwgk/2007-11/26/content_815498.htm
19. Liu, Z.: China's Carbon Emissions Report 2016. Report for Harvard Belfer Center for Science and International Affairs, Cambridge, MA (2016). https://dash.harvard.edu/handle/1/29916843
20. The Central People's Government of the People's Republic of China. Notice of the State Council on Issuing the National Environmental Protection "the Twelfth Five-Year Plan" (2011). http://www.gov.cn/zwgk/2011-12/20/content_2024895.htm
21. The State Council Information Office of the People's Republic of China. Environmental Protection Law of the People's Republic of China 2015 (2015). http://www.scio.gov.cn/32344/32345/32347/20150213/xgzc32645/Document/1399119/1399119.htm
22. The Central People's Government of the People's Republic of China. Notice of the State Council on Issuing the "the Thirteenth Five-Year" Ecological Environment Protection Plan (2016). http://www.gov.cn/zhengce/content/2016-12/05/content_5143290.htm
23. Ministry of Ecology and Environment of the People's Republic of China. During the "14th Five-Year Plan" period, the national urban ambient air quality monitoring points will increase to nearly 1,800 (2020). http://www.mee.gov.cn/ywdt/hjywnews/202001/t20200111_758804.shtml
24. The Central People's Government of the People's Republic of China. Notice of the National Development and Reform Commission and the National Energy Administration on Issuing the "Energy Production and Consumption Revolution Strategy (2016–2030) (2017). http://www.gov.cn/xinwen/2017-04/25/content_5230568.htm
25. Ministry of Commerce of the People's Republic of China Comprehensive Department. The vision and actions for promoting the joint construction of the silk road economic belt and the 21st century maritime silk road (2015). http://zhs.mofcom.gov.cn/article/xxfb/201503/20150300926644.shtml
26. China Outline #1519054: Clipart Library (2020). http://clipart-library.com/clipart/8ixrbkMbT.htm
27. Xie, Z., Zhang, N.: 21 Data Journalism Lab (2019). https://m.21jingji.com/article/20190812/herald/ec0af856d33510e1c75f4c85b1cc4faf.html
28. Dong, S., et al.: Study on the resource environment, economy and urbanization situation and green rise strategy in Central China. Resour. Sci. **41**(1), 33–42 (2019)

The Predictability and Analysis of American Stock Market

Songyuan Liu(✉)

St. Michael University School, 3400 Richmond Road, Victoria, BC, Canada

Abstract. The paper uses AR, MA, and ARMA models to analyze the return of the S&P 500 index, and forecasts the possible future return of the S&P 500 index. Based on the results of ACF and PACF, I find AR(9) or MA(12) is the best model. And when using the AIC, MA(2) is best. Furthermore, using the estimation results of AR(2), MA(12) and MA(2), I find the predictability of MA(12) are best, and the predictability of AR(9) is very close to MA(12), which could be used as a reference for investors' investments. However, The values of more than 2 steps of MA(2) are constant because the MA(q) model can predict q periods at most.

Keywords: American stock market · ARMA · Predictability

1 Introduction

The stock market is an important financing and investment market, occupying an important position in the economic development of a country. Since 1928, when Dow Jones Industrial Average was officially established, the sharp increase of the index from 100 to 26000 nowadays has symbolized the growth of America's national power. However, different from the other steady benchmarks, such as GDP, the stock market index responds to various events in a much greater magnitude. The major cause of which is human's irrational impulse when viewing a pessimistic investing environment. For example, the high unemployment rate, companies bankrupting, and government's emergent stimulus policies caused by the 2020 corona-virus pandemic created such a pessimistic attitude. As a result, it shrank most of the American stock index back to 2017's level, destroying the three-year effort of the country in a blink. Therefore, in the first half-year of 2020, successfully predicting the magnitude of the impact of the pandemic on the stock market could help the investors to dodge the collapse and follow the rise, resulting in considerable return.

How to analyze and predict the changes in stock market returns has become a very interesting topic and attracted lots of attention of research scholars. Econometric models can give us some rigorous analysis criteria and forecast results. Therefore, lots of literature use econometric models to analyze the stock market, and the autoregressive–moving-average (ARMA) model is one of the most widely used models. The general ARMA model was described by Peter Whittle (1951) [1], and it was popularized in the book by Box et. al (2015) [2]. For example, Gärtner and Wellershoff (1995) [3], Blazsek

C. Yuan et al. (Eds.): *Proceedings of the 4th International Conference on Economic Management and Green Development*, AEPS, pp. 75–79, 2021.
https://doi.org/10.1007/978-981-16-5359-9_10

and Mendoza (2016) [4], Rounaghi and Zadeh (2016) [5] used ARMA and its derivative models to analyze the US stock market. Hence, this paper also uses the autoregressive (AR) model, moving average (MA) model, and autoregressive moving average (ARMA) model to analyze American stock market, and forecasts the possible future return of the American stock market.

2 ARMA Model

To predict the future possible trend of the American stock market, the paper will analysis the S&P 500 index using three time series models (AR, MA, ARMA model), and then compare the feasibility of the respective model. ARMA model can be understood as a model that consists of AR and MA model.

An AR model predicts the future data based on the past. A general form for AR(p) model is:

$$X_t = \varphi_0 + \varphi_1 X_t - 1 + \cdots + \varphi_p X_{t-p} + a_t \tag{1}$$

A MA model predicts the future data based on the white noise of the past. A general form for MA(q) model is:

$$X_t = c_0 + a_t - \theta_1 a_t - 1 - \theta_2 a_t - 2 - \cdots - \theta_p a_{t-p} \tag{2}$$

The ARMA model is the combination of AR and MA model. A general form for ARMA(p,q) model is:

$$X_t = \varphi_0 + \sum\nolimits_{i=1}^{p} \varphi_i X_{t-i} + a_t - \sum\nolimits_{i=1}^{q} \theta_i a_{t-i} \tag{3}$$

In the function, the first term "$\varphi_0 + \sum_{i=1}^{p} \varphi_i X_{t-i}$" is the AR model and the second term "$a_t - \sum_{i=1}^{q} \theta_i a_{t-i}$" is the MA model, where a is the white noise. φ_0 is the constant term, and φ_i and θ_i are the coefficients (Fig. 1).

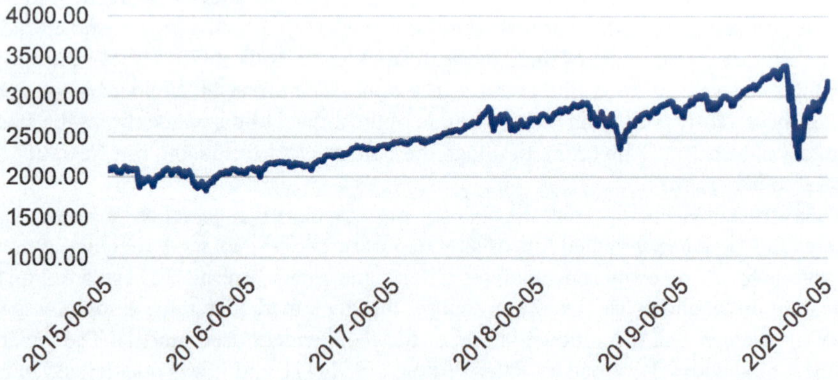

Fig. 1. S&P500 index

3 Empirical Result

3.1 Data

The paper uses the S&P 500 index from June 5th, 2015 to June 5th, 2020 as a reference to American stock market. The following table shows part of the S&P 500 index from June 5th, 2015 to June 5th, 2020.

3.2 Results

Order selection. The autocorrelation function (ACF) and partial autocorrelation function (PACF) of a stationary time series are useful tools for determining the order p of an AR model and the order q of an MA model. Based on Fig. 2, using the 5% significant level, I identifies an AR(9) or MA(12) model for the return of S&P 500 index. Some coefficients of the AR(9) or MA(12) model are restricted to be equal to zero based on the 5% significant level. The Table 1 and Table 2 are the estimation results.

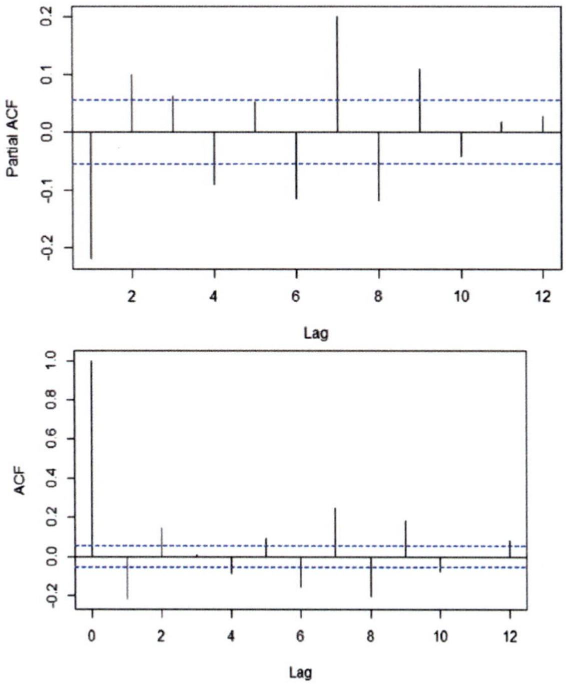

Fig. 2. ACF and PACF of the return of S&P 500 index. The dotted lines give approximate pointwise 95% confidence interval.

In addition to ACF and PACF, there are several information criteria available to determine the order p of an AR model and the order q of an MA model. The Akaike Information Criterion (AIC) proposed by Akaike (1973) [6] is a well-known criterion.

Table 1. The estimation results of AR(9).

AR(9) model	ar1	ar2	ar3	ar4	ar5	ar6	ar7	ar8	ar9
Coefficients	−0.130	0.081	0.052	−0.091	0.000	−0.082	0.184	−0.131	0.119
Standard error	0.028	0.028	0.028	0.027	0.000	0.027	0.029	0.026	0.029

Table 2. The estimation results of MA(12).

MA(12) model	ma1	ma2	ma3	ma4	ma5	ma6	ma7	ma8	ma9	ma10	ma11	ma12
Coefficients	−0.139	0.102	0	−0.099	0.047	−0.077	0.186	−0.132	0.129	−0.066	0	0.043
Standard error	0.028	0.028	0	0.028	0.029	0.030	0.028	0.026	0.030	0.031	0	0.029

Table 3. The estimation results of MA(2).

MA(2) model	ma1	ma2
Coefficients	−0.197	0.166
Standard error	0.028	0.029

Table 4. Predictability Results of AR(9).

AR(9) model	1	2	3	4	5	6	7	8	9	10
Predicted values	0.00641	0.00568	0.00183	−0.00118	0.00057	−0.00142	0.00548	−0.00235	0.00266	0.00034
Predicted errors (S.E.)	0.01130	0.01140	0.01143	0.01145	0.01150	0.01150	0.01153	0.01172	0.01181	0.01189
Average S.E	0.01155									

Table 5. Predictability Results of MA(12).

MA(12) model	1	2	3	4	5	6	7	8	9	10
Predicted values	−0.00758	0.00652	−0.00326	0.00192	0.00192	−0.00155	0.00503	−0.00313	0.00360	−0.00056
Predicted errors (S.E.)	0.01127	0.01137	0.01143	0.01143	0.01148	0.01150	0.01153	0.01172	0.01181	0.01190
Average S.E	0.01154									

Table 6. Predictability Results of MA(2).

MA(2) model	1	2	3	4	5	6	7	8	9	10
Predicted values	−0.00452	0.00411	0.00034	0.00034	0.00034	0.00034	0.00034	0.00034	0.00034	0.00034
Predicted errors (S.E.)	0.01170	0.01193	0.01208	0.01208	0.01208	0.01208	0.01208	0.01208	0.01208	0.01208
Average S.E	0.01203									

Hence, the paper also uses the AIC selection rule, which evaluates the function at the maximized likelihood. Using AIC and the function 'auto.arima' in the package 'forecast' in R, the best model is MA(2). The Table 3 is the estimation results.

Predictability. Using the estimation results of AR(2), MA(12) and MA(2) and the function 'predict' in the package 'stats' in R, I can get out-of-sample 10-step ahead forecast results of AR(2), MA(12) and MA(2). Tables 4, 5, 6 report the points and standard error (S.E.) of prediction. Based on Table 4, 5, 6 and average standard error (S.E.) of prediction, I find the predictability of MA(12) are best, and predictability of AR(9) is very close to MA(12). The values of more than 2 step of MA(2) are constant because the MA(q) model can predict q periods at most.

4 Conclusion

In order to analyze and predict the changes in American stock market returns, this paper uses the autoregressive (AR) model, moving average (MA) model, and autoregressive moving average (ARMA) model to analyze the return of the S&P 500 index, and forecasts the possible future return of the S&P 500 index. Based on the results of ACF and PACF, I find AR(9) or MA(12) is the best model. If I use the AIC, MA(2) is best. Using the estimation results of AR(2), MA(12) and MA(2), I find the predictability of MA(12) is best, and the predictability of AR(9) is very close to MA(12). The values of more than 2 steps of MA(2) are constant because the MA(q) model can predict q periods at most. If the lag period of the AM model is too short, it will limit the predictive ability of the model. Econometric models can give us some rigorous analysis criteria and forecast results, so comparing different models to maximize or minimize possible profit and loss is necessary and important.

References

1. Whittle, P.: Hypothesis testing in time series analysis, Almquist and Wiksell, Upssala (1951)
2. Box, G.E.P., Jenkins, G.M., Reinsel, G.C., et al.: Time Series Analysis: Forecasting and Control. John Wiley & Sons (2015)
3. Gärtner, M., Wellershoff, K.W.: Is there an election cycle in American stock returns? J. Int. Rev. Econ. Financ. **4**(4), 387–410 (1995)
4. Blazsek, S., Mendoza, V.: QARMA-Beta-t-EGARCH versus ARMA-GARCH: an application to S&P 500. . J. Appl. Econ. **48**(12), 1119–1129 (2016)
5. Rounaghi, M.M., Zadeh, F.N.: Investigation of market efficiency and financial stability between S&P 500 and London stock exchange: monthly and yearly forecasting of time series stock returns using ARMA model. . J. Physica A **456**, 10–21 (2016)
6. Akaike, H.: Information theory and an extension of the maximum likelihood principle. In: Petrov, B.N., Csaki, F. (eds.) 2nd International Symposium on Information Theory. Akademia Kiado, Budapest, pp 267–281 (1973)

Clustering of NASDAQ Stocks Based on Elbow Method and K-Means

Xuhuyang Guo(✉)

Mathmatics and Statistics, Southwest University, Tiansheng Street, Chong Qing, China

Abstract. As the most important component of the global stock market, the influential US stocks have always been the focus of research in the financial field . Faced with a large amount of data, it is not meaningful to study a single stock or several stocks. The goal of this article is to cluster all the NASDAQ stocks based on the stock prices in 2020, by converting a single-day stock price into a monthly daily return. We will do the clustering analysis with the K-Means algorithm. In the end, this article has successfully clustered 3264 stocks into 6 categories, laying a solid foundation for further analysis of the common nature of the stocks and predictions.

Keywords: K-means · Clustering · Stock cluster · NASDAQ · Elbow method

1 Introduction

As a pivotal part of the global stock market, the US stock market is also influencing the stock markets of other countries while its own volatility changes. Therefore, accurate analysis of the direction of the US stock market has become a vital part. Generally speaking, when the stock price is rising, most of the stockholders will choose to continue to add and increase the number of their positions. Once the stock drops, due to the high price of the stock price in the later period, the stockholders will be affected by the stock price and suffered heavy losses due to the fall. With the popularity of stocks and the growth of stockholders' experience, ordinary people's investment awareness has also continued to increase. They no longer blindly buy or sell, and gradually cultivate more awareness of the future trend of prices. However, the stock market is a product under the common influence of all kinds of information, and it is difficult to make accurate predictions. So how do ordinary shareholders choose the appropriate stocks and invest in stocks at the appropriate time? Therefore, mining valuable information from massive data and predicting future trends is necessary for stock investment [1].

Because of the important role the stock market plays in the financial system, scholars have done a lot researches on the stock market. Xiong Zhigang learned to use the K-Means method to achieve clustering research on data such as comprehensive stocks, M. Suresh Babu et al. compared the performance of K-Means, hierarchical clustering and reverse K-means three clustering algorithms, and proposed a new clustering method—HRK is also used to predict after the financial report is released based on the short-term trend of changes in the SME board [2], Zhi Tao Jian Ma studied the value of stock

C. Yuan et al. (Eds.): *Proceedings of the 4th International Conference on Economic Management and Green Development*, AEPS, pp. 80–87, 2021.
https://doi.org/10.1007/978-981-16-5359-9_11

investment based on cluster analysis and discriminant analysis. Base on the non-linear Granger causality test, Zhou et al. has analyzed the time-varying spillover based on the connection of the mainland China stock market, the US stock market, the French stock market, and many other main stock markets of the world [3], Yang et al. used the Granger-causality test to analyze the relationship of the volatilities of A and B-shares index [4]. Nanda et al. classified the stocks into clusters with a data mining approach [5]. Li proposed a clustering method by using a dynamic time warping, and has achieved a good result in the stock clustering [6], Rui Xiang et al. realized the analysis of technical indicators of stocks based on the K-Means algorithm [7]. Sougata Das and Palani-Rajan Kadapakkam realized the stock price clustering based on the algorithm era based on the changes of the stock price pricing standard [8].

This article will be divided into five parts. In the second part of this article, the cluster analysis will be briefly described. Also, the principle of the K-Means algorithm including the Elbow Method will be briefly explained. In the third part, the data and the clustering process and results of this paper are explained, and then the future work based on this paper is summarized in the fourth part, and finally the conclusion of the clustering results of this paper is given in the fifth part.

2 Models and Algorithms

With the development of the Internet and the increase in data volume, the application of cluster analysis in unsupervised learning has become more and more extensive. It can simplify data with complex structure during data processing, and discover the relationship between data items. Specifically, in business, cluster analysis can be used to discover different customer groups, and to find new potential markets through the characteristics of each group; in biology, cluster analysis allows humans to obtain knowledge of the inherent structure of the population by classifying animals and genomes; in e-commerce, cluster analysis analyzes customer characteristics and classifies customers with similar browsing behaviors, which can help companies to serve customers more personalized; in the insurance industry, cluster analysis facilitates industry analysis by classifying properties according to different locations, values, types and other factors. This article will apply cluster analysis to the analysis of the U.S. stock market. By clustering and grouping stocks with similar future trends, the characteristics of stocks can be analyzed more accurately and efficiently. The most popular and classic algorithm in cluster analysis is K-means algorithm.

2.1 K-Means Principle

As an unsupervised learning method, K-Means algorithm is widely used because its implementation is simple and the effect is relatively ideal. In addition to the most traditional algorithms, K-Means currently also has the initialization optimized K-Means ++ algorithm, the distance calculation optimized Elkan K-Means algorithm, and the optimized Mini Batch K-Means algorithm in the case of big data. This article will use the most traditional basic algorithm to cluster the US NASDAQ stocks.

For a given sample set, K-Means divides the sample points into different clusters, focusing on finding that the points in the cluster are as dense as possible, and the distance between the clusters is as far away as possible. In particular, suppose the given sample set is divided into k clusters $(C_1, C_2, \ldots, C_k)$, The goal of the next step is to minimize the error sum of squares SSE:

$$\mathrm{SSE} = \sum_{i=1}^{k} \sum\nolimits_{x \epsilon C_i} ||x - \mu_i||^2 \tag{1}$$

Where μ_i is the mean vector of the cluster C_i, sometimes called the centroid and the expression is $\mu_i = \frac{1}{|C_i|} \sum_{x \epsilon C_i} x$.

Since this problem is an NP problem, it is not easy to find it directly, so heuristic algorithm is used to iterate. The process is shown in Fig. 1:

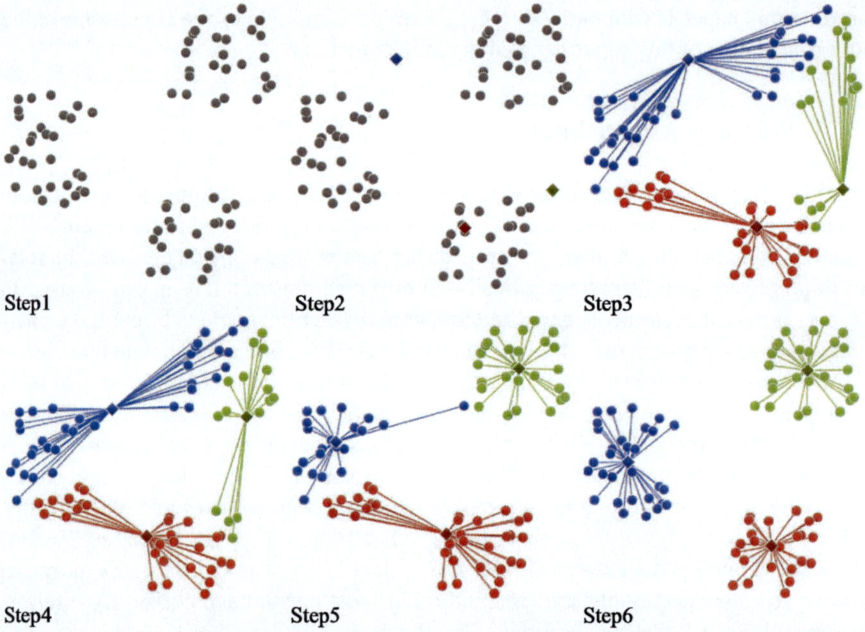

Fig. 1. The iterative process of K-Means algorithm

Step 2 and Step 3 are repeated in the iteration until the sample composition in each cluster no longer changes. In particular, when selecting random initial mass points, due to the uncertainty of the heuristic algorithm, the selection of the initial mass points may have a certain influence on the final result. In order to reduce the influence, the distance between the initial mass points should not be too close.

2.2 Elbow Method Principle

From the above, the most basic clustering idea of K-Means is to minimize the square error between the sample and the mass point as the objective function. People generally call the sum of the squared errors between the mass points of each cluster and the sample points in the cluster as the degree of distortions. Therefore, in terms of one cluster, if the distortions are lower, the connections between its internal members is closer. Conversely, the higher the distortions, the looser the internal structure. Although the degree of distortions will always decrease if the number of clusters is increasing, for data with a decided distinction, the degree of distortions will be greatly improved when it is reached to a determinate critical point, then the decrease speed will be slower, this certain point often regard as the point with better performance of clustering. And since its image resembles an elbow, it is named elbow method.

3 Data and the Result

3.1 Data

The data used in this article mainly comes from the yfinance library. This article selects all 3655 stocks in the NASDAQ constituent stocks, and collects their closing prices for each trading day from January 1, 2020 to August 31, 2020. Excluding the stocks that did not have a complete value in this August, the stocks analyzed in this article are 3264 NASDAQ constituent stocks. Due to the huge amount of data, this article focuses on the unit of month, where basic information such as stock prices is converted to daily return. The calculation formula is:

$$Daily\ return_n = \frac{Closing\ price_n - Closing\ price_{n-1}}{Closing\ price_{n-1}} \quad (2)$$

3.2 Result

After further sorting, this paper obtains the following data (see Table 1):

Table 1. The daily return data of obtained (showing the first 5 rows)

	2020–01	2020–02	2020–03	2020–04
AACG	−0.006115	−0.004328	−0.000849	−0.012146
AAL	−0.002742	−0.016844	−0.012576	0.001908
AAME	0.006368	0.010085	−0.005944	−0.007088
AAOI	−0.001343	−0.011395	−0.003674	0.017382
AAON	0.002900	0.002843	−0.002398	−0.000109
	2020–05	2020–06	2020–07	2020–08
AACG	0.012770	0.018242	−0.005585	0.005467
AAL	−0.004802	0.015665	−0.006113	0.008489
AAME	0.000358	0.006272	0.005435	0.004715
AAOL	−0.008035	0.010679	0.013666	−0.008335
AAON	0.006834	0.000634	0.004124	−0.001754

As for any given data set, the K-Means algorithm first needs to confirm the number of clusters, that is, the k value. This paper uses the elbow method to gradually cluster the collected effective information of all the constituent stocks and draws a graph of the k value and the WCSS value (where the WCSS value is the sum of squared errors SSE mentioned above). The k value at the inflection point of the graph is used as the k value of the final cluster.

Furthermore, K-Means needs to randomly select k centroids as the starting point of the iteration. The K-WCSS curve of this article is shown in the Fig. 2 below:

Since this article is derived from actual stock data and is interfered by many factors, there is not a very suitable inflection point in the figure, but the more appropriate k value can be taken as 6, 7, and 8. This article compares the clustering results, and finally decided to determine the number of clusters in this article into 6 categories. The first category includes stocks such as AACG, AFMD, BLFS, and the second category includes stocks such as GH, PFM. The third category just conclude one stock—DGICB, we can try to specify its character alone. The fourth category includes stocks such as TWOU, AXSM, and the fifth category includes stocks such as ADTN, CRVL, and the sixth category includes stocks such as ACCD and AEZS.

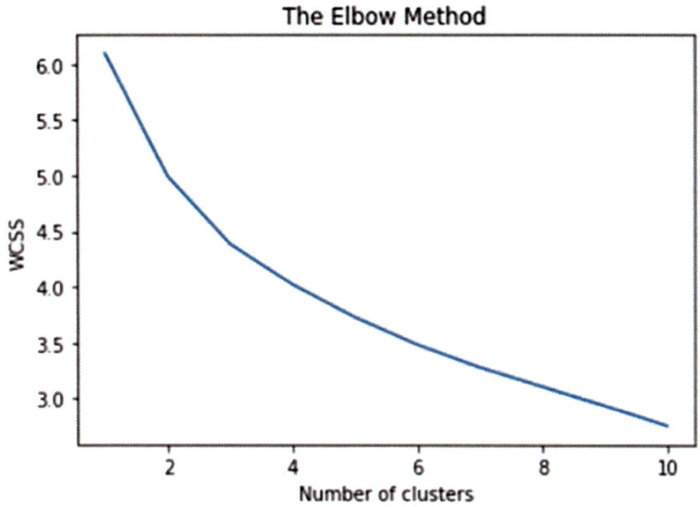

Fig. 2. K-WCSS (elbow-method) curve

4 Other Indicators and Future Work

In addition to using basic information such as the opening price, closing price and highest price of the basic stocks to cluster stock prices, you can also use certain indicators calculated based on the above information or other common information to cluster stocks. The important ones are:

4.1 Trading Volume

Trading volume is the total quantity of shares or contracts traded for a specified security. During a whole transaction, trading volume means the total number of shares transacted between buyers and sellers.

4.2 MACD—Moving Average Convergence Divergence

Moving Average Convergence Divergence (MACD) is an important index that computed by subtracting the 26-period EMA from the 12-period EMA. And the EMA is a special moving average. Compared to the normal, it puts more weight on the most recent data. The calculation formula is

$$EMA_{Today} = \left(Value_{Today} * \left(\frac{Smoothing}{1 + Days}\right)\right) + EMA_{Yesterday} * \left(1 - \left(\frac{Smoothing}{1 + Days}\right)\right) \quad (3)$$

and the common choice of smoothing is 2.

4.3 KDJ

KDJ (also named stochastic indicator) is also an important indicator. It is a really new and effective technical analysis indicator. It was firstly used to analysis future markets and then widely used for analyzing the short-term trend stock market. The specific steps are:

i. Calculate the RSV value. $RSV_n = \frac{C_n - L_n}{H_n - L_n} * 100$.

In the formula, C_n is the closing price on the nth day; L_n is the lowest price in n days; H_n is the highest price in n days.

ii. $K_n = \frac{2}{3K_{n-1}} + \frac{1}{3RSV_n}$

iii. $D_n = \frac{2}{3D_{n-1}} + \frac{1}{3K_n}$

iv. $J_n = 3 * K_n - 2 * D_n$

4.4 Market Capitalization

Market value indicates the sum of the dollar market value of the company's outstanding shares. Also, people call it the "market value". It's value equals with the multiple of the total number of shares outstanding by the company and the current market price of each share.

4.5 P-E Ratio

The price-to-earnings ratio (P-E ratio) is a kind of ratio that can used to value a company. It equals with the company's current share price relative to its per-share earnings (EPS). The P-E ratio relates a company's stock price to its earnings per share.

$$\frac{P}{E} Ratio = \frac{Market\ value\ per\ share}{Earings\ per\ share} \tag{4}$$

4.6 Price Change

In terms of the stock market, Price Change means a change in the company's security value or other kinds of asset to a higher or a lower level. It also refers to the difference of closing price between a trading day and the previous trading day.

The above indicators or information can be collected through Python stockstats or other methods. And based on the data in this article, the values of different indicators can be integrated with the data in this article to make a larger cluster, so that the clustering results can be similar to the stocks in each category from multiple perspectives. Analyze the place to achieve a better prediction effect. At the same time, the stocks in a certain category can be integrated for mixed investment based on this result, which greatly reduces the risk of loss.

5 Conclusion

This article successfully clustered all NASDAQ stocks that were normally traded from January 1, 2020 to August 31, 2020. From the clustering results, the daily return values of most stocks belong to one value. Interval, for this type of stocks, increasing the number of clusters is a measure. At the same time, you can single out the stocks of the largest category and do another cluster analysis. At the same time, there are some stocks that have jumped out of the main value range and are composed of relatively small groups. Of course, there are also several special stocks, which are in their own category or a very small number of them are combined into one category.

References

1. Zhou, Z., Chen, W., Zhang, Z.. Application of cluster analysis in securities investment[J]. J. Chongqing Univ. (2002).
2. Suresh Babu, M., Geethanjali, N., Satyanarayana, B.: Clustering approach to stock market prediction[J]. Int. J. Adv. Netw. Appl., 3(4)(2012).
3. Zhou, P., Li, Z.R.: Time-varying spillover between the mainland China stock market and the other global main stock markets base on the non-linear Granger causality test. Syst. Eng.-Theory Pract, 32(3), 466–475 (2012).
4. Yang, M., Yang, D.: An analysis on Granger casuality between A and B-shares index's volatility. Appl. Stat. Manag. **22**(1), 23–27 (2003)
5. Nanda, S.R., Mahanty, B., Tiwari, M.K.: Clustering Indian stock market data for portfolio management. Expert Syst. Appl. **37**(12), 8793–8798 (2010)
6. Liao, S.H., Chou, S.Y.: Data mining investigation of co-movements on the Taiwan and the mainland China stock markets for future investment portfolio. Expert Syst. Appl. **40**(5), 1542–1554 (2013)
7. Xiang Rui, W., Hualing, L.L., Li, Z.: Analysis of stock technical indicators based on K-Means clustering algorithm[J]. Computer Programming Skills and Maintenance **12**, 4–7 (2019)
8. Sougata, D., Palani-Rajan, K. Machine over Mind? Stock price clustering in the era of algorithmic trading[J]. North American J. Econ. Finance, 51(51) (2020).

The Research of the Environmental Measures Differentiation for Chinese Seaports

Minyou Qing(✉)

Management School, University of Liverpool, Brownlow Hill, Liverpool, UK

Abstract. The cross-border flow of capital is inseparable from the global trade linked by the supply chain, in which shipping with high capacity and low-cost features takes about 75% of trade transportation tasks, including bulk cargo, liquid cargo, and break-bulk cargo. While massive transportation activities create a high level of profits, the energy consumption and environmental pollution issues related to them have also attracted attention. Besides the direct transportation process, seaports as the hubs that connect the upstream and downstream of maritime trade, its impact on the supply chain sustainability also cannot be ignored, whether the internal operation or external logistics services. This paper uses the keyword index method to collect information on the environmental reports of Chinese seaports, identifies mainstream measures and application areas, and finds out that there is a general lack of sustainable-oriented management of suppliers through comparative analysis. It might be beneficial to port companies have a clear future environment-oriented development direction.

Keywords: Sustainability · Seaports · Keyword index · Environmental-oriented measures

1 Introduction

Along the distant coastline in China, with the construction of numerous modern ports, and the formation of five major coastal port areas, China has developed as one of the hubs of shipping service network all over the world. According to the Ministry of Transport of the People's Republic of China [1], there were 2,520 10,000-ton-class berths at Chinese ports in 2019, of which 2,076 above 10,000 tons berths in coastal ports, in which 24 seaports' annual cargo throughput more than 100 million ton, moreover, 18 seaports' container throughput over 2 million transmission extension unit (TEU). However, the massive port-centric business accompanies with unimaginable pollutants emissions and energy consumption, both of which, would affect the sustainability performance of seaports, even break surrounding natural ecological environment. In 2019, Qingdao Port, the fifth largest port in China in terms of annual cargo throughput, its comprehensive energy consumption volume reached 151 thousand tons of standard coal and discharged 166,150.1 tons of greenhouse gas emissions in the meantime.

It is of great significance for these port corporates to pay more attention to sustainability during their daily operation. The World Commission firstly proposed the concept

C. Yuan et al. (Eds.): *Proceedings of the 4th International Conference on Economic Management and Green Development*, AEPS, pp. 88–98, 2021.
https://doi.org/10.1007/978-981-16-5359-9_12

of Sustainability on Environmental Development, chaired by Mrs Brundtland in 1987. After this concept just been raised, many researchers realised the significance of policies or systems used to support it. Such as proposing specific policy for a particular coastal area [2], considering the sustainability-related marine regulatory system under the threat of capital [3], and even the European Union, also pointed out the policy issue about the conflicts between Transport and Environment [4]. 21st century, more articles started focusing on the impact of the technology innovation [5–7], including adopting the renewable energy [8, 9], the automative guide vehicles application [10–12] and the management innovation [13, 14]. What is more, the cases studies of seaports also become a popular research area. Hiranandani [15] compared the policies, practices, as well as driving and constraining forces within several ports. Schippera, Vreugdenhilab, and Jonga [16] developed a comparative methodology to assess sustainability performance. Both Santos, Rodrigues, and Branco [17] and Wagner [18] attempt to identify the most popular ports sustainability practices within European by comparing their environmental reports.

China launched its first related national strategy "The White Paper on China's Population, Environment, and Development in the 21st Century" in 1994. Although until 2004 China's port industry under the socialist market economy with Chinese characteristics just finished the institutional reform of separating enterprise from the administration, many experts have already devoted to researching the port environmental sustainability during the past years. According to the search result of literature related to port sustainability, from 1994 to 2020, there are 633 articles relevant to "Green Port" and 307 for "Port Sustainable Development". Most of them focus on the Container Terminal [19–21], Intelligent Port [22, 23], and Port Logistics [24–26]. However, the study on the differentiation of seaport sustainable development practices has not attracted much attention.

In order to clarify the differentiation about the environmental sustainability of Chinese ports, this article aims to demonstrate the features in environment protection efforts of these ports, through benchmarking their official sustainability reports. The primary research method is the Keywords Index Analysis, taking environmental-related actions as the keywords to reflect what kinds of approaches these seaports are undertaking, its result could be the reference to help these corporates make a better review of sustainability and identify what else direction could probe in the future.

2 The Environmental Sustainability of Chinese Seaports

2.1 The Disclosure of Environmental Information

This article takes dominant coastal harbours as research samples, which selected based on the level of their annual cargo throughput, to ensure the analysis more representative. According to 2019 official statistics, China has 17 ports ranked Top20 both in the list of cargo throughput and container throughput. The ports such as Shanghai Port, Ningbo Port, and Guangzhou Port have mighty comprehensive strength owning the massive volume of business in both two fields, while some seaports have a business propensity, like Tangshan Port, Qinhuangdao Port, and Zhanjiang Port (Table 1).

Table 1. 2019 Chinese seaport cargo throughput rank.

2019 Chinese Coastal Port Annual Cargo Throughput TOP 20	Ten thousand Tons	2019 Chinese Coastal Port Annual Container Throughput TOP 20	Ten Thousand TEU
Ningbo Zhoushan	112,009	Shanghai	4,330
Shanghai	66,351	Ningbo zhoushan	2,753
Tangshan	65,674	Shenzhen	2,577
Guangzhou	60,616	Guangzhou	2,283
Qingdao	57,736	Qingdao	2,101
Tianjin	49,220	Tianjin	1,730
Rizhao	46,377	Xiamen	1,112
Yantai	38,632	Dalian	876
Qinhuangdao	37,400	Yingkou	548
Dalian	36,641	Lianyungang	478
Huanghua	28,761	Rizhao	450
Shenzhen	25,785	Beibu Gulf	382
Beibu Gulf	25,568	Dongguan	368
Yingkou	23,818	Fuzhou	354
Lianyungang	23,456	Yantai	310
Zhanjiang	21,570	Tangshan	294
Xiamen	21,344	Quanzhou	258
Fuzhou	21,255	Zhuhai	256
Dongguan	19,032	Haikou	197
Zhuhai	13,838	Jinzhou	188

Source: The Ministry of Transport of the People's Republic of China.

On the basis of standard international practice and the document requirements of major exchanges, listed port companies need to announce their social responsibility behaviours and final results regularly (usually by annual). The main report type includes Corporate Social Responsibility Report (CSR), Environmental, Social, and Governance Report (ESG), and Sustainable Development Report (STD). Unfortunately, as shown in Table 2, 5 in these 17 seaports (Beibu Gulf, Dongguan, Fuzhou, Yantai, and Zhuhai) have never publicly released relevant reports, let alone disclose specific environmental data. As for the rest of seaports, based on the establishment time, capital restructuring, corporate listing and other factors, they have already begun successively issuing there reports in the past 12 years, and with CSR Report as the main presentation form [27–38]. Besides, it is worth noting that Qingdao Port, Guangzhou Port, Xiamen Port and Dalian Port have only started relevant work within the past five years. In contrast, Tianjin Port not only began to publish CSR reports in 2010 but also began to upload its ESG report synchronously in 2016. Overall, although most of Chinese port corporates are enhancing there sustainablility awareness, not all seaports will regard environmental protection and sustainable development as a fundamental area in their operations.

Table 2. Seaport report release status.

Seaports	Report Types	Initial Release
Shanghai	STD Report	2008
Ningbo zhoushan	CSR Report	2014
Shenzhen	CSR Report	2009
Guangzhou	CSR Report	2018
Qingdao	STD Report	2016
Tianjin	CSR Report/ ESG Report	2010/ 2016
Xiamen	ESG Report	2017
Dalian	CSR Report	2017
Yingkou	CSR Report	2010
Lianyungang	CSR Report	2009
Rizhao	CSR Report	2010
Tangshan	CSR Report	2012
Remark: STD Report: Sustainable Development Report CSR Report: Corporate Social Responsibility Report ESG Report: Environmental, Social, and Governance Report		

2.2 The Analysis of Environmental Practices

Classification of Environmental Practices The environmental reports would write abide by the required format and content according to the investigation of the environmental disclosure information in those seaports, but based on actual conditions, each company have different scopes of practice in the field of green sustainability. Refer to the application of keyword dynamic query software based on big data technology [39], this article used the keyword index method to probe the content of reports published in 2019. It would take the fixed chapter titles as core keywords. And then, from it to mine detailed environment-related measures as the potential sub-aspects. After analysing and statistics, the keywords would finally be constituted by two dimensions, including four Core aspects and twenty-three Sub-aspects extend from it. The keywords division is shown in Table 3.

Table 3. Keywords division.

Core Aspects	Subaspects
Low-carbon Operation	1.Energy 2.Emission 3.Equipment 4.Technique 5.Office Supplies
Port Construction	6.Vehicle 7.Infrastructure 8.Intelligent 9.Port Centralised 10.International Communication
Ecosystem Protection	11.Environmental Evaluation 12.Ecological Restoration 13.Environmental Monitoring 14.Pollution Control 15.Environmental Propaganda 16.Emergency Drills 17.Greening Maintenance 18.Biodiversity Conservation 19.Garbage Classification 20.Climate Change
Supplier Management	21.Management Institution 22.Purchasing Process 23.Environmental indicators

Firstly, Low-carbon Operation, including the sub-aspects of Energy, Emission, Equipment, a Technique, that most of the ports are paying attention to, and the Daily Supplies, which only been mentioned by Shanghai Port and Lianyungang Port. Secondly, Port Construction, all companies usually focus on the sub-aspects of Vehicle and Infrastructure in this area. A few economically stable corporates have already begun the exploration of the Intelligent, but here still has something in unique, the International Communication (Shanghai Port) and Port Centralised (Shenzhen Port) also been cared. Thirdly, Ecosystem Protection, owning the most ten sub-aspects, which might mean most of coastal port corporate tends to improve the environment through digging this field. Both Pollution Control and Environmental Propaganda are the most popular. Then, some seaports explore the Greening Maintenance, Environmental Evaluation, Environmental Monitoring, Emergency Drills, while only very few ports think about the Ecological Restoration, Biodiversity Conservation, Garbage Classification, and Climate Change. Lastly, Supplier Management, which might be an indirect field to affect corporate sustainability performance. As a result, just two seaports (Shanghai & Xiamen) considering it from the sub-aspect of Management Institution, Purchasing Process, and Environmental indicators (Fig. 1).

Frequency Distribution of Environmental Practices. No matter which core aspects these keywords belong to, Fig. 2 demonstrates the frequency of each of it appears in 2019 documents comprehensively. Here has eight sub-points attract more than ten harbours' attention simultaneously, while the field of climate change, international communication, and port centralised, each of which just been attended by a particular port. As many as 14 sub-aspects received a few attention, actually less than half ports. These might become the future development direction of the rest of ports which do not show their interest.

As for the scope of core aspects, most of the seaports could carry out their environmental plan within Low-carbon Operation and Port Construction. Because both operation and construction are the fundamental parts of port development, and it much easier for them to add green elements to improve it. Such as using LED light to substitute

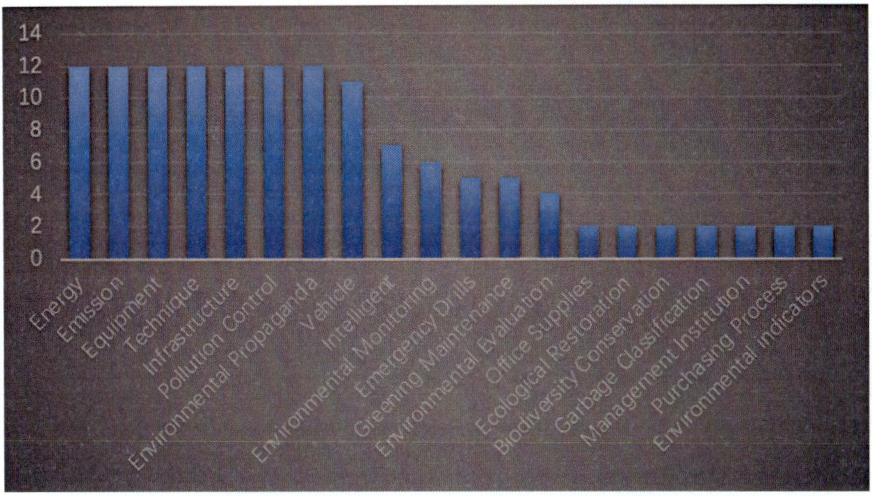

Fig. 1. 2019 seaports' report keywords frequency.

high-pressure sodium lamp (Tianjin Port, Xiamen Port, and Shanghai Port), adopting the Shore Power Technology to instead of diesel generator set (Shanghai Port, Ningbo Zhoushan Port, and Xiamen Port) Besides, Ecosystem Protection might be the most complicated direction. Guangzhou Port and Xiamen Port tend to protect the biodiversity, while Tianjin Port has already worried about the local climate change. Moreover, even the garbage classification (Qingdao Port and Xiamen Port) also get some attention. Based on the influence of various factors like investment budget, port layout, and geographical environment, it hard to let each company focus on a common topic, that might also be the reason why this core aspect exists as much as ten sub-points, and maybe more in the future.

The most apparent differentiation field is Supplier Management, with only two seaports conduct improvement in it. Maybe because of it related to the third-party companies, and would bring performance enhancement indirectly in a long-term, consequently has not arisen enough attention. However, each company in the same supply chain are never become isolated, the carbon footprint from their products as well as services would follow the chain spreading to anywhere in the world. Especially for the port corporate, which need to draw support from third-party companies, to extend its business and make itself become a more influential comprehensive physical distribution service provider.

Environmental Practices by Port. The average figure for Sub-aspects involved numbers is 11.58, but only five ports exceed it. The most comprehensive one is Shanghai Port, whose awareness of environmental sustainability has already beyond itself and starting to focus on the whole supply chain because of using the sustainable evaluation criterion to manage its suppliers. By contrast, both Lianyungang Port and Tangshan Port are the lowest (9 sub-aspects), which might owe to lack concern about Ecosystem Protection and Supplier Management.

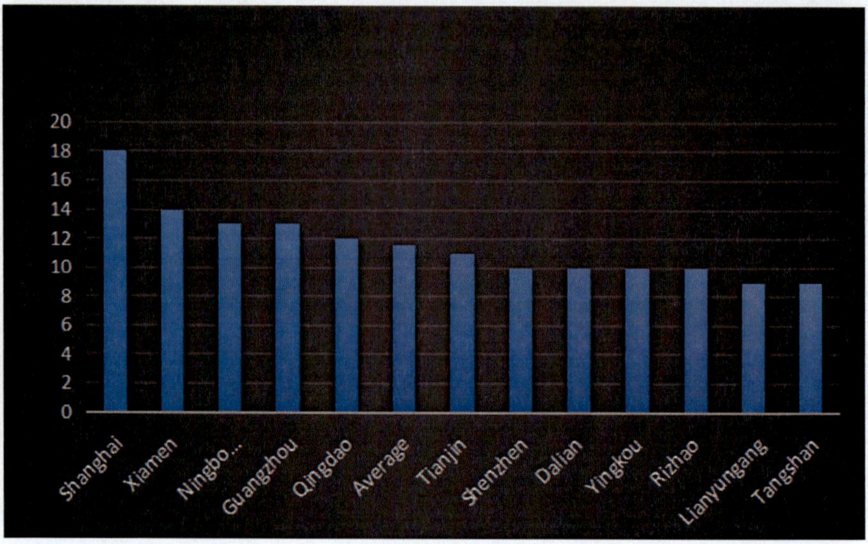

Fig.2. The number of sub-aspects for seaports.

To identify the consist status and structure feature of environmental measures in terms of helping each of them clear the structure reasonability of these measures and find out which direction should get more exploration in the future, we are taking stacked percentage column chart as the reference. According to the proportion of the sub-aspects contained in each core aspect in the total number of sub-aspects. If a port could take measures from all 23 sub-aspects during its operation, the ideal percentage composition should be Low-carbon Operation (LO): $5/23 = 21.7\%$, Port Construction (PC): $5/23 = 21.7\%$, Ecosystem Protection (EP): $10/23 = 43.5\%$, and Supplier Management (SM): $3/23 = 13.0\%$.

Figure 3 illustrates the composition proportion of the number of sub-aspects for each seaport. Then, we would calculate the sum of the absolute value (av) of the difference between real data and the ideal data, using the result to represent the level of structure reasonability for each ports' coverage compared with the theoretical perfect data structure.

After calculating and sorting the mark for each port, a smaller sum value might mean a more all-side and reasonable development in the field of environmental sustainability. Such as the port of Shanghai, Ningbo Zhoushan, Guangzhou, and Qingdao involve at least seven sub-aspects in total within Low-carbon Operation and Port Construction, as well as five in Ecosystem Protection. In contrast, the worst two ports (Shenzhen and Lianyungang) need to consider what kinds of approaches they could adopt to grow into an environmentally friendly company, especially for the Lianyungang Port, which nearly has not paid enough attention to other aspects except Low-carbon Operations.

What is more, the demonstration of Fig. 2 and Fig. 3 have dependency and consistency. It reflects the level of comprehensiveness for seaports green development and could illustrate the sustainability differentiation in each seaport, which represents that the

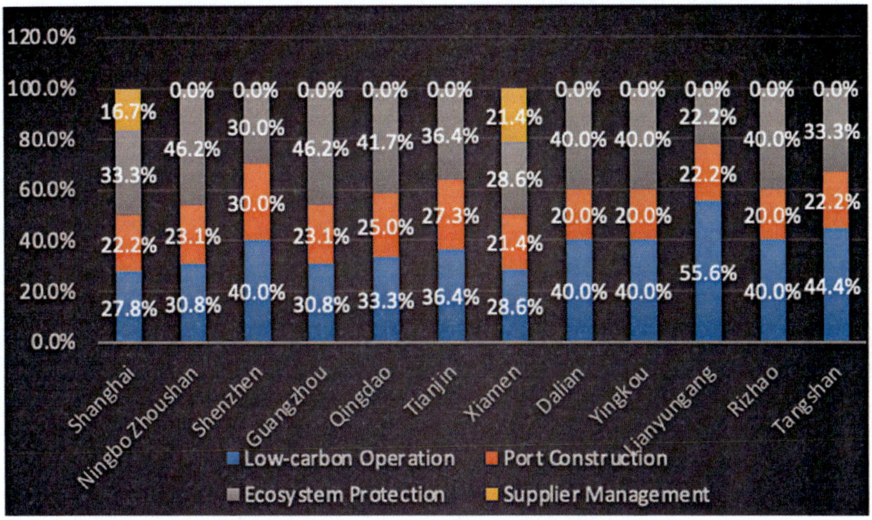

Fig. 3. Four core aspects in each port.

selection of keywords and the division of aspects have practical significance. However, we would not deny the meaning of some unique port green development structure, and perhaps it would bring better effects than the full-scale structure development pattern. For example, the Lianyungang Port pays more than half attention in a particular core aspect (Low-carbon Operation), if it divides its budget to develop other sub-aspects, the improvement of sustainability might not as good as current.

3 Discussion

This article has carried out a analysis around the keywords related to environmental sustainability in the report of port enterprises. Future research could combine other factors to make the result more realistic and reliable.

Firstly, the external factor, such as government policies. From a national perspective, the vast majority of large seaports or port areas are of strategic significance to the country's national defence, logistics supply chain, and foreign trade. Ports companies in China are often registered and listed with state-owned capital as the main body, even the predecessor of some of them were national institutions. In many cases, they will be directly affected by the local government and even the central government, and they have the responsibility and obligation to serve the development of national strategies. For example, the Tangshan Port committed to specialised transportation of bulk materials such as coal, grain, and steel; Tianjin Port develops as China's largest market for parallel imported cars. At the meantime, different commodities would undergo disparate handling processes and accompany with specific contamination, by which, each seaport has to give priority to its business relevant environmental improvement (Table 4).

Table 4. The sum of absolute value.

Port	av-LO	av-PC	av-EP	av-SM	Sum Value
Shanghai	6.1%	0.5%	10.2%	3.7%	20.5%
Ningbo Zhoushan	9.1%	1.4%	2.7%	13.0%	26.2%
Guangzhou	9.1%	1.4%	2.7%	13.0%	26.2%
Qingdao	11.6%	3.3%	1.8%	13.0%	29.7%
Xiamen	6.9%	0.3%	14.9%	8.4%	30.5%
Dalian	18.3%	1.7%	3.5%	13.0%	36.5%
Yingkou	18.3%	1.7%	3.5%	13.0%	36.5%
Rizhao	18.3%	1.7%	3.5%	13.0%	36.5%
Tianjin	14.7%	5.6%	7.1%	13.0%	40.4%
Tangshan	22.7%	0.5%	10.2%	13.0%	46.4%
Shenzhen	18.3%	8.3%	13.5%	13.0%	53.1%
Lianyungang	33.9%	0.5%	21.3%	13.0%	68.7%

Secondly, corporate capital might be a significant factor. This article focus on the coverage of environmental sustainability actions by conducting quantitative analysis theoretically. There is no doubt that the well-capitalised port like Shanghai Port and Ningbo Port could provide more budget in environment areas. According to Ningbo Zhoushan Port Annual Report (2019) and CSR Report (2019), between 2016 to 2019, the total net profit attributable to company shareholders is 11.2 billion Yuan, so it could expense 1.1 billion Yuan to invest its "Green Port Construction" project. As for the smaller scale seaport like Lianyungang Port which totals net profit attributable to company shareholders during the last four years, only 31 million in total (Lianyungang Port, 2019). It could only use limited funds to make environment-oriented improvements in the field of Low-carbon Operation and Port Construction. It might tough for this kind of company to afford a serial of Ecosystem Protection projects. What they could do might sustainably develop as much as possible under the premise of ensuring the continued survival for the enterprise.

4 Conclusions

Overall, as for the differentiation of environmental measures, through comparing these keywords, this paper finds that most of the Chinese coastal ports already considers the performance of environmental sustainability during their business, mainly emphasises the field of Low-carbon Operation and Port Construction. And then, the Ecosystem Protection, it also gets enough focus, although it owns various sub-aspects and would not involve all of them at the same time. Lastly, the more crucial thing is Supplier

Management, a topic almost be ignored by most of them. In the long term, sustainable supplier management, and making the whole supply chain network greener might be more meaningful. The four core aspects discussed in this article have its pros and cons, no one better than the other. No matter what measures the seaport takes to make itself more environmentally-friendly, according to its conditions to carry out environmental-related projects, and explore new sustainable development directions at the same time, it would become a better choice.

References

1. The Ministry of Transport of the People's Republic of China: National port cargo and container throughput in December (2019). http://xxgk.mot.gov.cn/2020/jigou/zhghs/202006/t20200630_3321297.html (2019)
2. Monyneath, V.: Policy and management proposal for sustainable development of coastal areas of Cambodia: Sihanoukville as a case study. World *Maritime University* Dissertations, p. 917. https://commons.wmu.se/all_dissertations/917 (1996)
3. Steinberg, P.E.: The maritime mystique: sustainable development, capital mobility, and nostalgia in the world ocean. Environ. Plan. D: Soc. Space **17**(4), 403–426 (1999)
4. Goulielmos, A.M.: European policy on port environmental protection. Global Nest: Int. J. **2**(2), 189–197 (2000)
5. Wiegmans, B.W., Geerlings, G.: Sustainable port innovations: barriers and enablers for successful implementation. World Rev. Intermodal Transport. Res. **3**(3), 230–250 (2010)
6. Acciaro, M., et al.: Environmental sustainability in seaports: a framework for successful innovation. Marit. Policy Manag. **41**(5), 480–500 (2014)
7. Attia, T.M.: Importance of communication and information technology and its applications in the development and integration of performance in seaports. Renew. Energy Sustain. Dev. **2**(2), 137–146 (2016)
8. Kotrikla, A.M., Lilas, T., Nikitakos, N.: Abatement of air pollution at an Aegean island port utilizing shore side electricity and renewable energy. Mar. Policy **75**, 238–248 (2017)
9. Yigit, A., Acarkan, B.: A new electrical energy management approach for ships using mixed energy sources to ensure sustainable port cities. Sustain. Cities Soc. **40**, 126–135 (2018)
10. Grunow, M., Günther, H.O., Lehmann, M.: Dispatching multi-load AGVs in highly automated seaport container terminals. Container Terminals and Automated Transport Systems. Springer, Berlin (2005)
11. Faulin, J., et al.: In: Sustainable Transportation and Smart Logistics. Joe Hayton (2019)
12. Aguiar, G., Oliveira, G., Tan, K., Kazantsev, N., Setti, D.: Sustainable implementation success factors of AGVs in the Brazilian industry supply chain management. Procedia Manuf. **39**, 1577–1586 (2019). https://doi.org/10.1016/j.promfg.2020.01.284
13. Beleya, P., et al.: Sustainability and green practices at Malaysian seaports: Contributors to the core competitiveness. Journal of Business Management and Economics **3**(3), 23–27 (2015)
14. Vaio, A.D., Varriale, L.: Management innovation for environmental sustainability in seaports: managerial accounting instruments and training for competitive green ports beyond the regulations. Sustainability **10**(3), 783 (2018)
15. Hiranandani, V.: Sustainable development in seaports: a multi-case study. WMU J. Marit. Aff. **13**(1), 127–172 (2013). https://doi.org/10.1007/s13437-013-0040-y
16. Schipper, C.A., Vreugdenhil, H., de Jong, M.P.C.: A sustainability assessment of ports and port-city plans: comparing ambitions with achievements. Transport. Res. Part D: Transport Environ. **57**, 84–111 (2017). https://doi.org/10.1016/j.trd.2017.08.017

17. Santos, S., Rodrigues, L.L., Branco, M.C.: Online sustainability communication practices of European seaports. J. Clean. Prod. **112**(4), 2935–2942 (2015)
18. Wagner, N.: Identification of the most important sustainability topics in seaports. Logist. Transport **34**(2), 79–88 (2017)
19. Fang, H.J., Luo, X.J., Zhou, W.F.: Analysis and prospect of environmental protection of automated container terminal. Port Waterway Eng. **9**, 9–13 (2016)
20. Wang, H.C.: Analyse the construction and future development of green, circular, low-carbon ports in container terminals. Technol. Innov. Appl. (18), 72 (2017)
21. Peng, P., Wang, Y.: Research on green design and evaluation of container terminals. Mar. Equip./Mater. Market. **6**, 95–96 (2020)
22. Cai, O.C., Yu, X.J.: Strategies to promote the green and sustainable development of international ports. Ship. Manag. (6), 1–3+25 (2017)
23. Wang, D.R.: High-quality construction of green and smart ports. Port Sci. Technol. (7), 23–24+52 (2020)
24. Niu, E.X., Meng, B., Shen, S.Y.: Evaluation model and empirical research on green logistics of port enterprises based on cloud model. J. Dalian Marit. Univ. **2**, 67–74 (2017)
25. Wu, X.F., Zhang, L.P.: Practical research on green crystals in China. J. Dalian Marit. Univ. (Soc. Sci. Ed.) **1**, 6–11 (2017)
26. Sun, L.: The original new path for the standardization of green and low-carbon development of port logistics. China Standard. **22**, 95–96 (2019)
27. Shanghai International Port (Group) Co., Ltd.: 2019 SIPG Sustainable Development Report. SIPG, Shanghai (2019)
28. Ningbo Zhoushan Port Co., Ltd.: 2019 Corporate Social Responsibility Report. Ningbo Zhoushan Port, Ningbo (2019)
29. Shenzhen Yantian Port Holdings Co., Ltd.: 2019 Corporate Social Responsibility Report. Shenzhen Yantian Port, Shenzhen (2019)
30. Guangzhou Port Company Limited: 2019 Corporate Social Responsibility Report. Guangzhou Port, Guangzhou (2019)
31. Qingdao Port International Co., Ltd.: 2019 Sustainable Development Report. Qingdao Port, Qingdao (2019)
32. Tianjin Port Development Holdings Limited: 2019 Environmental, Social, and Governance Report. Tianjin Port, Tianjin (2019)
33. Tianjin Port (Group) Co., Ltd.: 2019 Corporate Social Responsibility Report. Tianjin Port, Tianjin (2019)
34. Xiamen International Port Co., Ltd.: 2019 Environmental, Social, and Governance Report. Xiamen Port, Xiamen (2019)
35. Dalian Port (PDA) Company Limited: 2019 Corporate Social Responsibility Report. Dalian Port, Dalian (2019)
36. Yingkou Port Limited Company: 2019 Corporate Social Responsibility Report. YKPLC, Yingkou (2019)
37. Rizhao Port Co., Ltd.: 2019 Corporate Social Responsibility Report. Rizhao Port, Rizhao (2019)
38. Tangshan Port Group Co., Ltd.: 2019 Corporate Social Responsibility Report. TSPGC, Tangshan (2019)
39. Yu, Y.D., Yao, Y.G.: Application of keyword dynamic query software in relational database based on big data. In: 2020 International Wireless Communications and Mobile Computing (IWCMC), Limassol, Cyprus, pp. 1370–1375 (2020)

Comparing Portfolio Management Strategies in Factor Models

Aibo Wang(✉)

Business School, Imperial College London, Philbeach Gardens, London, UK
jw5520@ic.ac.uk

Abstract. Including using realized volatility to adjust portfolios, many portfolio management strategies can increase alpha values, Sharpe ratios and the utility gains of mean-variance investors. This paper uses the historical values of mean, standard deviation and variance of each factor to perform four different portfolio management strategies and tries to explore the most efficient one. It turns out that using mean and variance to adjust the original factor performs better in producing positively significant alphas. Among all asset pricing factors, the effect on the size factor *H-L* is the most remarkable, but our results vary within different periods.

Keywords: Portfolio management · Factor model · Mean-adjusted · Variance-adjusted

1 Introduction

Inspired from the previous study on volatility-managed portfolios of Alan and Tyler [1], this paper further explores more possible portfolio management strategies and tries to figure out the strategy that yields more utility gains for mean-variance investors. We collect asset pricing factors on the daily frequency and adjust them using their realized mean, variance and standard deviation of the past 22 trading days. After that, we regress the adjusted factors on those original factors to collect alphas and find the most positively significant ones. The impact of the financial crisis in 2008 is also considered in this paper and by setting the year 2008 as the breakpoint, we run two more sets of regressions within the time interval before and after 2008. It turns out that using realized mean and variance to adjust original factors produce the most positively significant alphas and the effect of the adjustment is most significant on the size factor *H-L*. In comparison with the work done by Alan and Tyler [1], which uses realized variance to adjust the original portfolio, our results are generally more positively significant.

The following paper begins with the review of the previous work done by different researchers in Sect. 2, followed by the description of data used in Sect. 3 and the empirical strategies in Sect. 4. Section 4 is divided into two parts where the first part details our portfolio adjustments and the second part summarizes the structure of our regression analysis. Section 5 contains the main empirical results and the last section is a brief conclusion of this paper.

C. Yuan et al. (Eds.): *Proceedings of the 4th International Conference on Economic Management and Green Development*, AEPS, pp. 99–107, 2021.
https://doi.org/10.1007/978-981-16-5359-9_13

2 Literature Review

This section begins with a summary of the research done by Alan and Tyler [1], followed by some previous perspectives on how to deal with investment risk in asset allocations and investment strategies. After that, some pieces of research related to the asset pricing models are summarized, including the three-factor model, five-factor model and q-factors model. This section ends with pieces of literature concerning the effect and value of volatility timing.

A volatility-managed portfolio is a trading strategy that scales monthly returns by the inverse of its realized variance and offers large risk-adjusted returns [1]. It yields positively significant alphas across many asset-pricing factors and economically, the utility gains obtained by mean-variance investors are much larger if they follow this volatility timing strategy. Incorporating realized variance can be implemented easily in real-time [1], but different methods of portfolio adjustments including other statistics are also worth investigating and comparing.

Risk is ubiquitous in portfolio investments and Risk Parity is a new strategy of asset allocation which uses risk to allocate resources across different components of portfolios. However, this strategy lacks sufficient justification and to desire diversification by risk is not enough [2]. Based on the idea of leverage aversion, Clifford et al. add a theoretical justification and provide broad empirical evidence to augment Risk Parity. In comparison with the historical performance of the value-weighted market portfolio, Risk Parity outperforms the market portfolio by 2.7% a year. Risk securities are also investigated by Andrea and Lasse in another research, where they suggest that many investors and funds overweight risky securities instead of using leverage [3]. Thus, they form the BAB (betting against beta) factor that holds low-beta assets and that shorts high-beta assets. Comparing with all standard asset pricing factors, the BAB factor outperforms all of them and yields positively significant risk-adjusted returns.

Many asset pricing models are established to explain the expected return on portfolios. The three-factor model of Fama and French explains the expected returns using *MKT-RF* (market excess return), *S-B* (size factor) and *H-L* (book-to-market ratio). These three factors can successfully capture much of the variation in the cross-section of average stock returns and explain the returns to portfolios formed on E/P, C/P and sales growth [4]. However, the three-factor model fails to explain expected returns on all securities and portfolios and Fama and French then come up with the five-factor model. The five-factor model further captures *R-W* (profitability factor) and *C-A* (investment factor) to help explain average returns on portfolios and it helps explain 71%–94% of the cross-section variance of expected returns [5]. Also inspired by investment-based asset pricing, the q-factor model that contains market, size, investment and probability factors is established by Hou et al. They suggest that well-specified models would produce regressions intercepts that are economically small and statistically insignificant from zero. Indeed, the q-factor model performs well in capturing many anomalies in the model of Fama and French [6].

Volatility in risk impacts the estimates of portfolio performance. Oliver et al. find that negative volatility-timing can inflate the unconditional momentum alpha [7]. Biased alpha can be resolved by using lagged loadings as instrument variables which produces lower estimates of conditional momentum alpha. The patterns of volatility are investigated by Tim et al. They disclose the strong similarities in unconditional distributions and structures of realized volatility patterns, and present a new risk model across global assets, which results in the improvement in out-of-sample forecasting performance [8]. They also find a new risk factor containing the information on the future volatility of individual assets. Beyond that, Jeff et al. analyze the economic value of volatility timing by considering a mean-variance investor allocating funds across four asset classes. They suggest that the predictability captured by volatility modeling is economically significant, which outperforms the unconditionally efficient static portfolios [9]. They further quantitatively analyze utility gains from volatility timing in portfolios, which are robust under different conditions and not restricted to short-horizon investors [10].

3 Data Description

There are two primary sources of data used in this paper. Daily asset pricing factors are collected from Kenneth R. French website on *MKT-RF*[1] (the excess return on the market), *S-B* (Small Minus Big), *H-L* (High Minus Low), *R-W* (Robust Minus Weak) and *C-A* (Conservative Minus Regressive), from July 1967 to June 2020. *MKT-RF*[1] refers to the value-weighted returns of all CRSP corporations in the US. *S-B* is the size factor that summarizes the difference in average returns of three small portfolios and three big portfolios. *H-L* represents the difference in average returns of two value portfolios and two growth portfolios. As an augment from the three-factor model, *R-W* (profitability factor) and *C-A* (investment factor) are added to help explain more cross-section variance of expected returns for portfolios [5]. Another set of data is collected from the website of Hou et al. From January 1967 to December 2019, collected factors are *ME* (size factor), *IA* (investment factor), *ROE* (profitability factor) and by subtracting the riskless rate of return from the return on the market portfolio, the excess return on the market portfolio *MKT-RF*[2] is calculated. The q-factors model of Hou et al. also outperforms the three-factor model on capturing many of the significant anomalies [6].

4 Empirical Strategy

4.1 Portfolio Formation

Our portfolios are constructed by adjusting them using the realized mean, standard deviation and variance of each factor. From Eqs. (1)–(3), these statistics are calculated based on a period of 22 trading days and they are used to adjust the factor on the next trading day. For instance, each factor on the 23rd trading day is adjusted by the mean, standard deviation and variance of that factor within the first 22 days.

$$Mean_t(f) = \frac{\sum_{i=21}^{i=1} f_{t+i}}{22} \tag{1}$$

$$Std_t(f) = \frac{\left(f_{t+i} - \frac{\sum_{i=21}^{i=1} f_{t+i}}{22}\right)}{\sqrt{22}} \tag{2}$$

$$Var(f) = \frac{\left(f_{t+i} - \frac{\sum_{i=21}^{i=1} f_{t+i}}{22}\right)^2}{22} \quad (3)$$

Using the mean, standard deviation and variance of each factor, we perform four types of portfolio management strategies: 1) dividing the original factor by its standard deviation, 2) dividing the original factor by its variance, 3) multiplying the original factor by its mean and then dividing it by its standard deviation, 4) multiplying the original factor by its mean and then dividing it by its variance. After making these adjustments, our sample now contains 14326 observations and 13318 observations respectively and then we group each factor into three groups, by its descending order. In Table 1, the group mean of the five-factor model is collected in Group A and the group mean of the q-factors model is collected in Group B.

As shown in Group A, for each factor after making adjustments, the adjusted mean values of the first group are always significantly larger than the other two groups and all of the mean values in the third group turn into negative. Among four types of adjustments, dividing the original factor by the variance of the previous 22 trading days produces the greatest mean values in the first and second groups, although the mean values of the third group are the most negative ones. Generally, multiplying the original factor by the mean value of the previous 22 trading days does not improve the group mean of that factor, but it decreases the absolute values of group means significantly so that the third group turns to be less negative, compared with the first and second types of adjustments. Similar results can be observed in Group B, where the second type of adjustment produces more preferable results in the first and second groups and multiplying the original factors by mean values improves the results of the third group.

4.2 Empirical Strategy

For each adjusted factor, as the dependent variable on the left-hand side of Eq. (4), we treat each original factor as the independent variable and run four time-series regressions of them.

$$f^*_{t+1} = \alpha + \beta f_{t+1} + \varepsilon_i \quad (4)$$

We limit the time horizon of our regression analysis from 2000 to 2019 and run three groups of regressions: 1) from 2000 to 2019, 2) from 2000 to 2007, 3) from 2008 to 2019. Regressions on different groups can evaluate the stability of our empirical results and 2008 is chosen as a breakpoint because the impact of the financial crisis is considered. Generally, if regression intercepts (alpha) are positively significant, then those adjusted factors could not be explained by the original factors and they yield positive income on the portfolios. Positively significant intercepts (alpha) also imply that Sharpe ratios relative to the original factors are increased and our adjustments expand the mean-variance frontier [1].

5 Empirical Results

We collect those alphas from regressions and summarize them in Table 2 and Table 3. Table 2 reports the regression results from 2000 to 2019. We can observe that before using mean values to adjust original factors, in the first row and second row, alpha values of *H-L*, *R-W*, *C-A* and *IA* are negative. For all positive columns, dividing by variance outperforms dividing by standard deviation since all positive alpha values in the second row are larger than those in the first row. As for the statistical significance, the alpha value of *C-A* divided by its variance is statistically significant at 10% level and the alpha values of *ROE* are both significant at 1% level. After using mean values to make adjustments, we observe that both market factors turn into negative while *H-L*, *R-W*, *C-A* and *IA* all turn into positive. The significance levels are also improved as in the third and fourth rows, *H-L* is significant at 1% significance level, *R-W* is significant at 10% significance level and 5% significance level respectively and importantly, *C-A* and *IA* in the fourth row are all significant at 1% significance level. The significance level of *ROE* remains the same with some decreases in the alpha values. Between the third type and fourth type of adjustments, the fourth type outperforms the third type since all alpha values are larger and there are three factors (*R-W*, *C-A* and *IA*) that are more positively significant. Therefore, we could conclude that using mean values to adjust original factors could produce more positively significant alphas and these results suggest that the adjusted factors produce more positive income on portfolios. Generally, dividing the original factors by their variance outperforms dividing the original factors by their standard deviations and the adjustments using mean values improve the performance of portfolios.

Table 3 reports the regression results after we apply the breakpoint. In Panel A, from 2000 to 2007, we see that if mean values are not used to adjust original factors, alpha values for both market factors, *C-A* and *IA* are negative and the alpha values of all factors, except for *ROE*, are not statistically significant. Dividing *ROE* by its variance produces more positively significant results comparing to dividing *ROE* by its standard deviation. After using mean values to make adjustments, the signs of *C-A* and the fourth row of *IA* turn into positive and the significance levels of *H-L*, *R-W* and *ME* are also increased as a result. As for the period from 2008 to 2019, we could generally observe more negative alpha values and lower significance levels. The alpha values of the *ROE* are not statistically significant except for the first row and we could still conclude that, between the third and fourth types of adjustments, dividing the original factors by variance still outperforms dividing the original factors by standard deviation, since it produces more positively significant results in *C-A* and *IA*.

Generally, we find that using mean values to adjust original factors yields more positively significant results and dividing original factors by the variance would outperform dividing them by the standard deviation. Therefore, we conclude that the fourth type of adjustment, multiplying the original factor by its mean and then dividing it by its variance, performs better in expanding the mean-variance frontier. However, the financial crisis in 2008 impacts the stability of our results significantly since in the Panel B of Table 3, those alphas are less statistically significant. Loss in the statistical significance may also be caused by the reduction of sample size.

Table 1. Group mean of adjusted asset pricing factors

	Ranking	Factor/std		Factor/var		Factor*mean/std		Factor*mean/var	
		Group A	Group B	Group A	Group B	Group A	Group B	Group A	Group B
Market factor	1st	1.153	1.122	1.849	1.717	0.200	0.200	0.271	0.261
	2nd	0.076	0.047	0.092	0.056	0.003	0.003	0.006	0.004
	3rd	−1.131	−1.157	−1.780	−1.769	−0.189	−0.189	−0.247	−0.242
Size factor	1st	1.169	1.164	3.278	3.040	0.150	0.146	0.410	0.374
	2nd	0.056	0.054	0.121	0.110	0.005	0.004	0.012	0.009
	3rd	−1.175	−1.163	−3.293	−3.084	−0.109	−0.111	−0.260	−0.253
Investment factor	1st	1.201	1.215	4.922	4.769	0.098	0.094	0.347	0.321
	2nd	0.029	0.050	0.088	0.151	0.003	0.002	0.012	0.007
	3rd	−1.123	−1.089	−4.652	−4.221	−0.079	−0.078	−0.268	−0.256
Profitability factor	1st	1.214	1.262	5.355	4.723	0.096	0.122	0.342	0.393
	2nd	0.031	0.101	0.106	0.300	0.003	0.005	0.009	0.015
	3rd	−1.098	−1.060	−4.687	−3.750	−0.076	−0.091	−0.260	−0.275
Book-to-market	1st	1.217		3.976		0.146		0.394	
	2nd	0.017		0.050		0.005		0.014	
	3rd	−1.125		−3.686		−0.108		−0.276	

6 Discussion and Conclusion

This paper collects daily pricing factors from the websites of Fama and French as well as Hou et al. Among four types of portfolio adjustments, we find that multiplying the original factor by its realized mean and then dividing it by its realized variance produces the most positively significant results, which expands the mean-variance frontier and increases the utility gains of mean-variance investors. This portfolio adjustment shows an improvement in the sign and the statistical significance level of alphas from the work done by Alan and Tyler [1], which only uses realized variance to make portfolio adjustment. Future studies could focus on other methods of portfolio management to improve the predictability on the stock performance and further increase Sharpe ratios. Also, it is worth including more detailed information of different companies to improve the prediction of expected returns, such as the culture and operating strategies of different companies in the stock market.

Table 2. Regression results (2000–2019)

		MKT-RF[1]	S-B	H-L	R-W	C-A	MKT-RF[2]	ME	IA	ROE
Factor/std	α	0.008	0.006	−0.003	−0.002	−0.012	0.005	0.005	−0.006	0.023***
		(0.008)	(0.006)	(0.009)	(0.008)	(0.008)	(0.008)	(0.006)	(0.008)	(0.008)
Factor/var	α	0.013	0.015	−0.018	−0.013	−0.095*	0.007	0.006	−0.048	0.121***
		(0.019)	(0.022)	(0.040)	(0.042)	(0.050)	(0.020)	(0.022)	(0.049)	(0.044)
Factor*mean /std	α	−0.003	0.002	0.008***	0.004*	0.002	−0.003	0.002	0.002	0.007***
		(0.003)	(0.002)	(0.003)	(0.002)	(0.002)	(0.003)	(0.002)	(0.001)	(0.002)
Factor*mean/var	α	−0.002	0.006	0.023***	0.010**	0.015***	−0.002	0.006	0.011***	0.018***
		(0.004)	(0.004)	(0.005)	(0.004)	(0.004)	(0.004)	(0.004)	(0.004)	(0.005)

Robust standard errors in parentheses.
*** $p < 0.01$, ** $p < 0.05$, * $p < 0.1$.

Table 3. Regression results

		MKT-RF[1]	S-B	H-L	R-W	C-A	MKT-RF[2]	ME	IA	ROE
Panel A: 2000–2007										
Factor/std	α	−0.003 (0.010)	0.015 (0.010)	0.015 (0.015)	0.005 (0.013)	−0.015 (0.014)	−0.004 (0.010)	0.015 (0.010)	−0.005 (0.015)	0.027** (0.013)
Factor/var	α	−0.002 (0.023)	0.028 (0.035)	0.054 (0.080)	0.062 (0.072)	−0.138 (0.088)	−0.005 (0.023)	0.025 (0.034)	−0.065 (0.087)	0.193*** (0.074)
Factor*mean/std	α	−0.006 (0.005)	0.007 (0.004)	0.011*** (0.004)	0.010** (0.005)	0.001 (0.003)	−0.005 (0.005)	0.008* (0.005)	−0.001 (0.003)	0.014*** (0.004)
Factor*mean/var	α	−0.004 (0.005)	0.010 (0.007)	0.041*** (0.009)	0.019*** (0.007)	0.011 (0.007)	−0.003 (0.005)	0.012* (0.007)	0.006 (0.007)	0.035*** (0.008)
Panel B: 2008–2019										
Factor/std	α	0.016 (0.012)	0.001 (0.007)	−0.019* (0.011)	−0.008 (0.007)	−0.001 (0.006)	0.013 (0.012)	−0.001 (0.007)	0.001 (0.007)	0.022** (0.010)
Factor/var	α	0.024 (0.028)	0.009 (0.029)	−0.085** (0.040)	−0.069* (0.041)	−0.023 (0.047)	0.017 (0.029)	−0.002 (0.028)	−0.002 (0.048)	0.084 (0.052)
Factor*mean/std	α	−0.001 (0.004)	−0.002 (0.002)	0.003 (0.003)	0.000 (0.002)	0.003* (0.001)	−0.002 (0.004)	−0.002 (0.002)	0.003* (0.001)	0.003 (0.002)
Factor*mean/var	α	−0.001 (0.005)	0.001 (0.005)	0.007 (0.005)	0.004 (0.005)	0.016*** (0.006)	−0.002 (0.005)	−0.001 (0.005)	0.014** (0.005)	0.007 (0.006)

Robust standard errors in parentheses.
*** $p < 0.01$, ** $p < 0.05$, * $p < 0.1$.

References

1. Moreira, A., Muir, T.: Volatility-managed portfolios. J. Finan. **72**, 1611–1644 (2017)
2. Asness, C.S., Frazzini, A., Pedersen, L.H., Aversion, L., Parity, R.: Finan. Anal. J. **68**(1), 47–59 (2012)
3. Frazzini, A., Pedersen, L.: Betting against beta. J. Finan. Econ. **111**(1), 1–25 (2014)
4. Fama, E.F., French, K.R.: Multifactor explanations of asset pricing anomalies. J. Finan. **51**, 55–84 (1996)
5. Fama, E.F., French, K.R.: A five-factor asset pricing model. J. Finan. Econ. **116**(1), 1–22 (2015). ISSN 0304-405X
6. Hou, K., Xue, C., Zhang, L.: Digesting anomalies: an investment approach. Rev. Finan. Stud. **28**(3), 650–705 (2015)
7. Boguth, O., Carlson, M., Fisher, A., Simutin, M.: Conditional risk and performance evaluation: volatility timing, overconditioning, and new estimates of momentum alphas. J. Finan. Econ. **102**(2), 363–389 (2011)
8. Bollerslev, T., Hood, B., Huss, J., Pedersen, L.H.: Risk everywhere: modeling and managing volatility. Rev. Finan. Stud. **31**(7), 2729–2773 (2018)
9. Fleming, J., Kirby, C., Ostdiek, B.: The economic value of volatility timing. J. Finan. **56**, 329–352 (2001)
10. Fleming, J., Kirby, C., Ostdiek, B.: The economic value of volatility timing using "realized" volatility. J. Finan. Econ. **67**(3), 473–509 (2003)

Statutory Power of WHO and Its Suggestions for Improvement Under Public Health Emergency

Jingnvying Su(✉)

Law School, Jilin University, Changchun, China

Abstract. WHO has taken a series of measures against the COVID-19, but they have been questioned by many parties, which is related to the marginalization of WHO as an international organization. This paper takes WHO's authority and behaviors under public health emergencies as the research object, and uses the methods of literature research and case study to explore the reasons why WHO has faded from the world public health authority from three aspects: WHO's international status, the effectiveness of relevant international agreements, and relevant cases in the past. At the same time, sort out emergency response behaviors and analyze the rationality of WHO's behaviors during the epidemic. On this basis, the article puts forward the following suggestions, including: Member States should respect the role of WHO in health emergencies, WHO itself should also standardize the election mechanism of functional departments, implementation procedures of related agreements and other systems, and reform the funding structure model. In the future, WHO can cooperate with non-official organizations to expand data collection channels, and its international agreements can also be linked with the domestic laws of sovereign countries, and the two can coordinate public health incidents.

Keywords: Public health emergency · WHO · Legal authority · System reform

1 Introduction

Since the outbreak of COVID-19, the World Health Organization (WHO), as the guiding and coordinating agency for health issues in the United Nations system, is responsible for integrating the epidemic data notified by countries to the organization and sharing information. However, things are not going smoothly. Tedros Adhanom Ghebreyesus, the director-general of the WHO said: "One of the biggest challenges we face is that too many countries affected by the novel coronavirus epidemic have yet to share data with WHO." Not only is it about information sharing, but the WHO's January warning that prohibits countries from issuing travel bans at will has few countries accepted. For example, US President Donald Trump announced in March that travel from the European continent was restricted [1]. Trump has repeatedly publicly criticized the WHO for not responding to the novel coronavirus in a timely manner. On May 29, he announced that

C. Yuan et al. (Eds.): *Proceedings of the 4th International Conference on Economic Management and Green Development*, AEPS, pp. 108–116, 2021.
https://doi.org/10.1007/978-981-16-5359-9_14

the United States would withdraw from the World Health Organization. On September 3, Morgan Ortagus, a spokesperson for the U.S. State Department, issued a statement stating that "WHO is not responding well to the COVID-19" and that "the organization refuses to reform to prove its independence" [2]. From the above facts, it is not difficult to see that WHO has been marginalized, and many member states have not fulfilled the obligations given to them by IHR. However, the international influence of WHO is still there, and the world still needs organizations with international influence to lead the global fight against the virus and guide public health. Therefore, it is necessary for WHO to make appropriate improvements and adjustments to deal with the current international situation.

2 The Status of WHO in International Law

On March 11, the WHO announced that the COVID-19 epidemic has constituted a global pandemic. As of September 12, the epidemic had affected 218 countries and regions around the world. People began to realize that the epidemic could no longer be controlled by a country's government, and at the same time they began to criticize the WHO's announcement for too late. In this epidemic, the WHO has continuously appeared in the public eye, broadcasting epidemic data to the world every day, and at the same time making suggestions or warnings to governments of various countries. However, the WHO did not play a leading role in this public health emergency. Many member states turned a deaf ear to WHO's recommendations, and even never fulfilled their obligations as a member state. They believe that the WHO's performance in the preparatory phase of this pandemic was negligent, and the WHO has not yet provided effective solutions. It can be seen that the status of WHO in the field of international public health has gradually declined.

2.1 The Establishment of WHO and Its Legal Nature

WHO is a specialized agency of the United Nations, not the main organ of the United Nations. On April 7, 1948, after the Constitution of the World Health Organization came into effect, the World Health Organization was declared established. The WHO incorporated the assets, personnel, and duties of the League of Nations' Health Organisation and the Office International d'Hygiène Publique, including the International Classification of Diseases (ICD) [3]. As of 2020, the World Health Organization has 193 member states and two associate members. But in fact, the powers that the WHO has are limited. According to the Constitution of the World Health Organization, the World Health Assembly is the highest authority of the WHO. It has the power to decide WHO policies and appoint the Director-General. However, Article 18 of the law also stipulates, the Health Assembly instruct the Board and the Director-General to bring to the attention of Members, any matter with regard to health which the Health Assembly may consider appropriate. The legal effect of words like "attention" is very weak. Similarly, the terms "recommendation", "communication," and "negotiation" are used in many places in IHR that regulate public health emergencies. Article 10 of the Regulation stipulates that after receiving information that may constitute a public health emergency of international concern, the WHO shall express its willingness to cooperate with the relevant state party.

However, if the state party does not accept the cooperation proposal, WHO can only encourage its cooperation. Therefore, in terms of public health emergencies, WHO's role is mainly to make temporary or long-term recommendations to the contracting parties, but Article 1 of the IHR stipulates that these recommendations are non-binding. To sum up, WHO has the decision-making power and executive power, but its power is still limited. The limited power makes its treaties frequently violated by member states, and its suggestions are not accepted by member states. All these make WHO gradually fade out of the international public health authority center.

2.2 Effectiveness of Relevant Agreements Signed by WHO

Articles 19 to 23 of the Constitution of the World Health Organization grant WHO the quasi-legislative power in the field of public health. So far, WHO has only concluded 1 convention (the Framework Convention on Tobacco Control) and 2 regulations (International Health Regulations and International Classification of Diseases). These international agreements belong to international public law, but WHO itself prefers to rely on non-legally binding resolutions and recommendations. These international documents are international soft law, that is, "soft law usually refers to international documents of principles, norms, standards or other statements other than treaties" [4]. The reason for this is that compared with domestic law, international law is not strong in legal effect, enforcement, and binding. Therefore, even if an international agreement is drawn up, the contracting parties will probably not comply. Once a contracting party violates the international agreement, the authority of WHO will also be hurt. These characteristics of international law have given rise to the nihilistic view of international law, that international law is not a real law, but such a view will lead to power politics in international relations and is even more detrimental to international cooperation. Take this novel coronavirus epidemic as an example. Although WHO has been criticized and ignored many times, the world needs a specialized international organization to lead the global fight against the epidemic. This is something that all sovereign countries cannot deny. Therefore, it should be considered that although the legal force of international law is not as strong as that of domestic law, it is determined by the nature of the international community. It is still a real law with certain legal effects.

3 Analysis of WHO's Behavior Under Emergent Public Health Events

The novel coronavirus epidemic is a public health emergency. At present, most of the international laws on emergency response involve the prevention of emergencies. Some countries directly focus on prevention as the principle, and the purpose is to achieve the effect of "using a small amount of money for prevention rather than spending a lot of money for treatment" [5]. The prevention here refers to relying on risk assessment, risk management and other risk regulation methods, striving to eliminate emergencies in the bud, that is, to prevent potential risks from evolving into emergencies in time, and then minimize the losses caused. Regarding risk, various social forms in various periods in human history are, in a sense, a risk society [6]. Modern society is under the background of globalization, and risks are also trending towards globalization. The above-mentioned reasons have prompted international organizations that transcend national boundaries to begin to assume the responsibility of risk regulation. As far as the public health field is concerned, WHO is professional, has the powers of risk assessment, monitoring, and making recommendations, and is a competent risk regulation body.

3.1 Modern Risk Society and Its Characteristics

Human beings have always been in a risk society. But in the contemporary era, the significance of risk to society is completely different from that of the past. On the one hand, modern social risks cover a wider range. That is to say, globalization has increased the interdependence of countries, societies, and people, and risks will affect more countries and regions than before, and the consequences will require more people to bear. On the other hand, the modern risk society has deeper risks, such as ecological deterioration and nuclear technology threats. Any kind of risk poses a serious threat to human survival and development. Once the risk regulation fails and the risk is transformed into an emergency, the harm to the world will be incalculable. In addition, modern social risks tend to shift from natural risks to man-made risks, from single risk consequence to multiple risk consequences. Take the novel coronavirus epidemic as an example. When the risk turns into a public health emergency, not only is the world's medical and health system destroyed, but the economic situation is also hit. International traffic has been severely restricted, and even food security problems have become more serious in some economically underdeveloped areas. Severe disasters exceed the processing capacity of early warning and monitoring, and catastrophic accidents have various impacts, making risk calculation impossible [7].

3.2 Modern Risk Society and Its Characteristics

There is a process for the occurrence and evolution of emergencies. There are many models in the world that can divide the specific stages of the above process. Among them, the PPRR model is the most widely used, including prevention, preparation, response, and recovery [8]. Emergency response activities can be classified according to the PPRR model. The first is "prevention". The purpose of prevention is to eliminate risks, avoid emergencies, or take measures in advance to reduce the hazards of emergencies if they cannot be avoided. In the early stage of the novel coronavirus epidemic, whether there will be human-to-human transmission of the COVID-19 requires a certain period of monitoring to determine. Monitoring data requires risk assessment to predict the possible negative effects of specific hazards. On February 28, the WHO raised the global risk level of the "Covid-19" epidemic from "high" to "very high", which is a risk assessment. On the basis of risk assessment, the subject of risk regulation should implement appropriate measures to prevent the occurrence of emergencies. All of these measures are "prevention". The response activity after prevention is "preparation", that is, to ensure that the affected area is prepared for the hazard. Taking this public health emergency as an example, preparations for nucleic acid testing, preparation of ventilators, and protective clothing for medical staff are all preparations. "Responding" refers to providing an effective response. The most widely adopted and most effective response to this epidemic is social isolation. The last emergency response activity is "recovery", which refers to providing recovery assistance to the affected areas, such as providing funds, material support and technical guidance.

3.3 Analysis of WHO's Emergency Response Behavior

The Director-General of WHO once introduced that WHO has done five aspects of work in response to the novel coronavirus epidemic. Among them, the three aspects of providing necessary medical equipment and training for frontline medical staff, mobilizing medical staff and accelerating the speed of vaccine development are fully meet the requirements of Article 13 of IHR, also meet the emergency response, and embody the international humanitarian spirit. However, other WHO work, especially the risk assessment of the novel coronavirus, has been seriously questioned and criticized. On January 23, WHO believed that the novel coronavirus epidemic did not meet the PHEIC standard, and the global risk level of the epidemic was also intermediate. On January 27, WHO admitted to the misjudgment and adjusted the risk level to high. It was not until January 30 that the WHO declared that the epidemic constituted an international public health emergency, but the number of confirmed cases at this time had increased tenfold. From the results of WHO's risk assessment, it is indeed not a reasonable emergency response behavior. One of the reasons for the failure of the assessment is WHO's major structural weakness, which relies on information provided by member states for risk assessment [9]. To confirm this point, North Korea has not reported any cases of coronavirus at present, and WHO is unable to verify it by itself. The root of WHO's weakness is that WHO as an international organization has too limited power, and it is also hindered by budgetary and political pressure. WHO also has other response behaviors and cannot judge the rationality of its behaviors. For example, the WHO does not approve or even oppose the implementation of travel bans by countries, but the countries ignore the warnings of the WHO. The reason for the inability to judge is that the positions of the two sides are different. WHO believes that international bans may hinder the transportation of required resources or delay aid and technical support. However, sovereign states believe that such restrictions are a policy that caters to public opinion and will indeed reduce the risk of domestic people contracting the novel coronavirus epidemic. International organizations and sovereign states consider different issues regardless of whether they are political or legal. However, the power of a sovereign state is stronger than that of an international organization. When the interests of both parties conflict, the authority of the international organization is often the result of damage. This is also an urgent need to be resolved in the formation and development of the law of international organizations.

4 Suggestions on Perfecting WHO's Statutory Power Under Public Health Emergencies

WHO is under great pressure and criticism in this epidemic, but at the same time, government officials and scholars from all walks of life are also calling on WHO to carry out reform in order to return to the authoritative center of public health in the world. Based on the previous analysis, WHO's own dilemma lies in two aspects: First, as an international organization, WHO has limited powers. The second is that WHO is considered to lack independence from its member states. WHO itself faces pressure from political powers, and because 70%–80% of its funds come from donations, it is also subject to pressure from major donor countries and even donor organizations. To solve these problems, WHO needs to readjust its related systems and improve related laws.

4.1 Respect and Improve WHO's Role in Responding to Specific (Emergency) Events

The novel coronavirus epidemic has once again confirmed that in the context of a modern risk society, infectious diseases cannot have a destructive effect in a certain fixed area, and no country can be immune to it. Only by starting from a global perspective and adopting consistent prevention and control actions through cooperation can countries prevent the spread of the disease. For this kind of cooperation to produce practical results, international public health organizations with world influence are needed to lead, provide professional advice, allocate rights and obligations in cooperation, and share information and technology. This is why, while WHO is constantly being questioned, many scholars still do not agree with the behavior of giving up or withdrawing from the WHO. WHO is the only global organization whose mission, scope and infrastructure are suitable for this, although it is in urgent need of reform and improvement. In fact, most sovereign countries know that the world urgently needs a well-functioning global health institution. According to Reuters, negotiations on WHO reform were held in early August, but because the United States still tried to lead the negotiations after it withdrew from the WHO, France and Germany have withdrawn from the negotiations [10]. Negotiations on the reform of the WHO should be held again, and member states should try their best to send representatives to participate in order to show the fairness of the negotiations. Representatives of member states should put forward suggestions on this WHO reform at the negotiation meeting. Various adjustment and improvement measures should be determined in the form of public international law. Member states should ensure that they respect the requirements and recommendations of the WHO when responding to public health emergencies, and at the same time, ensure that they fulfill their obligations as member states, and actively cooperate with WHO in the face of health emergencies. Only when these actions are implemented can the WHO carry out truly effective reforms.

4.2 Further Reform WHO's Operating Mechanism

The WHO does not have too much power but has been subject to interference by political powers, which has been criticized. In this reform, WHO should standardize its operating system so that its recommendations can be better implemented in all member states. At the same time, WHO should try to remove political factors, restore its scientific credibility, and make the overall operation of WHO more transparent and fair.

Innovate the Election System of WHO Functional Departments. During the novel coronavirus epidemic, the US government has been accusing the WHO of focusing on

China's interests. At the same time, many reports claimed that the current WHO Director-General Tedros was elected with the support of China and its group. In fact, the WHO functional department elections have always been full of political colors. It is said that in the early 1990s, Japan put pressure on small countries through trade actions and provided assistance to them in order to win support for the reappointment of its director general candidate. Historically, unwritten regulations also required five of the six WHO assistant directors to come from the United States, Britain, France, the Soviet Union and China [11]. Such excessive political involvement will undoubtedly affect WHO's leadership in the global fight against the novel coronavirus epidemic. Therefore, the adjustment of the election system of WHO's functional departments should also be the focus of this WHO reform. All member states should send representatives to conduct fair elections. If the leadership of each functional department continues to be the same person, strict review or prohibit the same person from continuously serving as the leadership of the same functional department. At the same time, the World Health Assembly maintains representative democracy in the 194 member states of the World Health Organization.

Standardize the Implementation Procedures of Relevant WHO Agreements. International agreements are soft law, and do not have actual enforcement power compared with the domestic laws of sovereign states. The WHO-related international agreements mainly stipulate the rights and obligations of WHO and member states, and seldom involve the responsibilities that both parties should bear if they violate international agreements. Therefore, in order to better implement the recommendations and policies formulated by the WHO, international agreements can increase the "cost" that member states should pay after breach of contract. For example, cancel the voting rights of the defaulting country in the next meeting or prohibit representatives of the defaulting country from participating in the election of members of the executive committee.

Promote the Change of WHO's Funding Structure. Currently, 70%–80% of WHO's annual funding comes from voluntary contributions. Powerful donor countries or organizations donate to WHO for specific purposes, in order to influence WHO's follow-up actions. This specific purpose is often not related to global health needs, but only related to the donor's own interests, which undoubtedly undermines WHO's planning in the health field. Therefore, WHO should lower the proportion of donations in the organization's funds each year and increase the share of long-term funding agreements [11].

4.3 Promote the Coordinated Development of International Affairs Governance

In the context of globalization, not only sovereign states need to cooperate with each other, but WHO as an international organization can also cooperate with other international organizations or civil organizations to broaden information and data collection channels and further expand its influence.

Collaborative Governance of Multiple Entities. In this epidemic, the performance of many non-official organizations was surprising, such as OWID and Johns Hopkins University, the latter's epidemic data was quoted by many countries. For a long time, the IHR infectious disease alert mechanism has had the problem of too limited information sources. It is difficult to control the crisis situation in a timely and effective manner by using official channels as the only channel to understand infectious disease outbreaks. Therefore, WHO can cooperate with other organizations, and WHO needs further screening of data and information transmitted by other organizations.

The Coordinated Governance of Domestic and International Law. At present, more and more sovereign countries have legislation on emergencies, such as China's "Emergency Response Act", the United States' "Disaster Relief and Emergency Assistance Act", and Japan's "Disaster Response Basic Law". The domestic laws of these sovereign countries have different focuses, but their pursuit of solving public health problems and maintaining public health is the same, which is also the legislative purpose of IHR. Therefore, these domestic laws can be governed in concert with international laws. When it comes to internal issues of each country, safeguard the sovereignty and constitutional authority of each country and implement relevant domestic laws. On international issues, follow international agreements and promote cooperation among countries to prevent disasters from spreading globally.

5 Conclusions

During the COVID-19 pandemic, WHO has been subject to controversy and doubts, and its lack of power and independence has been criticized by many parties. However, its influence in the world is still there, and the world needs global public health organizations to lead the world. Therefore, WHO should carry out reform instead of being completely marginalized. In order for the WHO's reform to play a substantial role, member states should respect the role of WHO in public health emergencies, and WHO itself should also improve its functions such as the election of functional departments and the implementation procedures of relevant international agreements. In the future, WHO can cooperate with other international organizations or non-official organizations to expand information and data collection channels. At the same time, deepen the connection between the relevant WHO international agreements and the domestic laws of sovereign states, so that domestic laws and international laws can coordinate public health incidents. Member states and international organizations cooperate well to build a community with a shared future for mankind.

References

1. Why has WHO failed to lead global solidarity in the fight against the epidemic. https://wsdigest.com/article?artid=4269
2. The US State Department has issued a statement on its next steps after withdrawing from the WHO. https://www.voachinese.com/a/US-WHO-withdrawal-update-20200903/5569997.html
3. Milestones for health over 70 years. https://www.euro.who.int/en/about-us/organization/who-at-70/milestones-for-health-over-70-years.

4. Shelton, D.: International law and relative Normativity. In: Evans, M.D. (ed.) International law, p. 166. Oxford University Press, Newyork (2003)
5. Osborne, D., Gabler, T.: Reinventing Government, p. 205. Shanghai Translation Publishing, China (1996)
6. Beck, U.: Wang Wulong compiled. "From Industrial Society to Risk Society (Part 1)". Marxism and Reality, Issue 3, pp. 26 (2003)
7. Beck, U.: Wang Wulong compiled. "From Industrial Society to Risk Society (Part 1)". Marxism and Reality, Issue 3, pp. 33 (2003)
8. Zhao, P.: The Problems of Administrative Law on Risk Regulation—with Perspective of Emergency Prevention, p. 20. China University of Political Science and Law, Beijing (2003)
9. What's Wrong With the World Health Organization. https://www.theatlantic.com/politics/archive/2020/04/world-health-organization-blame-pandemic-coronavirus/609820/
10. Exclusive: Germany and France quit WHO reform talks amid tension with Washington – sources. https://www.reuters.com/article/us-health-coronavirus-who-reform-exclusi-idUSKCN25329P
11. The World Health Organization can be reformed. https://www.ft.com/content/df72892c-8e19-11ea-af59-5283fc4c0cb0

How Does Overconfidence Affect Entrepreneurs at Loss

Yifu Liu(✉)

Beijing Royal School, Beiqing Street, Beiqijia Town, Beijing, China
liuyifu@st.brs.edu.cn

Abstract. This paper will review past economic models and through the analysis of mature economic and psychological models, this paper explains the rational man hypothesis and its deficiency in practical application. This paper further reviews the criticism of the rational man hypothesis in the past, and introduces behavioral economics and the overconfidence theory. The characteristics of overconfidence of entrepreneurs will be focused in this paper. Whether there is overconfidence in entrepreneurs and whether overconfidence will affect their entrepreneurial decision-making will be elaborated in this paper. It is expected to deepen the understanding of the entrepreneurs' own characteristics and provide a theoretical basis for the success of entrepreneurship.

Keywords: Behavioral economic · Overconfidence · Entrepreneurship · Management

1 Introduction

Nowadays, the entrepreneurial environment is complex and full of uncertain factors. The superposition of these factors has caused most of the entrepreneurial activities to generally end in failure. Not many new ventures can truly survive and grow in such a complex environment. Studies have shown that Chinese new ventures will face great challenges within 7 years of their establishment, and there will be the risk of exiting market competition. Among the new ventures, nearly 60% of the companies will exit the market within 5 years [1]. The success or failure of a business is related to many factors, such as the company's internal environment and external environment, but it is undeniable that the entrepreneurs' own characteristics, qualities, and abilities will play a decisive role in the entrepreneurial process.

The overconfidence of entrepreneurs is generally considered unfavorable and should be corrected. The overconfidence of entrepreneurs is rarely recorded in the literature, but in real cases, business can't stand without overconfidence from entrepreneurs: like in the early days of Apple's founding, many people did not think that Jobs and Wozniak would succeed, but the two entrepreneurs persisted and cultivated a technology start-up beyond people's imagination [2]. From this point of view, their irrationality or "overconfidence" in themselves makes entrepreneurial success.

C. Yuan et al. (Eds.): *Proceedings of the 4th International Conference on Economic Management and Green Development*, AEPS, pp. 117–124, 2021.
https://doi.org/10.1007/978-981-16-5359-9_15

Based on this, this article will address analysis of the concept of overconfidence and relevant effects of overconfidence on entrepreneurs, so as to deepen the understanding of entrepreneurs' own characteristics and provide a theoretical basis for entrepreneurial success.

2 Hypotheses and Theoretical Models

There is an essential monopolistic competitive market model in microeconomics. When the marginal revenue (MR) is lower than the short-term marginal variable cost (SRMC) [3], if the company continues to operate, the net profit will be negative. That is to say, if the company continues to produce products or provide services, it will gradually lose money, because the profit generated by a unit of product or service will not cover the unit cost. However, in reality, many entrepreneurs continue to operate the company even if they have reached the above-mentioned situation, not because they cannot calculate the specific MR and SAC or correctly figure out the critical point of production volume, but because the entrepreneurs believe that their companies will be successful in the near future. The factor that causes entrepreneurs to have this belief is overconfidence.

2.1 The Rational Man Hypothesis

Whenever we use any economic model to analyze economic phenomena from macro to micro perspectives, the economic man hypothesis proposed by Adam Smith or the rational man hypothesis that later evolved cannot be ignored. In the rational man hypothesis, all actions taken by all economic individuals are to choose the decision with the largest net-benefit after calculating costs and benefits, rather than emotionally. And all economic actions taken by all economic individuals are with the ultimate goal of maximizing their own economic interests. As a classic assumption in economics, the rational man hypothesis has been used for more than 200 years [4]. Until today, the rational man hypothesis is also a self-evident hypothesis in Western economics.

Economic individuals in the rational man hypothesis have the following four characteristics:

First, economic individuals are completely rational, and their decisions are optimized through rational analysis of final costs and benefits. As mentioned above, economic individuals will not be affected by any emotional factors in their decision-making. For example, whereas irrational people will give alms to beggars out of pity [5], rational people may give alms in consideration of the positive impact of alms on themselves, such as public relations and their own affinity.

Second is self-interest. The behavior of economic individuals is infinitely self-interested and avoids harm. Adam Smith pointed out in The Wealth of Nations, "The food and drink we need every day is not the blessing of the butcher, brewer or baker, but the self-interested plan." In Smith's view, human nature is selfish, and all economic phenomena are rooted in human's self-interested nature.

Third, economic individuals have stable and well-organized preferences and clarify their own preferences.

Fourth, the individual has the ability to calculate the net profit of each decision outcome. In the hypothesis, market information is consistent for every economic individual, and every individual can obtain and process it.

These assumptions help the processing of economic models. Ignoring the economic behavior of "disturbing individuals" in some economies, these assumptions able us to use a simpler way to derive some simple and logical conclusions or models such as market and the supply and demand curve. These conclusions or economic models can show economic principles more intuitively and use them to conclude economic laws. For example, by assuming that all economic entities in the market are rational people, two straight supply and demand curves can be drawn to help us understand how market prices are generated.

2.2 Simon's Bounded Rationality Theory

Herbert A. Simon proposed the concept of bounded rationality from two basic aspects – "The limit of rationality is seen from the fact that the human brain cannot consider the value and knowledge of a decision. And in all aspects of behavior, human reason works within the limits of the psychological environment."

The theory indicates that humans are animals that have demands, and their demands have a level of severity. Only when the needs of the lower levels are met, the needs of the higher levels will appear. These theories deny the traditional "rational man hypothesis" which regards economic interests as the only needs of human beings, making the "economic man hypothesis" move a big step towards reality and improving its ability to interpret reality (Fig. 1).

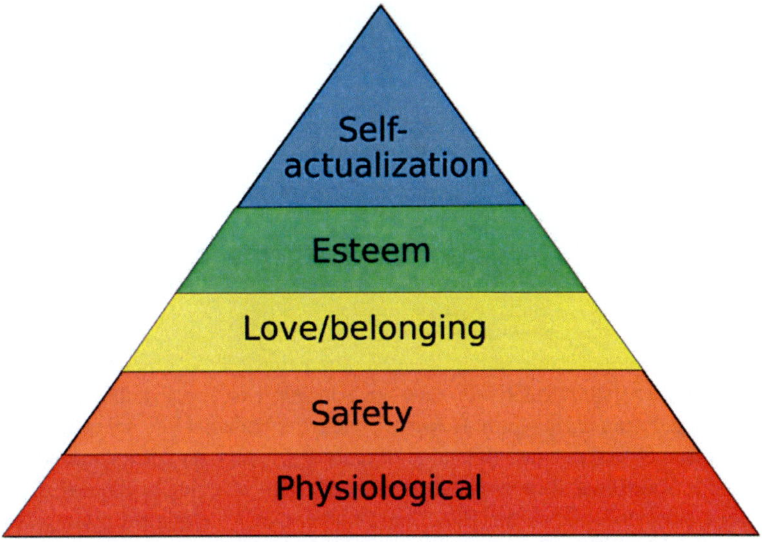

Fig. 1. The pyramid of human demands from Simon's bounded rationality.

2.3 Prospect Theory

Prospect theory was clearly put forward in 1979 by Daniel Kahneman (the 2002 Nobel Prize winner in Economic Sciences) and Amos Tversky. The characteristics shown by the choice made by individuals at risk are inconsistent with the basic principles of utility theory. Prospect theory is to describe people's performance in the process of decision-making when facing risks. They found that people's risk preference behavior was inconsistent when facing gains and losses.

Psychology professor Kahneman brought behavioral economics to the world. In fact, some people began to study behavioral economics in the 1950s, but the previous research was relatively fragmented. It was not until the 1970s that Kahneman and Tversky conducted extensive and systematic research in this field.

Prospect theory emphasizes that people's behavior is not only driven by interests, but also affected by a variety of psychological factors. Prospect theory effectively combines psychological research with economic research, reveals the decision-making mechanism under uncertainty, and opens up a new field of research.

The prospect theory points out that people make decisions based on the potential value of losses and gains, not the final result.

Kahneman concluded a series of experimental data and used mathematical functions to explain human behaviors (Fig. 2).

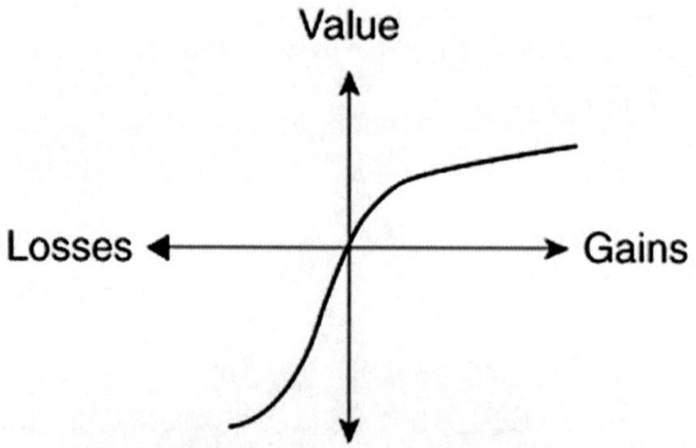

Fig. 2. The value curve of prospect theory.

The vertical axis represents the value one perceives in a circumstance and the horizontal axis represents the outcome of it. When the gain or loss of the result is relatively small, the value change is relatively slow, and during loss, the absolute value of issue increases quicker than during gain. However, when the gain and loss are relatively large, the rate of change declines, and the value of gain will nearly stay at a stable level while the value of loss still increases even in a slower rate. This function explains that people's sensitivity to loss and gain is different, and indicates that the painful feeling at loss is much greater than the happiness of gaining. In this function, when the horizontal axis result is larger and more inclined to profit, the value growth rate felt by the decision maker gradually decreases until it is a constant. When the result tends to be loss, even the rate of increase in loss felt by the tester has decreased, and loss still continues to increase in reality. There are three characteristics in the value function. They claim that first, most people are risk-averse when facing gains [6]; second, most people are risk-preferred when facing losses; third, people tend to be more sensitive to loss compared with gain. Therefore, people tend to be cautious and unwilling to take risks when faced with gains; but when faced with losses, they are willing and easy to take risks [7].

The vertical axis of the weighting function is the expected probability, and the dotted line draws the probability that investors consider in decision-making. If we consider the probability lower than 50% as a small probability, the probability higher than it is a bigger probability.

As can be seen from the figure, investors tend to believe that small probability events will occur (such as winning a lottery), while in fact, for events with higher probability (such as the stocks purchased are not immediately rising), investors prefer to believe that their chances of occurrence are lower than actual probability.

This function could explain why people tend to be irrational with small-probability events but faltering with relatively certain events (Fig. 3).

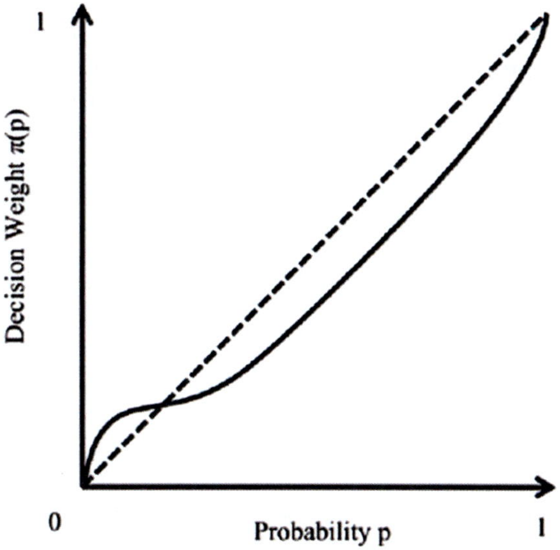

Fig. 3. The weighting function of prospect theory.

3 The Connotation of Overconfidence

Overconfidence is an external manifestation of individual psychology, which is considered as "irrationality" [8]. It originated from the related concepts of psychology at first, and then gradually developed and became an important content of behavioral cognition. In recent years, the concept of overconfidence has been given more and more attention. In psychology, overconfidence can be understood as a deviation or a mistake. This is mainly because, in a large number of psychological experiments, there will always be such a phenomenon, that experimental participants think that the correct probability is often higher than the real probability value. Therefore, psychologists use this phenomenon to describe overconfidence. This definition is the initial description of overconfidence by psychologists. It mainly occurred around the 1970s [9]. With further studies, the concept of overconfidence has gradually become clear. At present, the concept of overconfidence in academia can be divided into the following four aspects:

First is the wrong estimation of status. People tend to overestimate their ranking in the group and most people think they are better than average (shown by fund member experiments). Entrepreneurs may also overestimate their company's ranking among competitors and may think that they are "in the lead". Therefore, entrepreneurs will take it for granted that other companies must be suffering, which will help them to stick to the operation and believe that there will be fewer competitors and subsequent profits in the future.

Second, the overestimation of control: entrepreneurs tend to think that the company has control over the whole market. Entrepreneurs will think that their company has won a place in the market, and this "position" will bring profits to the company in the future. For example, entrepreneurs think that their previous losses are due to the top price, and have made their company out of the tight encirclement in the market to get a "place" (here is also a wrong estimation of the status). If it continues to operate, its absolute or dominant position in the market will bring long-term profits, so it should continue to operate even if it faces short-term losses. Entrepreneurs tend to think that the company has some control over its customers. Many entrepreneurs believe that even though the company is at a loss, the popularity, dependence, and recognition of customers have gradually accumulated in the past, so the control of customers will make the company "turn the corner" in the future.

Third is the wrong estimation of time. Many companies will inevitably suffer from cyclical effects, such as seasons, and some entrepreneurs will inevitably misestimate the timeliness of the impact of these cycles. If an entrepreneur mistakenly estimates the duration of this decline cycle and believes that the company can persist in resisting losses and usher in a "warm spring" in his estimated time, the company will continue to operate in the loss.

Fourth is the expectation effect. People overestimate the probability of a small probability event (Prospect Theory) just because the outcome is what they expect. If the entrepreneur has always believed that the reversal opportunity is around the corner, he must drive the company to operate or even expand.

3.1 The Influence of Overconfidence on Entrepreneurs

Overconfidence, as an unavoidable "nature" just like emotions of human, cannot be directly labeled as good or bad for entrepreneurs.

3.2 An Analysis of the Negative Impact of Overconfidence on Entrepreneurs

Geers and Ensley (2006) believed that most overconfident entrepreneurs have unrealistic expectations. They don't pay attention to some negative factors that others think, and make biased decisions on this basis. At the same time, these entrepreneurs tend to think that their ability is higher than those around them, and the constraints on those people are not applicable to themselves, which will lead to a more uncertain, unfamiliar, and complex environment for these entrepreneurs, which will cause more difficulties for entrepreneurship.

Gibson and Sanbonmatsu (2004) believed that overconfident entrepreneurs tend to regard the negative information in the public's minds to be positive, so they are less sensitive to it, and cannot analyze the reasons for the positive or negative factors of entrepreneurship. If entrepreneurs can maintain a reasonable level of rationality, their decisions will be more realistic [10]. Entrepreneurs will hold a relatively high or low confidence in their knowledge and ability, or in the future prediction of entrepreneurship, which leads to the increasing chance of entrepreneurial failure. Because entrepreneurs think that they have the basic ability to start a business successfully and are better than others, they will engage in high-risk entrepreneurial projects, which leads to an increasing failure rate.

To sum up, overconfident entrepreneurs are often considered to have unrealistic views and opinions. They will not value some recognized negative information, and even have a misunderstanding of their own ability. Those who are considered as the necessary conditions for the success of starting a business from a rational point of view, in their view, have already possessed or think that they cannot affect their own entrepreneurship. This situation leads to that they are easily encouraged by the so-called positive environment, which makes them blindly adhere to their own entrepreneurial behavior, and they will not retreat from the projects doomed to failure, thus resulting in a large amount of waste of resources and loss of potential competitive advantage. Therefore, the overconfidence of entrepreneurs will interfere with scientific judgment and decision-making, which will lead to the failure of new ventures and the doom of entrepreneurs.

4 The Influence of Overconfidence on Entrepreneurs

Kahneman and Lovallo (1993) believed that from the perspective of society as a whole, due to the constant sacrifice of these "overconfident" people who planted the seeds of success, the survival probability of entrepreneurial groups could increase in entrepreneurship. Therefore, the exploration of entrepreneurs' overconfidence can bring positive externality to entrepreneurial groups and bring greater social benefits. From the perspective of the whole society, due to the continuous attempt with overconfidence, people have effectively promoted the innovation of social technology and the progress of human civilization [11]. In addition to entrepreneurs, overconfident people also have a positive impact on some innovation projects. Moreover, many successful cases show that it is because of the hard management of these overconfident entrepreneurs that they can realize some projects which cannot be seen by outsiders, and achieve a certain number of amazing entrepreneurial spirit, creating an impossible entrepreneurial myth. Therefore, from this perspective, one of the advantages of overconfidence is to endow entrepreneurs with tenacious strength to complete the so-called "impossible tasks", and this background is also what entrepreneurs need to face the uncertain environment. Therefore, this paper argues that the benefits that overconfidence brings to entrepreneurs cannot be denied. More overconfidence and less innovation lead to failure, which requires entrepreneurs to find a balance according to their personal situation.

5 Summary

Traditionally, it is believed that the overconfidence of entrepreneurs plays a negative role in entrepreneurship. Entrepreneurs often succeed only when they are fully rational. However, in reality, many overconfident people are questioned at the beginning, but the effect achieved in the later stage is unexpected. Therefore, starting from the essence of overconfidence, this paper analyzes the relevant concepts and proves that overconfidence has a negative impact on entrepreneurship, but also has a certain positive effect. Therefore, entrepreneurs need a certain amount of overconfidence.

References

1. Wu, B., Knott, A.M.: Entrepreneurial risk and market entry. Manage. Sci. **52**(9), 1315–1330 (2006)
2. Trevelyan, R.: Optimism overconfidence and entrepreneurial activity. Manag. Decis. **46**(7), 986–1001 (2008)
3. Zhang, J., Souitaris, V., Soh, P., Wong, P.: A contingent model of network utilization in early financing of technology ventures. Entrepreneurship Theory Pract. **32**(4), 593–613 (2008)
4. Frank, J.D.: Some psychological determinants of the level of aspiration. . Am. J. Psychol. **47**(1), 285–293 (1935)
5. Griffin, D., Tversky, A.: The weighing of evidence and the determinants of confidence. J. Cogn. Psychol. **24**(2), 411–435 (1992)
6. Landier, A., Thesmar, D.: Financial contracting with optimistic entrepreneurs: theory and evidence. Working Paper, NYU Stern School of Business (2003)
7. Cooper, A.C., Woo, C.Y., Dunkelberg, W.C.: Entrepreneurs' perceived chances for success. J. Business Ventur. **3**(2), 97–108 (1988)
8. Stebro, T., Jeffrey, S.A., Adomdza, G.K.: Inventor perseverance after being told to quit: the role of cognitive biases. J. Behav. Decis. Mak. **20**(3), 253–272 (2007)
9. Van den Steen, E.: Rational overoptimism and other biases. Am. Econ. Rev. **94**(4), 1141–11518 (2004)
10. Hayward, M.L.A., Shepherd, D.A., Griffin, D.: A Hubris theory of entrepreneurship. Manag. Sci. **52**(2), 160–172 (2006). https://doi.org/10.1287/mnsc.1050.0483
11. Alvarez, S., Parker, S.C.: Emerging firms and the allocation of control rights: a Bayesian approach. J. Acad. Manag. Rev. **34**(2), 209–227 (2009)

Factors Affecting Development of Blockchain

Jiyin Shen(✉)

School of Finance, Shanghai University of International Business and Economics, 1900 Wenxiang Road, Shanghai, China

Abstract. Blockchain is a digital record of transactions made with cryptocurrencies. It is a newly occurred concept, whose appearance just started twelve years ago symbolized by the anonymous publication of the paper, Bitcoin: A Peer-to-Peer Electronic Cash System. There is limited research on this buoyant technology, thus this paper hopes to identify the factors that might have influence on the future development of blockchain to provide an explicit depicting of blockchain. To realize this goal, this paper identified three variables that may affect the future development of blockchain, including population acceptance, government policy, and technique support, and then constructed a model to describe development of blockchain.

Keywords: Blockchain · Development · Linear regression model

1 Introduction

According to Daniel Drescher, Blockchain is defined as a digital record of transactions made with cryptocurrencies [1]. To explain what Blockchain is, BTC, ETH, XRP, LTC and other cryptocurrencies can be regarded as Taobao, Beijing Jingdong Centuy Trade, Vipshop and other online trading platforms while Blockchain could be the analogy of Internet, providing the place for online trading. This is just a rough analogy, but also can help understand Blockchain superficially. As a newly invented technology, blockchain has limited information while it is blooming in multiple perspectives. Thus it is vital to identify the factors that might influence development of blockchain.

Blockchain officially appeared 12 years ago. During its development, technology relevant to it simultaneously boomed since its underlying technology is complicated and challenging, attracting individual elites and team to contribute to the relevant technology. With the inevitability explosion of blockchain, enterprise with foresight started their strategy correlated with blockchain and march into the blockchain projects. Government from multiple nations also started their regulations and supervision on blockchain since blockchain has inevitably break into routine business. Surrounded by these changes, public has pay attention on its newly occurred concept and are willing to learn more about blockchain, contributing to the prevalence of blockchain in recent years.

Against to this background, this paper would first identify the factors that might have influence on blockchain development. Then, after setting the variables, a linear regression model would be constructed and statistical test would be delivered to test the feasibility of the model. Finally, factors affecting blockchain would be ascertained.

C. Yuan et al. (Eds.): *Proceedings of the 4th International Conference on Economic Management and Green Development*, AEPS, pp. 125–132, 2021.
https://doi.org/10.1007/978-981-16-5359-9_16

2 Aim and Methodology

This paper hopes to find out the factors influencing development of blockchain by finding out factors intensifying or suppressing blockchain development based on its development process in the past years and construct a model using the variables. In the analysis, linear regression model would be applied via software R. This model will be constructed to exemplify the correlation between variables.

3 Analysis of Factors Affecting Development of Blockchain

Relevant analysis would include introduction on variable setting and model construction.

3.1 Variable Setting

Three factors that might influence the development of blockchain based on the development of blockchain, including acceptance from public, policy from government, and technique support, are selected as independent variables while the development of Blockchain is set up as dependent variable. All these variables are hardly to be quantified, thus indirect data would be referred to help build a model describing the development of blockchain, as introduced below.

(1) Dependent variable: development of blockchain

The development of Blockchain is described by the Blockchain financing quantity of Blockchain projects as shown below [2] (Table 1).

Table 1. Financing quantity and average amount of blockchain project in China from 2014 to 2018.

	2014	2015	2016	2017	2018
Financing quantity	10	6	26	54	19
Average financing amount (in hundred million/China yuan)	0.280	0.126	0.180	0.235	0.358

However, specific data in 2013 is missing and calculation is required. It would be calculated by multiplying the financing quantity to the ratio of number of Blockchain projects in 2013 and 2014. This is rational because in the initial stage of Blockchain projects development, where no enormous Blockchain project occurred, the financing quantity is proportional to the number of Blockchain projects. The calculated result of financing quantity is 4 hundred million China yuan and final version of blockchain development, represented by financing quantity, is shown below (Table 2).

Table 2. Blockchain development from 2013 to 2018.

	2013	2014	2015	2016	2017	2018
Blockchain development	4	10	6	26	54	19

(2) Independent variable: acceptance from public

Acceptance from public might affect development of Blockchain since it would influence the operation of Blockchain projects, and furthermore affect the potential development of it. There is no specific statistics to describe this variable, hence the acceptance from public will be described by the adoption rate of Fintech (Financial Technology), as shown below [3] (Table 3 and Fig. 1).

Table 3. Adoption rate of Fintech from 2015 to 2018.

	2015	2016	2017	2018
Adoption rate of Fintech	0.206	0.31	0.46	0.64

```
adpt=read.csv('adoption.csv')
adoption_rate=lm(Adoption_rate~Year, data=adpt)
summary(adoption_rate)
```

```
##
## Call:
## lm(formula = Adoption_rate ~ Year, data = adpt)
##
## Residuals:
##       1       2       3       4
##  0.0198 -0.0214 -0.0166  0.0182
##
## Coefficients:
##               Estimate Std. Error t value Pr(>|t|)
## (Intercept) -292.39180   24.33875  -12.01  0.00686 **
## Year           0.14520    0.01207   12.03  0.00684 **
## ---
## Signif. codes:  0 '***' 0.001 '**' 0.01 '*' 0.05 '.' 0.1 ' ' 1
##
## Residual standard error: 0.02699 on 2 degrees of freedom
## Multiple R-squared:  0.9864, Adjusted R-squared:  0.9796
## F-statistic: 144.7 on 1 and 2 DF,  p-value: 0.006839
```

Fig. 1. Linear regression model of adoption rate of Fintech.

However, adoption rate of Fintech in 2014 and 2013 were not counted. To attain the two missing data, a linear regression model describing the relationship between the adoption rate and year is set up as follows.

It could be viewed from the results that there is a linear relationship between 'year' and 'adoption rate' since the R-squared equals to 0.986369, which is approximate to 1, indicating the fitness of this linear relationship. Thus acceptance of 2014 and 2013 is able to be calculated referring to this relationship, which is 0.01042 and 0.041, respectively. Final version of acceptance from public is shown as below (Table 4).

Table 4. Public acceptance of blockchain from 2013 to 2018.

	2013	2014	2015	2016	2017	2018
Public acceptance	0.01042	0.041	0.206	0.31	0.46	0.64

(3) Independent variable: policy from government

Policy obviously has influence on the development since the regulation on Blockchain would encumber its development while policies encouraging Blockchain projects would have opposite effect. The variable of policy from government is capitalized by the number of policies published both by central government and local government to support or regulate the Blockchain development, the former of which is counted as 1 each while the latter is counted as −1 each. Among them, considering each policy varies in power and weighting, this paper utilized GDP, acquired in China National Statistics, as the foundation to weight policies that set the national policy weight as 1 and each policy issued independently by provinces or cities is weighted according to the percentage of the national GDP. For example, Beijing issued a policy to promote Blockchain development in 2016 and Beijing's GDP accounted for 3.47% of the national GDP that year, so the data of this policy in 2016 was 1*3.47%. To calculate the influence of policy in 2016, all policies issued in China in 2016 were needed to calculate the data of each policy and then added them together. As an analogy, the policy impact over the years can be calculated [4]. Since specific policy targeting on blockchain before 2016 is uncertain, influence weight of 2015 would be set as 1 because of positive attitude from government while influence weight of 2014 and 2013 would be set as −1 due to adverse attitude of government. The procedure and final result are shown below (Table 5).

Table 5. Calculation of policy influence power

	Number of policies			GDP/hundred million in China yuan			Percentage of GDP			Weight		
	2016	2017	2018	2016	2017	2018	2016	2017	2018	2016	2017	2018
China	2	3	1	740,061	820,754	900,309	1	1	1	2.00	3.00	1.00
Beijing	2	2	2	24,541	28,015	30,320	3.32%	3.41%	3.37%	0.07	0.07	0.07
Shanghai	0	2	2	26,688	30,134	32,680	3.61%	3.67%	3.63%	0.00	0.07	0.07
Guangzhou	1	2	1	20,004	21,503	21,500	2.70%	2.62%	2.39%	0.03	0.05	0.02
Shenzhen	1	1	1	19,300	22,438	24,222	2.61%	2.73%	2.69%	0.03	0.03	0.03
Chongqing	0	1	1	17,010	19,500	20,363	2.30%	2.38%	2.26%	0.00	0.02	0.02
Hangzhou	0	1	1	11,700	12,556	13,500	1.58%	1.53%	1.50%	0.00	0.02	0.01
Qingdao	0	2	0	10,193	11,037	12,002	1.38%	1.34%	1.33%	0.00	0.03	0.00
Guiyang	0	1	0	3,077	3,538	3,891	0.42%	0.43%	0.43%	0.00	0.00	0.00
Zhejiang	1	4	1	46,485	51,768	56,197	6.28%	6.31%	6.24%	0.06	0.25	0.06
Guizhou	1	1	3	11,734	13,541	14,806	1.59%	1.65%	1.64%	0.02	0.02	0.05
Jiangsu	0	6	2	67,008	85,901	92,595	9.05%	10.47%	10.28%	0.00	0.63	0.21
Shandong	0	2	1	76,086	72,678	76,470	10.28%	8.86%	8.49%	0.00	0.18	0.08
Jiangxi	0	2	1	18,364	20,819	21,985	2.48%	2.54%	2.44%	0.00	0.05	0.02
Guangxi	0	1	1	18,245	20,396	20,353	2.47%	2.49%	2.26%	0.00	0.02	0.02
							2013	2014	2015	2016	2017	2018
Influence of policy							−1	−1	1	2.20	4.44	1.67

(4) Independent variable: technique support

Technique support is represented by the number of blockchain relevant jobs [2], since once technique foundation has been made, the scale of practitioners would be one of the important factors to prompt the technique development (Table 6).

Table 6. Blockchain relevant jobs offered from 2016 to 2018.

	2016	2017	2018
Technique support	350	1990	6000

However, only statistics in 2016, 2017 and 2018 could be found while data in 2015 could be calculated through the increasing rate from 2015 to 2016. Since year 2013 and 2014 are the period when blockchain project has just started, the number of positions related to Blockchain is relatively steady and lower than that in 2015, both statistics in the two years are estimated as 100. Final version of technique support is shown below (Table 7).

Table 7. Technique support for blockchain.

	2013	2014	2015	2016	2017	2018
Technique support	100	100	350	350	1990	6000

3.2 Model Construction

A linear regression model describing correlation between blockchain development and acceptance from public, government policy and technique support was established using the software R.

By setting Blockchain development as dependent variable while acceptance from public, policy from government and technical support as independent variables, the linear regression model is constructed. The R-squared value equals to 0.911754, indicating that the fitness of this model is approximately equal to 91%, a relatively high value. Thus this model is applicable. Acceptance from public, policy from government and technique support are the three factors contributing to the development of Blockchain.

It could also be observed that each variable is positively correlated with development of blockchain, a reasonable phenomenon that satisfy the economic test. Since once public feels more confident about blockchain, investors and enterprises would have more motivation to develop blockchain project, believing that public would willing to accept and use these project. Promotion on technology also positively affects development of blockchain. With the progression on technique support, users are able to obtain a more impressed and comfortable experience, which apparently beneficial to the spreading of blockchain. Policy has always played significant role in every aspect in society, attributing to the great power owned by authorities (Fig. 2).

```
blck=read.csv('blockchain_development.csv')
blockchain_development=lm(blockchain_development~technology+acceptance+policy, data=blck)
summary(blockchain_development)
```

```
##
## Call:
## lm(formula = blockchain_development ~ technology + acceptance +
##     policy, data = blck)
##
## Residuals:
##       1       2       3       4       5       6
## -2.1771  6.1011 -9.3063  5.0031  0.5778 -0.1986
##
## Coefficients:
##               Estimate Std. Error t value Pr(>|t|)
## (Intercept)  17.663630   7.240895   2.439   0.1349
## technology    0.005066   0.004095   1.237   0.3416
## acceptance  -74.501594  53.196044  -1.401   0.2963
## policy       11.216799   3.175549   3.532   0.0716 .
## ---
## Signif. codes:  0 '***' 0.001 '**' 0.01 '*' 0.05 '.' 0.1 ' ' 1
##
## Residual standard error: 8.774 on 2 degrees of freedom
## Multiple R-squared:  0.9118, Adjusted R-squared:  0.7794
## F-statistic: 6.888 on 3 and 2 DF,  p-value: 0.1294
```

Fig. 2. Linear regression model of blockchain development.

4 Conclusion

Three factors, including acceptance from public, policy from government, and technique support, are variables contributing to development of Blockchain. To verify the correlation between blockchain development and the three factors, a linear regression model was constructed using the software R. Since both dependent and independent variable could not be quantified directly, indirect statistics were collected to describe the variables. Financing quantity of blockchain project was selected to measure blockchain development, number of jobs offered in blockchain industry was selected to measure technique support and popularity rate of Fintech was selected for acceptance from public. For policy from government, the influence from each policy was weighted by GDP proportion in the area the policy was introduced and annual influence was the summary of total influence from all policies.

It could be substantiated through the model that development of blockchain is affected by the three factors. Acceptance from public in the future is expected to increase while technical development tend to incline in the countries with supporting policies for blockchain, thus it could be speculated that blockchain would be boost in these countries.

5 Limitation on Analysis

Several limitations still exist from the previous analysis.

(1) The scale of data collected is limited in 6, which is under 30, indicating that the distribution does not satisfy normal distribution, weakening the accuracy of the modal. However, this is inevitable because blockchain is a new conception which was put forward in 2008. Relevant projects and census have just started a few years ago, highly restricting the collection of data.
(2) The data used to describe the dependent and independent variables, i.e. blockchain financing quantity of blockchain project, Fintech popularity rate, influence weight of policy from government and jobs related to blockchain offered, cannot totally represent the true situation in reality. This can also lead to misinterpreting of the model.
(3) There are only three factors in the model describing development of Blockchain. It is possible that certain elements which also deliver influence on blockchain development were not detected. Missing factors also impair the accuracy of the model.
(4) The data used might contain tiny errors itself. However, this kind of systematic error cannot be avoided.
(5) There might be better function form of the model but was not found out and used.

References

1. Drescher, D.: Blockchain Basics. Apress, Berkeley, CA (2017). https://doi.org/10.1007/978-1-4842-2604-9
2. Li, W.: A China Blockchain Development Report. Social Sciences Academic Press, Beijing, Beijing (2018)
3. Gulamhuseinwala, I., Hatch, M., Lloyd, J.: EY FinTech Adoption Index 2017. Retrieved February 5, 2019, from https://www.ey.com/Publication/vwLUAssets/ey-fintech-adoption-index-2017/$File/ey-fintech-adoption-index-2017.pdf (2017)
4. Zhu, L.: Summary and Interpretation of the Latest Policies for the Blockchain Industry in 31 Provinces and Cities of China in 2018. Retrieved February 5, 2019, from https://xw.qianzhan.com/t/detail/556/181024-a4689fe0, October 25, 2018

Research on Patented Drugs and Compulsory Licensing

Yunqing Luo(✉)

Georgia Institute of Technology, North Ave NW, Atlanta, USA

Abstract. Motivated by the COVID-19 drugs and vaccination competition, I want to discuss some core characteristics of drugs as a special good in the market. During this paper, we are hoping to discuss the uniqueness of drugs when it comes to the regulation policy: patented drugs and generic drugs. We will be able to understand how the companies and patentee get economic profits with high R&D costs. Because drugs are highly connected to people's well-being, I will also go over the current status of compulsory licensing of drugs and history. We will see how developing countries use the TRIP Agreement to improve certain disease situations with real examples.

Keywords: Patented drugs · Compulsory licensing · Generic drugs · R&D

1 Introduction

Like what Thomas Nagle said in his paper *Economic Foundations for Pricing:* Pricing, like most business decisions, is an art [1]. It is not a pure economic model as in real life, variables do not hold constant all the time. When it comes to pricing decisions, it is more likely to have an interdisciplinary learning: law, psychology, sociology, supply chain, etc. The broader the study covers, the more concise it is, but at the same time, more complicated and focused. In 2020, we had another outbreak of plague which has prompted us to focus more on pharmaceutical products. For most of us, this is a time when we focus on how different drugs and vaccines are invented and produced and cause all kinds of discussions and debates. In this paper, I try to focus on patents on pharmaceuticals, especially in combination with compulsory licensing.

2 R&D and the Specificity of Drugs

Drugs are considered as essentials, but their prices do not depend entirely on market supply and demand, but also on patent pricing, innovation cost, etc. The pricing system reflects several factors and is more complex than the average product. Even though the production of the product creates costs, the majority cost needs to be covered is the innovation costs. Pharmaceutical product is unique as it depends on uncertain innovation and research success, especially patent life doesn't depend on the investment cost.

C. Yuan et al. (Eds.): *Proceedings of the 4th International Conference on Economic Management and Green Development*, AEPS, pp. 133–137, 2021.
https://doi.org/10.1007/978-981-16-5359-9_17

Pharmaceutical companies, although highly correlated with social welfare, still aim to make a profit. There is evidence that the capital costs on the average R&D outlays of & 5.5 billion over 15-years, accounting for the cost of times and risks, are $17.2 billion [2]. This costly R&D process, however, doesn't mean there is definitely some product hitting the market. Failure on any trial phase can cause the failure of the whole innovation. For pharmaceutical companies, one successful product may need to cover multiple failing products' R&D costs. Another issue related to R&D costs is the production process is relatively cheap. Given the externalities of the technology, with no R&D costs and risks, pharmaceutical companies can just create a generic drug and sell them with a very low price. Without a doubt, the market will choose the generic drug and the innovator fails to cover the cost, and incentives for innovation are hampered, leading investors to be more cautious about investing in drug development.

3 Patented Drugs and Generic Drugs

In order to better study the patent compulsory licensing system, we first need to understand how the product uniqueness is defined: patented drugs and generic drugs, and how the current law protects the patentees' economic benefits. Patent is a form of intellectual property that gives its owner the legal right to exclude others from making, using, or selling an invention for a limited period of years in exchange for publishing an enabling public disclosure of the invention. In the drugs market, the number of years is 20 years. This means that the manufacturer has 20 years to cover all costs, or they will be ruled out of business. There is a common misunderstanding that the patent only begins when the product enters the market. However, the patent is effective long before public use to prevent potential copycat drugs. According to research, the profitable time frame after the FDA approval is typically between 7 and 12 years. Hence, most innovative drugs that are under patent are much more expensive than the older drugs that are no longer under patent. When the patent expires, competitor firms will submit an abbreviated new drug application (ANDA) to FDA to prove that its product is "bioequivalent", meaning it gets to the part of the body where the drug works at the same time and in the same amount [3]. This product is called "generic drugs". It works in the same way as the "innovative drug". Even though the FDA may still require certain clinical trials, it is much simpler than the innovative drugs trial. The profit margin is relatively bigger, considering there are no R&D costs. The innovator company loses its market share during this process due to loss of selling.

However, the mechanics of inventing generic drugs dictate that the results are only approximate, not duplicative. The FDA rules have a pretty big range. A generic's maximum concentration of active ingredients in the blood must not fall more than 20% or 25% above that of the brand name. Furthermore, drugs are composed of two components: active ingredient and additional ingredient. The FDA tends to have an easier regulation on the generic drugs' additional ingredients. In an effort to maximize profits, generic drug companies tend to use cheaper ingredients to manufacture these additional ingredients, which has proven to infect the dissolution of the drug.

Even though the generic drugs hit the patented drug hard, many pharmaceutical companies have another strategy to extend the lifetime of the patents. This idea usually involves making minor changes to the drug in order to create a new patent. Even though the minor changes don't change the overall effect of the drugs, the new patent successfully extends the exclusivity. This is called "evergreening" and has become a big problem as it doesn't help patients, but mainly helps the bottom line. The extension of the patent will create more revenues to help the company with more budgets to invest in new R&D. At some level, it is like a sophisticated generic drug produced by the same company while the patients still need to pay a high price.

4 Patent and Licensing

Licensing is a strategy for innovative companies to overcome the economic crisis. Considering the lengthy and costly R&D process, it is hard for an innovative company to cover all the cost from the very beginning. If the budget is low, smaller companies with relatively good technology often have to shut down in the middle of their innovation. The number of licensing transfers in the biotechnology industry is at least 10 times higher than in other industries (factors affecting pricing in patent licensing contracts in the biopharmaceutical industry). Valuation hence is important for these companies to find licensing partners and raise further capital. The commonly known factors of influencing licensing fee include deal value, royalty rate, upfront payment, etc.

4.1 There are Two Types of Licensing: In-Licensing and Out-Licensing

In-licensing refers to investment. Because of the high R&D cost, many pharmaceutical firms tend to find other investors to provide capital and increase the budget. This is even more common for small start-up companies. For those investors, they have positive expectations that those pharmaceutical products will bring economic benefits. However, this brings up a problem about orphan drugs. In order to maximize benefits, investors tend to invest in products that target large patient populations. There are limited innovation incentives for pharmaceutical firms to innovate products targeting rare diseases. Firms will find difficult retrieving money from products that create large amounts of value per patient but treat relatively few individuals [4]. As a result, many countries implement policies to encourage inventing these kinds of drugs: tax credits, longer market exclusivity, etc. Given the limited number of patients, it will also be almost impossible to have generic drugs, no matter how high the profit is.

Out-licensing is about having a partner to help target a specific market in order to have the best results. There is no specific type of partner, including legal firms, advertisement firms, etc. The reason the company out-license valued assets is because these assets have been developed over a number of years and, presumably, have the credentials, albeit at an early stage, to become products. While out-licensing has occurred over the years, this was because the out-licensed drug would not generate meaningful sales for a large company [5].

5 Patent Compulsory License System

When there are new drugs in the market, we expect there is an overall increase in people's health. However, due to the market reasons, not everybody has the ability to afford the corresponding medication to their diseases. To solve this problem, the government created another term: compulsory licensing. Compulsory licensing refers to the situation where the government allows others to produce patented drugs without the consent of the patent owner [6]. This is a requirement in the TRIPS Agreement to help improve public health. Because of the patent, the prices of some drugs are too high for most people to use. As the price becomes the only burden, the government will be able to have access to produce the drugs, especially those essential medicines. Many people argue that basic human rights like health, which includes access to essential medicines, should not be prevented by property rights. This Agreement is extremely helpful for developing countries to have essential medicines when they actually have no ability to innovate such medicines. However, some countries do not even have the capacity to produce these drugs, so there is another waiver for them to import these medicines.

This is widely used by countries in northern and southern hemispheres, and is not limited to medicines to treat cardiac diseases, AIDS, etc. In fact, a lot of countries issued licenses because of AIDS. Because of TRIPS, there is significant increase in production of antiretrovirals which help to decrease the prices. More people will be treated by these generic drugs. Despite these improvements, there are still big challenges. Patients can easily develop resistance to these generic drugs. Transforming to second-line treatments is too expensive for them to afford. Additionally, according to the TRIPS agreement, it is impossible to produce low-cost versions of patented antiretrovirals by 2016. This caused a conspicuous absence of compulsory licensing in the pharmaceutical sector, even after the South African authorities made a decision in 2003 to the effect that two foreign companies had refused to grant a patent license for essential HIV medicines, in violation of domestic competition laws [7]. Another example of compulsory licensing is Brazil with Efavirenz, a patented AIDS drug by Merck. In May 2007, Brazil issued a compulsory license for Efavirenz which is the first time for Brazil. Before this, the government just threatened to use compulsory licensing to force drug companies lowering the price. The price negotiations with Merck were partly induced by the steady increase in the market price of Efavirenz: in Brazil, the annual per-patient cost of treating patients with Efavirenz has risen from $1336 in 2004 to $2500 in 2006 [8]. The final issuance of compulsory license for Efavirenz was five years with a royalty rate of 1.5% to Merck [8]. Even though Brazil later found out they did not have the technology to produce Efavirenz, leading the government to import a generic version of the drug from India for two years, the case is still a success example for developing countries like Brazil.

Even though TRIPS helped overcome the patent barrier, this can cause certain misuses and possibly discourage companies from new innovations. A lot of governments are using this as a threat. A widely-used drug can create huge economic benefits while less competitive countries with weak IP regimes can easily lose it. For the patentee, TRIPS agreement interferes with their exclusive rights which may diminish their incentive to future innovations.

6 Conclusion

Drugs are not everyday products, but they are critical for people's lives with its close connection with people's well-beings. Along with the human welfare costs, scientists and governments are playing a critical role in improving the current drug shortage. Given the high spending during the R&D process and to encourage further study, it is not logical and impossible to get rid of patents. The market is free, but governments still need to consider human rights, which is why there are generic drugs, TRIPS agreements and compulsory licensing. These tools help developing countries improve consumers' access to drugs. However, this doesn't mean there is no need for improvement. Generic drugs tend to have weaker effects. Compulsory licensing is more likely to be used as a threatening tool than to actually be issued. The discussion on patented drugs will still continue.

References

1. Nagle, T.: Economic foundations for pricing. J. Bus. **57**, S3–S26 (1984)
2. Winegarden, W.: The economics of pharmaceutical pricing. Pac. Res. Inst. 2013–2015 (2014)
3. FDA. U.S.: Food and Drug Administration (2020). Accessed 21 September 2020
4. Garthwaite, C.: The economics of drug development pricing and innovation in a changing market. NBER Report. **3**, 19–24 (2018)
5. Lamattina, J.: Why is pharma out-licensing its compounds? [EB/OL]. Forbes 2020 (2020)
6. WTO | Intellectual property (TRIPS) - TRIPS and public health: compulsory licensing of pharmaceuticals and TRIPS [EB/OL]. Wto.org 2020 (2020)
7. Reichman, J.H.: Comment: compulsory licensing of patented pharmaceutical inventions: evaluating the options. J. Law Med. Ethics **37**(2), 247–263 (2009)
8. Bond, E.W., Saggi, K.: Compulsory licensing, price controls, and access to patented foreign products. J. Dev. Econ. **109**, 217–228 (2014)

How Welfare Policy Influence Firm Performance? Evidences from Chinese Listed Firms

Hong Han(✉)

Business Faculty, Macao University of Science and Technology, Macao, China

Abstract. The core competitiveness of the company's development depends on its employees, so motivating employees to make better contributions to the company, improve company performance and expand investor returns is the core concern of modern company management systems. Company welfare, as an important part of modern company management system and compensation system, plays an indispensable role in employee management. However, most of the current welfare research focuses on qualitative research, and the research on the financial value that welfare can bring is relatively few. In practice, some companies are skeptical about the role of welfare expenditure in their own development.

Keywords: The Chinese listed companies · Welfare policy of companies · Financial performance

Therefore, this paper mainly analyzes the relationship between welfare and financial performance of listed companies through empirical research and investor perspective. On the whole, the research shows that, the welfare improvement of listed companies is conducive to the improvement of their financial performance, which is beneficial to the development of the company and investors. Regarding the specific categories of welfare, the statutory welfare and non-statutory welfare of listed companies can promote the improvement of financial performance, but on this basis, too many other welfare categories are not conducive to the improvement of financial performance. It can be seen that welfare is necessary for the healthy development of the company, but not all the welfare contents play a positive role. Therefore, the key points that managers and investors should pay attention to the company's welfare structure and adopt a welfare that truly meets the need of employees, so that all welfare contents can bring about the improvement of financial performance and achieve the win–win situation between employees and the company.

1 Introduction

Modern company development to change the past as the traditional concept of human tools and gradually attached importance to the core position of talents in the development strategy of the company by various management strategies which encourage employees

C. Yuan et al. (Eds.): *Proceedings of the 4th International Conference on Economic Management and Green Development*, AEPS, pp. 138–153, 2021.
https://doi.org/10.1007/978-981-16-5359-9_18

to contribute to the company. As an important part of modern company management system and salary system, company welfare plays a substantial role in recruiting, utilizing and maintaining talents for the company.

However, the results of research on corporate welfare are diametrically opposed: on the one hand, welfare can improve employees' satisfaction, help employees to improve their organizational commitment and strengthen their investment in the organization. These can lead to the improvement of the corporate performance. On the other hand, it is argued that investment in employee welfare entails additional costs, reduces the profit earned by the company, and is an additional burden on the company's development.

Behind the two opposing research results, there are two different research perspectives: for one thing, from the perspective of human resource management, the competition of talents is the key to enhance the core competitiveness of the company. For another thing, from the financial point of view, the goal of financial performance is to improve the income of investors, while employee welfare will affect the allocation of company profits. Whether the return can exceed the investment, has undoubtedly become the focus of research in this field.

Therefore, the core question for listed companies and their shareholders is whether employee welfare is really conducive to the improvement of financial performance and able to improve shareholders' returns. If employee welfare can improve financial performance, then whether all kinds of benefits have a role in promoting financial performance, the importance of different benefit categories, how to configure their benefit categories to achieve financial performance improvement, these will become the research questions in this paper.

Based on the above research background, this study will comb the two research results and two research perspectives. Combining related knowledge of human resources management and financial management research ideas, this paper aims to find out how the welfare policy within firms impacts their financial performance by empirical analysis in the context of Chinese society. The innovation of this paper is as follows:

(1) Theoretically, there are relatively few studies on welfare in China at present, and the research direction mainly focus on the relationship between company welfare and employees instead of direct research on the relationship between company financial performance, so the relevant reference is limited. For the reason mentioned above, this study directly links welfare with financial performance, adopts the method of quantitative research, and uses financial indicators to measure the development level of listed companies. To a certain extent, it is a supplement to welfare related research.
(2) In the view of research methodology, the commonly used indicators for the study of corporate financial performance are: ROE [1, 2], ROA, Tobin Q [3], etc. This study takes the ratio of EVA to total assets (total asset EVA rate) as a measure.
(3) In practice, due to the existence of the above two opposite research results, the company's welfare practice lacks the corresponding basis and guidance. Taking the Chinese listed companies as a sample and combining the relevant contents of human resource management and financial management, this paper analyzes the impact of the welfare of listed companies on financial performance in the context

of Chinese economy and strive to find a suitable welfare mode for Chinese listed companies and even other enterprises.

2 Literature Review

2.1 Analysis and Definition of Related Concepts

Ding and Li proposed that corporate welfare is aimed at raising the level of themselves and family's quality of life by providing various non-monetary wages, postponement and other complementary services as the main form, based on employment relations, the mandatory laws and regulations of the state [4]. Wei believes that corporate welfare is one of the pillars of the compensation system, including statutory and non-statutory benefits. The goal is to help employees solve current and future life problems with a certain degree of company sentimentality [5].

Zhang and Yu [6] think that financial performance is a comprehensive reflection of all aspects of the company, including profitability, solvency, operating ability and development ability. Sun thinks that the performance level of the company is demonstrated in three aspects: the company's cash payment and debt repayment ability, the company's financial benefit income after the implementation of a certain category, and the company's current and future development situation and development prospects [7].

2.2 Theoretical Review

The complex human hypothesis, that is, the hypothesis of the contingency person, shows that human needs are subject to the changes of living conditions and the organizations they belong to. Therefore, there are great individual differences between people, and all the people cannot be simplified and generalized in one certain hypothesis. In the research on welfare, this type of management method has high value and great significance for reference——the characteristics of listed companies and employees are different, so that the benefits of company welfare will be different. Therefore, corporate welfare is special and worthy of further study. This is another way of thinking about and exploring on traditional welfare research.

Herzberg's two-factor theory divides incentives into two categories: hygiene and motivator. Hygiene factors are related to the external working environment, such as company policy, salary and reward bonus, working conditions and environment, etc., while motivation factors are closely associated with the work contents, such as the challenge and attraction of the work, good development prospects, etc. The welfare content of listed companies involves different factors: some parts of the welfare content as a hygiene factor to maintain employees' satisfaction, reduce the conflicts of organizations and their employees; the other parts of the welfare contents as an motivation factor encourage employees to use their skills to develop their talents and make contributions and returns to the organization.

The theory of corporate social responsibility points out that "corporate social responsibility means taking certain responsibility for the public interests of other interest groups and government representatives who are not shareholders, mainly refers to the tax benefits of creditors, suppliers, employees, users, consumers, local residents and government representatives" [8]. Ye Chenyi and Ye Chenyun [9] combed corporate social responsibility from the traditional concept of connotation, and judged whether the company guaranteed the interests of employees and provided corresponding welfare items from the perspective of corporate social responsibility, which was also an important basis for judging the compliance of the company's obligations, with certain compromise and contractual characteristics. In the research of company welfare, employees are the object of social responsibility. The company provides all kinds of benefits to its employees and protects the basic rights and interests of employees, which is an important embodiment of its social responsibility.

2.3 Related Research

In a study of 100 of the best companies in the United States and their performance (measured by ROA), the performance of the organization focuses on the study of financial performance, good company benefits bring good employee relationships, and play a positive role in financial performance [10]. From the point of view of cash holdings, the growth of employee benefits is an investment in employees, in favour of the employee organization's commitment to increase and contributes to increase in the company's cash holdings, both play a positive role in the company's financial performance [11]. From the perspective of corporate social responsibility, combined with the above reference to corporate social responsibility, a number of studies have shown that specific CSR investments can increase the level of return on investment [12]. Meanwhile, Corporate social responsibility is positively related to organizational commitment, and employees can perform through self-control without supervision. It is a loyalty to the organization, and that loyalty is stronger than commitment itself [13]. Corporate social responsibility to different stakeholders is positively related to corporate value, but the impact on company value is different [14]. From the perspective of employee satisfaction in human resource management theory, improved corporate welfare leads to increased employee satisfaction [15]. There is a strong positive relationship between employee satisfaction and investor return, Greater validation of the important role of human resources in the organization [12]. Employee welfare satisfaction has a significant impact on employee performance, Huang takes the employees of some companies in Northeast China as a sample, through a questionnaire survey and quantitative analysis of 236 employees in the company, it was found that the effect of satisfaction on employee performance was significantly positive [16].

Other studies show that employee benefits provided by companies have a negative effect on their financial performance. Generous employee benefits pose the risk of sudden declines in stock prices, a kind of damage to investors and shareholders, especially in labour-intensive companies. Although work benefits can increase the utility of work, there is no evidence that they can increase shareholders' wealth [17]. The traditional portfolio theory holds that employee welfare limits the choice of investors and results in a decrease in return on investment, and welfare itself is an expense of the company in the process of investment. Filbeck analyzes the relationship between different benefits, good working environment and employees' satisfaction and the return on investment [18], we conclude that the two are negative correlation, that is, the high level of employee welfare will lead to the decrease of the return on investment. From the perspective of corporate social responsibility, it is found that employee welfare is an investment in company capital, Markowitz & Harry thinks that any investment in social responsibility will make the performance of the company worse [19]. Corporate social responsibility is a non-profit because it provides benefits that outweigh its economic benefits [20].

3 Research Assumptions

3.1 Analysis on the Relationship between Welfare and Financial Performance of Chinese Listed Companies

According to the above-mentioned ideological development, theoretical basis and research analysis, the following assumptions are made:

H_0: There is a positive correlation between welfare and financial performance of listed companies in China.

Incentive theory provides a way to improve the relationship between listed companies and employees from the perspective of management and economics: Herzberg's two-factor theory holds that meeting the needs of "hygiene factors" related to the working environment can eliminate employee dissatisfaction and resistance, and "incentive factors" can stimulate individual ability and better contribute their talents to the company. As mentioned earlier, corporate welfare is a complex, including a large number of welfare contents. Some welfare categories can play a role in "hygiene factors", such as social insurance, and if a part of the equity incentive after achieving the goal becomes more challenging, it can play a "incentive role". The analysis of 100 most suitable companies in the United States found that good corporate welfare can improve the level of ROA and employees' satisfaction; the benefits enjoyed by employees can improve the company's cash holding level and make the company have good financial performance; corporate welfare can improve employee satisfaction, and employee satisfaction will lead to the improvement of company's financial performance. Zhuang used ROA as financial index test and listed company as sample to verify the positive relationship between welfare and company performance [21].

Based on the above analysis, we make H_0 assumptions. There is a positive correlation between welfare and financial performance of listed companies.

3.2 Analysis on the Relationship Between Different Welfare Projects and Financial Performance of Chinese Listed Companies

Corporate welfare is a complex including a variety of welfare contents. According to different standards, there are different ways of division, and the differences between countries and regions will also bring some differences. China's accounting standards No. 9 divides employee compensation into short-term compensation, post-employment benefits, dismissal benefits and other long-term employee benefits. The benefits provided by the company to the spouses, children, dependents, survivors of deceased employees and other beneficiaries are also employees' remuneration. Short-term remuneration specifically includes employee salaries, bonuses, allowances and subsidies, employee benefits, social insurance premiums such as medical insurance premiums, work injury insurance premiums and maternity insurance premiums, housing fund, union funds and worker education fund, short-term paid absence, short-term profit-sharing plan, non-monetary welfare and other short-term salary.

Therefore, combined with the classification of company welfare above, according to the actual contents of the financial statements of the database, the salary contents involved in the accounting standards are classified to determine the specific purpose of this study. The following four hypotheses are put forward:

H_1: There is a positive correlation between social insurance premiums and financial performance.
H_2: There is a positive correlation between housing fund and financial performance.
H_3: There is a positive correlation between labour union, education funding and financial performance.
H_4: There is a positive correlation between employee welfare and financial performance.
H_5: Additional benefits are positively related to financial performance.

Social insurance premiums are statutory benefits in the welfare classification and are provided in accordance with national regulations. It can eliminate the employee dissatisfaction which makes the employee to create the value better without worries. The social insurance premium is related to the employee's wage level, which is paid proportionally by the wage standard. When the employee's salary increases, the social insurance premium level will increase accordingly, and the "hygiene factors factor" will be converted into motivation factor under certain conditions. Gao pointed out in the study on social insurance that social insurance premiums are conducive to reducing employee mobility and are also positive for companies to reduce business risks in addition to providing protection for employees and families [22].

Housing fund is a form of protection to solve the basic housing problem of employees, which has certain financial attributes. Gao pointed out that the housing fund can promote the housing consumption of urban residents because of its function of reducing the cost of housing purchase [23]. Jin shows that housing fund has a positive impact on fairness, that is, housing fund is conducive to the realization of fairness among employees and the balance between efficiency and fairness of the company [24]. When employees feel fair, they are willing to serve the organization. Cheng showed that housing fund has a good appeal to employees: on the one hand, the interest rate of housing fund is lower than that of commercial housing loan. On the other hand, it does not have to pay personal income tax when it is withdrawn, equivalent to the employee's tax-free forward income [25].

Based on the above analysis, we make H_2 assumption. There is a positive correlation between housing fund and financial performance.

Labour unions are an extension of our national institutions in the company and a special form behind China's economic development. They are mainly used to serve employees and protect their rights and interests. They play an important role in building harmonious labour relations: promoting a democratic atmosphere of corporate management, enhancing employee participation, alleviating contradictions and problems between companies and employees, and protecting employee interests, especially in the prevention and occurrence of labour disputes [26]. Zhang selected 400 employees, labour union leaders and members, and company management of non-public companies in Beijing, Tianjin, Shijiazhuang, Qinhuangdao. Through analysis, it is concluded that labour unions have a positive impact on company performance, and the atmosphere of labour relations plays an intermediary role [27]. Education can improve personal ability, bring better competitive advantage, meet the needs of self-realization in Maslow's needs level and the pursuit of achievement in McLean's achievement theory. Ji pointed out that the improvement of education level will lead to the improvement of company performance and will also benefit the innovation of the company [28].

Based on the above analysis, we make H_3 assumptions. There is a positive correlation between labour union, education funds and financial performance.

According to Circular of the Ministry of Finance on Strengthening the Financial Management of Employee Welfare Expenses (No. 242 of the Financial Enterprise [2009]), the definition of employee welfare expenses is given by the exclusionary method, which means welfare benefits are exclusive of wages, wage subsidies, insurances, housing fund and staff education expenses. The article defines the scope of expenditure of employee welfare expenses by enumerating ways, including the expenses of staff and workers seeking medical treatment in the field on business, the expenses of staff recuperation, the expenses of the company subsidizing the internal canteen, the lunch expenses of the company that does not run the canteen for the staff, the heating expenses in accordance with the relevant regulations of the state, the expenses of preventing heat and lowering the temperature, etc. [29]. It can be seen that, unlike the above social insurance premiums, this kind of purpose benefits are freer and more diversified, more suitable for the company's own characteristics and the actual needs of employees, and there is no mandatory proportional requirement for employee benefits. Expenditure scale compared with social insurance premiums, relatively low financial pressure on the company.

Based on the above analysis, we make H_4 assumptions. There is a positive correlation between employee welfare and financial performance.

If employee welfare focuses on universal welfare needs, additional benefits are more flexible and higher. According to the content of Accounting Standard 9 and the classification of benefits, supplementary social insurance (including supplementary old-age insurance, supplementary medical insurance and commercial insurance), dismissal benefits, housing subsidies, benefits within one year, equity incentives, annuities, non-monetary benefits and profit-sharing schemes are attributed to other employee benefits. In categories that overlap with the foregoing benefits but are separately listed are not included. This kind of welfare content can better satisfy the employee's pursuit of self-worth and achievement. If the welfare category meets the employee's expectation, the incentive effect will be greater, and it will bring good financial performance more easily.

Based on the above analysis, we make H_5 assumptions. There is a positive correlation between additional benefits and financial performance.

4 Methodology

4.1 Sample Source and Selection

The data are derived from the CASMAR database of 3665 Shanghai and Shenzhen A-share listed companies for the years 2013–2017, with a sample time span of 5 years.

The content involved in the research hypothesis was analyzed and sample data were obtained on social insurance premiums, employee benefits, labour union and education funds, supplementary social insurance (including supplementary old-age insurance, supplementary medical insurance and commercial insurance), dismissal benefits, housing subsidies, one-year benefits, equity incentives, annuities, non-monetary benefits, employee incentives and welfare funds, profit-sharing schemes, asset-liability ratios, the proportion of cash flows to total assets and the equity ratio of the top ten shareholders. In order to ensure the accuracy of the data and results, 6047 samples were obtained by filtering and eliminating the missing data or content overlapping data of the same year.

4.2 Research Model

The following models are designed according to the theory mentioned above, drawing on Lin and Fu's research model design ideas [30] and Ben-Nasr and Ghouma model design methods [17]:

$$EVAAS = \alpha_0 + \beta_{it}X_{it} + \lambda_1 LLAB_{it} + \lambda_2 OPFI_{it} + \lambda_3 STOC_{it} + \varepsilon_{it} \quad (1)$$

X_{it} are all independent variables, the company benefits i the listed company in the t year are social insurance premium (LNINS), housing fund (HOU), labour union and education funds (UED), employee welfare (WELF) and additional benefits (OTWEL). The sum of benefits (TWEL) is composed of five items. The ratio of assets to liabilities (LIAB), the proportion of cash flow to total assets (OPFI), the proportion of the top ten shareholders' equity and (STOC) are the control variables of the whole model. ε_{it} is the error term.

Variable Interpretation. 1. Explained variables: according to Zhu [31] and Jiang et al. [32], the total asset EVA rate (EVA/total assets) was selected as the explained variables.

$$\text{total asset EVA rate} = \text{EVA/average of total asset} \quad (2)$$

The total asset EVA rate (EVAAS) is the ratio of the actual amount of the company's EVA to the total assets, which is a measure of how efficient companies use assets to create wealth, 2014). Dong showed that the total asset EVA rate (EVAAS) can reach 86.8% for the forecast accuracy of the listed company's financial situation for 2 years [33]. This index is to analyze the company's performance from the perspective of shareholder wealth growth level [34]. Chinese SASAC used economic added value to pilot central enterprises in 2010 also show the importance of EVA for financial analysis [35].

2. Explanatory variables: the main explanatory variables are the sum of the benefits of listed companies (TWEL), and the other explanatory variables are the components of the benefits: social insurance premiums (LNINS), housing funds (HOU), labour unions and education funds (UED), employee benefits (WELF) and additional benefits (OTWEL).

5 Empirical Analysis

5.1 Descriptive Analysis

From the descriptive statistics analysis table, it can be seen that the sample of listed companies selected in this paper has a low value-added ratio to total assets, and their financial performance is generally low, and even some listed companies have negative EVAAS (Table 1).

Table 1. Descriptive analysis.

Variable	Obs	Mean	Std.Dev.	Min	Max
EVAAS	6,047	0.001	0.040	-0.088	0.082
TWEL	6,047	18.08	1.436	12.82	25.74
LNINS	6,047	17.43	1.263	15.37	19.95
HOU	6,047	16.10	1.661	7.090	22.79
UED	6,047	14.94	1.500	12.05	17.70
WELF	6,047	16.19	1.319	13.86	18.74
OTWEL	6,047	14.31	2.195	10.34	18.20
LIAB	6,047	0.461	0.205	0.125	0.823
OPFI	6,047	0.039	0.05	-0.080	0.152
STOC	6,047	57.94	14.18	32.74	82.20

From the point of view of explanatory variables, the maximum value gap in descriptive statistics is large, which shows that the listed companies have a large gap in overall welfare expenditure and different kinds of welfare expenditure, and the difference in welfare level of different companies is obvious, especially housing fund and additional benefits expenses.

From the control variables, the debt level gap of listed companies is large. The proportion gap of net cash flow in total assets is large, which indicates that the net cash flow of some listed companies is negative, and the cash operation is tight. The average shareholding ratio of the top ten shareholders is 57.94, the maximum value is 82.20, that is to say, the degree of equity concentration is high.

5.2 Correlation Analysis

According to the above correlation analysis, in order to avoid the problem of insufficient accuracy of the results caused by the interaction between variables, the control variables are consistent in the above model design. It is reasonable to study the model design of each explanatory variable affecting the explained variable.

From the value of correlation coefficient, in each model, the correlation coefficient between explanatory variables and control variables is less than 0.6, and there is no problem of multiple collinearity.

The sign of correlation coefficient shows that there is a positive correlation between the total welfare (TWEL), social insurance (LNINS), housing fund (HOU), labour union, education funds (UED) and employee welfare (WELF) and the total assets EVA rate (EVAAS), while the correlation between additional benefits (OTWEL) and the total assets EVA rate (EVAAS) is negative (Table 2).

Table 2. Correlation table.

	EVAAS	TWEL	LNINS	HOU	UED	WELF	OTWEL	LIAB	OPFI	STOC
EVAAS	1									
TWEL	0.0915***	1								
LNINS	0.0851***	0.962***	1							
HOU	0.086***	0.921***	0.885***	1						
UED	0.115***	0.875***	0.869***	0.819***	1					
WELF	0.131***	0.886***	0.865***	0.790***	0.819***	1				
OTWEL	-0.030*	0.670***	0.614***	0.612***	0.593***	0.567***	1			
LIAB	-0.222***	0.388***	0.376***	0.356***	0.350***	0.344***	0.290***	1		
OPFI	0.373***	0.142***	0.131***	0.118***	0.142***	0.174***	0.095***	-0.192***	1	
STOC	0.201***	0.265***	0.230***	0.248***	0.232***	0.249***	0.198***	-0.032*	0.136***	1

5.3 Model Selection Analysis

This paper uses short panel data with time dimension less than the number of sections. The mixed regression model and fixed effect model are selected by F test, and then the fixed effect model and random effect model are tested by Hausman test.

The above test results show that the F test P values of the six models are 0.0000, and the original hypothesis is rejected at a significant level of 5%, indicating that the short panel data used in this paper have individual effects. A fixed effect model is selected between the mixed regression model and the fixed effect model. Then the Hausman test is used to test the fixed effect model and the random effect model. From the Hausman test results, we can see that the test P values of the six models are 0.0000, which also reject the original hypothesis. The panel data in this paper are more suitable for establishing fixed effect model. Based on the above analysis, the six models established in this paper are analyzed by fixed effect model.

5.4 Regression Analysis

According to the above analysis, the minimum value of R^2 is 0.276, the model has a good fit, and the explanatory variable can explain the explained variable. On the premise that the control variables in all models are consistent, the financial performance (EVAAS) of listed companies is significantly positively correlated with the sum of benefits (TWEL), which is consistent with the original hypothesis, indicating that, overall, the improvement of welfare of listed companies is beneficial to the improvement of financial performance. The financial performance of listed companies (EVAAS) is positively correlated with social insurance premiums (LNINS), consistent with the original hypothesis, which indicates that social insurance premiums can promote the improvement of financial performance of listed companies; The financial performance of listed companies (EVAAS) is significantly positively correlated with housing fund (HM), consistent with the original hypothesis, indicating that the level of housing fund payment is improved, which is beneficial to the improvement of financial performance; The financial performance (EVAAS) of listed companies is significantly positively correlated with labour union and educational funds (UED), consistent with the original hypothesis, indicating that education and labour union help for employees is beneficial to the improvement of financial performance; The financial performance (EVAAS) of listed companies is positively correlated with employee welfare (WELF), indicating that employee welfare plays a positive role in improving financial performance (Table 3).

5.5 Robustness Analysis

In order to make the empirical results more credible, this part tests the robustness of the empirical results. Choose the same analytical method and replace the performance of the listed company with the return on assets (ROA). The explanatory variables and control variables are consistent with the previous text and repeat the previous regression method. After passing the robustness test, the empirical results of this paper have not changed, indicating that the empirical results of this paper are reliable.

6 Findings and Recommendations

6.1 Conclusions

(1) On the whole, the welfare of listed companies in China is significantly positively correlated with their financial performance, which is consistent with the original scheme.
This paper verifies the theoretical contents mentioned above and some conclusions of domestic and foreign studies. From the perspective of management and economics, incentive theory holds that motivation for employees is helpful to improve employee satisfaction, so as to achieve the goal of good financial performance and further expand the development prospects of the company. Welfare will enhance the employee's sense of belonging and satisfaction to the company to a certain extent. Companies with good performance often have their own characteristics of employee welfare, which is obvious in the ranking of 100 most suitable companies in the United States.

(2) Specifically, the impact of subcategories of listed company benefits on financial performance.
Most of the welfare content is conducive to the improvement of financial performance, consistent with the original hypothesis: social insurance, housing fund,

Table3. Regression analysis.

	EVAAS (1)	EVAAS (2)	EVAAS (3)	EVAAS (4)	EVAAS (5)	EVAAS (6)
TWEL	0.004*** (-14.47)					
LNINS		0.004*** (-11.42)				
HOU			0.002** (-7.85)			
UED				0.004** (-6.38)		
WELF					0.005*** (-24.98)	
OTWEL						-0.000* (-3.71)
LIAB	-0.042*** (-24.22)	-0.041*** (-22.47)	-0.038*** (-18.80)	-0.041*** (-16.03)	-0.043*** (-41.16)	-0.028*** (-23.30)
OPFI	0.221*** (-42.07)	0.223*** (-46.3)	0.226*** (-48.66)	0.221*** (-46.3)	0.217*** (-35.44)	0.237*** (-43.57)
STOC	0.000*** (-13.77)	0.000*** (-13.63)	0.000*** (-15.23)	0.000*** (-18.11)	0.000*** (-12.16)	0.000*** (-13.32)
cons	-0.068*** (-12.14)	-0.070*** (-10.54)	-0.044** (-7.36)	-0.058** (-6.37)	-0.076*** (-15.09)	-0.002 (-0.51)
Time fixed	Yes	Yes	Yes	Yes	Yes	Yes
Company fixed	Yes	Yes	Yes	Yes	Yes	Yes
N	6047	6047	6047	6047	6047	6047
R^2	0.288	0.287	0.284	0.292	0.294	0.276

t statistics in parentheses, standard error in parentheses

* $p < 0.05$, ** $p < 0.01$, *** $p < 0.001$

labour union and education funds and employee welfare, the impact on financial performance is significantly positive, according with the original hypothesis. The theoretical basis and some contents of previous studies are verified.

Additional benefits are detrimental to the improvement of financial performance: additional benefits are significantly negatively correlated with financial performance. The conclusion of the empirical analysis, contrary to the view that the overall corporate welfare improves financial performance, supports the negative correlation research mentioned above. That is to say, the welfare structure of the company should be paid attention to instead of the amount. The study confirms that employee benefits may cause a sharp fall in stock prices to the detriment of investors and owners: because a wide range of spending on employee benefits affects the company's investment in other important elements that may add value to the company, and many employee benefits categories may become excuses that provide managers with a means to cover up their own problems or possible negative news about the company, both of which affect the company's financial performance [17]. A study of Huawei found that equity incentive has a more obvious role in retaining talent and stabilizing the team, but it has no significant effect on the value added of the company. They also found that equity incentive is lagging behind and the added value is "inverted U", that is, increase first and then decrease. The operability of equity incentives in the real economic context of China is worth exploring [36].

The research also shows that the design of listed companies for welfare purposes is relatively simple and cannot really meet the expectations of employees and the actual needs of employees. It is also an important reason why additional benefits expenses cannot effectively improve financial performance.

From the influence degree, in the analysis of the relationship between the welfare of listed companies on financial performance, the influence of employee welfare expenses is the greatest, which exceeds the influence of social insurance premium on financial performance. It is evident that because of the greater flexibility and adaptability of employee welfare compared with social insurance, the corresponding characteristics provided with time and events can stimulate employees to create benefits for the company. It also confirms that the company's welfare policy should be carried out according to the actual needs of employees and examine the company's welfare from the perspective of contingency.

6.2 Recommendations and Countermeasures

(1) Various forms of welfare realization are available, and the content of welfare is close to the real demand.
With reference to the current trend of research, a flexible welfare system was established. This is a "buffet" type of welfare form originated in the United States. Employees are selected according to their own needs within the company's defined welfare limits [37]. This form of welfare meets the different needs of employees of different genders, different ages, different levels and different backgrounds. The "one basket" welfare content is freely chosen to replace the unified welfare form. Under the premise of the same welfare content and welfare expenditure, the application form of welfare is diversified, and the efficiency of welfare utilization is improved.
(2) Design and application of cost control mechanism for welfare funds to further enhance welfare efficiency.
From the above analysis, we can see that the welfare expenditure, especially the large amount of social insurance and additional benefits expenses, has formed a great pressure on the listed companies, which is not conducive to the improvement of their financial performance. Therefore, it is imperative to establish the necessary cost control mechanism. Li found that 53.3% of companies think that the biggest difficulty of welfare policy lies in the cost problem, 46.7% have no effective cost

control mechanism, and lack of cost budget, accounting, decision-making and evaluation mechanism about welfare policy [38]. To establish a dynamic cost budget for welfare, carry out regular cost budget, prevent welfare policy expenditure and welfare content solidification, increase the economic burden of the company, welfare policy should pay attention to cost accounting. When welfare expenditure actually occurs, it should match the welfare category that has been determined in the current period. Because many welfare contents are non-monetary and under the premise of fair principle, the currency and non-monetary welfare should be reasonably converted, the actual operation process of welfare should be controlled, and every effort should be made to be feasible, matched and accurate. Evaluation of welfare policies, including employee satisfaction, cost consumption, performance improvement and other aspects, should be carried out to achieve a combination of short-term and long-term evaluation. At the meantime, exploring possible access to information on welfare policies, making the design of the welfare system more practical, balancing the needs of the company with the needs of employees and making welfare policies conducive to the improvement of financial performance are necessary.

(3) Welfare funds application monitoring system is established to avoid the misuse of welfare funds.
The misappropriation, abuse and unreasonable use of welfare funds are the reasons for the increase of welfare expenditure with unimproved company performance. The using direction of welfare funds should be strictly transparent by establishing a welfare fund tracking and supervision system from the budget allocation of welfare funds to the use and evaluation. At the same time, agency personnel, examination and approval personnel, implementation personnel, all should clearly define the post responsibility to create a welfare use fund line.

References

1. Wang, H., Song, T.: Empirical study on social responsibility and corporate performance of listed companies in China——empirical evidence from Shanghai 180 Index. J. Nanjing Norm. Univ. (Soc. Sci. Ed.), (02), 58–62+75 (2007)
2. Cai, Y.: Empirical study on corporate performance driven by social responsibility——empirical data from listed companies in Jiangsu province. Mark. Weekly (10), 91–93+102 (2018)
3. Li, K., Liao, Y.: An empirical analysis of the correlation between corporate social responsibility and financial performance. J. Tianjin Commer. Univ. **38**(05), 54–59 (2018)
4. Ding, G., Li, S.: Research on the innovation of corporate welfare system in China. Coop. Econ. Sci. Technol. (13), 26–28 (2012)
5. Wei, Y.: Research on welfare management of small and medium-sized companies in Henan province. Hum. Resour. Dev. (9), 72–74 (2015)
6. Zhang, Y., Yu, X.: Research on the correlation between financial performance and stock price of science and technology companies. J. Xiamen Inst. Technol. **26**(06), 38–43 (2018)
7. Sun, G.: EVA basis of the company's financial performance evaluation research. Acc. Stud. (14), 11–13 (2018)

8. Su, J.: The misunderstanding and improvement of corporate social responsibility in Chinese companies. J. Taiyuan City Vocat. Tech. Coll. (07), 177–178 (2018)
9. Ye, C., Ye, C.: Research on the relationship between social responsibility, value innovation and core competitiveness. In: The 12th (2017) China Management Academic Year, Tianjin, China (2017)
10. Fulmer, I.S., Gerhart, B., Scott, K.S.: Are the 100best better? An empirical investigation of the relationship between being a “great place to work” and firm performance. Pers. Psychol. **56**(4), 965–993 (2003)
11. Ghaly, M., Dang, V.A., Stathopoulos, K.: Cash holdings and employee welfare. J. Corp. Finan. **33**, 53–70 (2015)
12. Edmans, A.: Does the stock market fully value intangibles? Employee satisfaction and equity prices. J. Finan. Econ. **101**(3), 621–640 (2011)
13. Asrar-ul-Haq, M., Kuchinke, K.P., Iqbal, A.: The relationship between corporate social responsibility, job satisfaction, and organizational commitment: case of Pakistani higher education. J. Clean. Prod. **142**, 2352–2363 (2017)
14. Wang, X., Chen, H.: Research on the relationship between corporate social responsibility and corporate value based on stakeholders. Manag. Sci. **24**(6), 29–37 (2011)
15. An, S., Lin, D.: An empirical study on the impact of employee welfare satisfaction on job performance. China Bus. (33), 68–70 (2014)
16. Huang, He.: An empirical study on the impact of employee welfare satisfaction on employee job performance. Master Thesis, Northeast University (2010)
17. Ben-Nasr, H., Ghouma, H.: Employee WF and stock price crash risk. J. Corp. Finan. **48**, 700–725 (2018)
18. Filbeck, G.: MOTH Jones: do better places to work imply better places to invest? Rev. Finan. Econ. **10**(1), 57–70 (2001)
19. Markowitz, H.M.: Portfolio selection: efficient diversification of investment. J. Inst. Actuar. **119**(1), 243–265 (1992)
20. Carroll, A.B., Shabana, K.M.: The business case for corporate social responsibility: a review of concepts, research and practice. Int. J. Manag. Rev. **12**(1), 85–105 (2010)
21. Zhuang, P.: Research on the relationship between corporate welfare and corporate performance. Master Thesis, Capital University of Economics and Trade (2018)
22. Gao, X.: On the role of social insurance in human resource management of modern companies. Mod. Econ. Inf. (24), 1 (2014)
23. Gao, B.: China’s Urban Housing System Reform Research-Change, Performance and Innovation. Economic Science Press, Beijing (2017)
24. Jin, S.: A study on the impact of housing fund policy on income distribution——based on the analysis of payment-extraction path. Finan. Trade Res. **29**(12), 70–79 (2018)
25. Cheng, F.: On the negative impact of housing fund policy on personal tax fairness and suggestions for improvement. https://www.ntds.gov.cn/art/2017/2/24/art_1310_838361.html
26. Shen, Y.: On the role of company labour unions in building harmonious labor relations. Mod. Mark. (Next Ten Days) (01), 168 (2019)
27. Zhang, Q.: The impact of labour union practice on company performance. Hebei Co. (01), 23–24 (2016)
28. Ji, G.: Staff education funds, company innovation and company performance. Master Thesis, Nanjing University (2016)
29. Shen, J.: Discussion on some problems of employee welfare expenses. Employ. Secur. (24), 32–34 (2018)
30. Lin, Y.R., Fu, X.M.: Does institutional ownership influence firm performance? Evidence from china. Int. Rev. Econ. Finan. **49**, 17–57 (2017)
31. Zhu, R.: Research on performance evaluation system and incentive mechanism of senior managers in the company. Hum. Resour. Manag. (05), 180–182 (2014)

32. Jiang, Z., Yan, B., Sheng, C., Li, H.: Economic value creation, investment efficiency and macroeconomic growth-EVA methods and research on panel data of manufacturing industry in China and Beijing. Finan. Res. (11a), 118–128 (2007)
33. Dong, X.: Empirical study on financial early warning model based on EVA. Master Thesis, Chongqing University (2005)
34. Shao, Z., Chen, X.: A study on the valuation of company value based on the method of economic increase value——take Goliath as an example. Bus. Account. (01), 55–58 (2019)
35. Zhang, Z., Tang, Y.: Empirical analysis of performance evaluation of small household appliances companies based on EVA. J. Hubei Inst. Econ. (Humanit. Soc. Sci. Ed.) 16(01), 71–74 (2019)
36. Mo, X., Yu, S., Chen, C.: An empirical study on the relationship between equity incentive and corporate performance of listed companies. Financ. Econ. (Acad. Ed.), (14), 49 (2018).
37. Huang, R.R.: A brief talk on how companies should use flexible welfare system. Mall Mod. (06), 91–92 (2018)
38. Li, J.: Construction of welfare system of small and medium-sized companies under the background of cost-benefit analysis——take the investigation of 15 small companies in Hunan province as an example. Labour Secur. World (36), 26–27+30 (2016)

Overseas Experience of Top Management Teams and Firm Outcomes: Evidence from Mainland China

Ming Zhang(✉)

School of Economics, Shandong University, Shanda South Road 27, Jinan, China

Abstract. In emerging economies, overseas returnees from developed countries begin to take positions at all levels of firms. In this context, this paper investigates how overseas returnee executives affect firm outcomes. Using data on the real estate industry in China, I do multivariate analysis on the impact of the overseas experience of top management teams on firm performance and consider moderating effects of ownership properties and power distribution within TMTs. I find a positive association between the richness of overseas experience of TMTs and firm performance. Further analyses show the relationship holds for private and state-owned corporations and is unlikely to be driven by overseas returnee CEOs. A tentative explanation is TMTs with strong overseas experience may have advantages in raising funds internationally, which helps reduce capital costs and enhance firm performance.

Keywords: Overseas experience · Firm outcomes · Real estate industry

1 Introduction

A substantial number of talents choose to return to their home nations after studying or working in developed countries. For example, in China, the number of overseas-returned students has increased from 70 thousand in 2008 to 500 thousand in 2018.[1] As these overseas returnees participate in all levels of firms' organization, a growing body of studies in emerging economies pay attention to the role of overseas-returned executives, board directors, and entrepreneurs [1–4]. A relevant question in this context is whether executives with foreign experience exert a positive impact on firm performance. However, only a few studies answer this question and the findings on the impact of overseas returnee executives are mixed [5, 6]. Focusing on the real estate industry in China, this study contributes to current literature by investigating overseas returnee executives' role and giving new, tentative explanations.

Using data on real estate firms in China mainland, I find a positive association between TMT average overseas experience and firm performance. Further analyses also show this finding applies to both private- and state-owned corporations and is not driven

[1] Data from the Ministry of Education of the People's Republic of China.

C. Yuan et al. (Eds.): *Proceedings of the 4th International Conference on Economic Management and Green Development*, AEPS, pp. 154–165, 2021.
https://doi.org/10.1007/978-981-16-5359-9_19

by CEOs with overseas experience. One possible reason is that, TMTs with a stronger foreign experience may be better-informed of raising funds internationally and this advantage reduce the capital costs for firms.

The structure of the paper is as follows. The second section reviews the literature on TMT overseas experience. The next section provides information on methodology and data. The fourth is empirical results, followed by discussions and conclusion, the final section.

2 Literature Review and Hypothesis

In recent years, scholars began to explore the connection between overseas returnee directors and firm outcomes taking advantage of policy changes in emerging economies [3, 4]. Based on the discontinuity in talent introduction policies in China, a recent Chinese study [3] finds that, after recruiting more directors with overseas experience, firms generally experienced higher valuation, productivity, and profitability. Another study in Korea defines overseas experience as people getting a postgraduate degree overseas and focuses on the 100 largest Korean firms [4]. Their results show, following the introduction of new board regulations, firms with a higher proportion of returnee outside directors conformed to the new rules better and experienced a larger decrease of diversification and debt ratio. These studies attribute this contribution of overseas experience to overseas returnees' knowledge of superior management practice of foreign organizations, their connection in foreign nations, and their fringe position in the specific institutional environment [3, 4].

Other studies follow a similar logic and analyse the role of TMTs from the perspective of international activity engagement [7–9]. For example, internationally-experienced TMTs in Canada are more willing to develop foreign strategic partners [8]. In the context of China, firms with a TMT rich in international experience tend to choose a more aggressive way—to acquire a part control or the whole control of a foreign entity–to enter a foreign market and undertake more risks [7]. Firms with a strong overseas background of TMTs also facilitate the formation of international alliances [9], international diversification of the market [10], and internationalization behaviors [8]. While these empirical studies well support the contribution of overseas experienced executives to organizational outcomes, the direct connection between the overseas background of TMTs and firm outcomes are under-explored yet.

Very recently, another Chinese study [6] investigates Chinese manufacturing firms and finds overseas-returned executives improve firm performance, especially for non-state-owned firms, large firms, and firms located in eastern China. This literature finds a positive association between the overseas experience of TMTs and corporate outcomes in China. However, the study estimates on manufacturing firms, a broad scope of corporations, among which high-end, capital-intensive manufacturing firms, like electronics manufacturers, can be very different from low-end, labor-intensive firms in many aspects. Corporate governance structures may greatly vary within this broad sample and the estimations may suffer from biases [11]. Developing on this literature, this study focuses on a smaller scope of firms, the real estate industry, which guarantee similar governance structures within the sample. This study also considers a variety of control variables, including characteristics of directors [3, 4], and firm performance proxies, both market-based and accounting- based ones. In this way, I reduce omitted variable bias and measurement bias. The main hypothesis in this study is as follows:

Hypothesis: TMTs with stronger overseas experience can facilitate better current firm performance.

3 Methodology and Data

3.1 Data and Model

This paper uses panel data of all listed real estate companies in mainland China from 2014 to 2017. Data on the demographic characteristics and stock market information are from the Chinese Stock Market & Accounting Research (CSMAR) and Chinese Centre for Economic Research (CCER) databases. The foreign debt issuance data is from the Wind Financial Terminal. After deleting observations with missing values, the final sample is around 300. To remove extreme values, I winsorize all continuous variables at 1% and 99% level. The descriptions of the variables are in Table 1 below. To control for the impact of province-specific and time-specific factors, I further control dummies for province and years. I test my hypothesis by the following model (1).

$$Firm\ Perf_{it} = \beta\ oversea\ bkgd_{it} + \gamma\ control_{it} + \sum year_l + \sum prov_i + \varepsilon_{it} \quad (1)$$

3.2 Dependent and Independent Variables

This study adopts both market-based and accounting-based variables to proxy the current company performance. The market-based variables are stock return and abnormal return (by the market-adjusted model). Following the previous practice [12], the stock return and abnormal return are calculated from May to December to avoid the potential confounding price effects on stocks from corporate announcements. The account-based variables are EPS, EPS before interest and tax (EPSBIT), and ROE. In terms of dependent variables, I calculate the ratio of TMT members with foreign exposure, which represents whether a person has studied or worked in a foreign nation. Following the previous practice [6], I also use a foreign exposure dummy to serve a robust check.

3.3 Control Variable

At a firm level, consistent with previous literature [6, 13], I control for the state-owned enterprises, the size, equity multiplier, and fixed assets ratio of a firm. As the overseas experience of the board of directors correlates with that of TMTs and may influence firm performance [3, 4], I also control for their overseas background. Literature also shows that the ownership concentration [14], the executive share ownership, the education level [5, 12], the gender diversity [13, 15] and the average age of TMTs matter to operation of a firm. Thus, I control for shares of the largest shareholders, share ownerships of TMTs, and other demographic characteristics.[2]

[2] I do not control this average age and male ratio in the baseline regression due to a collinearity.

Table 1. Variable description

	Variable	Calculation
Firm performance	EPS	Earnings per outstanding share
	EPS before interest and tax	Earnings per outstanding share before interest and tax
	ROE	Earnings divided by all shareholders' equity
	Stock return	Changes in stock price from the beginning of May to the end of November[a]
	Abnormal return	Stock return minus index return during the same period
Overseas background	Average overseas experience	The ratio of TMT members with overseas degrees or working experience
	Overseas dummy	Dummy indicating at least one executive in TMTs with overseas experience
Firm	SOE	Dummy indicating a company owned by government or not
	Fixed asset ratio	The ratio of fixed assets to total assets of a company
	Equity multiplier	The reciprocal of equity ratio
	Ln(asset)	The log form of total assets of a company
Share	Management share	Total equity share owned by TMTs
Control	Largest shareholder's share	Top 1 shareholders' share
Other characteristics	Average overseas experience of the board of directors	The ratio of the directors[b] with overseas experience
	Financial background	The ratio of TMT members with working experience in financial agencies
	Average bachelor ratio	The ratio of TMT members whose highest degrees are a bachelor
	Less bachelor ratio	The ratio of TMT members without a bachelor degree
	Male ratio	The ratio of TMT male members

(*continued*)

Table 1. (*continued*)

	Variable	Calculation
	Average age of TMTs	The Average age of TMT team members
	CEO overseas experience	Dummy indicating a CEO with overseas experience

[a]Some corporations do not start to trade their stock at the immediate beginning of May and end at the immediate end of December. For these firms, I calculate the stock returns based on their actual starting and ending days prices and get their abnormal return subtracted from the index return for their corresponding periods.
[b]To avoid collinearity between variables, the calculation only based on directors who do not serve TMTs.

3.4 Summary Statistics

Table 2 is summary statistics of variables in this study. From the table, the stock and abnormal returns are both negative on average, which is consistent the poor stock market performance in China from 2015 to 2016. The statistics show that the ratio of companies with at least one overseas returnee executives is around 20% and the average ratio of overseas returnee executives is 3%. Surprisingly, the average ratio for overseas returnee directors (excluding those serving executives) is 16%, much higher than that of executive counterparts. Overall, the data suggests a low internationalization degree in the sample and most TMTs with positive overseas exposure may only have one or zero member with overseas experience.

Table 2. Summary statistics

Variable	Obs	Mean	Std. Dev.	Min	Max
EPS	346	0.25	0.366	−0.633	1.626
EPS before interest and tax	348	0.805	0.793	−0.528	3.739
ROE	342	0.067	0.105	−0.506	0.289
Stock return	339	−0.05	0.277	−0.607	1.2
Abnormal return	339	−0.091	0.29	−0.608	1.016
Average overseas experience	342	0.034	0.094	0	0.571
Overseas dummy	348	0.184	0.388	0	1
SOE	348	0.509	0.501	0	1

(*continued*)

Table 2. (*continued*)

Variable	Obs	Mean	Std. Dev.	Min	Max
Average financial background	348	0.047	0.114	0	0.75
Average age	348	47.297	3.295	37.333	55.8
Male ratio	348	0.815	0.153	0.429	1
Fixed asset ratio	348	0.022	0.029	0	0.163
Equity multiplier	321	3.777	2.004	1.139	10.544
Log(asset)	346	23.504	1.549	19.579	27.321
CEO oversea	348	0.072	0.259	0	1
Average overseas experience of directors	342	0.159	0.153	0	0.667
Share owned by management	348	0.475	2.178	0	17.316
Top 1 shareholders' share	342	38.147	15.334	7.12	76.95
Bachelor ratio	267	0.395	0.275	0	1
Ratio without a bachelor	267	0.1	0.207	0	1

4 Empirical Result

Table 3 shows the regression results in this study. From Panel 1 column (1), the overseas background of TMT members has a significant impact on current corporate performance, as measured by EPS. With the inclusion of more control variables, the coefficient stabilizes at 0.7. Specifically, all else being equal, a 1% increase in TMT overseas experience ratio leads to 0.007 units of increase in the earning per share of a firm on average. Alternatively, in a TMT with ten executives, one more executive with overseas exposure contributes to a roughly 0.07 unit increase in the earning per share. This finding is significant in both economical and statistical sense. This result is also consistent with the previous study [6], which find that a 1% increase in the ratio of top executives with overseas exposure will lead to 0.005 units of increase in ROA.

However, the baseline regression may ignore some important metrics, like the overall educational background of TMTs. To deal with potential endogeneity problems, more factors and proxies should be considered to test the robustness of the result.

4.1 Demographic Characteristics

As discussed in the previous section, a higher education level will add value to firm performance [5, 12]. I further control for education levels of TMTs by *lessbache* and

Table 3. Regression table

Panel 1: Baseline regression and regressions with extra control and consideration					
	(1) EPS baseline	(2) EPS Education	(3) EPS characteristics	(4) EPS SOE	(5) EPS Power
ave_oversea	0.7012***	1.0585***	0.6512***	0.6954***	0.8901***
	(0.2118)	(0.2740)	(0.2144)	(0.2430)	(0.2507)
ave_bache		0.0104			
		(0.1415)			
ave_lessbache		-0.0108			
		(0.2571)			
ave_age			0.0107		
			(0.0074)		
male_ratio			0.0491		
			(0.1362)		
ave_oversea*soe				0.0277	
				(0.5585)	
ave_ovesea*ceo oversea					-0.7055
					(0.5035)
ave_oversea_bod	-0.1986	-0.4217**	-0.1969	-0.1976	-0.2272
	(0.1388)	(0.1857)	(0.1387)	(0.1407)	(0.1400)
Other controls	X	X	X	X	X
Year & Province FE	X	X	X	X	X
N	319	186	319	319	319
r2	0.1699	0.2650	0.1771	0.1699	0.1756
Panel 2: Alternative proxies for overseas experience and firm performance					

(continued)

Table 3. *(continued)*

ave_overseas or overseas_dummy coefficients	(1) Educaiton	(2) Characteristics	(3) Ave_oversea*soe	(4) Ave_ovesea* Ceo_oversea	(5) Oversea_dummy	(6) Baseline
EPS	1.0585***	0.6512***	0.6954***	0.8901***	0.1387**	0.7012***
	(0.2740)	(0.2144)	(0.2430)	(0.2507)	(0.0558)	(0.2118)
EPSBIT	1.7156***	1.4099***	1.4009***	1.3555***	0.2815***	1.3467***
	(0.4985)	(0.3703)	(0.4192)	(0.4339)	(0.0963)	(0.3653)
ROE	0.0411	0.0830**	0.0038	0.0281	0.0210**	0.0849**
	(0.0427)	(0.0382)	(0.0424)	(0.0446)	(0.0099)	(0.0379)
Stock return	0.7246***	0.3986**	0.3426*	0.5943***	0.0837*	0.3700**
	(0.1965)	(0.1700)	(0.1920)	(0.1975)	(0.0442)	(0.1673)
Abnormal return	0.6122***	0.3391*	0.2665	0.5493***	0.0742	0.3106*
	(0.2309)	(0.1803)	(0.2036)	(0.2094)	(0.0468)	(0.1774)

Panel 3: SOE and power distribution term coefficient

Interaction term coefficient	ave_overseas * SOE	ave_oveseas * CEO_oversea
EPS	0.0277	-0.7055
	(0.5585)	(0.5035)
EPSBIT	-0.2553	-0.0329
	(0.9633)	(0.8715)
ROE	0.3822***	0.2124**
	(0.0974)	(0.0896)
Stock return	0.1290	-0.8329**
	(0.4406)	(0.3957)
Abnormal return	0.2070	-0.8865**
	(0.4671)	(0.4196)

Notes: Panel 1 of this table shows a baseline regression result and results after consideration of characteristics, power distribution, SOE heterogeneous effects, and education levels. Constants are not shown in the table. Panel 2 of this table shows the results after changing proxies of firm current performance and growth opportunities. Constant terms are not shown in the table. Panel 3 of this table shows the coefficients of interaction terms after considering SOE and Overseas CEO. Standard errors in parentheses. * $p < 0.1$, ** $p < 0.05$, *** $p < 0.01$.

bache[3] in column (2). I also control for the average age of TMTs and the male ratio[4] in column (3), as many previous studies point out that a lower average age and higher gender diversity may contribute to superior firm performance and correlate with overseas experience. Column (2) and (3) show, while the coefficients change due to changes in sample sizes after controlling for education levels, the overseas background still exerts a positive influence on firms' performance.

4.2 CEO Power and SOE Effects

One implicit assumption made by using the ratio of TMT members with overseas experience as the proxy for the whole team's overseas experience is that each member is the same influential to decision-making in TMTs. This assumption may not hold, however [16]. As average statistics ignore the power distribution within TMTs, this raises concern that the result is mainly driven by the CEO, the most powerful member in TMTs, not other less powerful members. To address the concern, I include an extra control variable, an interaction term between the demeaned *ave_overseas* and *CEO_oversea.* After considering this power distribution effect within TMTs, Panel 1 column (4) shows that the coefficient of experiences a small increase of 0.2 units and is still significant while the coefficient of the interaction term is insignificant. Thus, the result is not mainly driven by only powerful members.

Previous scholars find that non-state-owned firms gain more benefits from management characteristics, including overseas exposure and gender diversity, compared with state-owned counterparts [6, 13]. If this feature holds in this study, there should be a larger increase in the overseas background effect in private firms. To investigate the potential heterogeneous effect of ownership properties, I add an interaction term of *SOE* and demeaned *ave_oversea.* Panel 1 column (3) shows, however, the significance of overseas experience does not change, and the coefficient for interaction term is statistically insignificant. Put together, the pattern suggests the impact of overseas backgrounds applies to both state-owned and private- owned firms.

4.3 Alternative Proxies for Overseas Exposure and Firm Performance

To test the robustness of previous finding, I use a series of alternative proxies for overseas experience and firm performance. The result is shown in Panel 2 and Panel 3 of Table 3.

To begin with, I use other four other proxies for firm performance to regress again. The baseline model for these proxies shows consistent pattern in significant levels as shown in column (6). When I control for characteristics, CEO Power effects, SOE effects,

[3] The education level information of around 30% of TMT members are missing in my dataset. Using available members' information to construct average statistic will lead bias, as those executives with low-level degrees hold back their degree information on annual reports. Thus, to make best use of my data while ensuring accuracy of measurement, I leave out firms with less than 80% members' degree information available, and by doing this, the sample size decreases to around 200.

[4] Variance Inflation factor shows there is a strong correlation between male ratio, average age of TMT members, and log(asset). Thus, I not control these characteristics in baseline regressions and only control these in this part.

and education levels, the results are largely the same (at least three coefficients among the four proxies' regression are significant). The only exception is when I control for interaction term of *SOE* and demeaned *over_ave*, the two coefficients for ROE and abnormal return regressions are both insignificant. But overall, the results support the positive roles of overseas returnee executives on firm performance.

Then, I further explore if new proxies display different patterns for the moderation effects of ownership properties and CEO power effects in Panel 3. The examination on interaction terms, however, shows that SOE is unlikely to mediate the relationship between overseas experience and firm performance as four out of five regressions gives insignificant estimation results. What is interesting is CEO power interaction coefficients displays are statistically significant albeit different in their signs.[5] The mixed signs of CEO with overseas experience do not support a consistent mediation effects of CEO power either. Put shortly, my finding on overseas exposure of TMTs are applicable to both private- and state-owned corporations and are unlikely to be driven by CEO with overseas experience.

Finally, I replace average overseas experience of TMTs with an overseas dummy.[6] Panel 2 column (5) shows a significant positive role of TMTs with overseas exposure. Taking the result for EPS for example, firms with at least one member having overseas experience show a 13% higher EPS than those not, on average, while controlling for other factors.

5 Discussion and Conclusion

The positive association provides strong support for the main hypothesis. Consistent with previous literature, the study demonstrates that returnee executives, with their unique experience and knowledge learned from their overseas life, can contribute to better corporate outcomes. This finding applies to both private- and state-owned corporations and is unlikely to be only driven by CEOs with overseas experience. This indicates enterprises should create a tolerant firm environment that attracts talented overseas returnees, since their knowledge facilitate better firm performance. But one remaining question that has not been explored yet in this study is, what may be the channels though which executives promote firms' performance for this study?

Previous research mainly focuses on the exports and international alliance [9, 17] while a few focuses on fund-raising activities [3]. In the context of the real estate industry, I argue, the presence of returnees among TMTS may affect the corporation outcomes through supporting unfamiliar financing strategies or attracting overseas financers to serve their financial needs. Since 2003, regulations for foreign capital investment have gradually relaxed, with investment among increasing from 1.7 billion $ in 2003 to 76.6 billion $. Previous studies support that overseas background help raise funds, with their

[5] VIF are small enough for these two regressions. Thus, this is not a result of strong collinearity after controlling for interaction terms. One possible reason might be the data is from 2015 to 2017, which coincides the collapse of Chinese stock market.

[6] This replacement may not lose much information according to statistics summary, which shows the average statistics of average overseas experience within TMTs is only less than 4% and my data shows over half of corporations with positive average overseas experience has only one overseas returnee executive.

overseas experience a positive signal to international investors that corporation governance and decision-making process can be more effective, similar to those MNCs in developed nations [3]. This is important in drawing funds internationally, especially in an environment with weak investors protection and corporate governance. The participation of foreign institutional investors will stabilize the stock price of firms and facilitate better performance [18, 19]. In the same vein, in the recent years, as there are fewer restrictions on firm,[7] overseas-returned talents may take advantage of their knowledge or connections gained from developed nations [3, 4], where raising funds internationally is a commonplace. Having garnered experience from foreign study or work, they can propose constructive advice and enhance team's understanding, which means corporations can follow the trend of borrowing foreign money wisely. Foreign capital can be an extra powerful financing tool for these companies to meet their financial needs and reduce capital costs [20, 21]. Overall, previous literature shows strong evidence linking international financing activities and overseas backgrounds to firm performance, at least for the capital-intensive industry, like the real estate industry. However, with limited data, this study does not empirically examine these potential channels, the study of which may be beneficial to understanding the role of overseas executives.

Finally, this study is subject to following limitations. First, the study only supports a correlational relationship, not a causal relationship. Although many important control variables are considered, the problem of multi-causality typical in corporate finance research remains unsolved [16]. Some missing data, including education degree information, and other variables, including the independent director ratio, the board size, and the firm age, may also lead to omitted variable bias (OVB). Future studies can conduct a more convincing identification to eliminating reverse causality and OVB.

Second, while my focus on real estate firms can be helpful to understanding the role of overseas-returned top executives in corporate decisions and outcomes, the study based on this sample may not have the same explanatory power for other industries. The result may apply to high-technology corporations, for example. High-technology firms with good knowledge of the foreign world may benefit from increasing R&D investments [2], international networks [1] and willingness to develop partners [8]. But no evidence shows this will apply to manufacturing corporations with low internationalization degrees. As discussed in the previous section, real estate firms are capital-intensive and, different from most industries in China, have strong incentives to issue foreign debt and introduce foreign institutional investors. Thus, this finding may be more applicable to capital-intensive industries.

Finally, as updated upper echelons theories point out [16], two aspects that researchers tend to ignore are intra-TMT power distribution and TMT behavioural integration. Even though I investigate the powerful member, CEO, this study fails to explore the whole team's power distribution and structure. Also, different "behavioral integration" may exist in firms within a small scope of industry for many reasons. In some cases, TMT members only engage in bilateral relations with the CEO but have little to deal with other members [16]. Lack of consideration of these factors prevent this study making a stronger analysis. Including how TMT members interact with each into

[7] In terms of corporate foreign debt, according to statistics in the Wind Financial Terminal Platform in China, the number of foreign-currency-denominated debt issued in China (including those listed in Hong Kong) has increased nearly tenfold, from around 20 to over 150.

analysis, future studies may generate valuable insights on effective TMTs and efficient operation of firms.

Acknowledgements. I am indebted to professor Li in Shandong University for inspirations and advice. I also thank for professor Zhang and my classmate Miao for the data collection and writing. All remaining errors are mine.

References

1. Dai, O., Liu, X.: Returnee entrepreneurs and firm performance in Chinese high-technology industries. Int. Bus. Rev. **18**(4), 373–386 (2009)
2. Filatotchev, I., et al.: The export orientation and export performance of high-technology SMEs in emerging markets: the effects of knowledge transfer by returnee entrepreneurs. J. Int. Bus. Stud. **40**(6), 1005–1021 (2009). https://doi.org/10.1057/jibs.2008.105
3. Giannetti, M., Liao, G., Yu, X.: The brain gain of corporate boards: evidence from China. J. Finan. **70**(4), 1629–1682 (2015)
4. Lee, J.-H., Roberts, M.J.D.: International returnees as outside directors: a catalyst for strategic adaptation under institutional pressure. Int. Bus. Rev. **24**(4), 594–604 (2015)
5. Darmadi, S.: Board members' education and firm performance: evidence from a developing economy. Int. J. Commer. Manag. **23**(2), 113–135 (2013)
6. Zhang, C., Fu, P.: Overseas-returned executives and their roles in firm performance: evidence from China. Asia-Pac. J. Account. Econ. 1–10 (2020)
7. Ge, Y., Shen, S.: The influence of the management team's international experience on the internationalization behaviors of SMEs. Contemp. Econ. Manag. **33**(9), 53–58 (2011)
8. Reuber, A.R., Fischer, E.: The influence of the management team's international experience on the internationalization behaviors of SMES. J. Int. Bus. Stud. **28**(4), 807–825 (1997)
9. Lee, H.U., Park, J.H.: The influence of top management team international exposure on international alliance formation. J. Manag. Stud. **45**(5), 961–981 (2008)
10. Herrmann, P., Datta, D.K.: Relationships between top management team characteristics and international diversification: an empirical investigation*. Br. J. Manag. **16**(1), 69–78 (2005)
11. Haifa, S., Xiaoyi, W.: Review on studies of top management teams. J. Manag. Sci. China **6**(4), 82–89 (2003)
12. Cheng, L.T.W., Chan, R.Y.K., Leung, T.Y.: Management demography and corporate performance: evidence from China. Int. Bus. Rev. **19**(3), 261–275 (2010)
13. Liu, Y., Wei, Z., Xie, F.: Do women directors improve firm performance in China? J. Corp. Finan. **28**, 169–184 (2014)
14. Shahrier, N.A., Ho, J.S.Y., Gaur, S.S.: Ownership concentration, board characteristics and firm performance among Shariah-compliant companies. J. Manag. Gov. **24**(2), 365–388 (2020). https://doi.org/10.1007/s10997-018-9436-6
15. Rose, C.: Does female board representation influence firm performance? The Danish evidence. Corp. Gov.: Int. Rev. **15**(2), 404–413 (2007)
16. Hambrick, D.C.: Upper echelons theory: an update. Acad. Manag. Rev. **32**(2), 334–343 (2007)
17. Tyler, B.B., Steensma, H.K.: The effects of executives' experiences and perceptions on their assessment of potential technological alliances. Strateg. Manag. J. **19**(10), 939–965 (1998)
18. Genming, Z.: A study on the mechanism of the influence of qualified foreign institutional investors' shareholding on corporate performance. Commun. Finan. Account. **812**(12), 68–72 (2019)
19. Han, L., et al.: Do foreign institutional investors stabilize the capital market? Econ. Lett. **136**, 73–75 (2015)
20. Fei, G., Huixin, Y., Huimin, G.: Why do firms raise foreign currency denominated debt? Evidence from Chinese listed companies. J. Finan. Res. **453**(3), 137–154 (2018)
21. Fei, G., Panpan, Y., Huixin, Y.: Retrospect and prospect: the research of enterprise' foreign currency debt. J. Zhengzhou Univ. Aeronaut. **3**, 105–112 (2017)

Public Service Motivation and Prosocial Behavior Among College Students: The Mediating Role of Social Innovation

Hung-Yi Liao[1], Kang-Hwa Shaw[2(✉)], Zi-Shan Zhu[1], Si-Xin Huang[1], and Yu-Yao Song[1]

[1] Department of Human Resource Management, Shanghai Normal University, Shanghai, China
hyliao@shnu.edu.cn

[2] School of Management, Shandong University, Jinan, Shandong, China
kevinshaw99@sdu.edu.cn

Abstract. Based on motivation theory, this study investigated the relationship between public service motivation and prosocial behavior among college students, while further examining the mediating role of social innovation in that context. We conducted a questionnaire survey to collect relevant data from college students in China, thereby obtaining 125 valid responses for use in testing our hypotheses. Results showed that public service motivation and social innovation were both positively related to prosocial behavior, while social innovation mediated the relationship between public service motivation and prosocial behavior. We discussed the implications of the results, limitations, and future directions.

Keywords: Public service motivation · Social innovation · Prosocial behavior

1 Introduction

Due to the steady implementation of China's "One Belt, One Road" initiative and the continued advancement of the government-supported "community of shared future for mankind," it is increasingly important to cultivate prosocial behaviors among college students. Previous research has shown that public service motivations are a motivational orientation rooted in public institutions or organizations [1]. This further pertains to the formation of public policy, a commitment to public interest, compassion, self-sacrifice, and the patriotic spirit of Chinese culture. Public service motivation is therefore a highly valuable psychological resource. This study explored the relationship between public service motivation and prosocial behavior among college students in China, thereby highlighting the importance of both concepts.

Although previous studies have found that public service motivation triggers prosocial behavior at the individual level, there is relatively little empirical evidence in this regard [2]. It is also important to determine how public service motivation influences individual prosocial behaviors. Motivation theory posits that motivation is a behavioral tendency which influences certain individual activities that are directed toward

C. Yuan et al. (Eds.): *Proceedings of the 4th International Conference on Economic Management and Green Development*, AEPS, pp. 166–171, 2021.
https://doi.org/10.1007/978-981-16-5359-9_20

goals. This is important for social innovation, or the process by which innovative concepts and methods are used to identify business opportunities based on social problems, which are thereby addressed through an understanding of open-market spaces and social/commercial values at both the individual and organization levels. We assume that individuals with high levels of public service motivation are more willing to devote time and energy to social service activities (i.e., social innovation), while individuals who are highly oriented toward social innovation will exhibit more prosocial behaviors. We thus examined the relationships between public service motivation, social innovation, and prosocial behavior from a theoretical perspective. In this regard, we expanded the application of the theory of public service motivation and appropriately enriched the literature through new findings related to both social innovation and prosocial behavior.

In sum, this study used the concepts outlined in motivation theory to explore the relationship between public service motivation and prosocial behavior among college students, with a further focus on the mediating role of social innovation. In turn, these efforts clarified the internal mechanism of public service motivation. Our conclusions thus provide guidance for the development of prosocial education in the college setting. See Fig. 1 for this study's research model.

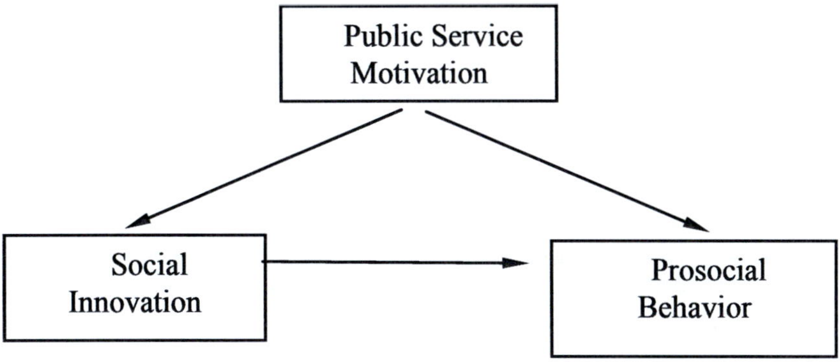

Fig.1. Research model

2 Literature Review and Hypotheses

2.1 Public Service Motivation and Prosocial Behavior

According to Perry and Wise [1], public service motivation is an orientation that is rooted in public institutions or organizations. Public service motivation can therefore be described as a belief and attitude that extends beyond the interests of individuals and particular departments. In this regard, public service motivation drives individuals to take corresponding actions at the appropriate times. It is also effective in serving the public and selflessly defending the motivations behind public interests. Indeed, Chen et al. [3] found that public service motivation positively influenced work engagement among civil servants at the grassroots level. Moreover, individuals who are active in the public sector and have high public service motivation tend to exhibit a greater willingness for engaging in social service activities when compared to private sector workers (Houston) [4]. In this paper, we further argue that public service motivation can trigger individual prosocial behaviors; that is, behaviors that meet social expectations while benefiting other groups and/or society. Esteve et al. [2] previously showed that public service motivation positively influenced prosocial behavior among students, while Piatak [5] claimed that individuals with high levels of public service motivation were more willing to volunteer in the public sector context. We therefore developed the following hypothesis: *H1: Public service motivation is positively related to prosocial behavior.*

2.2 Social Innovation and Prosocial Behavior

Based on its definition, social innovation links new ideas to create solutions that "improve living standards" (Mulgan et al.) [6]. The term also refers to the search for more effective and sustainable ideas that can solve social issues and/or create social value from existing solutions rather than focusing on personal value (Phills et al.) [7]. As important components of any social entity, individual social attributes determine whether a person will pay attention to social phenomena and/or problematic issues. Thus, social innovation can trigger individual prosocial behaviors. According to Bhatt et al. [8], social innovation aids entrepreneurs in focusing on target beneficiaries while influencing various entities to help solve social problems. Some social innovation projects have even stimulated social awareness among non-directly related subjects, who are nonetheless concerned about similar issues, including those who participate in volunteer teams or citizen groups with similar experiences (Garrone et al.) [9]. The generation and implementation of social innovation is therefore increasingly important for enterprises, citizens, and stakeholders who are interested in cultivating creativity, seeking to transform knowledge into commercial products and services, and solving social problems. Based on these factors, we developed the following hypothesis: *H2: Social innovation is positively related to prosocial behavior.*

2.3 Mediating Role of Social Innovation

Public service motivation is derived from individual inner demand, and is thus an internal driving force that influences the desire to serve others. More specifically, the goal is to benefit to others and society. On the other hand, the goal of social innovation is to promote human wellbeing, which makes it conducive to improving individual prosocial behavior. According to motivation theory, motivation is a behavioral orientation that influences individuals to engage in certain activities with a tendency to progress toward goals. Individuals with high levels of public service motivation will actively seek way to ensure their social participation while promoting cooperation. This often requires an awareness of the imbalanced relationship between existing social needs and the supply of social services, thereby stimulating the search for solutions aimed at existing social deficiencies. Individuals with this behavior can develop social services and products designed to improve the quality of social life while enhancing social innovation capabilities. Such behaviors are intended to meet social expectations in a way that benefits other individuals, groups, and/or society. Based on this, we developed the following hypothesis: *H3: Social innovation mediates the positive relationship between public service motivation and prosocial behavior.*

3 Methods

3.1 Participants and Procedure

This study's survey sample was comprised of college students from several universities in Shanghai, China. We examined the abovementioned hypotheses using data obtained from questionnaires that we distributed to these individuals online. Ultimately, this resulted in 125 valid questionnaires. Among respondents, 80.3% were female, while 48.5% were third-year students. Their majors mainly included those from the humanities (20.5%), with most students coming from areas outside Shanghai (73%). In terms of activity engagement, 51.6% participated in the "Summer social research activities," while 38.5% participated in the "College student innovation and entrepreneurship training program."

3.2 Measures

In addition to the frequency measurement of prosocial behavior, all items were assessed on a 5-point Likert scale ranging from 1 = strongly disagree to 5 = strongly agree. Public service motivation was measured using the 6-item scale developed by Bao and Li [10], with sample items including "For me, contributing to social welfare is very important." Social innovation was measured using the 8-item scale developed by Bulut, Eren, and Halac [11], with sample items including "I look for solutions to create political and social changes in society." Finally, prosocial behavior was measured using the 7-item scale developed by Oda et al. [12], with sample items including "When taking a train or a long-distance bus, I will help others put their luggage on the luggage rack." We also controlled for the effects of various demographic characteristics.

4 Analysis and Results

4.1 Reliability and Validity Analysis of Scale

This study conducted reliability analyses for all scales, with results showing Cronbach's alphas between 0.86–0.93 for each, thus indicating good reliability. Factor analyses were then conducted to test validity, with results show that the factor loadings of all items on each scale were between 0.67–0.95, while the cumulative explanatory variance of each scale was between 58.42%–68.00%. This indicated that all scales were of sufficient validity.

4.2 Correlation Analysis

As expected, public service motivation was positively correlated with prosocial behavior ($r = 0.42, p < 0.01$), while public service motivation was positively correlated with social innovation ($r = 0.52, p < 0.01$), and social innovation was positively correlated with prosocial behavior ($r = 0.37, p < 0.01$).

4.3 Hypothesis Testing

After controlling for the effects of demographic variables, public service motivation had a positive influence on prosocial behavior ($\beta = 0.49$, $p < 0.001$), thus supporting H1. Moreover, social innovation had a positive influence on prosocial behavior ($\beta = 0.38$, $p < 0.001$), which supported H2. We then used PROCESS macro (Model 4) for SPSS to test the statistical significance of these indirect effects. The point estimate for the indirect effect of public service motivation on prosocial behavior via social innovation was 0.14 (95% confidence interval: [0.01, 0.33]). Thus, social innovation mediated the relationship between public service motivation and prosocial behavior, which supported H3.

5 Discussion

5.1 Conclusion

This study found that public service motivation could positively influence prosocial behavior among college students via social innovation. This was based on three specific aspects. First, public service motivation showed a positive influence on prosocial behavior. Second, social innovation was found to enhance perceived prosocial behavior. Third, public service motivation positively impacted prosocial behavior via social innovation.

5.2 Management Implications

Based on the research results detailed in the previous sections, we propose the following two main suggestions. First, college students should be encouraged to develop public service motivation based on the concepts outlined in motivation theory. In specific regard to the management implications, colleges must encourage students to participate in public services, strive to build public service platforms, and organize public welfare activities. Moreover, colleges can work to assemble service or volunteer teams through existing student unions, clubs, and organizations. These newly formed teams can assist their communities by engaging with other institutions and organizations, such as local nursing homes and Red Cross societies. Colleges can reform scholarship and grant standards, incorporate public service participation into the scholarship evaluation system, and offer additional rewards to students who are active in public welfare activities and voluntary services.

Second, colleges should help students develop social innovations. More specifically, they should support students through information openness and knowledge sharing for the purpose of integrating and distinguishing large datasets that can help them discover the real needs of society. They can also promote social innovation by encouraging students to focus on more specific social problems and needs. In this way, students can more properly enter the frontiers of their respective professional fields, specifically equipped with the ability to comprehensively analyze related phenomena while continually improving their understanding of relevant social problems.

5.3 Limitations and Future Research

This study also had some limitations. First, we implemented a cross-sectional design, which does not allow researchers to establish causality. Future research should therefore adopt a longitudinal design involving multiple time periods, thus improving overall rigorousness. Second, we investigated the positive relationship between public service motivation and prosocial behavior via social innovation, but did not examine the moderator. As such, future research should consider the boundary conditions (e.g., individual values and school education).

Acknowledgements. This research was supported by the College Students' Innovation and Entrepreneurship Training Program (Grant No. 202010270046).

References

1. Perry, J.L., Wise, L.R.: The motivational bases of public service. Public Adm. Rev. **50**(3), 367–373 (1990)
2. Esteve, M., Urbig, D., Witteloostuijn, A., Boyne, G.: Prosocial behavior and public service motivation. Public Adm. Rev. **76**, 177–187 (2016)
3. Chen, W., Zhang, Y., Chen, G.: Linking professional identification to work engagement: the mediating role of public service motivation and the moderating role of self-efficacy. Hum. Resour. Dev. China **35**(2), 118–128 (2018)
4. Houston, D.J.: Walking the walk of public service motivation: public employees and charitable gifts of time, blood, and money. J. Publ. Admin. Res. Theory **16**(1), 67–86 (2016)
5. Piatak, J.S.: Public service motivation, prosocial behaviours, and career ambitions. Int. J. Manpower **37**(5), 804–821 (2016)
6. Mulgan, G., Wilkie, N., Tucker, S., Ali, R., Davis F., Liptrot, T.: Social silicon valleys - a manifesto for social innovation. The Young Foundation (2006)
7. Phills, J.A., Deiglmeier, K., Miller, D.T.: Rediscovering social innovation. Stanf. Soc. Innov. Rev. **6**(4), 34–43 (2008)
8. Bhatt, P., Ahmad, A.J., Roomi, M.A.: Social innovation with open source software: user engagement and development challenges in India. Technovation **52–53**, 28–39 (2016)
9. Garrone, P., Groppi, A., Nardi, P.: Social innovation for urban liveability. empirical evidence from the Italian third sector. Ind. Innov. **25**(6) 612–631 (2018)
10. Bao, Y., Li, C.: Measuring public service motivation: theoretical structure and scale revision. Hum. Resour. Dev. China **7**, 83–91 (2016)
11. Bulut, C., Eren, H., Halac, D.S.: Social innovation and psychometric analysis. Proc. Soc. Behav. Sci. **82**(9), 122–130 (2013)
12. Oda, R., Dai, M., Niwa, Y., Ihobe, H., Kiyonari, T., Takeda, M., et al.: Self-Report altruism scale distinguished by the recipient (SRAS-DR): validity and reliability. Japanese J. Psychol. **84**(1), 28–36 (2013)

Research on the Economic Model Construction and Development of Chinese Painting Industry

Wanghao Zou(✉)

Guangzhou Foreign Language School, Taoyu Road 79 Tianhe District, Guangzhou City, China

Abstract. Traditional Chinese painting is a kind of traditional painting form in China, and it is one of the important non-material cultural heritages in China. Unlike other painting products, traditional Chinese painting uses a "Chinese Calligraphy Brush Pen", which is dipped in water or ink and then painted on silk or rice paper. In the past, the public did not know much about it, and the industry was not taken seriously. However, in recent years, under the background of the government's policy of vigorously promoting the development of cultural industry and the rapid development of e-commerce, Chinese painting has ushered in new opportunities. Based on SWOT analysis, this paper analyzes the current strengths, weaknesses, opportunities and threats of the traditional Chinese painting industry by using all kinds of Chinese data and puts forward policy suggestions to prompt the development of the Chinese painting industry according to the analysis results.

Keywords: Chinese painting · SWOT analysis · Artworks · E-commerce · Culture

1 Introduction

Chinese cultural industry has fallen into a dilemma in the past few decades: China has a wide range of regions, different regions having distinctive levels of development of cultural industries, and many cultural products have not been valued and appreciated by the public. However, in recent years, the central government pays attention to the real problem and the dynamic trend in the combined development of cultural industry and other industrial integration and timely further launches a series of incentive type industrial policies; for instance, promote the deep fusion of cultural industry and the real economy, in order to achieve the world within the scope of "made in China" to "created in China" and realize the regional coordinated development of cultural industry [1]. In the face of government policies and the background of the modern times -- the combination of internet and finance, intangible cultural heritage has a new space for development. Among them, traditional Chinese painting as a typical kind of intangible cultural heritage, has ushered in the development opportunities. Helping Chinese painting develop industrial economic model can not only inherit Chinese traditional skills, but also promote the prosperity of other industries. Cultural tourism industry, for example, is a kind of cultural activities; in the process of cultural tourism project development, non-material

C. Yuan et al. (Eds.): *Proceedings of the 4th International Conference on Economic Management and Green Development*, AEPS, pp. 172–182, 2021.
https://doi.org/10.1007/978-981-16-5359-9_21

cultural heritage is an important development content; entrepreneurs reveal the cultural industry advantage through the innovation of intangible cultural heritage to attract a large number of tourists, improving the sale of cultural tourism product; it facilitates the construction of the brand of cultural tourism industry [2]. To better promote economic model construction and the development of the Chinese painting industry, this paper will be based on the SWOT analysis method, having a comprehensive combing present situation of economic advancement of traditional Chinese painting industry, analyzing its current development strengths, weaknesses, opportunities and threats, thus further putting forward the corresponding protection and development way, to better respond to and implement the 19th National Congress's great goal of "building a modern economic system".

2 SWOT Analysis

Theory of SWOT analysis model is put forward by potter. The four letters of SWOT represent respectively strengths, weakness, opportunities and threats. This analysis method is to sum up the industry internal resources, external environment and some other aspects synthetically. To move forward a single step, people can analyze industrial strengths, weaknesses, opportunities and threats. Through the analysis, enterprises can adopt the appropriate strategy to develop the industry, such as mergers and acquisitions to expand. Art industry, which is mainly composed of modelling art, as the sum of all kinds of cultural and economic activity, with the creation of art as the source, involving investment, exhibition, warehousing, trade, derivatives and other related fields of art, extends to the collection of consumption, investment and wide demand of all kinds of social groups [3]. The traditional Chinese painting as a cultural art good, possesses the spiritual attributes of culture and art and common qualities of goods meanwhile like aesthetic nature, the persistence of the hedging, the increment of value and etc. [4].

2.1 Strengths

National characteristic and the brand advantage. As one of the treasures of the Chinese nation and traditional art creation, the traditional Chinese painting is Chinese essential spiritual product. Compared to the western culture and other Asian countries' cultures, such as Japan and South Korea, the unique national characteristics of Chinese painting is real, and China has the Chinese painting industry in the international market of absolute advantage. According to the characters of the nationality of the traditional Chinese painting, we can apply the element of Chinese painting to packaging design of other products, promoting the organic combination of cultural elements and crafts design, strengthening the construction of independent brands, to fully embody the traditional Chinese painting culture, highlighting the national characteristics of other products, making the commodities packaging be full of national, regional and spatial information, reflecting the culture connotation and artistic conception. Thus, as a national brand, Chinese painting has rooted in the national cultural soil, fully mixing and giving a play to nationality, creating their own specific brand image and entering the international market, developing its own brand advantages.

Cost advantage. Creation of traditional Chinese painting has a particular trait that is of low cost and high value. The economic cost of Chinese painting's painting tools like pens, ink, paper and inkstone is low. However, the value of created works will vary based on the value of artists.; the higher ones, such as those made by Ji Zhao and Bohu Tang, are of great value, while the lower ones can also be used as works of art for ordinary household decoration [2]. Figure 1 is about the Chinese urban residents' CPI for entertained, educational and cultural goods and services. The Chinese painting's painting tools can be classified as CPI1 or CPI3. Figure 1 shows that CPI1 is the only curve that is always below 100; CPI3 numerical value is close to 100 and is the penultimate low curve. These evidences further confirm that Chinese traditional painting's raw material cost is indeed relatively low.

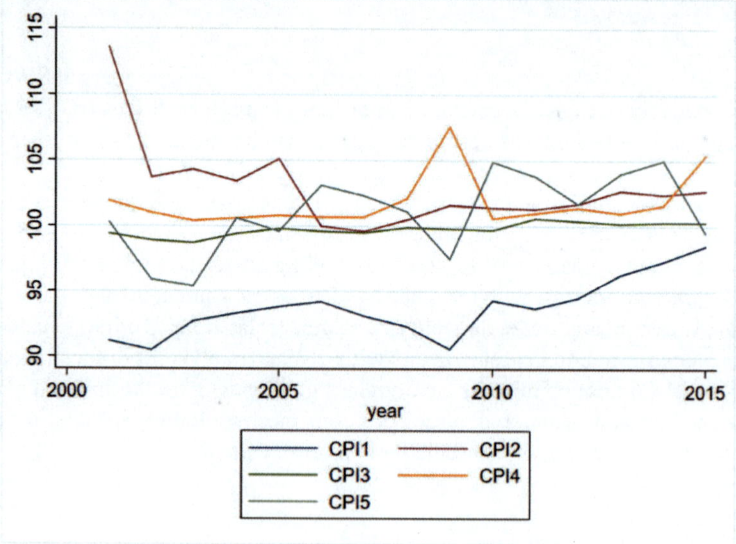

Fig. 1. Chinese urban residents' CPI for entertained, educational and cultural goods and services(previous year 100).

Annotation: CPI1: CPI for recreational and entertainment durables and services, CPI2: CPI for education, CPI3: CPI for cultural and entertainment articles, CPI4: CPI for books, newspapers and magazines, CPI5: CPI for travelling.

Source: National Bureau of Statistics of China.

Hedging function. From the economic point of view, the Chinese painting can be classified as break-even floating income financial products. For China with a long history, the history of Chinese painting is comparatively long, and the Chinese painting is consistent with the Chinese aesthetic, so for Chinese consumers, Chinese painting is more resonant with consumers than oil painting. Therefore, the current market hedging of traditional Chinese painting is better than the oil painting market [4]. By buying Chinese paintings, investors seek a way to preserve or increase their wealth [5]. Chinese painting artworks above the middle-end products can maintain and increase in value and can be widely promoted in the field of financial management; through this channel, more middle and high-end customers can be cultivated; on the one hand, it can give another way to customers to learn the artistic style of Chinese painting; on the other hand, it can promote the development of medium and high-end artistic products [6]. Therefore, the traditional Chinese painting has a strong hedging advantage.

2.2 Weaknesses

The lack of engaged in the personnel. Because of the traditional classical style of traditional Chinese painting, with the progress of science and technology and the development of the world, less and less staff works on it, and many people choose to work in the high-tech or fashion; even painters tend to create modern paintings. Comparing the number of personnel who works on Chinese and western painting, the number of painters and researchers occupied in traditional Chinese painting is gradually at a disadvantage [7]. In addition, the creation of Chinese painting is sometimes difficult since the required tools and techniques are complex, and it is inconvenient to modify. Thus, it is not easy to recruit a large number of staff with the foundation of traditional Chinese painting [8]. In table1, the number of employees in 19 urban units in Different industries in China is arranged from the largest to the lowest according to the average value. Table 1 shows that the employment of culture, sports and entertainment column is penultimately low and has very significant distinction with other units. At the same time, the standard deviation is the second from the bottom; that is, the number of employees is stable every year and almost does not rise. This further proves the personnel engaged in traditional Chinese painting is very little indeed.

Annotation: urban industries code corresponding vocations: job1: agriculture, forestry, animal husbandry and fishery, job2: the mining industry, job3: manufacturing, job4: production and supply of electricity, heat, gas and water, job5: the construction industry, job6: transportation, warehousing and postal services, job7: information transmission, software and information technology services, job8: wholesale and retail, job9: accommodation and catering, job10: finance, job11: the real estate industry, job12: leasing and business services, job13: scientific research and technical services, job14: water, environmental and public utility management industries, job15: residential services, repairs and other services, job16: education, job17: health and social work, job18: culture, sports and entertainment, job19: public administration, social security and social organizations.

Source: National Bureau of Statistics of China.

The marketing awareness of traditional Chinese painting is not strong, and the marketing channels are single. Among the main income sources of the Chinese painting museum, the sales income of the Chinese painting museum only accounts for a very small part of the total income, while the service income, rental income, exhibition income and training income result in the most of the income. It can be seen that the Chinese painting museum did not focus on the sale of Chinese painting, and marketing awareness is not strong. There are two possible reasons. First, the traditional Chinese painting museum, as a public institution allocated by the regional financial balance, has no business pressure. Second, the Traditional Chinese painting museum has been engaged in tourism

Table 1. China's urban industries employment(ten thousand people).

Variable	Obs	Mean	Std. Dev	Min	Max
job3	16	4015.65	798.1626	2980.5	5257.9
job5	16	1788.212	855.8482	833.7	2921.9
job16	16	1606.356	108.2243	1442.8	1736.5
job19	16	1459.644	199.937	1171	1817.5
job6	16	713.4688	107.0786	612.7	861.4
job8	16	682.175	158.5252	506.9	890.8
job17	16	677.95	154.0781	485.8	912.4
job2	16	537.5688	61.73244	414.4	636.5
job10	16	497.4875	121.842	353.3	699.3
job1	16	354.8063	87.01888	192.6	484.5
job4	16	340.375	42.13318	297.6	404.5
job12	16	338.7687	121.2037	183.5	529.5
job13	16	316.225	80.41089	221.9	420.4
job11	16	272.1063	127.8344	120.2	466
job7	16	238.1875	108.4688	116.8	424.3
job9	16	230.475	46.32644	172.1	304.4
job14	16	225.3625	38.05732	172.5	273.3
job18	16	135.7438	11.12702	122.4	152.2
job15	16	64.14375	9.574547	52.8	78.2

for many years, so it has no platform advantage in the field of art. In addition, traditional Chinese painting is a passive marketing model; the main marketing channels are limited, only such as waiting for customers or participating in the auction. Waiting for customers has very significant restriction; only familiar customers or experts come to purchase. Besides, the requirements of walk-ins are usually high. What they place favors is the national level famous painting, so this is unfavorable to a lot of broad provincial calligraphy and painting sale. The auction of the auction company is also too narrow for the crowd, merely aiming at people that are interested in Chinese painting and have intention to participate the buying of Chinese painting. It does not propagandize Chinese painting to more people who are unfamiliar with and do not comprehend Chinese painting. Hence, the market of Chinese painting cannot be expanded. For international trade, the sale of traditional Chinese painting is not dominant. China's export products with explicit comparative advantages are designs, handicrafts and musical instruments, and it has obvious disadvantages in sectors with high cultural creativity and knowledge content, such as audio-visual, visual art and publications [9]. Figure 2 shows the total export volume of Chinese products and the export volume of artworks. It can be seen from Fig. 2 that the total export volume is increasing year by year, but the export volume of artworks only accounts for a small part of the total export volume and fluctuates greatly.

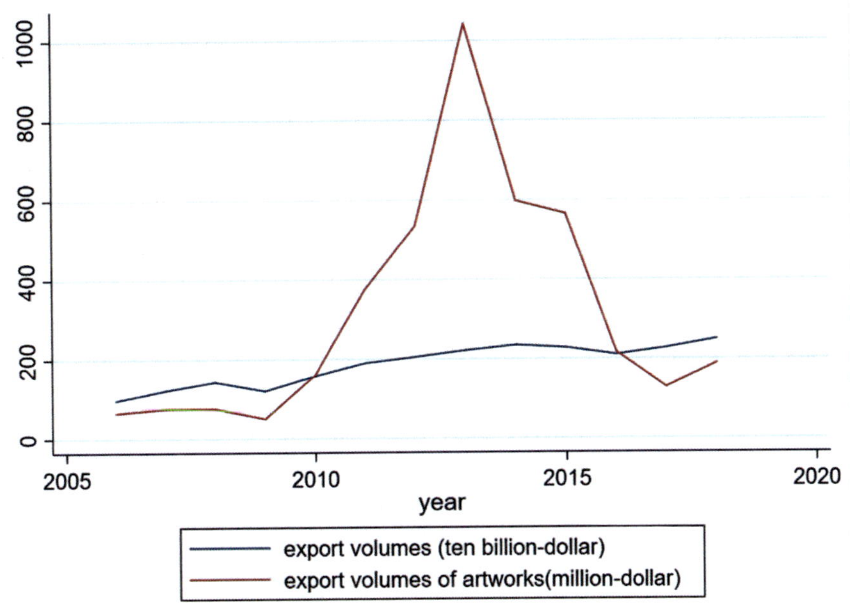

Fig. 2. Export volumes of China

Source: National Bureau of Statistics of China.

Difficult to preserve. One of the inherent disadvantages of the art collection industry is the problem related to preservation. Particular art collections require special ways of keeping them, and a small mistake may cause them to suffer a serious depreciation. The texture of Chinese paintings is rice paper. The properties of the paper are easy to be molded, decomposed in water, become ashes in the event of fire and bitten by insects and rats, so it is very difficult to preserve the classical precious paintings [2]. Also it is not easy for individual collectors to satisfy all the preservation conditions for different works, such as lighting and temperature requirements.

Low liquidity. The fourth weakness is that compared with securities and stocks, artworks have poor liquidity, inflexible trading time and flexible trading place, lack of guarantee measures and are prone to fraud [10].Generally speaking, when a collector buys fine art, he or she will not sell it quickly but in a few years, decades or even hundreds of years. He or she will wait for the appreciation of works of art and then flows the fine art into the auction market. Therefore, works of art are sometimes not equal to money. What is more, in the current market environment with irregular trading system, imperfect legal system and lack of integrity, sellers often have more information about art transactions, and art consumers are at a serious information disadvantage [11].

2.3 Opportunities

The development of e-commerce background. Since the birth of Guardian Online (auction company) in 2000, Chinese art e-commerce has experienced 20 years of development. Especially under the condition of Chinese art market into recession in 2012, art industry wants to obtain more development opportunities, via the art e-commerce, so the market of electric transaction rose 50% and has grown to 2.7% in the secondary market share; by 2014 the market trading amount has reached 4.5 billion yuan; Meanwhile, its share in the secondary market has increased to 6.7% [12]. Art e-commerce is a new mode of art trading on the Internet that has got rid of the traditional auction mode. There are many platforms at present, such as Bobao Art Network, HIHEY art network, Xianyu Auction and so on. The emergence of art e-commerce not only caters to the public's aesthetic demand and collection demand for artworks, but also helps contemporary artists to increase their personal popularity and enhance their market influence [13]. The bidirectional effect is the inherent advantage of the new model of online transactions. The linked attempt between the Chinese art industry and the internet will bring new development opportunities for the Chinese art market and even the world market [14].

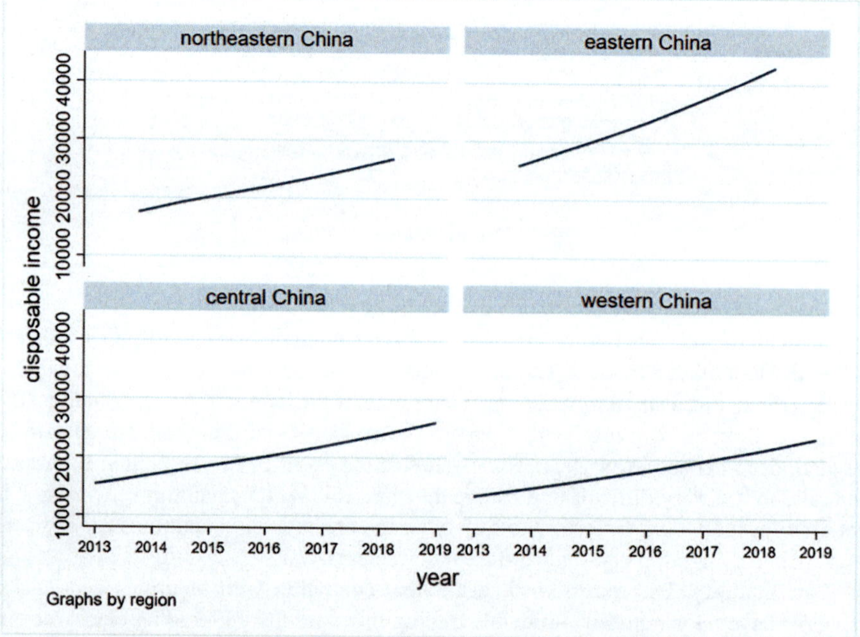

Fig. 3. Disposable Income Per Capita of Chinese residents (yuan)

Annotation: According to the division measure of the Chinese National Bureau of Statistics in 2011, the eastern region is divided into 10 provinces: Beijing, Hebei, Tianjin, Shandong, Shanghai, Jiangsu, Zhejiang, Guangdong, Hainan and Fujian; central region is divided into 6 provinces: Shanxi, Henan, Anhui, Jiangxi, Hubei and Hunan; the western region is divided into 12 provinces: the Xinjiang Uygur Autonomous Region, Gansu, Qinghai, Sichuan, Yunnan, Guangxi Zhuang Autonomous Region, Guizhou, Chongqing Municipality, Shaanxi, Tibet Autonomous Region, Inner Mongolia Autonomous Region and Ningxia Hui Autonomous Region; northeast China consists of 3 provinces: Liaoning, Jilin and Heilongjiang.

Source: National Bureau of Statistics of China.

The increase of public's income. Disposable income per capita is an important factor affecting the national consumption level. At present, China's disposable income per capita is increasing year by year, which has brought a great impact on the development of the Chinese art market [15]. Entering into the contemporary society, the public's demand for ceramic artworks and the scale of investment gradually increase with the gradual increase of income level; therefore, the ceramic art industry in JingDe town has ushered in a new period of development opportunities [16]. The same is for the demand for the traditional Chinese painting. At present, China's GDP per capita has reached 4,000 dollars, and the art market is ushering in a rare opportunity for development. As can be seen from Fig. 3, the disposable income per capita of the four regions in China is increasing gradually, including the western China which has the lowest disposable income per capita.

The government strongly promotes the development of the cultural industry. Under the trend of vigorous development of cultural industry, "cultural art + finance + Internet" has encountered unprecedented development opportunities [17]. As an important part of cultural industry, artworks can not only bring people the spiritual utility of culture, aesthetics and social status symbol, but also bring returns to investors in the form of assets [18]. The cultural characteristics of artworks themselves make the financialization of artworks not only driven by the art market and the financial market, but also in line with China's current requirement of vigorously developing cultural industry, which is supported by national policies [19]. For example, Wuhu cultural system was reformed, the leading group for cultural industry was established in Wuhu municipality, and it formulated the "suggestions about promoting the implementation of the great development and prosperity of socialist culture prosperity", "policies on accelerating the development of cultural industry", "the development plan of Wuhu cultural creative industry", "the relevant provisions about the promotion and support the development of cultural creative industry" and a series of supporting policies; adopting the way such as reward, discount and allowance to the digital animation, artworks production and other cultural industry projects can attract business investments, moderately promoting industrialization of "the Wuhu iron painting" [20].

The advancement of science and technology. Through the digital Chinese painting creation platform, Chinese painting is easy to copy, modify and save. First of all, the finished work can be reproduced indefinitely, unlike traditional Chinese painting that there is only one work in the end. In the process of creation, if there is a mistake, or a new layout is needed, the user only needs computer instructions to complete the modification work, but it is difficult to do in the traditional drawing part. The finished work created on the digital platform can be saved as a picture file, printed out or sprayed onto a scroll. Diversified preservation methods make the preservation of works last longer [2]. Although the platform still has some shortcomings, the overall advantages outweigh the disadvantages.

2.4 Threats

Low security. According to Jian Ma's book "The Economics of the Art Market", "the safety of works of art is crucial for both collectors and private collectors. Theft, fires and floods threaten the safety of works of art." Since the material of traditional Chinese painting is mainly rice paper, it will be easily damaged in the process of art trading, transportation and collection.

It is difficult to maintain the value. After the bold attempt of British railway pension funds, many of the newly established funds cannot turn art into an effective property category, and its reasons mainly lie in: to outsiders, the art market operation is full of mystery, and its lack of liquidity management also makes a lot of experienced funds managers take artworks as a threat to traditional funds so will not focus on [21]. In addition, Deloitte and ArtTactic jointly issued the "2016 art and financial report" of the research toward 53 private Banks, 14 family businesses, 126 professionals, 94 authoritative art collectors and 39 key leaders of art and financial enterprises, thinking that the lack of professional qualification standards, the lack of professional valuation personnel, difficulty in art valuation, low veracity became the greatest threat to the art market reputation, which to a certain extent, damaged the interests of the art managers, collectors and art professionals.

Copyright issues. The progress of science and technology is both an opportunity and a threat to the traditional Chinese painting industry. Unlike mechanical copying, downloading is a digital copy. Digital technology and 3D printing make the reproduction of artworks easily and directly threaten the copyright protection of artworks [22]. If pirated works are rampant, the public will take a skeptical attitude towards the Chinese painting market, which will lead to the decline of the reputation of the Chinese painting industry and the number of potential customers, thus seriously affecting the development of the Chinese painting industry.

3 Conclusion and Policy Proposals

According to SWOT analysis method, it can be seen from Table 2 that the strengths of Chinese painting industry are characterized by national characteristics, brand advantage, cost advantage and value preservation. Weaknesses include lack of personnel, weak marketing awareness, single marketing channel, high difficulty in preservation and poor liquidity. Opportunities consist under the background of e-commerce development, the improvement of people's income level, the vigorous development of cultural industry and the advancement of science and technology. The threats include low security, high difficulty of hedging and the copyright issue.

According to the strengths, weaknesses, opportunities and threats of the Chinese painting industry, the followings are some Suggestions to help the prosperity and development of the Chinese painting industry: (1) Public should be clear that the development of the traditional Chinese painting industry has disadvantages such as lack of personnel,

Table 2. SWOT strategy

SWOT	Content
strengths	national characteristics, brand advantage, cost advantage and value preservation
weaknesses	lack of personnel, weak marketing awareness, single marketing channel, high difficulty in preservation and poor liquidity
opportunities	under the background of e-commerce development, the improvement of people's income level, the vigorous development of cultural industry and the advancement of science and technology
threats	low security, high difficulty of hedging and the copyright issue

weak marketing awareness, single marketing channels, high difficulty in preservation and poor liquidity. It is also necessary to guard against threats caused by low security, high difficulty in maintaining value and the copyright problem. (2) Based on the background of e-commerce, utilizing media and internet finance to publicize the national characteristics of Chinese painting, spread national brands of Chinese painting, improve the popularity of painters, change passive marketing to active marketing, and sell Chinese painting products through multiple channels. (3) With the increase of national income, people should cultivate and publicize the value preservation characteristics of traditional Chinese painting, enhancing the national awareness that traditional Chinese painting can be used as a value preservation product and expanding the investment product market of traditional Chinese painting. (4) On account of the cost advantage of traditional Chinese painting, governments should guide relevant enterprises to increase the cost of materials on the mounting of Chinese painting to prevent theft, floods and fires and employ professionals to provide the authenticity of traditional Chinese painting identification, improving the mobility of traditional Chinese painting; in the process of transportation and collection, take preventive measures to avoid safety problems. 5. With the bonus of vigorous promotion of the development of the cultural industry by China, the government advocated to fundamentally support the traditional Chinese painting industry, such as raising the salaries of professionals and appraisers of the traditional Chinese painting industry and increasing the job supply associated with Chinese painting industry, encouraging the establishment of high-quality Chinese painting cram schools or extra-curricular activities to cultivate talented Chinese painting talents since childhood, legislating to limit infringement, crack down on piracy, and maintain the artistic value of works of art.

References

1. Sun, Z.J., Li, X.: Empirical research on spatial spillover effect and convergence form of cultural industry agglomeration. Chin. Soft Sci. **8**, 173–183 (2015)
2. Dai, H.H.: Design and implementation of a learning and creation platform for Chinese classical paintings based on WPF technology. Beijing University of Technology (2013)

3. Hua, J.: Development strategy of chinese art industry-towards the international vision of the "13th Five-Year plan" and China's path. J. Shanghai univ. Financ. Econ. **17**(05), 57–70 (2015)
4. Cheng, X.M.: Analysis on the impact of macroeconomic cycle on the price of art market. Yunnan University of Finance and Economics (2014)
5. Luo, C.R.: Contextual expression of loneliness and meditation. Hubei Institute of Fine Arts (2018)
6. Feng, X.S.: Study on marketing Strategy of TB Company's Traditional Chinese painting Artworks in Jilin district. Jilin University (2017)
7. Shi, S.: Two major contradictions in the transformation of modern Chinese landscape painting. Jiangnan University (2017)
8. Tang, X.X.: Research on world outlook setting of modern animation based on chinese mythological elements. Sichuan Normal University (2015)
9. Fang, Y., Yue, S.J.: Cultural trade between China and BRICS: comparative advantage and cooperation potential. Fujian Forum (Humanities and Social Sciences) **2**, 70–79 (2019)
10. Chen, L.Q.: Analysis on industrialization Trend of art collection industry. Market Modernization **23**, 242–243 (2015)
11. Wu, Y.P.: Research on the rights protection of art consumers. Yunnan University of Finance and Economics (2017)
12. Cai, S.H.: Development opportunity of art finance in china under the background of "Internet +." J. Fujian Financ. Manage. Cadre Instit. **01**, 19–23 (2019)
13. Chen, M.Y.: Research on online Trading of calligraphy and painting artworks under the background of "Internet +". Jiangsu University (2019)
14. Zhao, X.: Comparison between international art Market and Chinese art Market, pp. 236–241. Yunnan University of the Arts, Discipline Construction and Cultural Creativity (2015)
15. Gao, Y.: Research on the financialization of Chinese art. Xun Lu Academy of Fine Arts (2019)
16. Xiong, X.H.: Evaluation and strategy research on the development of financialization of Jingdezhen ceramic art market. Jingde Town Ceramic University (2016)
17. Li, Y.T.: 2015–2016 annual observation report on the status of cultural and creative industry. J. Henan Instit. Educ. (philosophy and social sciences edition) **36**(04), 38–43 (2017)
18. Shi, Y., Li, Y.: Chinese art investment returns – overall characteristics and masterpiece effect. Financ. Res. **12**, 194–206 (2013)
19. Xu, R.Z.: Research on the development status and problems of financialization of Chinese art. Southwest University of Finance and Economics (2013)
20. Ding, M.Y.: Research on the Industrialization development of Intangible Cultural Heritage. Anhui Normal University (2014)
21. Xu, J.X.: Comparative study of contemporary Chinese and western art and financial development. Shanghai University (2018)
22. Xu, W.G.: A New method of artistic creation. Chinese Art **03**, 52 (2016)

How Autarky and Specialization are Chosen Under COVID-19

Sheng Pan(✉)

University of Sheffield, Sheffield, South Yorkshire, UK

Abstract. With the spread of COVID-19, interactions between different countries are limited such as imports and exports. As a result, those industries that used to relied heavily on imports have been hit hard in some countries. For instance, at the first lockdown in some European countries, the shortage of medical supplies caused more unnecessary deaths. And the panic of closure forced people to buy a lot of necessities which lead to demand exceeds supply. Hence, those governments, such as France, are starting to reconsider whether to autarky in some industries. This paper is going to talk about whether a country should specialize or not under the situation as COVID-19, according to the actions from other countries.

Keywords: COVID-19 · Autarky · Specialization

1 Introduction

At the end of 2019, with the discovery of COVID-19, China decided to shut down Wuhan to contain the outbreak. As it continues to spread, many governments, such as Italy and the United Kingdom, have restricted unnecessary social contact to slow the spread of the virus.

As a result, the worldwide economy is suffering from a tremendous amount of pressure. Many countries rely heavily on imported goods for some industries, such as manufacturing, and due to the closing of many cities, many factories are prohibited opening during the COVID-19 pandemic.

As the disruption of the supply chain, adding pressure on the price. According to PR Newswire (2020), China has almost 85% of the total value of components used in smartphones at 2019 and it owns majority of important components in many industries. And the research shows those components' total value has been rose by about 3% by Chinese vendors in US during pandemic period. As a result, the price of those goods will have to increase in order to maintain the profits.

Moreover, with the navigation ban and logistics limited, import and export business becomes harder to achieve during the crisis period. It seems in some special situation which has difficulties in trade, specialization or autarky, should be reconsidered.

Although there are many benefits associated with free trade traditionally. For example, free trade encourages countries to specialize when they have a comparative advantage. Also, low tariffs and comparative advantage can increase exports, which could

C. Yuan et al. (Eds.): *Proceedings of the 4th International Conference on Economic Management and Green Development*, AEPS, pp. 183–193, 2021.
https://doi.org/10.1007/978-981-16-5359-9_22

boost GDP growth. And with more trade, the business competition will become more intense, curbing the formation of domestic monopolies, which might charge high prices. At the same time, the shortage of medical supplies happened among the COVID-19 in different countries like France. This shift in European thinking, the German Chancellor Angela Merkel said those things they must have should be developed in a very targeted manner. Hence, autarky might be mentioned more in the future.

Therefore, using our model can decide whether choose specialization or autarky under COVID-19. And when the probability of trade goes down, government would prefer to use resources to build a fully independent economy, which produces and sells all goods within this economy. As a result, countries can no longer pay most attention to the industries they are good at, and the resources need to be evenly distributed as people need that makes industrial diversification. Autarky not only can prevent the recession caused by COVID-19. For instance, the June 2020 Global Economic Prospects indicates there could be 5.2% drop in global GDP (World Bank 2020), which is the deepest recession in decades can be prevented under autarky. It also can protect domestic enterprises from the monopolization of foreign oligopolies. Hence, those domestic company can compete fairly. Also, it can rise the employment rate due to the decreasing of foreign companies, that can reduce the occupation of existing jobs by foreigners, and increasing of jobs for industrial diversification.

2 Literature Review

Irwin (2015) points out in 'Free Trade Under Fire' that free trade is always controversial. Under the free trade, the investment would increase due to investment choices are not limited to specific places; different choice means better allocation, which can reduce the risk of investment.

Moreover, free trade brings specialization across worldwide. It allows workers or firms to do what they are good at. According to The Wealth of Nations, Adam Smith (1776) thinks if one foreign country could offer higher quality goods and lower prices, then it is better to purchase them by some part of our industrial products. And then the efficiency of working can be improved. However, Irwin also mentioned that specialization would lead to many job losses. It finally raises the unemployment rate, even the pressure of deflation.

With limited resources within the country, producers have to choose what to produce to satisfy their people, therefore as David Ricardo (1817) said in 'On the Principles of Political Economy and Taxation', under a system of perfectly free commerce, each country naturally devotes its capital and labor to such employments as are most beneficial to each. This pursuit of individual advantage is admirably connected with the universal good of the whole". And when countries focus on producing those goods they are good at, which is called specialization. France has always been famous for its fashion industry, and many top clothing brands were born there. And New Zealand is well-known for its excellent agricultural products such as milk and wool. Whereas, specialization also means countries have to give up some industries which they are not good at. This might cause some problems such as shortage during in disaster. Therefore, there should be an equilibrium between specialization and autarky.

The Heckscher-Ohlin model is a model that finds the equilibrium of trade with different specialties and resources in two countries. It describes how a country should work and trade when they have limited natural resources through mathematical justification. One crucial target for a country is to make the economy flourish. And specialization allows workers to do the thing they are good at; therefore, this can reduce the waste of time and resources, which can be seen as potential outputs. Once countries have trade with others, they can sell the things they are good at and seek those they lack. Finally, the GDP goes up.

Whereas, if the trade cannot proceed normally, those countries are too dependent on international trade so they would be suffering shortages of goods which they choose not or less productive. As a result, this article is going to use the Heckscher-Ohlin model and think about specialization or autarky, which is a better choice under a trade or no trade condition.

3 Model Set up

Our model is based on the HO model, which finds the equilibrium between consumer's utility and Production-possibility frontier and compares them under different conditions, specialization, or not specialization. So we can get the best choice under game theory.

In addition, all numbers in the model are **assuming**.

3.1 Period 2: The HO Model, Taking the Production Technology as Given

Assuming there are two goods, c_1 and c_2, are supplied in countries A and B. In addition, both country and same limited resources. Therefore, the original endowment when they specialization in country A is $Y_1 = 1$, $Y_2 = 5$, and Y1 = 5, Y2 = 1 in country B. While in not specialization case, $c_1 = c_2 = \sqrt{}5.5$ at both countries. In this period, both countries have the **same decision** about whether specialization.

The consumers: Each country has a representative consumer, they have a utility of the form U, which is Eq. (2), and their budget constraint is Eq. (1).

$$\mathrm{c1} \times \mathrm{p1} + \mathrm{c2} \times \mathrm{p2} = \mathrm{W} \tag{1}$$

$$\mathrm{U} = \mathrm{c1} \times \mathrm{c2} \tag{2}$$

P is the price for goods, and W represents wage. Equation (3) follows the Lagrangian method maximize problem.

$$\Lambda = c_1 \times c_2 - \lambda \times (W - c_1 \times p_1 - c_2 \times p_2) \tag{3}$$

And after the first-order condition was taken, we finally get

$$\frac{c_1}{c_2} = \frac{p_2}{p_1} \tag{4}$$

Λ is the Lagrangian multiplier. Therefore, Eq. (4) states the profits of goods c_1 and c_2 are the same when they in one country due to $c_1 \times p_1 = c_2 \times p_2$.

$$c_1 = \frac{W}{2 \times p_1}, c_2 = \frac{W}{2 \times p_2} \tag{5}$$

When Eq. (4) is substituted into Eq. (1), which is budget constraints. Hence, $2 \times c_1 \times p_1 = W$, so that Eq. (5) can be found.

Equilibrium. If both countries decide **not to trade** with each other, then as we assuming $c_1^A = 1, c_2^A = 5$ and $c_1^B = 5, c_2^B = 1$. Hence, the price can be described as Eq. (6) and (7)

$$\frac{p_2^A}{p_1^A} = \frac{c_1^A}{c_2^A} = \frac{1}{5} \tag{6}$$

$$\frac{p_2^B}{p_1^B} = \frac{c_1^B}{c_2^B} = \frac{5}{1} \tag{7}$$

Also, if they choose to **trade**, we can have the market clearing level:

$$c_1^A + c_1^B = 6, c_2^A + c_2^B = 6 \tag{8}$$

$$c_2 = -c_1 + 6 \tag{9}$$

$$U' = -\frac{c_2}{c_1} \tag{10}$$

And the supply function will be Eq. (8), and then to maximize the utility, Eq. (10) shows the First order condition, the gradient of utility curve's tangent should be the same as market clearing level which is -1, hence:

$$\frac{c_1}{c_2} = 1 = \frac{p_2}{p_1} \tag{11}$$

$$c_1^A = c_1^B = 3 \tag{12}$$

$$c_2^A = c_2^B = 3 \tag{13}$$

Therefore,$c_1 = c_2 = \frac{W}{2 \times p_1} = \frac{W}{2 \times p_2}$. As a result, $c_1^A = c_1^B = c_2^A = c_2^B = 3$ and $p_1 = p_2$, when two country are trading.

$$c_1 = c_2 = \sqrt{5.5} \tag{14}$$

$$p_1 = p_2 \tag{15}$$

Also, country A and B might choose **not to specialize and not trade**, therefore, $c_1 = c_2 = \sqrt{5.5}$, and Price can be inferred from $\frac{W}{2 \times p_1} = \frac{W}{2 \times p_2} = \sqrt{5.5}, p_1 = p_2$.

3.2 Period 1: Choice of Production Technology

This time, one country makes a decision which is not specialization. Therefore, if country A not specialize, the endowments are:

$$\text{Country A}: Y_1 = Y_2 = \sqrt{5.5},$$

$$\text{Country B}: Y_1 = 5, \ Y_2 = 1.$$

Equation (1) states the budget constraint. Therefore, from Eq. (5), we can get those two goods in different countries in the details below:

$$W_A = p_1 \times \sqrt{5.5} + p_2 \times \sqrt{5.5}, W_B = p_1 \times \sqrt{5.5} + p_2 \times \sqrt{5.5} \tag{16}$$

$$c_1^A = \frac{W_A}{2 \times p_1}, c_2^A = \frac{W_A}{2 \times p_2}, c_1^B = \frac{W_B}{2 \times p_1}, c_2^B = \frac{W_B}{2 \times p_2} \tag{17}$$

Equilibrium. The market-clearing conditions are when they trade:

$$c_1^A + c_1^B = \sqrt{5.5} + 5, c_2^A + c_2^B = \sqrt{5.5} + 1 \tag{18}$$

Assume p_1 equals to 1, then:

$$\frac{W_A}{2 \times p_1} = \frac{W_B}{2 \times p_1} = \sqrt{5.5} + 5 \tag{19}$$

$$\frac{p_2}{p_1} = \frac{\sqrt{5.5} + 5}{\sqrt{5.5} + 1} \tag{20}$$

As a result, solving the Eq. (19), which p_1 equals 1, we can have p_1 to p_2 is (20).

$$c_1^A = \frac{p_1 \times \sqrt{5.5} + p_2 \times \sqrt{5.5}}{2 \times p_1} \tag{21}$$

$$c_1^A = 3.75(2\,decimal) \tag{22}$$

Putting Eq. (17) into c_1^A, then substitute (20) in it, the exact c_1^A can be found, and the same way for c_2^A.

$$c_2^A = 1.71(2\,decimal) \tag{23}$$

Finally, with the value of c_1^A andc_2^A, go back to Eq. (16), hence, c_1^B equal to $\sqrt{5.5} + 5$ minusc_1^A, also for the c_2^B:

$$c_1^B = 3.60(2\,decimal), c_2^B = 1.64(2\,decimal) \tag{24}$$

Whereas, if they decide not to trade, they can only have the endowments as:

$$\text{Country A}: Y_1 = Y_2 = \sqrt{5.5}$$

$$\text{Country B}: Y_1 = 5, \ Y_2 = 1$$

3.3 Comparison

The total expected utility of both countries is the combination of utility that when they like to trade and not trade under the situation, which is specialization or not specialization. P is the probability they might choose to trade.

$$E(U) = p \times (c_1^T \times c_2^T) + (1 - p) \times (c_1^{NT} + c_2^{NT}) \quad (25)$$

As a result, we can get a table below shows all different utilities as a prisoner's dilemma type.

	B specialization	B not specialization
A specialization	$E(U_A) = p \times 9 + (1 - p) \times 5$ $E(U_B) = p \times 9 + (1 - p) \times 5$	$E(U_A) = p \times 5.9 + (1 - p) \times 5$ $E(U_B) = p \times 6.41 + (1 - p) \times 5.5$
A not specialization	$E(U_A) = p \times 6.41 + (1 - p) \times 5.5$ $E(U_B) = p \times 5.9 + (1 - p) \times 5$	$E(U_A) = 5.5$ $E(U_B) = 5.5$

Fig. 1. Game theory

To compare these results, we can find Eq. (26), which states when B country tends to specialization. Country A will choose to specialization if **p is greater than 0.16**.

$$p \times 9 + (1 - p) \times 5 > p \times 6.41 + (1 - p) \times 5.5 (p > 0.16) \quad (26)$$

However, when **p is smaller than 0.56**, which means the probability of trade is less than 0.56, then it shows it is better not to specialization for country A when B is not specialization.

$$p \times (5.9) + (1 - p) \times 5 > 5.5 \, (p > 0.56) \quad (27)$$

Therefore, there are three cases from above, [0, 0.16), [0.16, 0.56] and (0.56, 1]. And for each interval, different Nash equilibrium can be found.

For case 1, when p-value is smaller than 0.16, then both countries finally prefer not to specialize according to Eq. (26).

In case 2, if $0.16 < p < 0.56$, country A or B would have the same decision as to the other one. For example, when A chooses to specialize, B would specialize as well.

And once p is greater than 0.56, the Nash equilibrium is to specialize for both countries through Eq. (27).

Also, the social planner should specialize when p is above 0.125 normally. What they actually do, whereas, is when p is higher than 0.16.

4 Conclusion

Additionally, there are some limitations in this model. The original endowment in both countries are not the data from the real world due to we cannot get the numbers without any specialization and trade. And the productivity may change during the period 'work from home' which also might affect the predictions in the model.

Undoubtedly, as any disease that human have conquered, the Covid-19 will be overcome. But the effects of the it will not disappear with the virus. The impact of it will depend on whether people believe it will last or happen again which is the probability in the model we built.

A Mathematical Appendix

Here we present the details of the results that appeared in the text due to the text part show the idea, and we will calculate more in detail here.

Period 2

$$c_1 \times p_1 + c_2 \times p_2 = W.$$

$$U = c_1 \times c_2.$$

We have a budget constrain and a utility function. Therefore, we can use lagrangian to solve it.

$$\Lambda = c_1 \times c_2 - \lambda(W - c_1 \times p_1 - c_2 \times p_2).$$

When we have the first derivative, we can get:

$$\frac{\partial \Lambda}{\partial c_1} = c_2 - p_1 \times \lambda = 0$$

$$\frac{\partial \Lambda}{\partial c_2} = c_1 - p_2 \times \lambda = 0$$

So we have:

$$c_2 = p_1 \times \lambda \text{ and } c_1 = p_2 \times \lambda.$$

As a result,

$$\frac{c_1}{c_2} = \frac{p_2}{p_1}$$

$$c_1 \times p_1 = c_2 \times p_2$$

substitute it into budget constrain function,

$$2 \times c_1 p_1 = W = 2 \times c_2 p_2 = W$$

$$c_1 = \frac{\mathrm{W}}{2 \times p_1} \text{ and } c_2 = \frac{\mathrm{W}}{2 \times p_2}$$

Equilibrium

In the period 2, both countries have the same choice about specialization or not. Hence, the endowments in period 2 are $Y_1 = 1, Y_2 = 5$ in country A and $Y_1 = 5, Y_2 = 1$ in country B. If they choose to specialization but not trade. So due to $\frac{c_1}{c_2} = \frac{p_2}{p_1}$, we can get:

$$\frac{p_2^A}{p_1^A} = \frac{c_1^A}{c_2^A} = \frac{1}{5}$$

$$\frac{p_2^B}{p_1^B} = \frac{c_1^B}{c_2^B} = \frac{5}{1}$$

Whereas if they choose to trade, then

$$c_1^A + c_1^B = 6,\ c_2^A + c_2^B = 6$$

And the demand equation as a function of c_1 is:

$$c_2 = -c_1 + 6$$

The equilibrium when they trade is when the gradient of the utility curve's tangent is the same as the gradient of market-clearing level:

$$-1 = \frac{dc_1}{dc_2} = -\frac{c_2}{c_1}$$

$$c_1 = c_2$$

Because of $\frac{c_1}{c_2} = \frac{p_2}{p_1}$, as a result,

$$p_1 = p_2$$

and

$$c_1^A = c_1^B = c_2^A = c_2^B = \frac{W}{2 \times p_1} = \frac{W}{2 \times p_2} = 3$$

If they both decide to not specialization, when they trade:

$$C_1^A + C_1^B = 2\sqrt{5.5},\ C_2^A + C_2^B = 2\sqrt{5.5}$$

$$W_A = W_B = \sqrt{5.5} \times p_1 + \sqrt{5.5} \times p_2$$

substitute $c_1 = \frac{W}{2 \times p_1}$ and $c_2 = \frac{W}{2 \times p_2}$ into it:

$$\frac{W_A}{2 \times p_1} + \frac{W_B}{2 \times p_2} = 2\sqrt{5.5}$$

Also, as we know $W_A = W_B$, then it becomes to

$$\frac{W_A}{2 \times p_1} + \frac{W_A}{2 \times p_2} = 2\sqrt{5.5}$$

$$W_A = 2\sqrt{5.5} \times P_2$$

$$\sqrt{5.5} \times p_1 + \sqrt{5.5} \times p_2 = 2\sqrt{5.5} \times P_2$$

$$p_1 = p_2$$

when they not trade, $c_1 = c_2 = \sqrt{5.5}$, so

$$\frac{c_1}{c_2} = \frac{p_2}{p_1} = 1$$

Period 1

In this period, one country would prefer to specialization, and the other would not. Therefore, if country A decides to not specialization, the endowments are:

$$Country\, A : Y_1 = Y_2 = \sqrt{5.5}$$

$$Country\, B : Y_1 = 5, Y_2 = 1$$

Substitute them into the budget constrain. We can get $\sqrt{W_A} = p_1 \times \sqrt{5.5} + p_2 \times \sqrt{5.5}$ and $W_B = 5 \times p_1 + 1 \times p_2$.

Equilibrium

When they trade, the market-clearing conditions are:

$$c_1^A + c_1^B = \sqrt{5.5} + 5, c_2^A + c_2^B = \sqrt{5.5} + 1$$

As we already know $c_1 = \frac{W}{2 \times p_1}$ and $c_2 = \frac{W}{2 \times p_2}$, put these into the market clearing conditions and assume p_1 equals to 1:

$$\frac{W_A}{2 \times p_1} + \frac{W_B}{2 \times p_1} = \sqrt{5.5} + 5$$

$$\frac{\sqrt{5.5} \times p_1 + \sqrt{5.5} \times p_2}{2 \times p_1} + \frac{5 \times p_1 + 1 \times p_2}{2 \times p_1} = 2\sqrt{5.5}$$

$$\frac{p_2}{p_1} = \frac{\sqrt{5.5} + 5}{\sqrt{5.5} + 1}$$

Hence, we can get c_1 and c_2 if we substitute $\frac{p_2}{p_1}$ into $c_1 = \frac{W}{2\times p_1}, c_2 = \frac{W}{2\times p_2}$.

$$c_1^A = \frac{\sqrt{5.5}\times p_1 + \sqrt{5.5}\times p_2}{2\times p_1}$$

$$c_1^A = \frac{\sqrt{5.5}}{2} + \frac{\sqrt{5.5}}{2}\times\frac{p_2}{p_1}$$

$$c_1^A = 3.75(2\ decimal)$$

Also, use the same way, we can get c_2^A

$$c_2^A = \frac{\sqrt{5.5}\times p_1 + \sqrt{5.5}\times p_2}{2\times p_2}$$

$$c_2^A = \frac{\sqrt{5.5}}{2}\times\frac{p_1}{p_2} + \frac{\sqrt{5.5}}{2}$$

$$c_2^A = 1.71(2\ decimal)$$

Due to $c_1^A + c_1^B = \sqrt{5.5} + 5\ and\ c_2^A + c_2^B = \sqrt{5.5} + 1$, we can find c_1^B and c_2^B:

$$c_1^B = \sqrt{5.5} + 5 - c_1^A,\ c_2^B = \sqrt{5.5} + 1 - c_2^A$$

$$c_1^B = 3.60(2\ decimal)\ and\ c_2^B = 1.64(2\ decimal)$$

Comparison

From the equation $p\times 9 + (1-p)\times 5 > p\times 6.41 + (1-p)\times 5.5(p > 0.16)$ and $p\times 5.9 + (1-p)\times 5 > 5.5(p > 0.56)$, we can simply get three intervals, [0,0.16], (0.16,0.56) and [0.56,1].

When p¡0.16, then according to $p\times 9 + (1-p)\times 5 > p\times 6.41 + (1-p)\times 5.5(p > 0.16)$, neither country B specialize or not, country A would choose not to specialize. And vice versa. As a result, the Nash equilibrium is they both not specialization.

In case 2, (0.16, 0.56). Both country would have a same decision due to $p\times 5.9 + (1-p)\times 5 < 5.5(p < 0.56)$, therefore, if A not specialization, B would do it as well. and when A decide to specialize, B should follow this action as $p\times 9 + (1-p)\times 5 > p\times 6.41 + (1-p)\times 5.5(p > 0.16)$.

And if p is greater than 0.56, then $p\times 5.9 + (1-p)\times 5 > 5.5(p > 0.56)$ and $p\times 9 + (1-p)\times 5 > p\times 6.41 + (1-p)\times 5.5(p > 0.56)$. Hence, the benefits of specialization will always higher than not to specialize. Finally, the Nash equilibrium is they both specialize.

In addition, if the probability of trade is higher than 0.125, which means the benefit of specialization is bigger than when both countries do not specialize (i.e., $p\times 9 + (1-p)\times 5 > 5.5$), the social planner should specialize. However, according to the Game Theory, they actually specialize when p is above 0.16.

References

Prnewswire.com: Impact Of COVID-19 On The global manufacturing industry (2020) https://www.prnewswire.com/news-releases/impact-of-covid-19-on-the-global-manufacturing-industry-2020-301042150.html. Accessed 4 May 2020

World Bank: The global economic outlook during the COVID-19 pandemic: a changed world (2020). https://www.worldbank.org/en/news/feature/2020/06/08/the-global-economic-outlook-during-the-covid-19-pandemic-a-changed-world. Accessed 17 Dec. 2020

Irwin, D.: Free Trade Under Fire 4E. Princeton University Press, New Jersey (2015)

Smith, A., Cannan, E.: The Wealth Of Nations, 5th edn. MethuenCo. Ltd, p.Chapter II, London (1904)

Ricardo, D.: The Principles Of Political Economy And Taxation, 3rd edn. John Murray, London (1821)

Relationship Between Emotional Intelligence and Subjective Economic Well-Being

E. A. Sergienko[1], E. A. Khlevnaya[2], T. S. Kiseleva[3], E. I. Osipenko[4](✉), and A. A. Nikitina[4]

[1] Institute of Psychology, Russian Academy of Sciences, Moscow, Russia
[2] Department of Financial Management, Plekhanov Russian U of Economics, Moscow, Russia
[3] International Centre Creative Technologies of Consulting, Moscow, Russia
kiseleva@mc-ktk.ru
[4] Emotional Intelligence Lab, Skolkovo, Moscow, Russia
{osipenko,nikitina}@eilab.ru

Abstract. The objective of this study was to investigate the relationship between ability emotional intelligence (EI) and subjective economic well-being (SEW). It was hypothesized that ability EI positively correlates with SEW. EI was measured with the Russian-language Emotional Intelligence Test (EIT), which is conceptually based on MSCEIT. SEW was measured with the Subjective Economic Well-being Questionnaire. The sample consisted of 243 Russian working adults (61 men, 182 women, mean age = 36.1, aged between 18 and 67 years). The results indicated that higher levels of emotional intelligence statistically significantly correlated with higher levels of subjective economic well-being. In particular, the ability to use emotions for problem solving, the ability to manage emotions, and the experiential domain of EI were key emotional intelligence factors associated with SEW. This finding indicates the importance of the effect of EI on people's subjective economic well-being for their overall well-being.

Keywords: Emotional intelligence · Ability · Subjective economic well-being · EIT

1 Introduction

Emotional intelligence (EI) can be defined and conceptualized as an ability consisting of 4 branches: identifying, using, understanding, and managing emotions in oneself and others [1], and can be measured with testing methods [2].

Research of the last couple of decades shows that EI is associated with various indicators indirectly related to the performance and psychological well-being of a person. For example, the higher the level of EI, the higher the productivity in the workplace [3]. Employees with a high level of EI have higher levels of job satisfaction, loyalty to the organization and show lower turnover [4]. The higher a person's EI level in college, the higher is their salary level 10–12 years after starting a career [5]. EI is

C. Yuan et al. (Eds.): *Proceedings of the 4th International Conference on Economic Management and Green Development*, AEPS, pp. 194–200, 2021.
https://doi.org/10.1007/978-981-16-5359-9_23

also associated with effective macro-level decision-making [6] and team effectiveness in high-cost management environments [7].

Moreover, EI is positively associated with well-being, coping skills, competence, and negatively associated with perceived stress, which suggests that heightened feelings of control and emotional competence help people adopt proactive and effective coping strategies, which, in turn, increases their subjective well-being [8]. Researchers also find empirical confirmation of a positive relationship between EI and the components of psychological well-being - self-esteem, life satisfaction and self-acceptance [9].

Based on the findings of previous research on the importance of EI in well-being, in our work we investigate the relationship of EI with indicators of subjective economic well-being of a person. The category of subjective economic well-being (SEW) is not one of the deeply developed scientific concepts. This is largely due to the fact that for a long time the very concept of "economic well-being" was considered the prerogative of only economic science, where it is viewed as synonymous with the objective characteristics of wealth, material, financial, and social conditions of human life in general [10]. The term "subjective economic well-being" entered the parlance of psychological science thanks to the work of Strümpel [11] and Campbell [12]. SEW is very often considered as an aspect of the general well-being of a person directly related to money and material means, and as one of the main substructures of the quality of life [13].

Previous research suggests that economic well-being largely explains differences in overall life satisfaction. Moreover, SEW may be a more accurate indicator of a person's perception of their economic well-being than objective values such as GDP per capita [14].

In our work, SEW is defined as an integral psychological indicator that expresses a person's attitude to her current and future material well-being. SEW is an essential element of a person's subjective well-being. This means that SEW acts, on the one hand, as a constituent part, a component of general subjective well-being (or satisfaction with life), and on the other, as an independent factor of its determination, a source of a person's general well-being, the value of which changes at different periods of her life [15].

At the moment, there is not enough research data considering the relationship between EI and SEW. At the same time, both SEW and EI are important indicators and possible predictors of the general well-being of a person and society as a whole.

Thus, the novelty of the proposed study consists in the analysis of the role of EI in the SEW of a person, as an indicator of her psychological health and adaptability to modern living conditions. In addition, this study sheds light on the psychological factors of economic well-being and economic behavior of people that make up the human resources of a society's economy. We believe that the level of EI is interconnected with the SEW indicators of a person, while the degree and specificity of these relationships remain poorly understood.

2 Methods

The study was conducted online during the summer of 2020. The sample consisted of 243 Russian adults (61 men and 182 women, all employed) aged 18 to 67 years (mean age = 36.1 years), who were randomly invited to participate in the study via social media.

2.1 Measuring Emotional Intelligence

As a method for measuring emotional intelligence, we used the EIT (Emotional Intelligence Test) [16], a Russian methodology based on the MSCEIT (Mayer-Salovey-Caruso Emotional Intelligence Test) model with a number of reasonable changes for the Russian sample. The EIT methodology allows to determine the general level of development of EI, as well as the level of development of its 4 components: identification of emotions, use of emotions in problem solving, understanding of emotions, and conscious management of emotions. EIT has a high degree of reliability (Alpha Cronbach of general level of EI = 0.93), structural and factorial validity ($\chi 2 = 39.87$, RMSEA = 0.027 (0.001; 0.044), CFI = 0.99), and the theoretical validity of the method is confirmed by rather high correlations of indicators of the EIT methodology and the MSCEIT methodology (at the level of generalizing indicators rs = 0.51).

2.2 Measuring Subjective Economic Well-Being

To measure SEW, the "Subjective economic well-being" method was used, which was developed by V.A. Khashchenko [17]. The methodology includes 22 questions with suggested answer options. All items of the questionnaire were tested for discrimination - the ability of individual items on the questionnaire to differentiate participants with respect to the minimum and maximum measurement results.

The method measures five individual factors, as well as a general SEW index:

- The first factor - the economic optimism/pessimism index (EOI) reflects an optimistic or pessimistic assessment of external and internal conditions for the growth of material well-being.
- The second factor - the economic anxiety index (EAI) is an assessment of the severity of negative emotional states in connection with financial and material problems.
- The third factor - the index of the subjective adequacy of income (IAI) is an assessment of the accordance of the amount of income with the requests and needs of the individual.
- The fourth factor - the financial deprivation index (FDI) reflects the degree of wealth or lack of financial resources.
- The fifth factor - the index of current family well-being (IFW) reflects subjective assessments of the family's financial situation.

The integral (general) SEW index shows the level of a person's satisfaction with their material and economic indicators - an integral psychological indicator that expresses a person's attitude to the material aspects of life and her life position in the sphere of material aspirations and consumption. The questionnaire has been experimentally tested for reliability and design validity. Reliability coefficients were calculated separately for each scale and for all items of the questionnaire as a whole. All the obtained coefficients turned out to be high, which indicates the internal consistency of the items. The coefficient of reliability of all items of the questionnaire was acceptable and at the level of 0.73.

3 Results

Descriptive statistics and test results for normality of the EIT by section, branch, domain and total EI score are presented in Table 1. These data show that the distribution has negative skewness for almost all scales. That is, according to the data obtained, there is a small number of respondents who received a very low score, while symmetrically high scores are not found in the sample. This may be due to the fact that the scale is more sensitive at the low end and not sufficiently sensitive at the high end. Checking for normality due to this asymmetry shows a significant difference from the normal distribution on all scales, except for section 9b.

Descriptive statistics and the results of testing for normality on the scales of the SEW methodology are presented in Table 2. The distribution of points on the scales of this method differ from normal (at the level of $p < 0.05$) in all cases, except for the integral index. At the same time, for the EOI, IFW, and FDI scales, there is a negative, left-sided asymmetry, and for the IAI and EAI scales, there is a positive one (right-sided).

Spearman's rank correlation coefficient was used to assess the relationships between the emotional intelligence and subjective economic well-being variables. This is due to the fact that the distribution of many of the variables included in the analysis differed from normal.

The correlations between the results of the EIT and the scales of the SEW method are shown in Table 3. As can be seen from the table, there are quite a few significant positive correlations between the results of the two methods, but all of them are rather weak (from 0.130 to 0.283). Branch 1 significantly positively correlates with the EOI and FDI scales. Branch 2 - also with the EOI and FDI scales, as well as with the general index. Branch 3 - only with the FDI scale, and branch 4 - with the EOI, IFW, EAI scales and the general index. The experiential domain of EI significantly positively correlates with the EOI and FDI scales, as well as with the general index. The strategic domain - with the FDI scale and the general index. Finally, the overall EI score is significantly positively correlated with the EOI, IFW, EAI scales and the general index.

4 Discussion and Conclusion

The results of the statistical analysis showed that EI and subjective economic well-being are positively correlated on many indicators. In particular, the subjective degree of sufficiency of financial resources, the subjective optimistic assessment of external and internal conditions for the growth of material well-being, as well as the general SEW index had a large number of positive correlations with EI abilities.

Specifically, people with higher levels of emotion identification, emotion use, emotion understanding, experiential and strategic domains, and general EI had higher subjective levels of financial wealth. People with higher levels of emotion identification, emotion use, emotion management, experiential domain, and general EI levels had a more subjectively optimistic assessment of the external and internal conditions for increasing their material well-being. And people with higher levels of the ability to use emotions, emotion management, experiential and strategic domains, and general EI had a higher integral SEW index.

Table 1. Descriptive statistics and test results for normality for the EIT methodology.

Scale	Average (st. dev)	95% confidence interval of the mean	Asymmetry	Shapiro-Wilks test significance (test for normality)
Section 1	0.362 (0.050)	[0.356 0.369]	– 0.773	$p < 0.001$
Section 2	0.448 (0.061)	[0.441 0.456]	– 0.958	$p < 0.001$
Section 3	0.453 (0.083)	[0.443 0.464]	– 0.344	$p = 0.003$
Section 4	0.358 (0.049)	[0.351 0.364]	– 0.613	$p < 0.001$
Section 5	0.291 (0.040)	[0.285 0.296]	– 0.743	$p < 0.001$
Section 6	0.371 (0.063)	[0.363 0.379]	– 0.495	$p = 0.001$
Section 7	0.427 (0.063)	[0.419 0.435]	– 0.409	$p < 0.001$
Section 8	0.418 (0.067)	[0.410 0.427]	– 1.292	$p < 0.001$
Section 9a	0.316 (0.025)	[0.312 0.319]	– 0.746	$p < 0.001$
Section 9b	0.301 (0.055)	[0.294 0.308]	– 0.141	$p = 0.176$
Branch 1 Identifying emotions	0.321 (0.024)	[0.318 0.324]	– 0.768	$p < 0.001$
Branch 2 Using emotions	0.417 (0.051)	[0.411 0.424]	– 0.975	$p < 0.001$
Branch 3 Understanding emotions	0.401 (0.044)	[0.396 0.407]	– 0.330	$p = 0.015$
Branch 4 Managing emotions	0.383 (0.045)	[0.378 0.389]	– 1.074	$p < 0.001$
Experiential domain	0.355 (0.028)	[0.352 0.359]	– 0.938	$p < 0.001$
Strategic domain	0.393 (0.035)	[0.388 0.397]	– 0.731	$p < 0.001$
Overall EI	0.365 (0.025)	[0.361 0.368]	– 1.053	$p < 0.001$

Thus, we can talk about a positive relationship between the levels of development of EI as an ability and SEW. This finding is consistent with previous research on relationships between EI and other types of well-being (see Introduction), which can be explained by the fact that SEW is an essential element of a person's overall subjective well-being.

As can be seen from the results of our research, the ability to use emotions for problem solving, the ability to manage emotions, and the experiential domain of EI are key emotional intelligence factors associated with SEW. This finding indicates the

Table 2. Descriptive statistics and test results for normality for the SEW methodology.

Scale	Average (st. dev)	95% confidence interval of the mean	Asymmetry	Shapiro-Wilks test significance (test for normality)
EOI	19.506 (3.226)	[19.099 19.914]	−0.323	p = 0.004
EAI	14.473 (4.430)	[13.913 15.033]	0.284	p = 0.013
IAI	12.823 (2.213)	[12.543 13.103]	0.244	p < 0.001
FDI	19.420 (3.047)	[19.035 19.805]	−0.418	p < 0.001
IFW	14.967 (2.840)	[14.608 15.326]	−0.232	p = 0.006
Integral (general) index	81.189 (11.132)	[79.783 82.596]	−0.122	p = 0.751

Table 3. Correlations of the EI and SEW scales.

EIT scale	EOI	EAI	IAI	FDI	IFW	Integral (general) index
Branch 1	r = 0.165, p = 0.010	r = 0.042, p = 0.514	r = 0.025, p = 0.697	r = 0.155, p = 0.016	r = 0.022, p = 0.731	r = 0.125, p = 0.051
Branch 2	r = 0.168, p = 0.008	r = 0.059, p = 0.362	r = 0.065, p = 0.310	r = 0.235, p < 0.001	r = 0.027, p = 0.678	r = 0.157, p = 0.014
Branch 3	r = −0.036, p = 0.581	r = 0.006, p = 0.927	r = 0.057, p = 0.375	r = 0.128, p = 0.046	r = 0.014, p = 0.834	r = 0.060, p = 0.354
Branch 4	r = 0.201, p = 0.002	r = 0.149, p = 0.020	r = 0.107, p = 0.097	r = 0.119, p = 0.064	r = 0.163, p = 0.011	r = 0.227, p < 0.001
Experiential domain	r = 0.207, p = 0.001	r = 0.071, p = 0.268	r = 0.055, p = 0.391	r = 0.247, p < 0.001	r = 0.031, p = 0.635	r = 0.177, p = 0.006
Strategic domain	r = 0.111, p = 0.084	r = 0.095, p = 0.140	r = 0.110, p = 0.086	r = 0.180, p = 0.005	r = 0.121, p = 0.060	r = 0.194, p = 0.002
Overall EI	r = 0.217, p = 0.001	r = 0.114, p = 0.076	r = 0.094, p = 0.143	r = 0.268, p < 0.001	r = 0.065, p = 0.314	r = 0.228, p < 0.001

importance of the effect of EI on people's subjective economic well-being for their overall well-being.

The scientific significance of the results obtained consists in expanding the understanding of the role of EI as an internal resource in achieving SEW, as well as subjective well-being in general. In the future, EI should be further explored as an important predictor of various aspects of human well-being, which must be taken into account in consulting, psychotherapeutic practice, organizational management in order to solve professional and personal problems.

Acknowledgements. This work was supported by the Russian Foundation for Basic Research, project 19-013-00085.

References

1. Salovey, P., Mayer, J.D.: Emotional intelligence. Imagin. Cogn. Pers. **9**(3), 185–211 (1990)
2. Mayer, J.D., Caruso, D.R., Salovey, P.: Emotional intelligence meets traditional standards for an intelligence. Intelligence **27**(4), 267–298 (1999)
3. O'Boyle, E.H., Jr., Humphrey, R.H., Pollack, J.M., Hawver, T.H., Story, P.A.: The relation between emotional intelligence and job performance: a meta-analysis. J. Organ. Behav. **32**(5), 788–818 (2011)
4. Miao, C., Humphrey, R.H., Qian, S.: A meta-analysis of emotional intelligence and work attitudes. J. Occup. Organ. Psychol. **90**(2), 177–202 (2017)
5. Rode, J.C., Arthaud-Day, M., Ramaswami, A., Howes, S.: A time-lagged study of emotional intelligence and salary. J. Vocat. Behav. **101**, 77–89 (2017)
6. Ali, A.M., Jarboui, A.: CEO emotional intelligence and firms' financial policies. Bayesian Netw. Method Contemp. Econ. **8**(1), 5–24 (2014)
7. Farh, C.I., Seo, M.G., Tesluk, P.E.: Emotional intelligence, teamwork effectiveness, and job performance: the moderating role of job context. J. Appl. Psychol. **97**(4), 890 (2012)
8. Por, J., Barriball, L., Fitzpatrick, J., Roberts, J.: Emotional intelligence: its relationship to stress, coping, well-being and professional performance in nursing students. Nurse Educ. Today **31**(8), 855–860 (2011)
9. Carmeli, A., Yitzhak-Halevy, M., Weisberg, J.: The relationship between emotional intelligence and psychological wellbeing. J. Manag. Psychol. **24**(1), 66–78 (2009)
10. Osberg, L., Sharpe, A.: New Estimates of the Index of Economic Well-Being for Selected OECD Countries, 1980–2007. Centre for the Study of Living Standards, Ottawa (2009)
11. Strümpel, B., Gurin, G., Curtin, R.T.: Economic Incentives, Values and Subjective Well-being, 1971–1974. Inter-university Consortium for Political and Social Research (1984)
12. Campbell, A.: Subjective measures of well-being. Am. Psychol. **31**(2), 117 (1976)
13. Diener, E., Biswas-Diener, R.: Will money increase subjective well-being? Soc. Indic. Res. **57**(2), 119–169 (2002)
14. Cracolici, M.F., Giambona, F., Cuffaro, M.: The determinants of subjective economic well-being: an analysis on Italian-Silc data. Appl. Res. Qual. Life **7**, 17–47 (2012)
15. Easterlin, R.A.: Life cycle happiness and its sources: intersections of psychology, economics, and demography. J. Econ. Psychol. **27**(4), 463–482 (2006)
16. Sergienko, E.A., Khlevnaya, E.A., Vetrova, I.I., Kiseleva, T.S.: Development and psychometric testing of the method of measurement of emotional intelligence (EIT). Kazan Pedag. J. **3**(122), 114–117 (2017)
17. Khashchenko, V.A.: Subjective economic well-being and it's measurment: building of a questionnaire and it's validation. Exp. Psychol. **4**(1), 106–127 (2011)

Comparing Environmental Impact of Various Energy Sources Powering Data Centres's at Indian Candidate Locations

B. Hari Raghavendran(✉), Shivansh Agarwal, and P. Srinivasan

Department of Mechanical Engineering, Birla Institute of Technology and Science, Pilani, India
{f20180514,f20171044,psrinivasan}@pilani.bits-pilani.ac.in

Abstract. While major 21st century developments are data-intensive, the use and storage of such high volumes of data are processed through datacentres. Though Datacentres (DC) are typically high-power consumption applications, the percentage of energy consumption is even higher in developing economies such as India. This paper studies the net environmental impact of energy supply to such datacentres, at four candidate locations, by examining their net Green House Gas Emissions per kWh and land usage per kW. The study analyses various sources of energy such as thermal, wind, and solar plus battery systems to provide a comprehensive view of environmental impacts caused by the power supply.

Keywords: Energy · Environmental impacts · Datacentres · GHG emissions · Land use

1 Introduction

One of the most unique factors of major 21st century developments have been its reliance on data. As the number of internet-connected users increases exponentially, the data generated necessitates the need for tremendous volumes of data storage. While, highly secure Data Centres (DC) have been set up to facilitate the storage of such high-volume data, their energy consumption has been a prime cause of concern. DC's consume large amounts of energy to run, accounting for about 3% of America's power consumption. The U.S. Department of Energy's Lawrence Berkeley National Lab reports that data centres in the United States use 70 billion kW hours (kWh) of electricity per year. Globally DC's consume energy derived from various sources, accounting for nearly 2% of total Green House Gas Emissions (GHG), roughly equal to the GHG emissions of the entire airline industry.

India is considered to be one of the fastest-growing markets for DC's in the world due to multiple reasons such as the growing number of users, data localisation policies, etc., The Indian DC market is currently valued at $1.25 billion, with estimates expecting it to grow at 11.4% Compound Annual Growth Rate (CAGR). While, the Mumbai continues to be a significant site of DC's in India, other IT cities such as Bangalore, Hyderabad, and Chennai have multiple planned investments in line. While the corporation's race to

C. Yuan et al. (Eds.): *Proceedings of the 4th International Conference on Economic Management and Green Development*, AEPS, pp. 201–208, 2021.
https://doi.org/10.1007/978-981-16-5359-9_24

reduce their carbon footprint in a bid to achieve carbon neutrality and reduce climate change impacts, it becomes imperative to focus on high-energy consuming applications such as DC's. While data centres globally consume about 3% of energy totally produced, the fraction is much higher in developing countries like India, where the per-capita power consumption is considerably low. Therefore, it becomes imperative to establish newer DC's in India, that are environment friendly. While India is a country that predominantly sources energy from combustion of fossil fuels such as coal, the growth rate of renewable energy systems is considerably higher than the growth rate of such conventional sources. Therefore, this study analyses the four-candidate location's GHG emissions per unit of energy consumed from various sources such as Sub-critical and super-critical thermal power plants, to renewable energy sources such as Wind & Solar. While the latter part of this study analyses the impact of land usage on using the above-mentioned energy sources, as land usage is the primary cause of concern in high population density countries such as India.

2 Methodology and Calculations

2.1 Thermal Power

One of the significant sources of power generation in India is thermal energy, accounting for about 62.22% as of July 2020. While coal continues to be the primary source of thermal energy accounting for nearly 86.23% of all thermal-based power generation, it can be primarily classified into sub-critical and super-critical technology based on the operating pressure of working fluid. While super-critical power plants account for 10.6% [1] of coal-based power generation, the remaining is obtained from Sub-critical units. While advanced super-critical technology is more efficient at an exergy efficiency of 46.13% [2] compared to the 30.41% [3] average efficiency obtained from sub-critical units, the relatively higher costs of super-critical technology has been the primary constraint in its expansion. Accounting for more than half of India's power generation, it becomes imperative to analyse the feasibility of using coal-based thermal power for Indian datacentres.

While the candidate locations of Mumbai and Chennai uses coal derived primarily from their parent states of Maharashtra and Tamil Nadu respectively, the cities of Bangalore and Hyderabad uses coal from Orissa and West Bengal due to the lack of abundant coal mines in their parent states of Karnataka and Telangana respectively. Since coal composition used in different candidate locations varies considerably, the following coal composition data were used to calculate the emission levels [1] (Table 1):

The above-mentioned coal composition's Gross Calorific Value is used to calculate the various emissions based on the specific coal composition (SCC), which is defined as the mass of coal required to produce 1 kWh of energy, and the net impact of Green House Gases (GHG's) is calculated in terms of CO_2 equivalent accounting NOx's impact on climate change to be 298 times of CO_2. The net GHG emissions incurred during the

Table 1. Coal composition in mass percentage and Gross Calorific Value (GCV) in MJ/kg.

Coal composition	Mumbai	Bangalore	Hyderabad	Chennai
	Maharashtra	Orissa	West Bengal	Tamil Nadu
C	39.7	33.5	44.5	19
S	0.4	0.6	0.5	0.8
N	1.5	1.5	1.4	1.6
VM	29.7	29	31.5	23.5
Ash	21	29.5	17	4
GCV	22.08	19.89	24.28	11.46

manufacturing phase of thermal systems are neglected as they comprise of an insignificant volume compared to the emissions during operations. While 80% of ash is disposed of as fly ash, an Electrostatic Precipitator (ESP) efficiency of 99.9% [1] is assumed to calculate the level of Particulate Matter (PM) emissions. The SCC and various emissions per kWh are given below in Fig. 1.

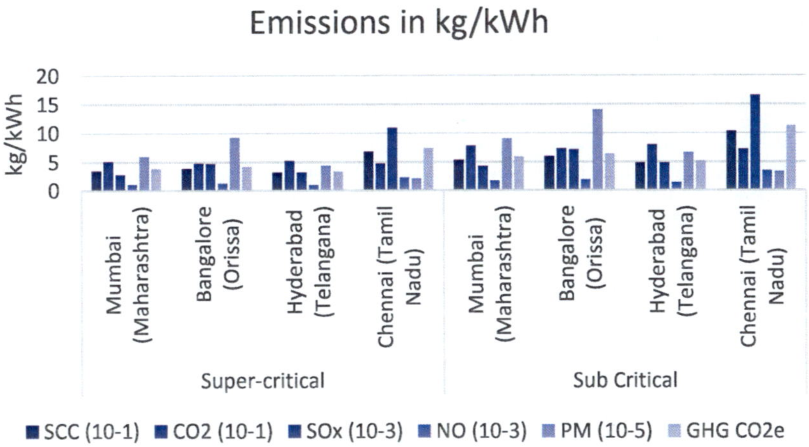

Fig. 1. SCC and net emissions per kWh in candidate locations (origin of coal used)

2.2 Wind Energy

While wind energy is touted to be one of the cleanest sources of energy, their availability varies significantly over time and demography. Therefore, for standardisation of comparison, mean wind speed of parent states at a height of 100 m from ground level recorded in 2019 [4] was accounted for and studied in detail. The recorded mean wind speeds at 10% and 20% of the windiest areas were used to calculate the power output per turbine based on the power curve, assuming the use of Suzlon S128 wind turbines, one of India's most efficient and reliable wind turbines. The usage of a battery-based energy storage system is neglected in this case, as wind energy systems are typically connected to the grid. As the systems are connected to the grid; export and import occurs over a wide range of demography, thus reducing the variability in generation. While the nominal net GHG emission for wind based power generation is accounted at 4.97 g CO_2eq/kWh for a 3 MW onshore wind turbine, the same is adjusted for the current study based on the work by Y.Wang & T.Sun [5]. The average mean wind speeds, and power output per turbine are given below in Table 3 (Table 2):

Table 2. Suzlon S128 wind turbine properties.

Parameters	Value
Rated power	2.7 MW
Rated wind speed/Cut-in wind speed	9.5 (m/s)/3.0 (m/s)
Rotor diameter/Swept area	129 m/13,070 m^2
Cut-out wind speed	30 m/s

Table 3. Mean wind speeds and power output per turbine.

Mean wind speed	10% Windiest area (m/s)	Power output (kW)	20% Windiest area (m/s)	Power output (kW)
Mumbai (Maharashtra)	6.30	725.25	6.02	621.59
Bangalore (Karnataka)	6.76	916.66	6.53	817.58
Hyderabad (Telangana)	6.00	614.54	5.82	553.21
Chennai (Tamil Nadu)	7.93	1533.68	7.16	1105.73

2.3 Solar Energy

Solar energy has grown to be one of the top-most contenders of renewable energy supply due to its ubiquitous nature, though the exact solar irradiation levels vary over time and demography. To account for the demographic variations, the Direct Normal Irradiation (DNI) at the candidate locations are used. To account for the seasonal variations, the minimum daily DNI was used assuming a conversion efficiency of 40% for state-of-the-art technology and dual-axis tracking. While battery storage based on LiFePo4 cells were used to counter the intermittency of daily variations, the excess power generated in peak DNI seasons is assumed to be exported for other applications and not considered to be part of the study. The following Table 4 lists the Minimum daily DNI in kW/sq.m per day sourced from Solar Atlas Data [6].

Table 4. Minimum daily DNI at candidate locations [6].

	Mumbai	Bangalore	Hyderabad	Chennai
Min. daily DNI (kWh/sq.m)	1.031	1.523	1.561	2.407

Based on the above data, the minimum area required to generate 24 kWh per day using Solar Photo Voltic (PV) cells were calculated. Based on the calculated values and on average hourly solar irradiation data sourced from SolarAtlas [6], the battery independent operation time were found out. While the GHG emissions of solar power was estimated to be 20 g CO_2eq/kWh by Kato and Murata [7], for similar irradiation conditions. The net solar PV panel area and battery capacity required are given below in Table 5.

Table 5. Solar PV area and battery capacity required [6].

City	PV area (sq.m)	Battery off-time	Battery on-time	Battery capacity (kWh)
Mumbai	58.196	07:52	16:36	15.265
Bangalore	39.396	07:18	16:43	14.601
Hyderabad	38.437	07:12	16:27	14.765
Chennai	24.927	07:07	16:13	14.912

While the solar PV cells are low-pollution sources of power generation, even the cleanest and most sophisticated energy storage solutions such as Lithium-Ion batteries have a considerable amount of GHG emissions over their lifespan. Though the energy density of such batteries is constantly being improved indirectly reducing the GHG emission per unit of energy storage, the net GHG emissions for lithium iron phosphate battery is 12.7 kg CO_2eq/1000 kWh capacity as calculated by Y. Liang et al [8].

3 Results

3.1 Green House Gas Emissions

One of the primary concerns of humanity in the 21st century is global warming. While the single most important factor contributing to global warming is the release of CO_2, other gaseous emissions such as NOx also contribute considerably to global warming. Though most GHG are absorbed and offset by the environment in 100 years, CO_2 stays in the atmosphere for more than a thousand years, causing concerns over irreversible damage. While the impact of CO_2 is long term, it has a relatively low intensity in short-term impact per unit of emissions compared to other GHG's such NOx, therefore the net Global Warming Potential (GWP) of all emissions are measured based on their equivalence to CO_2. The comparison of GHG in terms of net CO_2eq/kWh, is presented in Fig. 2 below.

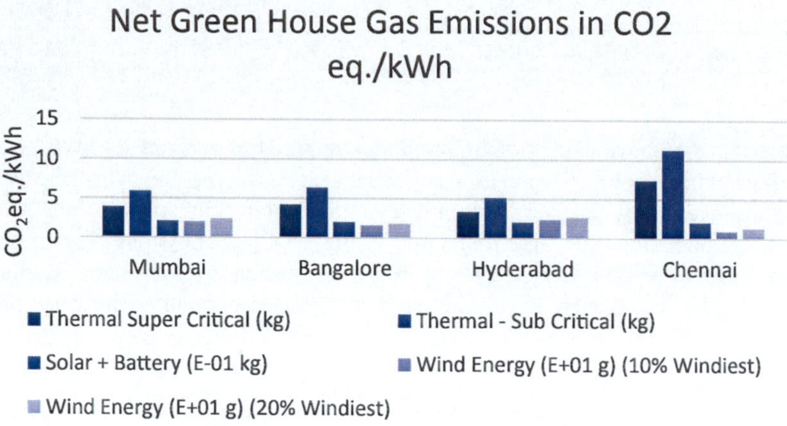

Fig. 2. Net green house gases emission in CO_2eq/kWh.

3.2 Land Usage

While GHG emissions is one of the most critical parameters to measure environmental impact, the next major parameter is land usage. While most conventional power sources are compact, renewable energy sources are known to use extensive land areas. While the land usage for thermal power plants is derived based on the study conducted by the Government of India [9], the land usage for solar power is based on the direct PV cell lay-up area. Wind-based energy systems use a conservative estimate of 0.75 hectares of land for a 2.6 MW generic wind turbine based on various studies conducted by the Government of the United States. The net land usage required to produce 1 kW of consistent power supply, in presented in Fig. 3 below.

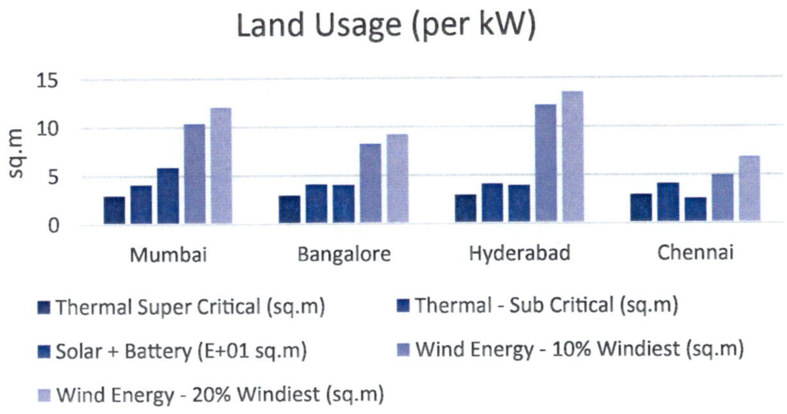

Fig. 3. Land usage to produce 1 kW of consistent power supply.

4 Conclusion

As the growth of data centres in India becomes imminent, it is imperative to consider the environmental impacts of such developments. The comparative analysis of the same is as below:

- The most compact renewable energy site was Chennai both in Solar and Wind Energy systems, occupying about less than 50% compared to Mumbai across all renewable sources.
- While Hyderabad had the most land extensive wind-based energy systems, Mumbai had the largest land usage for solar power primarily due to its higher latitude.
- While the best-case scenario of solar energy in terms of GHG was found in Bangalore, it was just marginally lower at 96% of emissions at Mumbai.
- While the use of thermal sources is not preferred due to average emissions crossing 1000% of renewable sources, they occupy about 20% space as compared to wind energy systems, and only about 4% compared to solar energy systems.

The study, therefore, concludes that wind energy is the best source of renewable energy both in terms of GHG emissions and land usage, if available and transmitted state-wide. Whereas solar energy is the least polluting form of energy source if localisation is demanded either by policies or law.

References

1. Srinivasan, P., Shekhar, A.: Internalizing the external cost of gaseous and particulate matter emissions from the coal-based thermal power plants in India. Part. Sci. Technol. (2020). https://doi.org/10.1080/02726351.2020.1815256
2. Rashidi, M.M., Aghagoli, A., Ali, M.: Thermodynamic analysis of a steam power plant with double reheat and feed water heaters. Adv. Mech. Eng. **6**, 940818 (2014)

3. Rajper, M.A., Memon, A.G., Harijan, K.: Energy and exergy analysis of 210 MW Jamshoro thermal power plant. Mehran Univ. Res. J. Eng. Technol. **35**(2), 265–274 (2016). ISSN 2413-7219
4. Global Wind Atlas. https://globalwindatlas.info
5. Wang, Y., Sun, T.: Life cycle assessment of CO2 emissions from wind power plants: methodology and case studies. Renew. Energy **43**, 30–36 (2012)
6. Global Solar Atlas. https://globalsolaratlas.info
7. Kato, K., Murata, A., Sakuta, K.: Energy pay-back time and life-cycle CO2~ emission of residential PV power system with silicon PV module. Prog. Photovolt. **6**(2), 105–116 (1998)
8. Liang, Y., et al.: Life cycle assessment of lithium-ion batteries for greenhouse gas emissions. Resour. Conserv. Recycl. **117**, 285–293 (2017)
9. Review of land requirement for thermal power stations, Central Electricity Authority, Govt. of India, September 2010

Estimation of Customer's Repayment Date Based on Machine Learning Methods

Hongliang Li(✉)

Southwest Jiaotong University School of Mathematics, Chengdu, China

Abstract. In the company's development process, finance factoring becomes one of the vital links to protect from the shortage of funds. The finance factoring companies need to build a model to estimate the customer's repayment date based on history data. In recent literature, machine learning techniques have been widely applied in finance prediction field. 2466 samples about customer's repayment were acquired Kaggle website. Then we constructed Random Forest model and Support Vector Machine (SVM) model and compared their results. It found that Random Forest model shows plenty of advantages and it can choose some essential features to help us make some appropriate plans. Using the results of the model, based on the existing data set, it is finally found that the feature 'RepeatCust' and 'countlate' are the main factors affecting the late payment of customers. Through this model, the capital of financial factoring companies can be better managed so that the enterprise can achieve capital balance and ultimately maximize the company's earnings.

Keywords: Finance factoring · Payment date · Features · Random Forest · Support Vector Machine

1 Introduction

Finance factoring is growing rapidly in recent years, aiming to provide assistance to small and medium-sized enterprises (SMEs) that have financial problems. The so-called factoring is comprehensive financial services including trade financing, receivables collection and bad debt guarantee for receivables generated from goods sales or service contracts signed by enterprises [1–3]. Unpaid invoices are one of the major challenges to decide whether the factoring companies need to accept this task. In real life, it is easy to link unpaid invoices with the credit rating of a company or customer, and credit is often inseparable from the prediction of financial risks. Thus, it is quite necessary to set up a standard which can help the factoring companies to reduce the risk as much as possible. Most statisticians and scientists, etc. hold a point of view that is to build a predictive model. This model should best present the current situation perfectly and enable to analyze the necessary factors affecting financial risks as much as possible and predict the future size of the risk. However, with the continuous development of the information age, machine learning as an efficient and scientific method is widely used

C. Yuan et al. (Eds.): *Proceedings of the 4th International Conference on Economic Management and Green Development*, AEPS, pp. 209–221, 2021.
https://doi.org/10.1007/978-981-16-5359-9_25

in various big data analysis problems, especially in the field of financial risk analysis. Machine learning uses its own speed to process big data and high prediction accuracy to make the results more realistic.

The importance of factoring is to help enterprises to increase cash flow in a more rapid way of financing. Meanwhile, factoring can also increase the working capital of enterprises. Besides, it allows businesses to increase their purchasing power without affecting their current loan book, increase their borrowing power, and increase their access to supplier discounts. As a result, the demand for funds has increased rapidly due to the continuous growth of the business of SMEs. In addition to being the best financing method for SMEs to grow their business, finance factoring can also reduce their collection risk [4, 5]. Therefore, financial factoring business is an indispensable task for small and medium-sized enterprises to operate normally.

There are a large number of models and algorithms for credit risk assessment. At the beginning, some experts would like to predict risk in the traditional way, such as Markowitz model and the mean-semivariance efficient frontier model (E-SV, Ballestero 2005) [6, 7]. As a result of data mining in handling big data is easy to find and extract the useful information. Machine Learning as the means of data analysis and processing is gradually applied in risk assessment and forecasting. Some people have a preference for the K-Nearest Neighbor algorithm (KNN). KNN algorithm is used to extract the key variables, and the regression can also be used to predict [8–11]. Other experts especially favor the use of Random Forest algorithms to solve this type of finance prediction problem. It is mainly to rank the importance of each feature and select the key factors affecting the late payment [12–14]. In addition, Support Vector Machine (SVM) algorithm has a great advantage in dealing with classification issues [15–17]. Other algorithms can also use known data to accurately predict financial factoring risks by means of classification or regression, especially logistic regression [18, 19]. The financial credit risk here is specifically to predict whether the enterprise will delay the payment of loans.

In this paper, it is going to choose a suitable model which enables to screen different important variables about finance factoring. In Sect. 2, the correlation statistical results of the data set are obtained, and the correlation analysis is used to eliminate the irrelevant variables and select some characteristics with high correlation. Based on the previous preliminary analysis, Random Forest model and SVM model are selected to achieve the results (Sect. 3 and Sect. 4). Then we compare the results of the first two models and choose the one that performs better in terms of speed, results presentation, and accuracy. Moreover, Sect. 5 shows some conclusions through the final result and how to improve the model or choose a better model to optimize the final result.

2 Data Research

There are 2466 samples deriving from the data from Kaggle website (www.kaggle.com). The initial characteristics of the dataset are twelve. It should be noted that the data set has no missing values. Therefore, the step of dealing with missing values can be omitted. And by observation of the data set, outliers found no centralized or white noise in the data. So the data is relatively clean, easy to handle. Then the best choice for processing the exist data is to convert some character - type text variables to numbers such as features 'Customer ID', 'Disputed', 'PaperlessBill'. For 'Disputed', since its answers are only Yes or No, we can naturally use 0, 1 instead of whether or not. Meanwhile, it is convenient to analyze data through doing the same operation to the feature 'DaysLate'. For literal variables that express the meaning of time, like 'InvoiceDate', 'DueDate' and 'SettledDate', we convert them to time. To create a new character, the feature 'DueDate' is converted into quarterly data. Here is a variable that records the number of times the customer appears, called 'RepeatCust'. However, 'countlate' is the time to delay payment on behalf of the customer. In this way, we artificially construct two very important variables from the original data in the data set. Moreover, remove irrelevant variables based on what some variables actually mean, for example, 'SettledDate'.

Table 1. Statistics of different features for data set

	Mean	Std	Min	25%	50%	75%	Max
countryCode	1.75	1.4	0	1	2	3	4
customerID	50.39	28.85	0	25.25	50	76	99
invoiceNumber (E+09)	4.98	2.88	0	2.53	4.96	7.49	9.99
InvoiceAmount	59.9	20.44	5.26	46.4	60.56	73.765	128.28
Disputed	0.23	0.42	0	0	0	0	1
PaperlessBill	0.51	0.5	0	0	1	1	1
DaysLate	0.36	0.48	0	0	0	1	1
countlate	4.41	5.36	0	0	2	7	32
RepeatCust	9.04	8.39	0	2	6	15	32

Through Table 1, it is easily to find that most of results in the feature 'Disputed' are zero (No), more customers choose to renew the deferred repayment period. The longest time that the customer completely repays accounts is 75 days. The feature Customer ID indicates that there are 99 different customers in this data set and nearly a quarter of them pay in excess of the specified time. It is easy to get from the table that the time of late payment represented by feature 'countlate' is 32 days at most, which is nearly 1 month. But the average is only about 4 days, indicating that the data span is very large and fluctuates greatly. The relatively large variance of the feature 'countlate' confirms this from another perspective. The 75% quantile of 'RepeatCust' indicator is 15, but the maximum value is 32, which is more than twice the former, and the variance is as large as feature countlate. Therefore, the data of this indicator is also small and there are more data, and the span between the data is also big (Fig. 1).

Correlation analysis is made for these indicators. From the correlation coefficient between feature 'RepeatCust' and feature 'DaysLate' on the figure is 0.66, it can be concluded that 'RepeatCust' and 'DaysLate' are closely related. Given the meaning of the previous set of 'RepeatCust', it's easy to see why this is closely related to 'DaysLate'. With this heatmap, it is clear that 'countLate' ranks second with a correlation coefficient of 0.46. Judging from the statistics provided by this heatmap, there is no multicollinearity

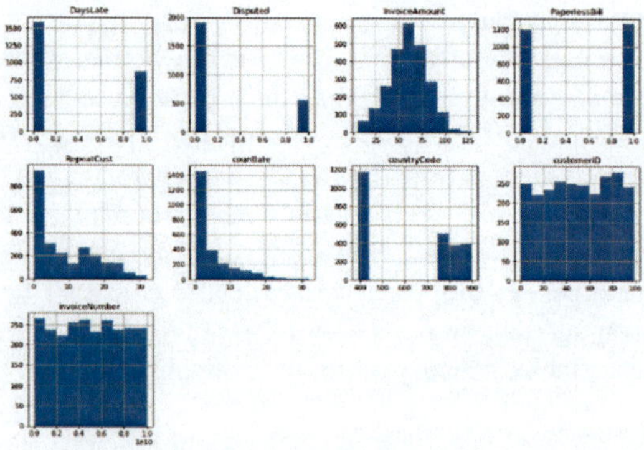

Fig. 1. Histogram of characteristic information

in all variables. This brings us great convenience so that we can start building models directly. It means it is not necessary to filter characteristics. By correlation analysis, features 'RepeatCust', 'countlate' and 'Disputed' provide a huge support to predict the final result (Fig. 2).

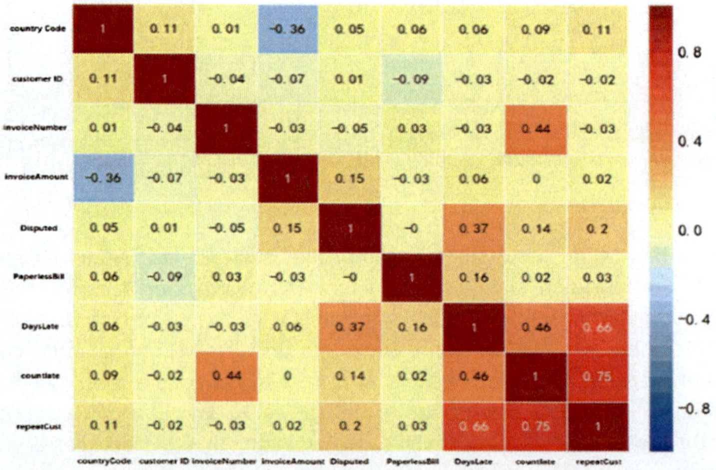

Fig. 2. Correlation of different characteristics

3 Models

Machine learning is an important method to solve the problem of data mining. Its basic idea is to use a large number of training data to solve the decision function of classification or regression problems, and then make decisions on unknown samples. However, because the specific data mining problems are very different, and there are different requirements for solving time and storage cost, so the machine learning algorithm used is also completely different. This part mainly summarizes the principle of machine learning algorithm used to solve the problem of customer's repayment date.

3.1 Support Vector Machines (SVM)

SVM, one of the machine learning models of supervised learning, effectively solves classification and regression problems. It has gained a mass of satisfaction from different people like scientists and researchers. Its popularity owns to two aspects: on the one hand, SVM enables to predict a precise result; On the other hand, the model is based on a more complete and beautiful mathematical theory. To classify two categories, hyperplanes are regarded as a tool to clearly divide two sets of sample points.

The mathematical expression of hyperplane is:

$$w^{\mathrm{T}}x + b = 0$$

where w is the normal vector to the hyperplane, b is the displacement term which refers to the distance between the hyperplane and the origin.

In the given training sample set $D = \{(x_1, y_1), (x_2, y_2), \ldots, (x_m, y_m)\}, y_i \in \{-1, +1\}$.

For $\forall(x_i, y_i) \in D$, there is:

$$\begin{cases} w^{\mathrm{T}}x_i + b \geq +1, \ y_i = +1 \\ w^{\mathrm{T}}x_i + b \leq -1, \ y_i = -1 \end{cases}$$

The distance between two separation boundaries is defined as the margin:

$$\gamma = \frac{2}{\|w\|}$$

And the points located on the boundary of the interval are called support vectors (Fig. 3).

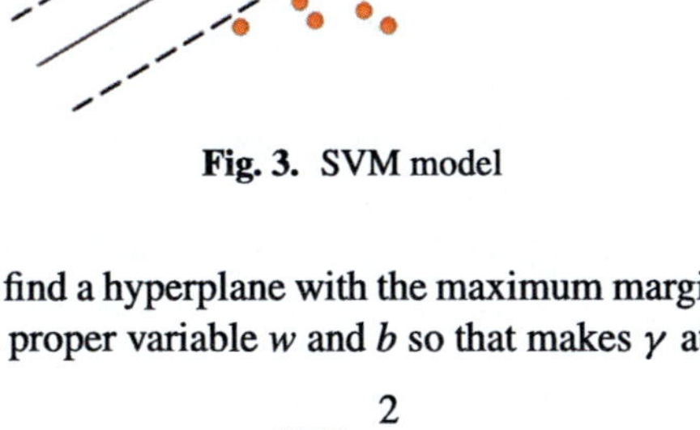

Fig. 3. SVM model

The next task is to find a hyperplane with the maximum margin as much as possible, it also means to find a proper variable w and b so that makes γ attain the most value.

$$\max_{w,b} \frac{2}{||w||}$$

$$s.t.y_i\left(w^{\mathrm{T}}x_i + b\right) \geq 1, \quad i = 1, 2, \ldots, m$$

To simplify the operation, it is a quite convenient method to get the minimum of $||w||^2$. Lagrange multiplier method can be used to solve the dual problem, thus the model can be solved.

However, in practice, it is often difficult to find a good hyperplane to completely separate samples of different classes. Furthermore, if a hyperplane happens to meet the requirement, the result will appear overfitting leading to inaccurate forecasts. The way to mitigate this problem is to allow individual points of dissatisfaction. Therefore, the concept of soft margin is introduced. As a result, the mathematical description of the soft interval SVM can be defined:

$$\min_{w,b,\varepsilon_i} \frac{1}{2}||w||^2 + C\sum\nolimits_{i=1}^{m} \varepsilon_i$$

$$s.t.y_i\left(w^{\mathrm{T}}x_i + b\right) \geq 1 - \varepsilon_i$$

$$\varepsilon_i \geq 0, \quad i = 1, 2, \ldots, m$$

The corresponding solution can also be obtained by using Lagrange multiplier method.

So let's assume that the sample space is linearly separable, and now we're constructing a kernel function that maps the original sample space to a higher dimensional sample space, so that we can find a proper hyperplane, and fortunately, if the dimension of the property is limited, there must be a higher dimensional space where the sample is separable. The expression $\kappa(\bullet, \bullet)$ usually stands for the kernel function. Then, there are some common types of kernel functions as follows:

Linear kernel	$\kappa(x_i, x_j) = x_i^T x_j$.
Polynomial kernel	$\kappa\left(x_i, x_j\right) = \left(x_i^{\mathrm{T}}x_j\right)^d$, $d \geq 1$ is degree of polynomial.
Gaussian kernel	$\kappa(x_i, x_j) = \exp\left(-\frac{\left\|x_i - x_j\right\|^2}{2\sigma^2}\right)$, $\sigma > 0$ is the width of it.

4 Random Forest

If the features of the sample points are non-numerical features in the classification problem, there is no direct use of Bayesian classification, support vector machine, k-nearest neighbor method to train the sample set with numerical characteristics and establish a classifier to solve the classification problem. At this time, there are two options: one is to transform the non-numerical features into numerical features, and then solve the problem. In general, the information in non-numerical features will be lost, and the dimension of sample point features will be increased, and the time complexity of solving classification problem will be increased; secondly, non-numerical features can be directly used to solve classification problems by using Decision Tree method.

Decision Tree algorithm is a kind of common supervised machine learning method, which mainly uses recursive process to help people complete the final decision. Generally, a root node, several internal nodes and leaves together form a decision tree. Random Forest uses samples to randomly generate the required decision trees, and then uses the judgment results of each decision tree to determine the calculation results of the final model.

The Random Forest algorithm needs to Bootstrap the sample and take the sampling method with put back for the sample data. Categorical variables should also be selected at random. The advantage of this operation is that it can effectively avoid the problem caused by wrong data.

Assuming that there are *N* samples and *M* categorical variables in the training set. In order to successfully construct a decision tree, the primary step is to select *N* samples randomly from the samples set that are put back. Then select *k* categorical variables in a random way. Next, repeat *n* times to generate *n* decision trees to form Random Forests. At last, collect every result from each decision tree before define the final result by voting (Figs. 4 and 5).

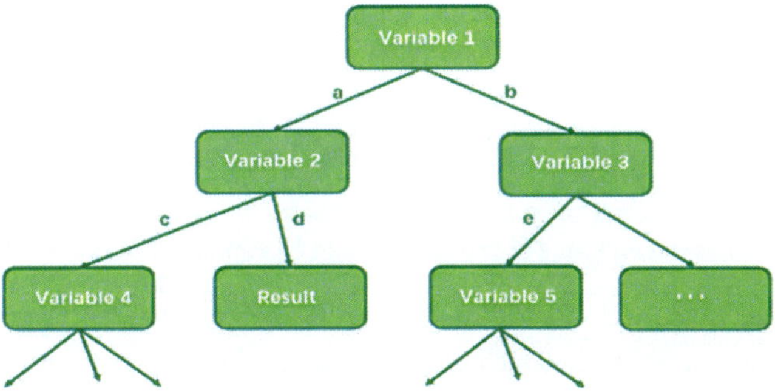

Fig. 4. Decision tree structure

It is obvious that the decision tree is an essential part of the Random Forest. So the quality of the decision tree will affect the effectiveness of the Random Forest. The goal of the decision tree is to learn a classification tree that has strong classification ability and generalization performance for unknown samples. To achieve it, we need to introduce some judgment basis to assist us. Information entropy is an important indicator in the decision tree, reflecting the purity of the sample set. Let's assume that the sample set D has N categorical variables, and the ratio of the i-th sample to the total sample is p, then the information entropy of the set is

$$I(D) = -\sum_{i=1}^{N} p_i \log_2 p_i$$

Define information gain as the difference in entropy of two categorical variables:

$$\Delta I = I(p) - \sum_{c} P_c I(p_c)$$

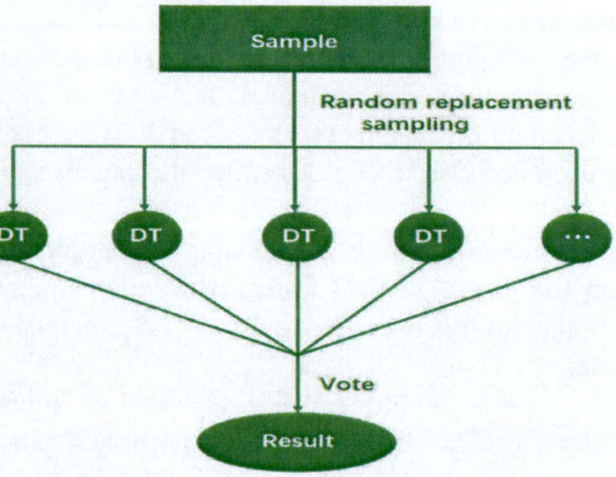

Fig. 5. A flow chart of a Random Forest

where p_c is a child of the P node, P_c is the probability of splitting to p_c. In internship applications, the information entropy will be biased to take more values in the classification variables as the basis for division. To avoid the impact of this bias on the model classification, an evaluation index of information gain rate is added.

$$I\_ratio = \frac{\Delta I}{I(D)}$$

During the model, by means of pruning has the ability to reduce the risk of over-fitting. The two most common methods of pruning need to be explained below: pre-pruning and post-pruning. Pre-pruning is to evaluate each node before dividing it into a decision tree. If current division enables to enhance efficiency hardly in the decision tree, the node will be marked as a leaf node which is not split again. Post-pruning is to find non-leaf nodes that can be replaced by leaf nodes after finishing the whole decision tree. If the new left nodes make the decision tree better, keep these left nodes.

5 Results and Discussion

In fact, there are some necessary parameters need to be decided both Random Forest and SVM. In Random Forest algorithm, the number of decision trees (n_estimators) is an important parameter and the value is 100 after several tests. Moreover, when building the optimal decision tree model, the largest number of features (max_features) and minimum number of leaves (min_sample_leaf) are also paid more attention. The max_features considers the number of features when limiting branching, any feature that exceeds the limit is discarded, here it is set to 20. On account of the smaller leaf can reduce the risk that the model gains noise in train data set, the min_sample_leaf appears its importance and it takes 50.

In SVM model, choosing a proper kernel function is one of the key procedures, the result finally confirms Gaussian kernel function through comparing all kinds of kernel functions. Penalty coefficient (C) means the tolerance of error. The higher the value, the more likely the over-fitting phenomenon will occur and will cause the model to fit well, but the prediction results are not ideal. In the same way, the smaller C is, the greater the possibility of underfitting. So it is settled after many trails (Table 2).

Table 2. The comparison results of the two models

Model	Accuracy	Precision	Recall	f1_score	Roc
SVM (linear)	0.55	0.36	0.38	0.37	0.51
Random Forest	0.83	0.74	0.79	0.76	0.82

After comparing these two models, it is clearly that Random Forest model performs better than SVM model on the whole. Since the Random Forest performs well in all the indicators in the above table, it is more accurate to choose this model to predict. The results of Random Forest model are prior to that of SVM due to characterization of dataset.

In the above histogram, a variable called relative importance is constructed, which can be understood as the score or importance of each indicator that affects the final result. Then it can be defined as the ratio of the importance of each indicator to the importance of the largest indicator. In addition, relative importance can highlight which features may be most relevant to the target, and vice versa, which features are the least relevant (Fig. 6).

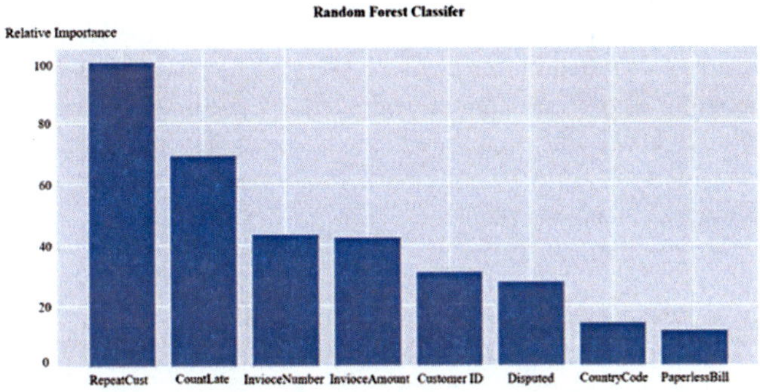

Fig. 6. Importance of indicators

So the feature 'RepeatCust' is of great importance among these features, and 'countlate' becomes the second most important with around 70%. It can be seen from here that these two features are more related to the prediction results. 'InvioceNumber' and 'InvioceAmount' has equal status in prediction. It is clearly that feature 'PaperlessBill' has a little relationship with the payment time. To sum up, from the above picture, we can get the characteristics closely related to paytime, which provides a clear goal for future predictions.

6 Conclusions

Finance factoring is fairly useful for banks or other companies to decide whether the customer will delay the date of payment. The result based on the data from Kaggle indicates that decision tree model is more suitable for this issue after applying the SVM and decision tree model. It also makes sense factors influencing deferred payment. Through the results of the Random Forest model, features 'RepeatCust' and 'CountLate' are important factors to determine whether the customer is delayed or not. It is consistent with the preliminary conjecture in the correlation analysis. The reason for the low importance of feature 'PaperlessBill' may be that there is no substantial relationship between the way to pay the loan and the final choice to postpone the payment, which is also consistent with our life cognition.

Certainly, the result will be more reasonable and reliable if the data set includes more factors or features. We can also make some changes in the model itself to make the results more credible. First of all, try to make some improvements to the Random Forest model, say the choice of more advanced sophisticated Random Forest model, for example, AdaBoosted decision tree models [20]. SVM also has many improved methods, which can use weighted SVM model [16], or it can be combined with some heuristic algorithms like genetic algorithm [17]. It is a good idea to make some attempts to other new models, such as K-means, XGBoost and LightGBM [21–23]. Besides, these may also be mixed together several models, constituting the hybrid model, the literature suggests the effect is sometimes mixed much better than a single model [24–27]. This also provides a new direction for us to optimize the model.

With the advent of the era of big data, data mining problems will become more and more boring and complex. In the future, in order to make the solutions of the two data mining problems studied in this paper better practical, we should constantly try to improve the existing algorithms, further improve the performance of relevant machine learning algorithms in classification accuracy and speed, and study the solutions to the problems such as data set skew, labeling bottleneck, over classification, etc. And try to expand the original data mining problem.

Appendix

See Table 3.

Table 3. Partial data presentation.

countryCode	customerID	PaperlessDate	invoiceNumber	InvoiceDate	DueDate	InvoiceAmount	Disputed	SettledDate	PaperlessBill
391	0379-NEVHP	4/6/2013	611365	1/2/2013	2/1/2013	55.94	No	1/15/2013	Paper
406	8976-AMJEO	3/3/2012	7900770	1/26/2013	2/25/2013	61.74	Yes	3/3/2013	Electronic
391	2820-XGXSB	1/26/2012	9231909	7/3/2013	8/2/2013	65.88	No	7/8/2013	Electronic
406	9322-YCTQO	4/6/2012	9888306	2/10/2013	3/12/2013	105.92	No	3/17/2013	Electronic
818	6627-ELFBK	11/26/2012	15752855	10/25/2012	11/24/2012	72.27	Yes	11/28/2012	Paper
818	5148-SYKLB	8/28/2013	18104516	1/27/2012	2/26/2012	94	Yes	2/22/2012	Paper
897	8690-EEBEO	12/5/2012	23864272	8/13/2013	9/12/2013	74.69	No	9/9/2013	Electronic
770	4460-ZXNDN	6/27/2013	27545037	12/16/2012	1/15/2013	75.06	No	1/12/2013	Paper
770	3831-FXWYK	3/8/2013	28049695	5/14/2012	6/13/2012	80.07	Yes	7/1/2012	Paper
897	7654-DOLHO	4/4/2012	32277701	7/1/2013	7/31/2013	48.33	No	7/26/2013	Electronic

References

1. Sopranzetti, B.J.: The economics of factoring accounts receivable. J. Econ. Bus. **50**(4), 339–359 (1998)
2. Fu, J.C., Zhang, H., Li, K.C.: Discussion of factoring based on mathematical finance. J. Beijing Jiaotong Univ. (Soc. Sci. Ed.) **2**, 15 (2010)
3. Auboin, M., Smythe, H., Teh, R.: Supply chain finance and SMEs: evidence from international factoring data (2016)
4. Cela, S., Shkurti, R., Hilaj, B.: Factoring as the short term finance for SME and possibility of its application in Albania. Int. J. Econ. Perspect. **7**(3) (2013)
5. Klapper, L.: The role of factoring for financing small and medium enterprises. The World Bank (2005)
6. Alexander, G.J.: From Markowitz to modern risk management. Eur. J. Finance **15**(5–6), 451–461 (2009)
7. Pla-Santamaria, D., Bravo, M.: Portfolio optimization based on downside risk: a mean-semivariance efficient frontier from Dow Jones blue chips. Ann. Oper. Res. **205**(1), 189–201 (2012)
8. Galindo, J., Tamayo, P.: Comput. Econ. **15**(1/2), 107–143 (2000)
9. Imandoust, S.B., Bolandraftar, M.: Application of k-nearest neighbor (KNN) approach for predicting economic events: theoretical background. Int. J. Eng. Res. Appl. **3**(5), 605–610 (2013)
10. Henley, W.E., Hand, D.J.: A k-nearest-neighbour classifier for assessing consumer credit risk. Statistician **45**(1), 77 (1996)
11. Mukid, M.A., Widiharih, T., Rusgiyono, A., Prahutama, A.: Credit scoring analysis using weighted k nearest neighbor. J. Phys. Conf. Ser. **1025**, 012114 (2018)
12. Tang, L., Cai, F., Ouyang, Y.: Applying a nonparametric random forest algorithm to assess the credit risk of the energy industry in China. Technol. Forecast. Soc. Change (2018)
13. Malekipirbazari, M., Aksakalli, V.: Risk assessment in social lending via random forests. Expert Syst. Appl. **42**(10), 4621–4631 (2015)
14. Lin, C.D., Peng, G.L.: Application of random forest on selecting evaluation index system for enterprise credit assessment. J. Xiamen Univ. (Nat. Sci.) **2**, 11 (2007)
15. Wang, G., Ma, J.: A hybrid ensemble approach for enterprise credit risk assessment based on support vector machine. Expert Syst. Appl. **39**(5), 5325–5331 (2012)
16. Yu, L., Yao, X., Wang, S., Lai, K.K.: Credit risk evaluation using a weighted least squares SVM classifier with design of experiment for parameter selection. Expert Syst. Appl. **38**(12), 15392–15399 (2011)
17. Zhou, J., Bai, T.: Credit risk assessment using rough set theory and GA-based SVM. In: 2008 The 3rd International Conference on Grid and Pervasive Computing – Workshops (2008)
18. Vikulov, V., Butrin, A.: Risk assessment and management logistics chains. LogForum **10**(1) (2014)
19. Laitinen, E.K.: Predicting a corporate credit analyst's risk estimate by logistic and linear models. Int. Rev. Financ. Anal. **8**(2), 97–121 (1999)
20. Kim, S.Y., Upneja, A.: Predicting restaurant financial distress using decision tree and AdaBoosted decision tree models. Econ. Model. **36**, 354–362 (2014)
21. Zhang, M., Zhou, Z.: A credit rating model for enterprises based on projection pursuit and k-means clustering algorithm. J. Risk Anal. Crisis Response **2**(2), 131–138 (2012)
22. Wang, Y., Ni, X.S.: A XGBoost risk model via feature selection and Bayesian hyper-parameter optimization. arXiv preprint arXiv:1901.08433 (2019)
23. Sun, X., Liu, M., Sima, Z.: A novel cryptocurrency price trend forecasting model based on LightGBM. Finance Res. Lett. (2018)

24. Wang, M., Yu, J., Ji, Z.: Credit fraud risk detection based on XGBoost-LR hybrid model (2018)
25. Tsai, C.-F., Chen, M.-L.: Credit rating by hybrid machine learning techniques. Appl. Soft Comput. **10**(2), 374–380 (2010)
26. Ghodselahi, A.: A hybrid support vector machine ensemble model for credit scoring. Int. J. Comput. Appl. **17**(5), 1–5 (2011)
27. Zhu, Y., Zhou, L., Xie, C., Wang, G.-J., Nguyen, T.V.: Forecasting SMEs' credit risk in supply chain finance with an enhanced hybrid ensemble machine learning approach. Int. J. Prod. Econ. (2019)

The Influence of Social Media Community Marketing on Brand Loyalty

–Take the Marketing Strategy of Xiaomi Mobile as An Example

Zihan Fang(✉)

Michael Smurfit Graduate Business School, University College Dublin, Pingleyuan Road, Beijing, China
zihan.fang@ucdconnect.ie

Abstract. With the development of e-commerce and instant messaging, community marketing is on the rise. Understanding consumer's behavior in the community is particularly important for the development of enterprises. To understand how the consumers' decision-making models work under community marketing strategy, this article takes Xiaomi's marketing model as an example and explores how community marketing affects consumers' brand loyalty by analyzing consumers' decision-making models after entering the community. The article also analyzes the reason why Xiaomi faced a declining sale in the later stage of development, which we could use for reference that companies can not only rely on internet marketing, but also value the importance of offline promotion.

Keywords: Consumer's decision · Brand community · Social media · Community economy · Xiaomi

1 Introduction

With the rise of the Internet and social media, the mode of interaction between brands and consumers is undergoing substantial changes. As a brand-new marketing model in the context of the Internet, community marketing can stand out in the fiercely competitive e-commerce environment and be adopted by brands. Therefore, exploring how the community marketing model affects consumers decision making can enable brands to better understand customer relationship management and help companies improve and refine their marketing strategies.

2 The Basic Theory of Community and Community Economy

2.1 Community Economy

Sociologically defined, a community is a group of people who interact with each other and live in a common area. Later, it was often referred to a group of people who share common values or group cohesion due to a common geographical relationship. American

C. Yuan et al. (Eds.): *Proceedings of the 4th International Conference on Economic Management and Green Development*, AEPS, pp. 222–228, 2021.
https://doi.org/10.1007/978-981-16-5359-9_26

historian Burstein first proposed the concept of consumer community in 1974. Since then, the meaning of community has changed from geographically gathered relationships between people to social relationship related to the use of a particular product or brand. On this basis, the concept of a brand community is derived: a special, not restricted by geography community based on a set of structured social relationships between consumers using the same brand.

Community economy refers to the use of communities by companies to discuss products and reduce the cost for consumers to obtain information. Products and consumers are no longer purely functional links—emotional links of image, culture, etc. Since then, consumers have established emotional trust in the brand. Eventually, achieve value consensus and product consumption.

2.2 Social Media and Community

With the rapid development of the Internet and communication technology, the online community has undergone various forms such as forums, Weibo, WeChat, and short video platforms. Social media has achieved progress from static graphics to dynamic video, from a single type to multiple types. Driven by Internet technology and smartphones, social media has burst out with unparalleled energy, and the information it spreads has become the content that people browse and screen every day.

The emergence of social media has provided a platform for people to communicate and interact in real time. Make the communication time interval, geographical restrictions, and other constraints continue to weaken. To meet their needs for emotional communication and social identification, people with the same values or hobbies can completely break through the limitations of time and space-based on social tools and can freely aggregate and build relationships. Therefore, many interests in certain areas in the itinerary the core functional community. At the same time, with the strengthening of community users' initiative, satisfaction, and sense of belonging after joining the community, it has further promoted the community's activity and development.

Social media has accumulated a fairly stable number of users, and its business ecosystem is becoming more and more complete. According to the China Internet Network Information Center report, social platform advertising has dramatically increased the arrival rate and conversion rate of advertising information under the features of socialization, video, and intelligence, and the market share of social advertising continues to expand. Compared with traditional media, social media has many advantages in disseminating information. First, social media information dissemination is characterized by transparency, which is conducive to public supervision of information. Second, consumers can use social media to understand consumers' actual needs, understand consumers' opinions on products, and further improve product quality. Third, it is conducive to help enterprises better expand sales channels. Fourth, social relations are freer, and consumers have more choices. Fifth, the broad masses of the people can have a more comprehensive understanding of large enterprises. Sixth, users can interact and communicate with each other.

2.3 Community and Branding

Muniz and O'Guinn (2001) proposed that the brand community respond to a social assembly centered on a particular brand, emphasizing the relationship between consumers and consumers who travel based on the use, emotion, and connection of a specific brand relationship.

According to American marketing scholar McAlexander, brand community is based on multiple relationships between consumers and brands, consumers and products, consumers and companies, consumers and other consumers. Based on the Internet and social platforms, companies and consumers interact in real-time, inspiring consumer participation and collaborative initiatives to achieve co-creation of brand value. Brand community marketing integrates relationship marketing, emotional marketing, experience marketing, and word-of-mouth marketing, reshaping the relationship between brands, communities and consumers, building a new marketing model of community interaction.

Operators of brand communities can connect with community users through community activities, publish media content related to brands and products, guide community users to generate a large number of sharing, comments and other behaviors, and at the same time carry out dynamic tracking and precision of the community Positioned big data marketing function, based on the historical behavior data generated by community users to analyze their product needs and targeted product information push. In this process, the community operator obtains revenue while meeting the needs of the community users, bringing into play the commercial value of the community and realizing the community economy.

3 The Impact of the Community Marketing Model on Consumers' Decision-Making Behaviors—Improving Brand Identity and Brand Loyalty

When analyzing consumer decision-making behavior, it is usually divided into three components according to the customer life cycle theory: input-processing-output.

Consumers' decision-making process of deciding whether to purchase a product or service is essential for most brands, and companies often use relationship marketing to influence the consumer's decision making process and enhance consumer awareness of their products and services. Dependence improves consumer loyalty, and ultimately form a closed-loop relationship. The establishment of a brand community is a form of relationship marketing. It can improve customer satisfaction and achieve customer loyalty by identifying potential consumers, maintaining existing consumers, and retaining lost consumers. The brand community can help companies achieve one-to-one communication with customers, which can increase consumer participation, understand their exact needs, improve products and services quality, and improve the relationships with customers. Increased consumer recognition and loyalty to the brand value, thus affecting the processing and output links in the consumer decision-making process, making consumer decision-making form a closed loop (Fig. 1).

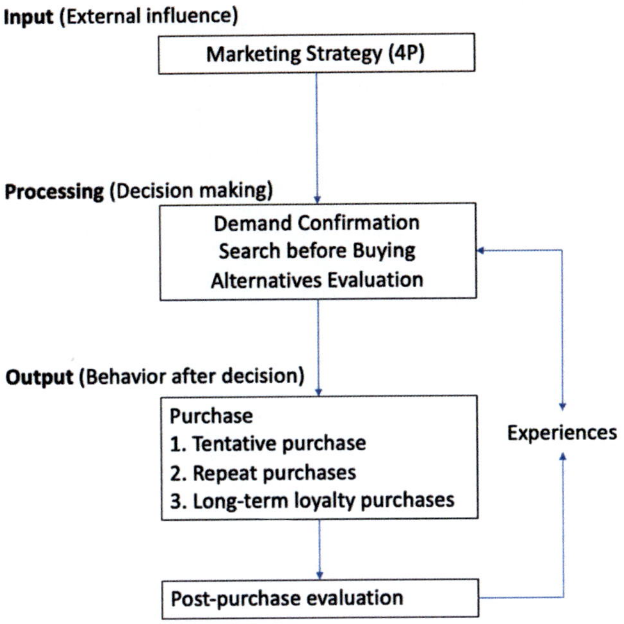

Fig. 1. Consumer decision-making model.

4 A Case Study of the Community Marketing Model Taking Xiaomi as an Example

4.1 Introduction on Xiaomi

Xiaomi was founded in 2010 and initially focused on developing mobile phone operating systems and app applications. Later, it built an ecological product chain with smartphones, software, hardware, and Internet information technology as the leading products, as well as related product accessories and peripheral items. In 5 years, the brand value has increased 180 times. In 2014, Xiaomi's sales in the domestic mobile phone market exceeded the top two Apple and Samsung. This fantastic result is inseparable from its consistent Internet thinking and unique living marketing model. At the beginning of its establishment, Xiaomi took "birth for fever" as the original intention of the brand, regarded users as participants in creating brand value for the enterprise, established an online community, and called on users to actively participate in the company's design, research and development, and production. And throughout the marketing campaign.

Before releasing its first mobile phone in August 2011, Xiaomi communicated with users about user needs every week and developed an operating system with users. Because of interaction and communication between brands and users, a close social relationship was formed. Before the mobile phone was released, it had accumulated 500,000 users. Six months before the release of the first Xiaomi mobile phone release, the Xiaomi community, an online communication group around the Xiaomi brand, was officially launched.

In addition to regular user interaction and information sharing, the Xiaomi community organizes and organizes the Xiaomi City Meeting event at regular intervals. The staff will publish notices of related events in the forum in advance and set up an event chat group so that all registered participants Active people gathered together. These members have the same hobbies and interests. This approach makes the community users participating in the event increase their sense of belonging to the enterprise and allows more users to join the promotion activities of Xiaomi products. They established an emotional bond between community users.

4.2 Xiaomi's Community Marketing Model

By building a brand community, the new customer relationship management process established by Xiaomi is (Fig. 2):

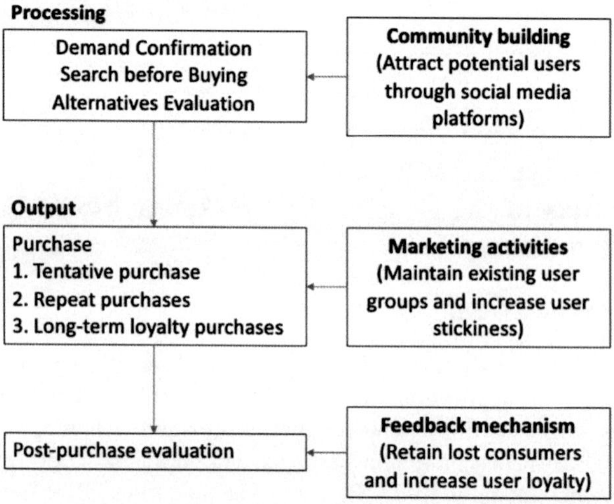

Fig. 2. Xiaomi's community marketing model

The marketing strategy of the Xiaomi community is divided into the following three steps.

1. Cultivate new users and identify potential consumers. The construction of the brand community, which plays a role in the consumer decision-making process, through the brand's market positioning, around the characteristics of potential users, when customers search before purchase, to attract potential customers to establish initial customer relationship is established; through community interaction, members form common brand awareness and sense of belonging and identify with their community members.

2. Activate community activities to maintain existing consumers. Xiaomi community will regularly hold some feedback activities for old users, such as free tickets for product launches, product priority purchase rights, etc. Maintain a long-term and good relationship with consumers.
3. Maintain loyalty and retaining lost consumers. In the Xiaomi community, there are multiple channels to receive user feedback. Any problems encountered by users in the use of products can be connected to online customer service through Weibo and WeChat. Xiaomi will regularly promote product improvement plans and feedback solutions to solve problems for users. Maximum consumer convenience.

4.3 Problems at the Present Stage of Xiaomi

As a brand that started to attract consumers through the Internet and innovative community marketing methods in mainland China's mobile phone market, it took only three years since it entered the market in 2011. It became the brand with the largest market share (13.3%) at that time (Statista 2020), and due to its novel marketing model and marketing content that fits the consumer's psychology at that time, has made its brand and products have a higher spread rate and sought after by young consumers.

However, in the Internet era, where information spreads widely and quickly, the model of community marketing is straightforward to be imitated and surpassed by competing products. Many brands such as Huawei, OPPO, vivo have followed the marketing model of Xiaomi and added innovative marketing content that combined their own brands and current social hot spots. At that time, Xiaomi's marketing plan was not adjusted in time according to the social hot spots of the time. Still focusing on the marketing strategy around mobile phone enthusiasts, while other brands such as Vivo and OPPO, by inviting traffic stars to endorse, focus on promoting the beautiful appearance of the mobile phone and the powerful camera function, so that the brand has a broader broadcast surface. At the same time, due to these with the rise of the brand, Xiaomi is no longer one of the few low-cost, cost-effective brands on the market. At the same price, newer community marketing methods and content have made these brands attract a broader base of consumer groups.

2016, is the beginning of the decline in sales of Xiaomi mobile phones. In the 2016 annual mobile phone report released by Toutiao, it was mentioned that the consumption of Xiaomi mobile phones had dropped significantly, and the loss of Xiaomi users to Vivo and OPPO is a typical phenomenon, which is concentrated in the younger age group of 18–23 years old, who are more interested in selfies. However, the report also pointed out that 56% of Xiaomi's mobile phone users' source of replacement is still the brand. Compared with other brands, user loyalty is relatively high, reflecting the role of community marketing in promoting user loyalty. According to the latest data in 2019 (Counterpoint), although Xiaomi's market share is still 11%, the Huawei brand's share has reached more than 35%, and Xiaomi's market share can only rank fourth.

Another major reason for the decline in Xiaomi's market share is that it relied on Internet communication and sales channels at its inception, while ignoring or even deliberately reducing its offline stores. In the 2016 mobile phone report, it was pointed out that in the third, fourth and fifth-tier cities with a larger base of mobile phone users, users are more concerned about more extensive sales channels, mobile phone maintenance, and are not sensitive to mobile phone parameters. According to relevant news in 2016, the Vivo and OPPO brands' offline retail outlets reached more than 200,000 at the time, and Huawei was 80,000, mostly distributed in third- and fourth-tier cities. These brands have more than 1,000 franchise stores, while Xiaomi has only 40,000 offline retail points and 41 specialty stores, and they are concentrated in second and third-tier cities.

5 Conclusion

In the process of effective use of community marketing, Xiaomi corresponds relevant products to individual users in need of internal services and publicity through the relevant Xiaomi dedicated forum or network and makes uses the relevant network interaction platform. In the meantime, through the channels of Weibo, WeChat and other related popular information release and dissemination, while effectively attracting the attention of the masses, through the directly connected network interaction platform, it can timely and effectively consult with customers for online services and answers, effectively maintain the user experience and Perfection, to a certain extent, has played a role in shared promotion. In the practice of this approach, but also objectively changed the relevant publicity and interaction, from the previously widespread and comprehensive publicity and promotion, to the target customers of the targeted precision needs. In the process of promoting precision marketing, while reducing marketing and promotional costs, it greatly improved the influence of the product and the promotion effect.

However, the rapid decline of Xiaomi in the later period was due to its failure to iterate its marketing strategy with the development of society and the Internet in a timely manner and excessively focused on Internet community marketing, while ignoring the offline mode of traditional marketing, which caused other competition. Brands have seized a larger market share.

This article analyzes the role of community marketing in consumer decision making, from the perspective of customer management, analyzes the impact of community marketing on the input and processing of consumer decision-making processes, and derives the community marketing model for Consumers' brand loyalty has a significant role in promoting, thus forming a closed-loop conclusion of the relationship between users and brands. At the same time, combined with the reasons for the early success of Xiaomi mobile phones and the later decline, it is hoped that it can provide certain practical significance and reference value for the development of the brand.

References

1. Jin, S.: The communication characteristics and business model of "community economy". Mod. Commun. **237**(4) (2016)
2. Xue, H., Wang, X.: An empirical study of the impact of transcendent customer experiences in brand community on consumers' brand loyalty. In: Proceedings of the Sixth Annual Conference of Shanghai Social Sciences (2008)
3. Huang, X.: The integration of social economy, e-commerce and new marketing model. J. Commer. Econ. **14**, 75–77 (2019)

Community Epidemic Prevention and Control Based on Statistical Analysis and KLR Signal Analysis

Cao Jiarui and Wang Shuqi(✉)

School of Management and Engineering, Nanjing University, 22 Hankou Road, Nanjing, China

Abstract. This article mainly studied the social epidemic control at the grass-roots level. We analyzed the role of community work in epidemic management, and explored the relationship between community work and the local epidemic situation, the living conditions and psychological states of residents. Through the AHP and cluster analysis of the data, we found that the work intensity of communities is positively correlated with local epidemic prevention and control. At the same time, we found that there was also a positive correlation between people's satisfaction with community work and the intensity of community work. This meant that the mainstream attitude of the masses towards epidemic prevention and control is to actively cooperate and strictly implement, which strongly proved that China's grass-roots management policy put the personal safety of residents in the first place, and fully complied with People's will and sentiment. Finally, taking China as an example, we analyzed and predicted the development trend of the epidemic situation in the United States with the help of KLR early warning model, which is common in the field of economics.

Keywords: Social governance · Statistical analysis · KLR analysis

1 Introduction

The outbreak of Covid-19 has a huge impact on the country's social stability, economic development, basic people's livelihood, industrial structure and other aspects, and has brought personal safety loss, trade and population flow constraints in the global scope. In terms of public health events in human history, it usually takes several months or even years for different regions to completely get rid of the impact of epidemic situation and restore market order and people's normal life according to their own conditions. Taking the impact of 2009 influenza A (H1N1) as an example, under the background of the financial crisis and the rising unemployment rate, the flu seriously damaged human security and the global economy, causing many countries to fall into a long-term downturn. These historical experiences showed us the far-reaching impact of the epidemic on human society and warned us of the importance of comprehensive prevention and control. In the past studies on infectious diseases (such as SARS, hepatitis B, etc.), the influence of environmental factors on the epidemic of infectious diseases was generally

C. Yuan et al. (Eds.): *Proceedings of the 4th International Conference on Economic Management and Green Development*, AEPS, pp. 229–235, 2021.
https://doi.org/10.1007/978-981-16-5359-9_27

studied. Xiao Xiong et al. (2013) used the geographically weighted regression model to study the infectious diseases in Sichuan Province, and drew the conclusion that population density affects the epidemic scale, and pointed out that whether there are prevention and control measures will directly have a significant impact on the infection situation. We also showed the obvious difference through simulation method (Fig. 1 and Fig. 2). At the beginning of the study, we also made a similar hypothesis based on a wide range of infectious diseases such as COVID-19. [1].

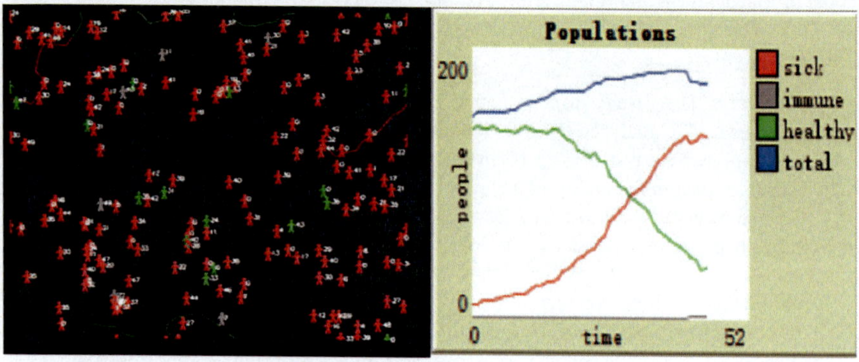

Fig. 1. Population infection without epidemic control measures.

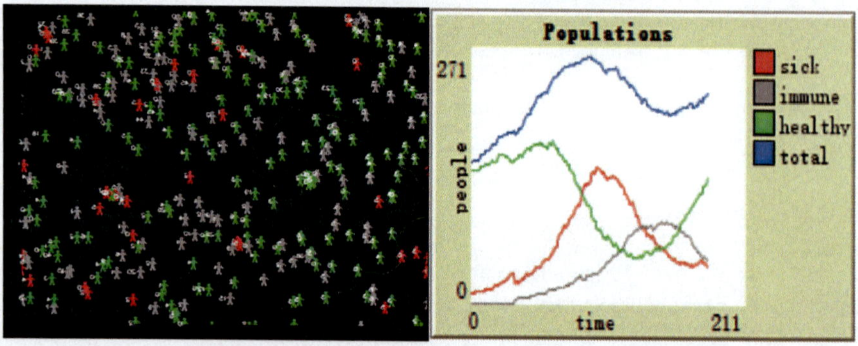

Fig. 2. Population infection with epidemic control measures.

Based on the needs of the project, through an online questionnaire survey, offline structured interview, data crawler method, and literature method, we collected important information including residents' evaluation of community prevention and control, grass-roots staff's work strength, and the epidemic situation in local areas. More than 300 questionnaires and network information about population migration and mass sentiment were obtained. To ensure that the follow-up data analysis work can be carried out normally, we standardized and regularized the original data, and set control questions such as "residents' views on community work before the epidemic" to eliminate the irrelevant variable impact brought by the local community working conditions.

2 Data Processing and Analysis

2.1 Pearson Analysis of Outbreak and Aggregation Pattern

Our preliminary guess was that the higher the density of people gathering, the more serious the infection situation. Thanks to the transparency of epidemic control data in China, the public regional dispersion of epidemic situation and population flow were mainly obtained from Baidu migration and the platform of National Health Commission. Through investigation and analysis, we made a connection between the aggregation mode and the epidemic distribution, and the results are as follows (Table 1). It can be seen that there is a positive correlation between the density of aggregation and the infection of the epidemic, but the correlation is not obvious, which is not consistent with our preliminary conjecture.

The reasons for this are as follows: (1) There is little correlation between the density of people and the severity of infection. Although the density of people without aggregation is low, there was a large number of people flowing in every community during the Spring Festival. Therefore, the correlation is far less obvious than we thought. (2) When the spread of the epidemic has been paid attention to, under the strict management policy and the strong implementation of the community in China, the strict management measures were basically implemented in all parts of the country. Even in the areas with high population density, people's contact opportunities were basically reduced in the first time after the outbreak, and the correlation was weakened.

The discovery of this firstly shows that our strict regulation has an important effect. Otherwise, it is not difficult to imagine that Pearson correlation coefficient will not only be weak correlation. Therefore, reducing the contact between people and carrying out necessary physical isolation is an effective way to prevent the spread of the virus. Strengthening the management and control of personnel flow and encouraging the wearing of masks are the correct measures for epidemic prevention. The impact of personnel flow on the epidemic situation is very obvious [2].

Table 1. Analysis on the mode of aggregation and community infection.

	Aggregation and management mode during epidemic period
Community infection	0.026*
$^*p < 0.05$ $^{**}p < 0.01$	

2.2 Analysis of Community Epidemic Situation and Community Disinfection

In the process of fighting the epidemic, the community has become the basic unit of fighting the epidemic at the grassroots level. Timely disinfection of public areas such as elevators and stairs in the community is an effective means to prevent the spread of the epidemic. Through the analysis of the community epidemic situation and community disinfection situation, we got the results consistent with expectations. Generally speaking, better sanitized communities have fewer infections. However, the community without disinfection measures will have a slightly less number of infected people. The main reason is that once there are people infected, the community's attention to the area will be greatly increased, which also shows that China's grass-roots work will make real-time adjustments according to the situation. However, we noticed that the communities without disinfection measures account for more than 10%, which is a little high. Although some communities can't disinfect the community due to personnel, equipment and other restrictions, the grass-roots staff should pay attention to the implementation of disinfection measures to avoid secondary infection and other events in the community.

For the areas with disinfection measures, we found that more than 50% of them disinfected many times a day, which shows that grass-roots personnel attach importance to the prevention and control of epidemic situation and their conscientious and responsible work at ordinary times. It also explains the reason why the epidemic situation in China was reduced relatively quickly. It is necessary to make joint efforts of a large number of grass-roots personnel behind the anti-epidemic line and the conscious cooperation of most people home isolation. According to the chi square distribution results, it is wrong that disinfection measures have no effect on the infection situation in the community, it can be concluded that disinfection has curbed the severity of infection to a certain extent (Table 2).

Table 2. Results of Cross (chi square) analysis of community aggregation pattern distribution.

	Life gathering and management mode during epidemic period		χ^2	p
	Yes	No		
Community infection	85.71	80.67	2.049	– 0.359
	8.57	6.31		

* $p < 0.05$ ** $p < 0.01$

2.3 Analysis of Community Epidemic Situation and Community Travel Restriction and Isolation

Community isolation data were mainly obtained by questionnaire. Most of the communities began to have blockade measures three to eight days after the closure of Wuhan City, and there were no communities that did not take any measures. Given China's huge population and industrial system, there is no doubt that community action is efficient if the region is blocked orderly and effectively while maintaining stability.

We also noticed that about 26% of the districts were closed late. Excluding the influence of various factors such as special areas, we still thought that the proportion was too high. Then we looked up and analyzed the local government management documents, and found that there was a significant correlation between the closing time of communities and the issuing time of government control documents, which also explained why some closing time was too late. For example, the Heilongjiang and Qinghai samples, which account for about 2% of the total samples, basically chose eight days after the closure of Wuhan among the closing time options. Moreover, the time for the two provinces to release the first level response to major health events was later than the national median time (1.24) (Heilongjiang 1.25, Qinghai 1.26). On the one hand, it showed that China's government was effective. One the other hand, it reflected the top-down way of policy implementation requires the government to take measures to guide the work in time, and the grass-roots managers need to cultivate initiative, find problems and solve problems in time.

In addition to the implementation time, we also conducted a corresponding survey on the implementation of the community. From the feedback of the questionnaire, as long as the measure are implemented, they are basically strictly enforced, the proportion of loose measures is very low, and there were no community took no measures.

2.4 Summary of Community Epidemic Management

From the survey results, it is easy to find that in this epidemic situation, the recognition and satisfaction of the masses for community work has increased significantly, with the average satisfaction increased from 3.46 to 3.84. Especially in the areas without cluster management, they were also helped and managed by the surrounding communities during the epidemic period, reflecting the popularity and effectiveness of grass-roots management ability in China. We find that most people thought that the more strict the community control measures, the higher the degree of satisfaction, which also showed that the popularization of public health knowledge in China was very obvious.

At the same time, through the analysis of the opinions and praise of the masses, we found that most of the people thought that the community was worthy of praise, and there were few places that needed to be improved. Only about 8% of the people said that they needed to be improved.

There were aspects worthy of praise, including the delivery of masks, disinfection and universal education. There were also aspects that needed to be improved, basically focused on the aspect of insufficient strength, which showed people's need for high-intensity and effective epidemic prevention measures.

3 Analysis Based on the Theory of KLR Early Warning Model

3.1 Comparison of KLR Signal Analysis Method with Common Early Warning Models

There have been many models for crisis early warning. From logit model, signal analysis and other mainstream analysis ideas to the neural network model ANN proposed by Nag [3] and binary tree proposed by Ghosh [4], the effect is not nearly the same. This paper focuses on the KLR model and compares the most widely used FR and STV early warning models. The FR model is a probabilistic model developed by Frankel and Rose [5] based on a large number of real sample data from developing countries. STV is a cross-sectional regression model constructed by Sachs et al. [6], which focuses on the causes of similar crises, so as to more accurately analyze some variables that have greater impact on each type of crisis. Although the three models have their own characteristics and emphases, according to the conclusion of Andrew [7] and others using the data of Asian crisis, the test rate of KLR model is the highest among the three models. Therefore, we can draw a preliminary conclusion that the KLR model has the highest accuracy among the mainstream crisis warning models.

The formula is expressed as follows:

$$I_t = \omega_t(\Delta\alpha_t)/\alpha_t - (1 - \omega_t)(\Delta\mu_t)/\mu_t \tag{1}$$

Based on the results of previous surveys, we considered the importance and stress level. at refers to the proportion of infected people in the total population and the weighted mortality rate in the t month. μt refers to the medical equipment reserve in month t (assigned according to the importance of medical equipment, such as ECMO value 10, ventilator value 7, etc.). It is the quantitative embodiment of the country's pressure indicators, ωt is the weight mainly reflected by epidemic prevention measures, including whether to carry out isolation measures, whether to carry out large-scale disinfection, and the state of public opinion. Data can be obtained through government documents, community surveys and social network crawlers such as Facebook.

It will increase with the increase of infection rate or decrease of Medical Reserve, which shows the pressure of anti-epidemic in the country. When it is higher than the set value, it is generally believed that there is a serious public health crisis in the country in t. This quantitative way to measure the health crisis provides the basis and direction of numerical analysis for early warning, with strong operability and readability, but this measurement will also make the national economy and other factors ignored, which has a certain impact on the accuracy of the measurement.

3.2 KLR Index Threshold and Early Warning Effect Analysis

For the n early warning indicators, we will change the hazard index into a binary variable s, whose value is 1 or 0. Let X be the value of a certain index in a certain period, then P(x) represents the cumulative probability distribution value of X, and P′(x) represents the threshold value of warning signal. When P(x) > P′(x), s value is 1, and an early warning signal is sent. On the contrary, if it is less than or equal to the set threshold value, s value is 0. The formula is as follows:

$$\begin{cases} S = 1;\ P(x) > P'(x) \\ S = 0;\ P(x) \le P'(x) \end{cases} \tag{2}$$

According to the analysis of the current infection rate and mortality rate, we found that both the infection rate and the mortality rate were below the threshold through the analysis of the current infection rate and mortality rate. However, we also need to pay attention to the fact that after April 8 and October 1, the pressure value of the infection rate in China slightly rose. According to the news information, it should be related to transnational staff and asymptomatic infected people, which worth national vigilance.

Based on the results of this study, we realized that it's important to further strengthen the execution and management scientificity of grass-roots organizations and enhance the organizational ability and execution of grass-roots organizations, which is the emphasis and difficulty of epidemic control.

References

1. Xiong, X., Changhong, Y., Ke, T., Hongyan, H., Tao, Z., Xiaosong, L.: Application of geographically weighted regression model in spatial analysis of infectious diseases. China Health Stat. **6**, 833–841 (2013)
2. Zhang, H., Xu, J.: Modern Psychology and Education Statistics, 3rd edn. Beijing Normal University Press (2009)
3. Nag, A., Mitra, A.: Neural networks and early warning indicators of currency crisis. Reserve Bank of India Occas. Pap. **20**(2), 183–222 (1999)
4. Ghosh, S.R., Ghosh, M.A.R.: Structural vulnerability and currency crises. International Monetary Fund (2002)
5. Frankel, J.A., Rose, A.K.: Currency crashes in emerging markets: An empirical treatment. J. Int. Econ. **41**(3–4), 351–366 (1996)
6. Sachs, J., Tornell, A., Velasco, A.: Financial crises in emerging markets: the lessons from 1995. National bureau of economic research (1996)
7. Berg, A., Pattillo, C.: Are currency crises predicatable. IMF Working Paper No. WP/98/154 (1998)

Capital Structure of New Energy Automobile Industry

Analysis Based on Tesla and Nio

Lin Zhao(✉)

Illinois Institute of Technology, 35th Street, Chicago, IL, USA

Abstract. This paper examines the capital structure of two firms in new energy automobile industry to get a clear understanding of the propriety of the two firms' capital structure and the feature of the industry. It turns out that as a nascent industry, new energy automobile industry is not one that enables firms to grow fast at the beginning. A typical feature of the industry is the low profit margin at start-up or even expanding stages, which makes the firms' external financing vital. However, for especially start-up firms, it is relatively hard to finance by equity or long-term debt, therefore, they relied much on short-term debts or more on other external funding like government subsidies. As the firms grow larger, they are able to finance by equity and long-term debts, but the limited profit margin inhibits them to get high ratings, which restraints their debt financing. Consequently, equity becomes the best choice. Yet as the high volatility in stocks, the firms still should pay more attention to their debt ratio to make sure their capital structure is stable enough to sustain their operation.

Keywords: Capital structure · New energy automobile industry · Financing · Cost of capital · Tesla · Nio

1 Cost of Capital Analysis

1.1 Models and Mechanism

The CAPM model is implemented to estimate a firm's stock risk to the market and calculate the expected return of the firm to break even. To make the research more accurate, 30-year treasury bond rate rate is taken as a risk-free rate, meanwhile, risk premium is the United States' risk premium from Aswath Damodaran's "Country Default Spreads and Risk Premiums" [1], and default spread is estimated according to the ratings of the firms in each year.

$$\mathrm{Er_i} = \mathrm{R_f} + \beta_\mathrm{i}(\mathrm{Er_m} - \mathrm{R_f}) \tag{1}$$

Weighted Average cost of capital is calculated to estimate the optimal capital structure that saves the most cost of capital for the firms. For firms in New energy Automobile industry, there are nearly no profit to activate tax benefits, hence the cost of capital keeps

C. Yuan et al. (Eds.): *Proceedings of the 4th International Conference on Economic Management and Green Development*, AEPS, pp. 236–246, 2021.
https://doi.org/10.1007/978-981-16-5359-9_28

growing when there are more debts. However, though vague, the trend of the raise and the comparison between the two firms can still help to estimate a propriate capital structure that cost the firm lower capital expense with debt financing.

$$\mathrm{WACC} = r_{\mathrm{debt}}\left(\frac{D}{D+E}\right) + r_{\mathrm{equity}}\left(\frac{E}{D+E}\right) \tag{2}$$

1.2 Cost of Capital Analysis for Tesla

Overall Risk Profile 1. Volatility

The highest close price of TSLA in five years is 498.32 and the lowest close price is 28.73. The volatility of TSLA's stock is 57.98 (Fig. 1).

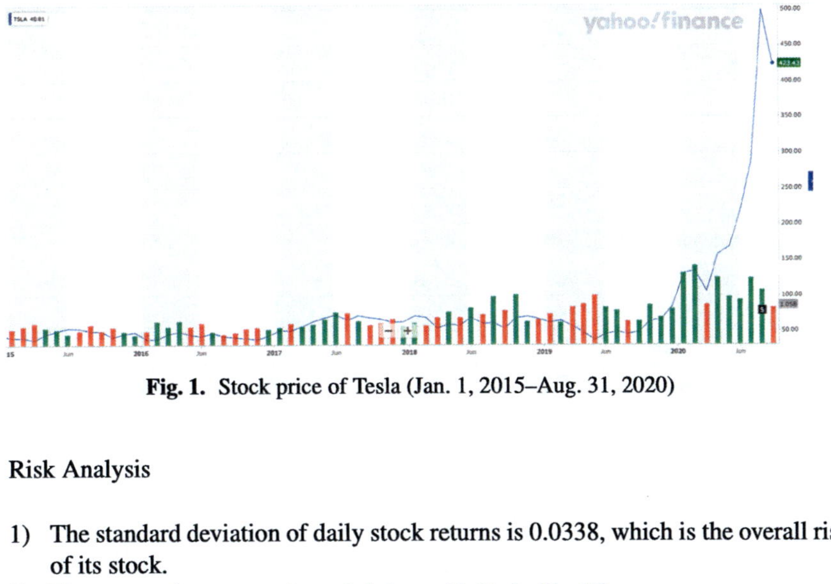

Fig. 1. Stock price of Tesla (Jan. 1, 2015–Aug. 31, 2020)

2. Risk Analysis

1) The standard deviation of daily stock returns is 0.0338, which is the overall risk of its stock.
2) There are a few competitors fighting with Tesla like Nio.
3) Tesla has a 5:1 stock split after close of trading on August 28, 2020 [2].

Expected Return (Cost of Capital) - CAPM Model Tesla has negative EBIT in the past 5 years, therefore the marginal tax rate for the past 5 years is assumed to be 0.

Betas are calculated by the regression of 5 years' (2015–2019 monthly) stock returns and 2020's (daily) stock returns separately. Since Tesla has debt, these betas are levered beta.

Default spread is estimated according to the rating of the firm in each year. The latest rating of Tesla is B2 by Moody's in 2020.

Since Tesla's stock price increased sharply in the year of 2020, two databases are used to calculate the cost of capital for Tesla. One is to use the 5 years' average data

Table 1. Past five-year and 2020's cost of capital for Tesla.

	Million	2015–2019 average	notes	2020.8.31	Notes
a	Debt (BV)	10238	Average	14095	2020.6
b	Stock price dollar	54	Average	498	2020.8.31
c	Shares	792	Average	932	2020.8.31
d	Equity (MV)	43239	b*c	464340	b*c
e	D/E	23.68%	a/d	3.04%	a/d
f	D/A	19.14%	a/(a + d)	2.95%	a/(a + d)
g	Marginal tax rate	0.00%	Negative profit	10.00%	2020.6
h	30-year Treasury rate (risk free)	2.80%	Average	1.54%	2020
i	Equity risk premium	5.23%	Open source of US equity risk premium	5.23%	Open source of US equity risk premium
j	Beta	116.94%	2015–2019 daily regression	133.67%	2020.1–8.31 daily regression
k	Unlevered beta	94.55%	j/(1 + (1–g)*e)	130.12%	j/(1 + (1?g)*e)
l	Cost of Equity	8.92%	h + i*j	8.53%	h + i*j
m	Default spread	4.96%	5 years average	4.21%	Rated B2 by Moody's
n	Cost of debt pretax	7.77%	h + m	5.75%	h + m
o	Cost of debt aft tax	7.77%	n*(1–g)	5.18%	n*(1–g)
p	Cost of capital	8.70%	l*d/(a + d) + o*f	8.43%	l*d/(a + d) + o*f

Notes: Financial data is sourced from the firm's financial report. Stock prices are from Yahoo! Finance [3]; Shares outstanding and 30-year Treasury bond rate are from macrotrends [4]; Equity risk premium is from Aswath Damodaran's NYU stern page [1].

that indicates its past performance; The other one is to use the latest 2020's data (2020 ended August 31) that better indicates its future performance.

According to the results in Table 1, the 2020's beta is greater than the past five years' beta, which means that the volatility of stock returns in 2020 is greater than the volatility of stock returns in the past 5 years, compared to the market.

When comparing the alpha with rf(1-beta), the excess returns are 0.5% and 1.5% in the past 5 years and in 2020, which indicates that Tesla's stock performed 1.5% better than expected and has a greater excess return in 2020 than in the past.

$$\begin{gathered}\text{Returns of Tesla (5years)} = 0.0256\% + 1.17\,\text{Returns of Nasdaq 100} \\ \text{Alpha} = 0.000256,\ \text{levered beta} = 1.1694,\ \text{R square} = 17.76\%\end{gathered} \tag{3}$$

$$\text{Returns of Tesla (2020 ended August 31)} = 0.98\% + 1.34\,\text{Returns of Nasdaq 100}$$

$$\text{Alpha} = 0.009845,\ \text{levered beta} = 1.3367,\ \text{R square} = 32.04\% \tag{4}$$

According to the results above, the 2020's cost of capital for Tesla is 8.43, which is 0.27 lower than the cost of equity in the past 5 years. This is basically due to the issuance of stocks and increase of stock prices that brought the firm's market value of equity up, causing it to be almost 10 times the market value of equity in the past 5 years. Therefore, without largely issuance of debt, the sharp increase of equity lowered the debt ratio of the firm and its cost of equity. The 2020's expected return of Tesla, to break even, is 8.43% (Table 2).

Table 2. Optimal cost of capital estimation for Tesla-past five years.

Debt to capital	Levered beta	Cost of equity	Aft-tax cost of debt	Cost of capital
0	94.55%	7.75%	0.00%	7.75%
5%	99.53%	8.01%	5.20%	7.87%
10%	105.06%	8.30%	6.31%	8.10%
15%	111.24%	8.62%	7.01%	8.38%
20%	118.19%	8.99%	7.95%	8.78%
25%	126.07%	9.40%	11.00%	9.80%
30%	135.08%	9.87%	11.44%	10.34%
35%	145.47%	10.41%	14.14%	11.72%
40%	157.59%	11.05%	17.92%	13.80%
45%	171.92%	11.80%	17.92%	14.55%
5 years cost of capital				8.70%

Note: Tesla has negative EBITs for the past 5 years, therefore, when calculating its cost of debt based on the 5 year average data, the default spread is estimated according to the rating of the firm, which is B3 for its current debt ratio around 20%, and increase or descend one level when its debt ratio goes down or up.

Model Optimal Capital Structure Estimation-Past Five Years Tesla as a whole has not earned money for the past 5 years, which means it has negative EBIT all along, therefore it is not possible for them to have an interest coverage ratio greater than 0. This makes the theoretical optimal debt ratio to be 0, that the firm has not got enough profit to cover its debt yet, so it is better not to issue any debt in this period of business.

However, although the cost of capital keeps increasing as the debt ratio increases, when the debt ratio is under 25%, the increase of cost of debt is below 0.5%, which shows a slower increase than when the debt ratio is above 25%. Therefore, if the debt ratio of the past 5 years went above 25%, the cost of capital will go much higher that the firm will need more effort to keep its returns increase in a faster speed.

Optimal Capital Structure Estimation-2020 On the other hand, since Tesla started to earn money in the year of 2020 according to its quarterly report in June, this estimation

Table 3. Optimal cost of capital estimation for Tesla-2020 ended August 31.

Debt to capital	Levered beta	Cost of equity	Aft-tax cost of debt	Cost of capital
0	130.12%	8.35%	0.00%	8.35%
5%	136.28%	8.67%	8.77%	8.67%
10%	143.13%	9.03%	11.59%	9.28%
15%	150.78%	9.43%	14.99%	10.26%
20%	159.39%	9.88%	14.99%	10.90%
25%	169.15%	10.39%	14.99%	11.54%
30%	180.30%	10.97%	16.58%	12.65%
35%	193.17%	11.64%	16.66%	13.40%
40%	208.19%	12.43%	16.66%	14.12%
45%	225.93%	13.36%	16.66%	14.84%
2020 cost of capital				8.43%

Note: Since the half year financial statements are unaudited and the depreciation amount is not listed separately as to cost of revenue, here EBITDA is used to calculate the interest coverage ratio, which can only be more reserved, and it is estimated to be the two quarter's EBITDA in 2020 multiply by 2, which is 1220 million.

is conducted separately to see when Tesla has its capability of paying part of the interests by itself, what will be the optimal capital structure for it.

What can be indicated by Table 3 is that in the 2020's condition, Tesla has better keep its capital structure or still lower its debt ratio if possible, because although the firm is capable of paying part of its interest in the first half of 2020, it still cannot pay for its full interest even with the lowest debt ratio at 5%.

Conclusion. The increase of stock price in 2020 made the debt ratio of Tesla dropped sharply from 19.14% to 2.95%, which effectively adjusted the capital structure of the firm to a reasonable level.

The comparison of the two estimation shows that the cost of capital for the latest condition of Tesla turns out to be more sensitive than in the past. Therefore, Tesla needs to pay more attention to the increase of its debt ratio than before.

1.3 Cost of Capital Analysis for Nio

Overall Risk Profile 1. Volatility.

The highest price of Nio since its issuance is 20.44 and the lowest price is 1.32. The volatility of Nio's stock is 3.34 (Fig. 2).

2. Risk Analysis.

1) The Standard deviation of daily stock returns is 0.0658, which is the overall risk of its stock.

Fig. 2. Stock Price of Nio (Sep. 12, 2018–Aug. 31, 2020)

2) There are strong competitors fighting with Nio like Tesla.

CAPM Regression and Interpretation Nio is a start-up firm whose initial public offering at the NYSE is on September 12, 2018. Since market value is used in the model, in the calculation and estimation of Nio's capital structure, 2 years data is used as database, which is data from 2019 to 2020 (2020's beta is estimated according to its stock price until August 31, 2020). All the calculations are based on the average of this database.

Since Nio has a negative EBIT all the time and as a start-up firm it has not reach the stage to earn money and adjust its spread yet, its default spread is estimated to be the highest of 15.12% in the calculation. As the average 30-year treasury rate of 2019 and 2020 is 2.06%, the constant default spread of 15.12% made the cost of debt for Nio to be a constant rate of 17.18% in any capital structure.

The result of regression on Nio's return from January 1, 2019 to August 31, 2020 is:

$$\text{Returns of Nio} = 0.4288\% + 0.8799\,\text{Returns of NYSE}$$
$$\text{Alpha} = 0.004288,\ \text{levered beta} = 0.8799,\ \text{R square} = 5.08\% \quad (5)$$

According to the result above, only 5.08% of Nio's risk comes from the market. When comparing the alpha with rf(1–beta), the excess return is 0.18% in the estimation period, which indicates that Nio's stock performed 0.18% better than expected.

Expected Return (Cost of Capital) - CAPM Model According to Table 4, the cost of capital for Nio is 7.85%, which is also the expected return of Nio if it is expected to break even, and 0.58% lower than Tesla.

Optimal Capital Structure Estimation Since Nio is a start-up firm who does not have positive EBIT to pay its interest, the cost of debt is assumed to be the highest of 17.18% at any capital structure. Then the cost of capital will grow when the debt ratio increased, therefore it is clear that the optimal finance source for these start-up firms is equity or

Table 4. Cost of capital for Nio.

	Million	2019–2020	Notes
a	Debt (BV)	1704	Average
b	Stock price dollar	12	Average
c	Shares	1165	Average
d	Equity (MV)	13419	b*c
e	D/E	12.70%	a/d
f	D/A	11.27%	a/(a + d)
g	Marginal tax rate	0.00%	Negative profit
h	30-year Treasury rate (risk free)	2.06%	Average
i	Equity risk premium	5.23%	Open source of US equity risk premium
j	Beta	87.99%	2019–2020.8.31 daily regression
k	Unlevered beta	78.08%	j/(1 + (1–g)*e)
l	Cost of Equity	6.66%	h + i*j
m	Default spread	15.12%	The highest spread of 15.12%
n	Cost of debt pretax	17.18%	h + m
o	Cost of debt aft tax	17.18%	n*(1-g)
p	Cost of capital	7.85%	l*d/(a + d) + o*f

Notes: Financial data is sourced from the firm's financial report.

other external financing sources without mandatory interest repayment. However, the estimation is still done to be used as a comparison to Tesla for industrial analysis (Table 5).

Table 5. Optimal capital structure estimation for Nio.

Debt to Capital	Levered beta	Cost of equity	Aft-tax cost of debt	Cost of Capital
0	76.04%	6.14%	0.00%	6.14%
5%	80.04%	6.36%	17.18%	6.90%
10%	84.49%	6.60%	17.18%	7.66%
15%	89.46%	6.86%	17.18%	8.41%
20%	95.05%	7.16%	17.18%	9.17%
25%	101.39%	7.50%	17.18%	9.92%
30%	108.63%	7.89%	17.18%	10.68%

2 Financing Source Comparison and Assessment

2.1 Compare Cost of Debt, Cost of Equity and Cost of Capital

Table 6. COD, COE, COC comparison between Tesla and Nio.

(million)	Nio	Tesla	
	2019–2020	2015–2019 average	2020 ended Aug. 31
Debt (BV)	1704	10238	14095
Equity (MV)	13419	43239	464340
D/A	11.27%	19.14%	2.95%
Beta	87.99%	116.94%	133.67%
Cost of Equity	6.66%	8.92%	8.53%
Cost of debt aft tax	17.18%	7.77%	5.18%
Cost of capital	7.85%	8.70%	8.43%

It is illustrated in Table 6 that the cost of equity for Nio is much lower than that of Tesla in any estimation period. However, since Nio has not issued any bond yet and its main financing sources are mandatory debts and government subsidies, the cost of debt for Nio is estimated to be much higher than Tesla, who has already issued debts and has a latest rating of B2. Even so, the cost of capital for Nio is much lower than it of Tesla, which is mainly because of the higher beta of Tesla that represents a higher volatility of its stock returns compared to the market.

According to a research by Aswath Damodaran [5], the total market cost of capital (without financial) until January 2020 is 6.9% based on the estimation of 5878 firms. The cost of capitals for these new energy automobile firms turned out to be much higher than the market cost of capital, which indicates that it cost much for these new energy automobile firms to finance, especially from debt.

2.2 Compare the Interest Expense for One Unit of Debt

Apart from the estimation of the cost of capital according to CAPM model, a more direct way to estimate the cost of debt for the two firms is to calculate their interest expense for one unit of debt. The equation used in this part is listed below:

$$\text{Interest Expense per unit of Debt} = \frac{\text{Total Interest Expense}}{\text{Debt}} \quad (6)$$

Where: Debt = ST Debt + Current Portion LT Debt + LT Debt (Capitalized Leases included).

The interest expenses per unit of debt are 4.66% and 3.52% for Tesla and Nio separately, which turned out to be different as our theoretical estimation of cost of debt.

In this case, Nio actually has a lower interest expense for each unit of its debts. This may due to the government subsidies of Nio that required no mandatory repayments in the past. However, as discussed in its annual report, with the government reducing such charitable fund, Nio will be facing heavier burdens in the future.

2.3 Compare the Balance Sheet

A proportion comparison is done to compare the proportion of different assets and liabilities of the two firms in 2019.

Nio and Tesla are two typical firms in the new energy industry. Compared to Tesla, Nio is a start-up firm that issued its stock publicly only since 2018. The characteristics of Cost of Equity, Debt and Capital of the two firms can serve as a reference of the cost of capital of different phases of the firms in the industry: Start-up(Nio), Rapid Expansion(2015–2019T) and high Growth(2020T) (Fig. 3).

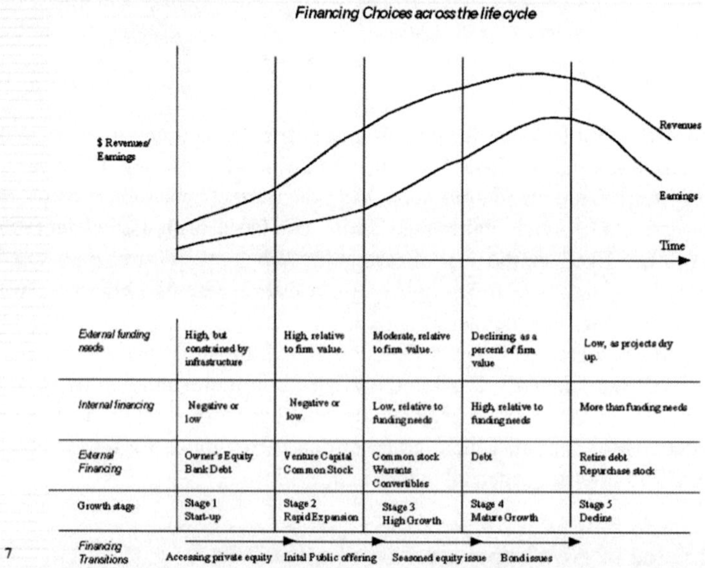

Fig. 3. Financing Choices across the life cycle

It is shown from the estimation before that for a relatively new industry, the two firms are much more levered than they are supposed to be. However, the financing choice of the two firms are relatively consistent with the regular financing choices of their life cycle. According to their balance sheet, Tesla favor equity and long-term debts, whereas Nio has more short-term debt and meanwhile relied on government subsidies largely.

3 Conclusion and Suggestions

Although the treasury rate decreased largely in 2020, the volatility of tesla's stock returns raised and thus the cost of equity did not change largely. However, the cost of debt dropped sharply in 2020 since both the two factors determining it decreased in this year: default spread, which is caused by the increase of rating recently, and risk-free rate, which is estimated by the treasury bond rate. Meanwhile, debt ratio dropped sharply due to the increase of stock price in the recent year and issuance of stock which caused the market value of debt of Tesla more than ten times of its historical five-year average debt. Due to the factors above, Tesla's 2020 cost of capital decreased to a lower rate merely due to the dramatic rise in stock price, which is not necessarily long-lasting.

3.1 Characteristic of the Industry: Limited Internal Financing

Since both Tesla and Nio are in the condition of not being able to pay for its own interests and their cost of debt keeps increasing or constant at the highest rate, they are not enjoying any saving in debt in their stage of business. Both firms are operating based on their debts and equities, and their internal financing, profit, is rather limited.

3.2 Industry's Financing Preference

From the characteristics of the two firms analyzed, the trend of financing sources preference in new energy automobile industry as they grow is indicated as: short-term debts with government subsidies, equity, long-term debts. The difference of financing preferences of the two firms indicates the advances of each financing source in different life cycle of this industry.

For relatively start-up firms, since it is rather difficult for them to get long-term debt from institutions, their choices of financing are limited to short-term debts or more equity buyers if possible; Whereas for expanding firms, as their reputation spread and rating increased, long-term debts are better to substitute for short-term debts to save their costs of debt, and equity becomes a more efficient way to finance.

However, since it takes long to gain profit in this industry, firms find it hard to cover its interests in a long period of time. The external financing source needs to be reliable and long-lasting. In Nio's case, it mostly relied on the government subsidies, which are gradually reducing recently. As to Tesla, it mainly depends on their shareholders and the innovative concepts they are establishing to draw the attention of more shareholders. In both the cases, the supplementary financing source is not that reliable, and it may be more appropriate for them to finance with less mandatory debts as a backup to any possible upheaval.

Compared with the Retail of Automobile industry or Green & Renewable Energy industry [5], the cost of capital in this New energy Automobile industry is much higher, mostly due to their high cost of debt which may be the consequence of this nascent industry's feature of slow and low profit that limited their ratings.

4 Insufficiency and Future Work

Weighted average cost of capital was used to estimate the optimal capital structure in this paper. However, since the two firms both have negative EBIT as a whole, tax is not taken into consideration in this paper. In this case, there cannot be an absolute optimal cost of capital from the calculation. This paper merely discussed the relative optimal cost of capital and the comparison between two firms to draw a sketch of the feature of financing in different stages in the industry. There still should be some modifications to the research such as adding tax into it to make it more accurate, as there are profits in some branches of the firms that they can enjoy some tax benefits from them. Instead of cost of capital, other criterion can also be used to estimate optimal capital structure like cash flows and firm value. Therefore, more research should be done to make it more precise.

References

1. Damodaran, A.: Country Default Spreads and Risk Premiums (2020). http://pages.stern.nyu.edu/~adamodar/New_Home_Page/datafile/ctryprem.html
2. Globe Newswire: Tesla Announces a Five-for-One Stock Split (2020). https://ir.tesla.com/press-release/tesla-announces-five-one-stock-split
3. Annual financial reports and 2nd quater financial reports of Tesla and Nio; Yahoo!Finance
4. Macrotrends: https://www.macrotrends.net/2521/30-year-treasury-bond-rate-yield-chart
5. Damodaran, A.: Cost of Capital by Sector (US) (2020). http://people.stern.nyu.edu/adamodar/New_Home_Page/datafile/wacc.htm
6. Joshua, R., Joshua, P.: Investment Banking-Valuation, Leveraged buyouts, and Mergers & Acquisitions, pp. 124–131 (2009)
7. Figure 1, Figure 2 data source: Yahoo! Finance. Figure 3: https://slideplayer.com/slide/5707356/

The Dilemma Faced by the RCSC of China and Suggestions for Improvement

Yaping Gong(✉)

Shandong Zhucheng Experimental Middle School, Shandong, China

Abstract. The legal status and mission of the RCSC of China have been confirmed by law. In terms of the organizational structure, the RCSC of China is composed of national and local Red Cross Societies. Existing problems include but are not limited to defects in the governance structure and operation mechanism, unclear legal position, the relationship with the government that needs to be optimized, and the lack of transparency in operation. To reform and improve the RCSC of China, the following work is necessary. First is to improve the governance structure and operational mechanism of the RCSC. Second is to clarify the nature of the social organization. Third is to optimize the relationship between the RCSC and the government. And finally, the transparency and openness need to be improved.

Keywords: RCSC · Governance structure · Sociality · Transparency · Independence

1 Introduction

In February 2020, the COVID-19 pandemic broke out nationwide, centered in Wuhan. As an important rescue force, the Red Cross Societies of Hubei Province and Wuhan city participated in the action against the epidemic. While playing an important role, they also exposed a series of problems, including but not limited to: first, the medical rescue materials were not timely and effectively distributed, resulting in a large number of materials piled up in the warehouse and could not be delivered to the hospitals in urgent need [1]. Second, the distribution of relief materials was unfair. Some private hospitals closely related to the Red Cross received more materials, but those not close to them and undertook heavier tasks received very few supplements, which was obviously against the impartiality of rescue. Third, medical materials were casually distributed to government agencies. According to a news report on the Internet, a government staff took away a box of surgical masks without going through relevant procedures. After the occurrence of such problems, people began to question the public welfare and professionalism of the Red Cross. We can't help but ask, as an organization with a long history and aims to save the dying and heal the wounded, why the RCSC of China (hereinafter referred to as "RCSC") has so many problems [2]. On this basis, the paper tries to analyze the problems and shortcomings of the RCSC on the basis of its characteristics and current situation, and puts forward corresponding suggestions for improvement, which is of

C. Yuan et al. (Eds.): *Proceedings of the 4th International Conference on Economic Management and Green Development*, AEPS, pp. 247–253, 2021.
https://doi.org/10.1007/978-981-16-5359-9_29

great significance for improving its society image and giving better play to its role and function.

2 Current Situation and Characteristics of the RCSC

2.1 The Legal Status of the RCSC

In order to ensure the operation of the RCSC, the Standing Committee of the National People's Congress formulated the Law of the People's Republic of China on RCSC (hereinafter referred to as "the Law") in 1993 and amended it in 2017. This law is specific for the establishment and operation of the RCSC, so it is the authoritative basis for its status in China. Through theoretical analysis, it can be seen that the Law stipulates the nature and function of the RCSC. According to the Law, the RCSC is a unified Red Cross organization in the people's Republic of China, which establishes the unity and uniqueness of it [3]. The nature of the RCSC is a social relief group engaged in humanitarian work. From a legal perspective, the Law stipulates that the RCSC must abide by the Constitution and laws, follow the basic principles established by the International Red Cross and Red Crescent Movement, and carry out its work independently in accordance with the Geneva Conventions and its additional protocols to which China is a party, as well as the constitution of the RCSC. Under the above provisions, the RCSC is a social organization with independent legal status, which is clearly defined and empowered by the Law. Therefore, from the legal point of view, the RCSC is a social organization rather than a government institution, which is not only based on the domestic law, but also the international law.

2.2 The Main Mission of the RCSC

Article 11 of *the Law* clearly stipulates the responsibilities of the RCSC. It aims at developing humanity, fraternity and dedication, protecting people's life and health, and promoting the cause of peace and progress. The original intention and legal status determine that its main mission and responsibilities are divided into two parts at home and abroad [4]. Internally, the RCSC is responsible for carrying out work of rescue and disaster relief and establishing the Red Cross emergency rescue system, carrying out training of emergency rescue, promoting the knowledge of emergency rescue, disaster prevention and health care, organizing volunteers to participate in on-site rescue, promoting blood, body and organ donation, participating in the related work of hematopoietic stem cell donation, and organizing the volunteer service and Red Cross youth work. Internationally, the RCSC should participate in the international humanitarian relief work, publicize the basic principles of the International Red Cross and Red Crescent Movement and the Geneva Conventions and their additional protocols, complete the tasks entrusted by the government in accordance with the basic principles of the International Red Cross and red Crescent Movement, carry out its work in accordance with the relevant provisions of the Geneva Conventions and their additional protocols, and assist the government in the development of international humanitarian relief related to their responsibilities [5]. The mission and responsibilities of the RCSC determine that it is required to have professional ability, relative independence and public welfare, otherwise it will not be able to truly complete the tasks above.

2.3 The Internal Structure of the RCSC

According to the introduction on the official website, the RCSC was founded in 1904. After its establishment, the RCSC has been engaged in activities of rescuing refugees, wounded soldiers and disaster victims [6]. It has actively worked to alleviate people's suffering from wars and natural disasters, and has participated in international humanitarian relief activities. After the founding of the People's Republic of China, the RCSC carried out a consultative reorganization in 1950. Premier Zhou Enlai presided over and revised its constitution. In 1952, the RCSC resumed its legal seat in the International Red Cross and Red Crescent Movement. According to the law, the RCSC has been established at the national level. It represents the RCSC in the international community. Local Red Cross Societies at or above the county level are established in accordance with their administrative regions, and full-time staff shall be provided according to the actual work needs. Red Cross Societies may be set up in national industries according to their needs. In other words, the RCSC in China consists of the general association, local branches and industry branches. At present, under the general association, there are also Red Cross Societies in various administrative regions, corps and railway exclusive Red Cross Societies, as well as the Red Cross Societies of Hong Kong and Macao special administrative regions. The general association has six internal organizations (Office, Relief and Rescue Department, Finance Department, Organization and Publicity Department, Liaison Department, and Department of Party-Related Affairs (Discipline Inspection Commission), and nine directly subordinate units are set up (Service Center of the RCSC, Training Center of the RCSC, Disaster Preparedness and Disaster Relief Center of the RCSC, Foreign Aid Material Supply Station of the RCSC, Hematopoietic Stem Cell Donor Database Management Center, China Organ Donation Administrative Center, Career Development Center, Chinese Red Cross Foundation, and Press Agency, etc.). Red Cross Societies at all levels shall be established with the approval of the government at all levels and supported by government funds [7]. At the same time, in terms of business, Red Cross Societies at all levels are guided by a higher level and implement a double track management system. The highest organ of power of the RCSC is the National Congress of Members. Its executive body is the Council of the RCSC. It has specialized committees, internal organs of the general association and units directly under the association. The members of the RCSC are composed of local Red Cross Societies, industrial Red Cross Societies, grass-roots organizations and organizations of special administrative regions.

3 Problems and Deficiencies of the RCSC

Through the investigation of the organizational structure and operation of the RCSC, it can be seen that there are many problems and deficiencies in the Red Cross system at present, including the following four aspects:

3.1 Defects in the Governance Structure and Operation Mechanism

At present, there are defects in the governance structure and operation mechanism of the RCSC, which leads to its lack of execution force and action. First of all, the governance structure of the RCSC is incomplete. The smooth operation of large-scale social organizations depends on the establishment of a modern governance institution composed of an efficient council, an executive body and a board of supervisors [8]. However, current governance structure of Red Cross Societies at all levels is unreasonable. Take its general association as an example. Although it has a council of about 200 members, most of the directors are officials of state organs, who have neither sufficient energy nor professional ability to carry out their work. The executive committee is composed of executive vice presidents, full-time vice presidents and secretary general, but most of them are senior state administrative officials. In addition, the Red Cross Societies at all levels have not set up a board of supervisors, which is a relatively large drawback for a governance institution. Secondly, the superior departments have no power to appoint personnel and supervise financial affairs and projects. Local Red Cross Societies follow the orders of governments at their levels, and there is no relationship of supervision and management between the upper and lower levels of Red Cross Societies, which cannot form an effective joint force. Finally, the operation system lacks efficiency and incentive mechanism. There is no incentive in the personnel system. Once the personnel are assigned to the RCSC, there is no chance of promotion. The presidents at all levels are sent from upper organizations, and middle-level cadres and managers are almost impossible to have promotion nor to have opportunity to communicate with other institutions. In terms of wages and benefits, no incentive mechanism has been established. Employees cannot be motivated to work with a fixed salary as civil servants.

3.2 Unclear Legal Position

At present, the legal position of the RCSC is still lack of clarity, making it a government agency, unable to play its initiative and independence. According to Article 2 of the Law of the RCSC, the RCSC is a unified Red Cross organization of the People's Republic of China and a social relief organization engaged in humanitarian work. Also, Article 10 stipulates that the RCSC shall have the status as a legal person of a social organization. The legal status of the RCSC is of great significance to its operation and construction. Currently, the law makes it clear that the RCSC is a social organization, but it does not specify the independence and sociality of the RCSC. Therefore, the RCSC has been alienated as a subordinate organization of the government, which has a negative impact on its development. The unclear legal position of the RCSC makes itself becoming a subordinate organ of the government, and its sociality and independence are not reflected, thus affecting the professionalism and autonomy of the RCSC.

3.3 The Relationship with the Government Needs to Be Optimized

Theoretically, as a social organization, the RCSC should be independent of the government. Keeping a moderate distance from the government can fully reflect its professionalism. However, in reality, there are problems in the relationship between the RCSC and the government in China. The most important problem is the lack of independence of the RCSC. In China, it is not only a public welfare institution but also a government institution, that is, it has administrative characteristics, and its staff basically participate in public management, which is equivalent to civil servants. Since the RCSC has become a government organization, its operation and style are similar to the government, and contrary to the nature of social organizations [9]. As a government institution, the RCSC cannot express its independence and professionalism, because it undertakes a large amount of administrative responsibilities and is used to obeying the orders of the government.

3.4 Lack of Transparency in Operation

As a public welfare organization, the survival and development of the RCSC need the trust and support from the public, which requires it to improve the transparency of actions and actively accept the supervision of the public. However, in reality, the RCSC in China has serious deficiencies in transparency. From foreign experience, the Red Cross Societies in other countries attach great importance to the disclosure of account books and accept public supervision, while the RCSC is lack of transparency at present. First of all, it doesn't make a comprehensive disclosure of its financial situation, such as the donations it received and the use of funds. Second, it doesn't actively promote public participation, and some projects do not widely accepted participants from the public. The lack of transparency is likely to breed corruption and lead to a lack of public trust in the Red Cross. Third, the staffs lack the awareness of enhancing transparency and openness.

4 Suggestions on Reforming and Improving the RCSC

In view of the problems existing in the RCSC and combined with foreign experience, the reform and improvement of the RCSC should do the following work:

4.1 Improve the Governance Structure and Operation Mechanism of the RCSC

Theoretically, it is necessary to modify the governance structure and operation mechanism of the Red Cross to improve its operational ability and professionalism. In order to improve the governance structure and operation mechanism of the RCSC, the following points should be emphasized. First, it needs to strengthen the specialization and professionality of the board of supervisors and the board of directors, reduce the number of part-time staff and add full-time personnel. Red Cross Societies at all levels should actively promote personnel reform, greatly reduce the number of government official members, and increase the number of full-time staff. Second, the relationship between the upper and lower levels of the RCSC should be clear [10]. On the basis of ensuring the independence of the RCSC at all levels, the supervision and guidance of the RCSC at a higher level to a lower level should be strengthened. The Red Cross Societies at higher levels shall conscientiously perform their duties and promptly correct the illegal and violation practices of the Red Cross Societies at lower levels. Third, an incentive mechanism should be established to improve the enthusiasm of internal personnel. It can introduce the concept of enterprise management, establish the incentive mechanism in line with the characteristics of the RCSC, and reward those who are professional and have high efficiency. Although the RCSC is a public welfare organization, it cannot only focus on dedication, but also need to meet the needs of the staff in order to ensure the sustainability of the activities.

4.2 Clarify the Legal Nature of the RCSC

In view of the current unclear legal nature of the RCSC, it is suggested that the nature and independence as a social organization of the RCSC should be clearly stipulated in the future revision of the Law on RCSC, so that it can truly return to the attribute and status of a social organization. Only by establishing the social organization attribute and status, can the administrative nature of the RCSC be weakened and its public welfare and autonomy be embodied. Moreover, clarifying the legal attribute of the social organization can provide legal guarantee for the reform. To clarify the nature of the RCSC, the key is to confirm its sociality, independence and autonomy, and to reduce other subjects' interference in its affairs.

4.3 Optimize the Relationship Between the RCSC and the Government

To establish a scientific relationship between the RCSC and the government is the key to improve the work of the RCSC. In view of the ambiguous relationship between the RCSC and the government, we should emphasize on independence, and focus on the government's support and supervision of the RCSC. First of all, the RCSC and the government should be independent, instead of relying on each other. The government shouldn't treat the RCSC as its subordinate and should respect its independence and sociality. Secondly, the RCSC should build a cooperative relationship with the government, supplementing government functions and roles. They can make up for each other's shortcomings or give full play to their own advantages. Finally, the government should strengthen the supervision and guidance of the RCSC. On the basis of respecting its independence, the government should strengthen the supervision and guidance, correct its illegal activities, guide it to strengthen its own construction, and provide corresponding policies and financial support.

4.4 Enhance the Transparency and Openness of the RCSC

The transparency and openness of operation is of great significance to optimize the self-restraint and regulation of the RCSC and enhance public trust. Therefore, multiple measures should be taken to improve the transparency and openness of the RCSC. First, through legislation, the Red Cross should be required to disclose its financial status and related activities. Second, under government supervision, financial transparency should be regarded as an important indicator to evaluate whether the RCSC is qualified. From the perspective of the RCSC itself, it should establish a sound mechanism of financial openness and business matters publicity, strengthen its openness to the public and actively accept supervision. Third, it should pay attention to actively absorbing the participation of the masses, so that the general public has the opportunity to understand the reality of its work. On the one hand, it can give the public an opportunity to participate in public welfare activities and enhance the sense of achievement [11]. On the other hand, it can also enhance the public's recognition of the work of the Red Cross. Fourth, the RCSC at higher levels should strengthen supervision and inspection on the openness and transparency of finance and affairs of the lower branches, and urge them to strengthen their publicity work.

5 Conclusion

The RCSC has its own characteristics. But its strong administrative feature has become an obstacle to the fulfillment of its responsibilities. In the future, the RCSC should be independent in the direction of de-administration, mainly returning to its sociality and public welfare, and sorting out its relationship with the government. It should be a social organization independent to the government, supplement the government's role and be supervised by the government. It is believed that after reform, the RCSC can better play its role as a social organization and enhance its professionalism and operational capability, instead of being so dissatisfying as the RCSC of Hubei Province in prevention and control of COVID-19 epidemic.

References

1. Jiang, H.: Legal consideration on improving the supervision mechanism of China's red cross. J. Guangxi Admin. Cadre Inst. Polit. Law 06 (2013)
2. Sun, Y.: Soft power and hard power of the construction of the RCSC. Theory Constr. 01 (2018)
3. Ge, D.: The RCSC should transform to a statutory body. Soc. Sci. Weekly 002 (2013)
4. Xu, G., Chi, Z.: Historical trace of the development of chinese red cross since the foundation of the new China – a review with chinese red cross constitution as its path. Jiangxi Soc. Sci. 4 (2009)
5. Xu, X., Duan, W.: Research on building an internal control in the RCSC. Contemp. Account. 14 (2020)
6. Tan, Z.: Non-governmental organizations' participation in public crisis management – taking the RCSC of Hubei province as an example. Young Soc. 20 (2020)
7. Wang, J.: Transformation and development of the RCSC under the new circumstance. J. Party School of C.P.C Qingdao Municipal Committee Qingdao Admin. Inst. 04 (2017)
8. Pan, M.: An exploration on way of reformation of the RCSC from "Multiple Governance Subjects' 'Publicity' Management Model". Admin. Law 08 (2016)
9. Liu, X.: Nonprofit organizations credibility crisis and rebuild – taking the RCSC as an example. Forward Pos. 08 (2015)
10. Sun, Y.: Analysis of the supervision and management mechanism of the RCSC. J. Liaoning Admin. Coll. 02 (2015)
11. Zha, X.: Sociological thoughts on the supervision of the RCSC. J. Guangxi Admin. Cadre Inst. Polit. Law 04 (2014)

The Impact of Financial Derivatives on Economic Growth: Implications for Financial Risk Management

Weiting Liu(✉)

Birkbeck College, University of London, Malet Street, Bloomsbury, London WC17HX, UK

Abstract. Derivatives market is important to many aspects of the financial system and the economy, but only limited research has shed light on its relationship to economic growth, let alone any studies that consider the influence of risk measures. Based on 32 quarters (2012Q3 to 2020Q2) of data from six countries and taking risk factors into account, this paper uses a panel vector autoregression to investigate the dynamic causality between economic growth and the financial derivatives market. The results indicate that there is significant unidirectional relationship from the factors of financial markets to the economic growth. Meanwhile, the results show that trade openness and inflation also have significant positive effects on economic growth. However, to investigate how risk factors, for instance, volatility and value at risk, would affect the model, further methodology needs to be introduced.

Keywords: Financial derivatives · Economic growth · Granger-causality test · Risk measures

1 Introduction

The financial derivatives market has contributed to the improvement of financial systems and plays an important role in economies. While there has been much literature in the past examining the relationship between economic growth and financial development. (Such as in, e.g., Levine 2005; Lucas 1988; Khan 2001; Pagano 1993) [1–4], the studies of dynamic relationship between financial derivatives market and economic indicators seem to be insufficient due to the limited research [5]. There are even fewer studies exploring the relationship between financial derivatives market and economic growth from the perspective of financial risk management. One of the most essential functions of Financial derivatives is that they can be used as a tool to control risk [6], investors and financial institutions can achieve their purpose of risk sharing, risk transfer and hedging with rational utilization of derivatives. However, due to its many types including the use of non-standardized contracts(Over-the-counter traded derivatives) and its own leverage, unreasonable use and management of financial derivatives will cause serious consequences: The financial crisis occurred during 2007–2008 is the biggest shock and it damaged the worldwide financial system unprecedentedly since the 1931, and it still

C. Yuan et al. (Eds.): *Proceedings of the 4th International Conference on Economic Management and Green Development*, AEPS, pp. 254–263, 2021.
https://doi.org/10.1007/978-981-16-5359-9_30

challenges the understanding of liquidity production and liquidity risk management from both financial institutions and regulators [7]. Financial institutions in bank sector pursued abnormal profit by utilizing collateralized debt obligation (CDO) and selling credit default swap(CDS),leveraging their potential profit and the uncertainty at the same time [8]. Due to the insufficient investigation about housing bubble and the lack of risk management, they were unaware of the situation that they had been facing huge exposure to troubled mortgages and they lacked the financial strength to cover it. Similarly, after the coronavirus outbreak in 2019, WTI crude prices plunged since March, 2020, crashing to minus 37 dollars on 20 April [9], the long holders of oil futures were sacked. Risk events in the derivatives market do not seem to be well controlled because of any advance in economy and the derivatives market. Therefore, studying how the implementation of risk management would affect the causality between financial derivatives markets and economic growth is very important and it needs to be solved further with more profound researches, which could be started by examining the causal and dynamic relation with data for those risk measures or risk control indicators of derivatives instruments such as Value at risk (VaR) and volatility.

Over the past 2 decades, on the other hand, financial risk management has experienced much transform and greater revolution and it is a very board concept: It can either be a strategy of profit-making institutions and non-profit organizations, or the policy of monitoring the whole derivatives market in a country [10]. As a result, the effect of its implementation is difficult to be quantified, compared and analyzed with other macroeconomic data. Even though the quantified risk control indicators are feasible to be analyzed, the results might not persuasively prove that any change of causality between derivatives market and economic growth is affected by the risk management.

My research, adopting similar methodology by Vo et al. [5], would firstly apply a panel vector autoregression and analyze the dynamic relationship between economic growth and financial derivatives markets using data from major countries. I will also try to introduce variables with risk measurement indicators such as volatility, VaR value, etc.

2 Literature Review

In the past, few studies have focused on the relationship between the risk factors of financial derivatives and economic growth. Several studies have attempted to analyze the link between derivatives markets and economic growth. Haiss and Sammer (2010) discuss the functions of derivatives through three channels: volume, risk, and efficiency. They derived these three channels to answer questions related to the links between financial development and economic growth in the US and how the channels worked in changing the nexus of them [11]. Using time-series data, Sendeniz-Yüncü et al. (2018) specialized in analyzing the roles and functions of stock index future markets and how the markets had contributed to economic gain in 32 advanced and emerging countries. The results of this study show that there are 22 countries, mostly low-income countries, where the stock index futures market has a unidirectional impact on the country's economic growth [12]. These studies seem to be contradictory: Haiss and Sammer (2010) concluded that there was only a weak correlation between economic growth and financial sector(especially derivatives markets) [11], while Sendeniz-Yüncü (2018) produced the result to prove that futures markets, under impeccable functioning, underpin for larger risk-sharing, therefore making it plausible for institutions to work on projects that are relatively higher risk but more lucrative, which helps improve economic growth [12].

On the other hand, some of effort has been expanding in studying the nexus between risk factors and growth. Obstfeld (1992) developed a time-series model examining the relationship between international risk sharing and expected consumption growth and he concluded that international risk sharing can help produce tremendous welfare gains through its positive effects on expected consumption growth [13]. Devereux and Smith (1994) studied how international risk sharing affect economic growth in a country in which the growth rates are endogenous and they found that if relative risk aversion exceeds one, growth and saving rates will also be reduced when the rate-of-return risk is diversified away [14]. Notwithstanding dissimilar circumstances, this kind of diversification always increases well-being. Acemoglu and Zilibotti (1997) proved that using diversified portfolios to decrease financial risk allows individuals and financial institutions to invest in high-return projects which would have a positive impact on economic growth [15]. Krebs also illustrated that reducing the change of specific risk in firms would increase the total investment return and growth. Angeletos (2003) and Storesletten et al. (2004) studied entrepreneurial risk and idiosyncratic risks respectively [16, 17]. Their results both show that a good control of these risks, i.e., either improving risk sharing or reducing economic agents' risks could be beneficial to economic growth. Although these studies agree with a theory that the improvement of financial risk management can help achieve the economic boom, none of them focus specifically on risk of financial derivatives or the markets and most of them are on the theoretical side. As a result, we can see that there is a conspicuous vacancy on empirical research examining the causality between financial derivatives markets and economic growth using data of risk factors along with other macroeconomic data.

To develop an empirical analysis related to this, the effect of financial risk management must be quantified, as they can be compared and analyzed with macroeconomic indicators. Also a proper model needs to be reasonable used.

Fortunately, many previous researches have developed models studying the causality among factors of financial development and those of economic growth. Using data averaged from 1960 to 1989, King and Levine (1993) conducted simply a cross-state analysis and they also developed a coalescent cross-nation and time-series study using data over 30 years since 1960s to test the correlation between the average level of different financial area and economic growth [18]. Arestis and Demetriades (1997) criticized King and Levine's work because their causal interpretation is based on a fragile statistical basis [19]. Instead, Arestis and Demetriades constructed vector autoregressive (VAR) procedures along with cointegration tests to examine the finance-growth relationship. Ouyang and Li (2018) conducted a PVAR approach to estimate the endogenous connection between financial development and economic growth in China [20]. In a closer relation to my topic, Rodrigues, Schwarz and Seeger (2012) developed an EGARCH model and also used GMM estimation to analyze the effect of institutionalized derivatives transaction on economic growth [21]. More recently, Vo et al. (2019) conducted an empirical analysis estimating the essentiality of derivatives markets to economic growth in China, Japan, US and India [22]. They built a vector error correction model (VECM) and tested whether variables have Granger-causality. In the same year, Vo studied a similar topic with a different model. They introduced a panel vector autoregression (PVAR) model to investigate dynamic relationships among GDP, derivatives and other macroeconomic variables.

The studies did find empirical evidence to support the nexus among economic growth and development of various financial sectors, but as Haiss and Sammer raised questions related to the impact of derivatives on economy (such as, who would benefit from derivatives other than the whole economy, in what ways derivatives and the market affect the economic growth), it is inevitable that further studies in the field of financial derivatives need to be done.

3 Methodology

The purpose of this study is to create a proper model that investigates the links between economic growth and financial derivatives, more specifically the stock index futures. The study applies macroeconomic factors along with variables of risk measurement. The regression is expressed as followed:

$$\mathrm{LnGDP_{it}} = \alpha_0 + \alpha_1\mathrm{OPEN_{it}} + \alpha_2\mathrm{INF_{it}} + \alpha_3\mathrm{IFPQA_{it}} + \alpha_4\mathrm{IFQEV_{it}} + \alpha_5\mathrm{IFRQA_{it}} + \alpha_6\mathrm{IFRQS_{it}} + \alpha_7\mathrm{VaR_{it}} + \varepsilon_{\mathrm{it}} \tag{1}$$

Where $OPEN_{it}$ is the market openness of a country (= total exports/total GDP), and INF_{it} is the inflation rate, $IFPQA_{it}$ is the quarterly moving average of each country's main stock index futures. Similarly, representing the indicators of a nation's major stock index futures, $IFQEV_{it}$ is the quarterly trading volume, $IFRQA_{it}$ is the quarterly mean, and the $IFRQS_{it}$ is the quarterly volatility. $IFVaR_{it}$ is quarterly value at risk (VaR) of main stock index futures. i and t stand for nations and the time scaled in quarterly respectively.

Regression (1) shown above illustrates the causality among the logarithm of GDP, macroeconomic factors and variables of futures (including risk factors). By applying a PVAR model, we can capture the dynamic behavior among variables. All variables can be treated as endogenous variables. Although VaR can be controlled by policies and rules using different level of confidence, $IFVaR_{it}$ is still an endogenous variable because it is the key variable that needs to be explained further.

Following Vo et al. [5], this study applied a PVAR model, where there is a set consisted of all the variables. To avoid bias estimation coefficients, the PVAR for this study, which the generalized method of moments (GMM) is utilized, is as follows:

$$\theta_{it} = \alpha \sum_{s=1}^{n} A_s\theta_{i,t-s} + N_i + D_{N,t} + \varepsilon_t \tag{2}$$

Where θ_{it} is constituted by a set of 8 variables, namely the economic growth (logarithm of GDP),market openness, inflation rate, and the moving average, trading volume, mean, volatility, value at risk (VaR) of major stock index futures. s stands for the serial correlation of the LnGDP and i is the number of cross-sectional nations. Taking the fixed effects of the nation into consideration in the Eq. (2), N_i is introduced. As a nation-specific dummy variable, $D_{N,t}$ is used to control the quarter impact. Using the PVAR approach, every nation and time t are considered to have the quarterly structure.

4 Data

This study uses data resources from a domestic website https://cn.investing.com/, which contains historical and real time information of stock futures, options and foreign exchange from major exchange markets.

This study selects data from China, Japan, South Korea, Germany, the United Kingdom, and the United States for 32 quarters from 2012Q3 to 2020Q2 for the variables of gross GDP, market openness, inflation rate, and quarterly moving average, trading volume, mean, volatility and value at risk of major stock index futures.

I will use CSI 300 index futures as Chinese major stock index futures, Nikkei 225 index futures as Japanese major stock index futures, KOSPI index futures as Korean major stock index futures. I will also choose DAX30 index futures, FTSE 100 index futures and S&P 500 Index Futures to represent the major stock index futures of Germany, the UK and the US respectively.

The value at risk (VaR) of each stock index futures is generated and calculated by Monte Carlo simulation. Firstly, I would apply the method to generate a random number between 0 and 1 on each trading day for every quarter, calculating Logarithmic rate of return. Then the mean and standard deviation (volatility) of each quarter's rate of return is derived. Assuming that the rate of return is subject to the normal distribution of the corresponding mean and standard deviation, I would use the Normiv function to simulate the normal distribution for the rate of return. Next, calculating the fluctuation range of each trading day and sorting the fluctuation range in ascending order. In general, VaR adopts a 95% confidence interval. Therefore, the trading day's fluctuation range is obtained by the number of trading days in each quarter times 0.95. Taking the absolute value of the range and we derive the value at risk for the quarter. Table 1 evinces the details of descriptive statistics.

Table 1. Descriptive statistics of the data

Variables	Obs	Mean	Std. dev.	Min	Max	p1	p99	Skew.	Kurt.
LNGDP	192	10.669	0.737	9.735	12.29	9.741	12.281	1.079	3.073
Openess	192	0.008	0.005	0	0.02	0	0.019	0.137	2.155
Inflation	192	1.416	0.883	−0.657	4.967	−0.499	4.267	0.293	3.885
IFPQA	192	6801.029	6318.386	236.297	23006.728	239.468	22644.265	1.011	3.033
IFQEV	192	6.413	8.939	0	62.654	0.017	45.067	3.189	15.421
IFRQA	192	0	0.001	−0.005	0.006	−0.004	0.003	−0.444	7.491
IFRQS	192	0.011	0.005	0.003	0.034	0.004	0.03	1.371	5.637
IFVaR	192	0.025	0.012	0.007	0.076	0.008	0.07	1.235	4.995

5 Results

Main empirical tests are conducted using data selected. First, I applied panel Fisher unit-root test to check whether the variables are stationary. Unit-root test is done again after the first-order difference. Then panel cointegartion test, model stationary test (inverse roots of AR Characteristic polynomial) and Granger-causality test are went through.. Lastly, to derive the coefficients of the PVAR model, I have conducted variance decomposition for $\boldsymbol{LnGDP_{it}}$. The empirical results are listed as follows.

Since some of the variables failed the stationary test, the variables were differentiated first-order. The results show that all variables are stationary after the first-order difference. The cointegration relationship among GDP and other variables exists because p value is less than 0.05 in cointegration test. Further testing reveals that the roots of the characteristic equations converge within the unit circle, which means that the model is stable and there are convergent solutions to the differential equations.

Table 2. Granger-causality test

Null hypothesis	Obs	F-statistic	P value	A/R
A does not Granger Cause B	**192**	—	$\mathbf{p \geq 0.05}$	**Accepted**
A:IFRQAB:LNGDP	192	3.80677	0.024	Rejected
A:LNGDP B:IFRQA		0.03594	0.9647	Accepted
A: IFPQA B:OPENESS	192	3.26757	0.0403	Rejected
A:OPENESS B:IFPQA		0.55975	0.5723	Accepted
A:IFVAR B:OPENESS	192	13.1366	5.00E−06	Rejected
A:OPENESS B:IFVAR		0.3632	0.6959	Accepted
A:IFPQA B:INFLATION	192	9.76876	9.00E−05	Rejected
A:INFLATION B:IFPQA		2.3305	0.1001	Accepted
A:IFQEV B:IFPQA	192	9.59883	0.0001	Rejected
A:IFPQA B:IFQEV		0.6388	0.5291	Accepted
A:IFRQA B:IFPQA	192	8.50195	0.0003	Rejected
A:IFPQA B:IFRQA		0.93642	0.3939	Accepted
A:IFRQS B:IFPQA	192	0.14838	0.8622	Accepted
A:IFPQA B:IFRQS		3.25127	0.041	Rejected

Note: From the Granger-causality test above, there was significant unidirectional Granger-causality from the quarterly mean of stock index futures to economic growth; From the quarterly moving average to the openness; From the VaR of stock index futures to the openness; from the quarterly moving average to the inflation rate; From the quarterly moving average to the quarterly volatility; From the quarterly mean to the quarterly moving average; From the quarterly trading volume to the quarterly moving average. Granger-causality tests between other factors(such as openness and LnGDP, Inflation and LnGDP) have been done and their relationships are bidirectional

The two-variable causality test proposed by Granger can determine how useful some variables are in predicting other variables, and thus provide a basis for the choice of variables in the PVAR model. As shown in Table 2, some of the proxy variables of financial derivatives reject the original hypothesis that the proxy variables of financial derivatives are not Granger causes of LNGDP at 1% or 5% significance level, while other original hypotheses are accepted, indicating that there is a one-way Granger causality nexus between financial derivatives(stock index futures in this case) and economic gain(GDP growth), which shows that financial derivatives have a significant impact on GDP growth, while the growth has no significant effects on financial derivatives.

Table 3. LnGDP variance decomposition

Period	LNGDP	OPENESS	INFLATION	IFPQA	IFQEV	IFRQA	IFRQS	IFVAR
1	100	0	0	0	0	0	0	0
2	60.5959	22.3381	0.3011	1.3173	0.0791	7.3905	7.3587	0.6194
3	52.5461	27.2695	0.1450	1.3324	0.4049	8.3641	5.8018	4.1361
4	47.1822	27.9393	0.2875	1.8718	0.5853	10.2352	5.2344	6.6643
5	45.9121	27.5283	0.5600	1.7989	0.5825	9.9098	5.5216	8.1868
6	44.6617	26.6965	1.0793	3.8907	0.6105	9.6137	5.5088	7.9388
7	43.1540	24.4422	1.0091	9.0651	0.5867	9.0992	5.2080	7.4358
8	45.2797	22.1896	0.8989	10.6317	0.6149	8.3609	5.4059	6.6184
9	47.7759	19.8212	0.7899	9.3443	0.9960	8.1251	6.8302	6.3174
10	46.7219	18.6242	0.7508	8.8926	1.2326	8.5742	8.2075	6.9962

Table 3 shows the variance decomposition results of LnGDP, which indicates how macroeconomic factors and variables of stock index futures affect the GDP. The results show that market openness and the quarterly moving average have a large impact on GDP, and the impact is largest in the lagged period of four periods, while the impact of inflation rate and the quarterly trading volume is smaller. The impact of the quarterly moving average and the quarterly risk value are reflected in the long-term impact, which is more obvious when the lag period is longer, while the impact of the quarterly volatility of futures is larger in the immediate period and the long-term period, and is non-linearly distributed.

Table 4 evinces the test results of the panel regressions, which show that market openness, the quarterly moving average of futures indices, and inflation have more significant positive effects on economic growth.

Table 4. Panel regression of LnGDP

	FE	PCSE
Openess	−62.4033***	−105.3148***
	(00000)	(00000)
Inflation	0.00973	0.2755***
	(0.1120)	(00000)
IFPQA	9.06e−06***	−0.00002 ***
	(0.0020)	(00000)
IFQEV	−0.0013**	−0.0037
	(0.0300)	(0.38)
IFRQA	−11.2162***	10.6915
	(0.0060)	(0.565)
IFRQS	−1.5744	−1.2186
	(0.4190)	(0.930)
IFVaR	−0.3853	4.4588
	(0.6330)	(0.450)
CONST	11.1449***	11.1786***
	(0.0000)	(00000)
Observations	192	192
R-squared	0.3572	0.5441
No. of countries	6	6

6 Conclusion

This study, adopting mainly the quarterly data for six major countries, investigates how the performances of PVAR model for estimating the causality among GDP and other variables. Notwithstanding financial risk management is of great significance to financial derivatives and economic growth, it seems difficult for the model to explain how the risk indicators affect the gross domestic product of a country, hampering the governments or financial institutions to put forward the effective and enduring method of risk management that can enliven the derivatives market.

However, my research did find existing evidence among economic growth, financial derivatives, and financial risk management factors. First, based on the tests of Granger Causality, LnGDP variance decomposition and panel regression—the PVAR estimation, the findings reveal that trade openness, quarterly moving average of the futures index and inflation rate are greater positive correlated with economic growth than other variables. Second, the direction of Granger causality among financial indicators and GDP internationally is confirmed based on selected data: Apart from the one-way causality from the quarterly average of stock index futures to LnGDP, these six nations experienced bidirectional causal relationship among economic growth, risk factors of financial derivatives market and other macroeconomic variables.

As the results of estimation does not plausibly embody the essentialities of financial risk factors, the model has to be refined and the risk measures might be taken as, in any cases, the control factors to examine the nexus between financial derivatives and economic growth such as futures, options and swaps in the markets again.

References

1. Levine, R.: Finance and growth: theory and evidence. Handb. Econ. Growth **1**, 865–934 (2005)
2. Lucas, R.E.: On the mechanics of economic growth (1988)
3. Khan, M.S., Ssnhadji, A.S.: Threshold effects in the relationship between inflation and growth. IMF Staff Papers **48**(1), 1–21 (2001)
4. Pagano, M.: Financial markets and growth: an overview. Eur. Econ. Rev. **37**(2–3), 613–622 (1993)
5. Vo, D.H., Van Nguyen, P., Nguyen, H.M., Vo, A.T., Nguyen, T.C.: Derivatives market and economic growth nexus: policy implications for emerging markets. North Am. J. Econ. Financ. 100866 (2019)
6. Kolb, R.W., Overdahl, J.A. (eds.): Financial Derivatives: Pricing and Risk Management, vol. 5. Wiley, Hoboken (2009)
7. Cornett, M.M., McNutt, J.J., Strahan, P.E., Tehranian, H.: Liquidity risk management and credit supply in the financial crisis. J. Financ. Econ. **101**(2), 297–312 (2011)
8. Brunnermeier, M.K.: Deciphering the liquidity and credit crunch 2007–2008. J. Econ. Perspect. **23**(1), 77–100 (2009)
9. Sorkhabi, R.: The oil price crash of 2020: causes, consequences and historical context. Geol. Today **36**(4), 140–145 (2020)
10. Chen, C.W., Gerlach, R., Lin, E.M., Lee, W.C.W.: Bayesian forecasting for financial risk management, pre and post the global financial crisis. J. Forecast. **31**(8), 661–687 (2012)
11. Haiss, P.R., Sammer, B.: The impact of derivatives markets on financial integration, risk, and economic growth. Risk Econ. Growth (2010)
12. Şendeniz-Yüncü, İ., Akdeniz, L., Aydoğan, K.: Do stock index futures affect economic growth? Evidence from 32 countries. Emerg. Mark. Financ. Trade **54**, 410–29 (2018)
13. Obstfeld, M.: Risk-taking, global diversification, and growth, No. w4093. National Bureau of Economic Research (1992)
14. Devereux, M.B., Smith, G.W.: International risk sharing and economic growth. Int. Econ. Rev. 535–550 (1994)
15. Acemoglu, D., Zilibotti, F.: Was prometheus unbound by chance? Risk, diversification, and growth. J. Polit. Econ. **105**(4), 709–751 (1997)
16. Angeletos, G.M., Calvet, L.: Idiosyncratic production risk, growth, and the business cycle (No. w9764). National Bureau of Economic Research (2003)
17. Storesletten, K., Telmer, C.I., Yaron, A.: Cyclical dynamics in idiosyncratic labor market risk. J. Polit. Econ. **112**(3), 695–717 (2004)
18. King, R.G., Levine, R.: Finance and growth: Schumpeter might be right. Q. J. Econ. **108**(3), 717–737 (1993)
19. Arestis, P., Demetriades, P.: Financial development and economic growth: assessing the evidence. Econ. J. **107**(442), 783–799 (1997)

20. Ouyang, Y., Li, P.: On the nexus of financial development, economic growth, and energy consumption in China: new perspective from a GMM panel VAR approach. Energy Econ. **71**, 238–252 (2018)
21. Rodrigues, P., Schwarz, C., Seeger, N.: Does the institutionalization of derivatives trading spur economic growth? (2012). SSRN 2014805
22. Vo, D.H., Huynh, S.V., Ha, D.T.T.: The importance of the financial derivatives markets to economic development in the world's four major economies. J. Risk Financ. Manag. **12**(1), 35 (2019)

Blockchain in Supply Chain: Great Potentiality for Perfecting Logistics Information Transmission

Yujiao Qiu(✉)

Department of Supply Chain Logistics, Soochow University, Suzhou, China

Abstract. In modern society with the rapid development of technology, the application of blockchain shows great potential. Currently, the supply chain logistics model has a series of problems in information transmission, which cannot meet the increasing demand of the industry. However, blockchain technology can effectively make up for these defects. Blockchain which based on de-centralization technology ensures the security and reliability of information, thus greatly improving the transparency and efficiency of the supply chain. In the future, the combination of blockchain technology and IoT can also further boost the development of the supply chain. However, in the current practical production, there are still some problems between the blockchain and the supply chain, making the degree of integration of the two in the application layer is not high. This paper aims to discuss the application potential and current challenges of blockchain technology in supply chain logistics in the future and to provide ideas for the application research of intelligent logistics system based on blockchain.

Keywords: Blockchain · Supply chain · Logistics · Information · Decentralization

1 Introduction

In today's fast-growing global information age, the application of emerging technologies has become a particularly important and urgent issue in economic society. blockchain technology is certainly one of them.

Blockchain technology is essentially a decentralized database [1], a model for the application of new computer technology that Bitcoin has made widely known. Since Nakamoto came up with Bitcoin, the blockchain technology behind Bitcoin has had huge potential applications in all walks of life. Bitcoin, for example, is a new type of electronic currency not issued by a central bank such as the Bank of England but controlled by a decentralized network of computers. The network relies on cryptography and other technologies to manage the supply of Bitcoins and track who owns them. Hence, Bitcoin is also called cryptocurrency. The bank records the customer's balance in a ledger and Bitcoin uses a ledger as well, but it is maintained by a decentralized network of computers known as a distributed ledger. When a new batch of entries is added to the distributed

C. Yuan et al. (Eds.): *Proceedings of the 4th International Conference on Economic Management and Green Development*, AEPS, pp. 264–272, 2021.
https://doi.org/10.1007/978-981-16-5359-9_31

ledger, they contain a reference to the previous batch of entries so that all participants can verify the true origin of all entries in the ledger themselves. These batches are called blocks and the entire collection is a blockchain. No Bank or other single third party records the ledgers and verifies transactions and no single entity controls the ledgers. Therefore, blockchain technology was created to transfer value, especially in the context of digital currency Bitcoin. For its part, the main use of blockchain includes payments and other financial transactions. Blockchains implement peer-to-peer transactions by eliminating the need for a trusted intermediary to verify transactions, a role that is necessary when peers do not know or trust each other.

Therefore, according to blockchain's essence, its basic characteristics can be summarized as followed: First, decentralization. In the whole blockchain network nodes, they can be classified as accounting, which completely avoids the operation of the center. During the process of data, it is no longer dependent on the central node and each client can record, update, store, and transfer data in real-time, making the information more open and transparent. Second, openness. This is for the blockchain common chain because the information of the common chain can be read and written by anyone as long as it is the node of the whole network system where nodes with accounting rights can be carried out. This kind of network storage data is based on the cryptology principle. Through the blockchain network in the system, data can be updated in time on the sharing platform, which improves the speed of information circulation and transparency of enterprises [2]. Third, tamper-proof. Each node on the blockchain can receive a complete copy of the database from the system in the form of a subdatabase. It is invalid for a node to tamper with the database unless it can control more than 51% of the nodes in the whole system at the same time. Otherwise, it will not affect the data of other nodes [3]. Fourth, anonymity. The anonymity of blockchain is a cryptographic technology used to protect privacy, such as deterministic wallet, coin mixing mechanism, ring signature technology, zero-knowledge proof, etc., which greatly guarantee the secrecy of blockchain information. Fifth, traceability. The mechanism of the blockchain is that the last block has a hash value for the previous block, just like a hook, which can only be hung up if the hash value in front is identified. It is a complete chain. This makes it easy to query the data because the block is uniquely identified. So, consumers can authenticate the products they consume based on the seller and the logistics data in each link.

Thus, the advantage of the blockchain embodies that it creates transparency and security for digital transactions because data is classified and encrypted. It also breaks the traditional trading model of centralized control with trusted third-party organizations at its core. Neither information nor funding is dependent on the presence of intermediate entities, which makes the blockchain more reliable and stable [4]. The blockchain, for example, offers a new trading and payment system for financial markets [5].

Due to the advantages of blockchain technology (consensus mechanisms, timestamps, etc.), this technology can be applied to most industries such as health care and energy markets. Recent research, however, has focused on distributed ledgers and financial transactions [6]. By contrast, there are not so many papers focusing on supply chains and blockchain. Nevertheless, as the supply chain structure is more and more complex, the stakeholders are more and more diverse and hard to tackle. There are still some problems to be solved, such as the imperfect response mechanism of the supply chain, the uncoordinated links of storage, purchase, transportation, and distribution, and the imbalance of supply and demand caused by the non-transparent information of the whole supply chain. Centralized Management Systems also face corruption, fraud, and tampering in the supply chain. Under the environment, blockchain technology is a necessity. On the one hand, blockchain technology not only provides a good solution for identity management in the supply chain [7], but also realizes mutual or multi-party trust due to its local decentralization and non-destructive modification [8]. On the other hand, the application of blockchain technology can improve efficiency and security, increase transparency, and reduce costs [9]. Meanwhile, blockchain can be combined with IoT technology to realize product tracking, traceability, and real-time measurement [10].

2 Role of Blockchain in Solving Deficiencies in Traditional SCM

2.1 Opaque Information

In the supply chain, the information among various main bodies is often difficult to share transparently, which results in the asymmetry of information between the main body. Furthermore, trust problems that arise from the asymmetry of information will circumscribe the supply chain to improve its efficiency. The circumstance may be on account of its low level of information, but also often due to business game between upstream and downstream [11]. If the information can be shared transparently, it will help companies to avoid the monopoly of a central information source and to better select suppliers and measure the performance of the corresponding development plan. Or more specifically, companies can track product life cycles and assess green sustainability potential by providing visual historical data and characteristics of product transactions (source, quantity, quality, time). These can greatly reduce the company's risk and cost. When it comes to costs, among the three types of transaction costs (transportation, management, and information), the main cost type facing supply chains today is information cost. Therefore, enterprises need to solve how to share information with their partners under different visibility to ensure the authenticity, integrity, and invariance of data and to minimize the negative impact of trust issues on the supply chain echelon.

Aiming to solve basic problems of information transparency, traditional logistics chooses to build a modern information base platform and an integrated supply chain management information system to strengthen the information flow efficiency among enterprises. However, faced with how to master the scale of information sharing and how to limit the information capacity, traditional logistics enterprises rely more on the government and industry authorities to formulate the corresponding supervision policy and management system for external constraints. Also, in order to achieve information sharing, traditional logistics enterprises choose to use some incentives like "bulk discount", "pricing discount" etc. to promote members on the chain actively opening information.

But platforms of information sharing tend to make information centralize in one center, such as an ERP system. While easy to navigate, it's harder to keep a stream of information relatively intact when such platforms are under attack. Besides, government mechanisms are less flexible when some of the companies' private information needs to be protected or encrypted in time. What's more, incentive mechanisms are often designed to redistribute profits based on information sharing or to carry out price incentives but it is also easy to cause problems concerned about uneven distribution or unreasonable pricing, leading to unobvious incentive effect.

The application of blockchain technology makes up for the above problems. From the perspective of the openness of blockchain, it can better eliminate the trust asymmetry and coordinate activities between both sides more effectively. Besides encrypting the private information of each party in transactions, the blockchain's data is open to all, making the system highly transparent. From the perspective of the decentralization of the blockchain, to achieve the information equivalence and improve the trust degree of members in the supply chain, the distributed database of the decentralization of the blockchain allows all the participants to record and share at the same time. It not only ensures the reliability and integrity of the transaction data information but also makes information open to upstream and downstream enterprises. Thus, information such as demand changes can be fed back to every subject in the chain in real-time and every enterprise can keep abreast of the latest progress of logistics. For example, raw material producers, manufacturers, distributors, and retailers can share storage information and sales information with the help of a logistics information system constructed by blockchain technology. This will improve the authenticity of point-of-sale (Pos) or point-of-purchase (POP) information. Under the "decentralized" supply chain coordination, each link's data platform will become a pedigree of distributed public accounts. Through the blockchain consensus algorithm, each node reconstructs the trust mechanism with the "peer-to-peer" network protocol, which transforms the initial distrust among the subjects into the trust of the blockchain mechanism itself. On this basis, the mode of centering on the traditional logistics company and transferring logistics information at different levels will also be transformed to a new mode of transferring logistics information with no hierarchy and sharing of information [12]: merchants, users, logistics enterprises, and other multi-cooperation agents will jointly construct and update information data in an equal capacity. Moreover, it effectively circumvents the possibility that logistics data information is manipulated by a central subject and ensures the visibility and reliability of information flow.

2.2 Unsafe Logistics Transmission

Nowadays, the chain of a supply chain is getting longer and longer and some problems, like data fraud and loss, personal information leakage, appear frequently in the process of information transmission. For example, if some data is not integrated into the whole supply chain information system or the full channel participation of the information system is not realized [13], it will be hard to pinpoint exactly where the problem is. Also, during the process of logistics transmission, guaranteeing whether a party in the commodity has provided true and reliable information is difficult to realize. Thus, when products are found to have problems of quality or safety, it will have difficulty judging the subject of liability and tracing back the source of problems.

Based on the overall direction of the supply chain, in face of data fraud and loss, the main bodies of the traditional supply chain are often unable to avoid such problems in advance. Companies generally conduct constraints through business models, such as withholding performance fines upon discovery of false information. Also, to avoid information disclosure, individuals are generally forced to conduct self-restraint by establishing incentive mechanisms or sign confidentiality agreements before cooperation. However, none of these methods can directly identify the subject of problems. Instead, all parties just take the initiative to ensure information security through after-action punishment or positive incentive, which makes each subject hide the fluke mentality of evading responsibility and buck-passing behavior. Subdivided into specific logistics links, most of the domestic logistics and express companies provide a mobile phone to inquire a tracking number so customers can inquire about the arrival time of the goods at each transfer station. But until the end customer receives the goods, they know little about the specific transportation of goods. Once the goods are found to have problems and the customer wants to trace back afterward, accurate monitoring information will often be missing. Even if not missing, there is the possibility of tampering and hiding.

Blockchain uses distributed accounting to link supply chain information together in a complete and orderly manner after accurately recording the information of each node. This interlocking feature makes it very difficult to change the contents of a blockchain but easy to trace, which makes the information traceable and tamper-proof. It also guarantees the authenticity of the data recorded on the chain. Also, if someone intentionally enters false information in the first place, it will be received by all parties as information on all nodes is updated simultaneously and the evidence of fraud will be permanently recorded. This makes counterfeiting more expensive. Therefore, the possibility of source data fraud is greatly reduced. These characteristics of blockchain overcome opportunism behaviors such as the responsibility shifting of participants in traditional logistics operations and provide an effective guarantee for the coordination of information in the supply chain. It also improves the degree of trust and transaction efficiency and provides reliable traceability for the responsibility division of supply chain subjects. In Singapore, for example, 2016 Fresh Turf is using blockchain technology to build a network of lockers that connect to lockers, transfer logistics, and help users track parcels. Because the parcel can be uniquely identified on the blockchain, the transaction trust problems such as the difficulty of adducing evidence and accusing and accusing can be solved [15]. Also, the anonymity of blockchain technology can be achieved through cryptography on each node to achieve selective encryption. It largely makes information "selective visible". For example, the "zero-knowledge proof", which is very promising in blockchain technology, cannot only fully prove that they are the legal owners of certain rights, but also do not leak out the relevant information, further ensuring the security of information.

2.3 Asymmetric Logistics Information and Inefficient Cooperation

The supply chain covers different stages such as raw material procurement, transportation, production, and storage. Each stage has different levels of suppliers, so the extension and complexity of the supply chain make it difficult to improve the efficiency of information transmission in the first place. And under the original business model, paper-based waybills were often more secure because electronic versions will be tampered with and impossible to be authenticated. It can't have the same effect as the paper document with an official seal [15]. But the paper order also brings about low efficiency of information sharing and inaccurate information transmission, which aggravates the bullwhip effect and increases the communication cost. Coupled with the surge in emergent problems of supply chain logistics during the outbreak of coronavirus, the issue of inaccurate information transmission and high communication costs has become increasingly prominent. Therefore, how to transfer information quickly and accurately to reduce inefficient communication in the supply chain is an urgent problem to be solved.

To solve the problem of communication efficiency, parties in the supply chain will centralize the related data of different levels through platform integration to the lean supply chain. But once the integration is not completely realized, troubles like link paralysis or missing consolidated information will emerge. Although enterprises encourage using electronic documents to save time and cost, there is still no strong technical guarantee of the reliability of the information.

Since the outbreak of the epidemic, the supply chain logistics business, which relies on centralized information management, has been seriously interrupted. Many processes in emergency logistics and each process is related to form a community of destiny, so there is a risk that the whole emergency supply chain will be paralyzed. Moreover, in this epidemic, the information is updated quickly, involving many issues and being highly intrusive but platforms cannot handling with centralized processing of data immediately. So, information asymmetry or even mistakes will occur during information transmission [16]. However, if we can use the distributed bookkeeping principle of blockchain, many problems can be easily solved. On the one hand, we can help the upstream and downstream enterprises to establish a secure distributed bookkeeping. Since the information in bookkeeping is open to all trading parties, monitoring each other is feasible. On the other hand, through the "smart contract" technology of blockchain, the content of the agreement between enterprises is recorded in the form of code on the account book. Once the terms of agreement come into effect, the code will be automatically executed to realize the self-verification, self-transmission, and self-management of transaction data. For example, when a buyer transacts from a supplier, a contract can be created on the chain. The contract is to send the payment to the supplier when the logistics data indicates that the goods have arrived at the destination. As soon as the arrival message is sent, the money will be transferred automatically. Besides, the blockchain technology can make data be backed up in each node of cooperative subjects so that the integrity of the information system will not be affected even if data of any single node is damaged. The intelligent contract mechanism of blockchain enables the logistics system to be automatically executed by referring to the agreed standard process without being affected by external conditions. Under this mechanism, there is no need to repeat communication between logistics entities, and unnecessary intermediate links are eliminated. It also enables the free operation and change of information among nodes, which effectively avoids the waste of resources caused by repeated processing and transmission of information and reduces the cost of information processing.

3 Prospect and Challenges for the Integration of Supply Chain and Blockchain

In the future, blockchain can provide greater scalability for the original supply chain and form a larger scale of supply chain management, without worrying about the management chaos between transactions and the tedious audit required by internal systems and processes. In this way, more legitimate supply chain entities will participate in sharing regardless of the geographical location of the members, thereby eliminating any single point of failure. The location flexibility of blockchain greatly increases the time efficiency of interaction, making it a new way of global collaboration [17].

In addition, on this basis, the supply chain is expected to open up a new business model that can further integrate and analyze information flow, business flow, and logistics based on the combination of IoT and blockchain with wider operability. Take the storage link of the supply chain as an example. In terms of business flow, the technology means of "IoT + blockchain + big data" can realize the warehouse control form integrating transaction risk management, liquidity management, and financial platform docking which can endue real property to the chattel in the warehouse. In terms of logistics, combined with RFID, radar and other IOT technology, the operation process of each link in the supply chain process is further reasonably. In terms of information flow, the key node data of participants are registered through blockchain to achieve the purpose of intelligent goods control and guarantee the authenticity of data.

Although the combination of blockchain and supply chain will bring huge advantages, in the current real economy, the degree of integration of blockchain technology in supply-chain related industries is not very high. The reasons that affect, explicitly or implicitly, the landing of blockchain in the supply chain are as follows.

(1) Technical difficulties
For the supply chain industry, it takes a lot of investment to be able to put these combined technologies to work. Besides, it is very difficult to ensure the data privacy of all parties in the sharing of data. And there are very few people left to work on and continue to work on blockchain technology. Therefore, the "scale" effect of blockchain technology has not been formed yet.

(2) Blockchain's dilemma
The current blockchain technology has encountered an "impossible triangle", that is, it cannot achieve scalability, decentralization, security at the same time. For example, as the information of unit blockchain on the supply chain becomes more, it needs to improve its malleability, but the cost is to make the algorithm simpler and easier to verify, so it cannot guarantee security. Also, there is no specific reference application for cross-chain interaction between different underlying architectures of blockchain or between blockchain and non-blockchain. On the most important "transaction performance", the throughput of blockchain is still extremely low. The performance problem is still the shortboard of the running blockchain, which makes the technology of the blockchain unable to link up with the real economy. When blockchain technology is booming and all parties put into the blockchain, there is also a need to balance the benefits of the mechanism.

(3) Imperfect environment and supporting system of blockchain
The implementation of blockchain technology is faced with the problem of cognitive environment and ecological supporting facilities. On the one hand, the public's understanding of blockchain technology generally stays at the stage of Bitcoin and the blockchain itself is obscure and difficult to popularize. On the other hand, the blockchain technology still needs to be improved in all aspects, especially in terms of the solution of extensibility (TPS), the guarantee of the authenticity of data,

the certainty of execution mechanism of intelligent contract, and the protection of intelligent contract by law (contractual loophole). The hysteresis makes it difficult for blockchain to be implemented on a large scale in the field of the supply chain.

(4) Immature incentives
Adding incentives to ensure that as many valuable participants as possible are willing to join the database is a big indicator, known as uplink data. The uplink data means that existing stakeholders or groups may have to be broken up, like a chain of medical records or medical records. But if there is no data injection or little data on the blockchain, network benefits will not form. Maersk and IBM, for example, kept the database growing by getting other big shipping companies to participate in their blockchain-related global shipping projects. However, not all companies have the charisma of Muskie or the purchasing power of Walmart to negotiate with their suppliers to persuade them to participate in their blockchain projects. Another incentive is to link the costs and benefits of information sharing, such as supply chain financing, but the credit of small and medium-sized enterprises still needs to be further judged. Of course, if you can issue tokens directly to the early participants, it will attract more people later on. With token as the carrier, digital rights and assets can be realized, which makes the value circulation more efficient and transaction cost lower. However, it is difficult for some enterprises related to supply chain logistics to have the technical support to issue digital currency, and relevant laws in China do not clearly define the illegal issue of digital currency. Therefore, the failure of coming up with an incentive mechanism to attract participants in the early stage is one of the reasons for the lack of better development of blockchain in the supply chain.

4 Conclusion

Under the background of blockchain development, the paper focuses on how to solve the pain points in the current supply chain through blockchain technology and analyzes the advantageous prospects of blockchain technology and some existing problems. Based on the basic characteristics of blockchain, this paper introduces the successful application of blockchain in the limitation of the traditional supply chain and does a summary. This paper aims to analyze and solve existing contradictions through in-depth communication between the two fields and takes an insight into the future of blockchain technology in the supply chain field.

References

1. Salah, K.: Blockchain for AI: review and open research challenges. IEEE Access **7**, 10127–10149 (2019)
2. Ruan, X.: A research on financial problems of supply chain finance model based on embedded blockchain technology—taking ant financial as an example. Shanxi Nongjing **274**(10), 168–169 (2020)
3. Feng, T.: An agri-food supply chain traceability system for China based on RFID & blockchain technology. In: 2016 13th International Conference on Service Systems and Service Management (ICSSSM), Kunming, pp. 1–6 (2016), https://doi.org/10.1109/ICSSSM.2016.7538424

4. Yli-Huumo, J.: Where is current research on blockchain technology? - A systematic review. Plos ONE **11**(10) (2016)
5. Ikeda, K., Hamid, M.N.: Applications of blockchain in the financial sector and a peer-to-peer global barter web. Blockchain Technol. Platforms Tools Use Cases **111**, 99–120 (2018)
6. Lee, J.H., Pilkington, M.: How the blockchain revolution will reshape the consumer electronics industry. IEEE Consum. Electron. Mag. **6**(3), 19–23 (2017)
7. Alam, M.: Why the auto industry should embrace Blockchain. In: CarTech (2016)
8. Bi, Y., Wang, C.: Analysis of the access system of photovoltaic power station based on photovoltaic power. Agric. Plant. Hybrid **4**(31), 95–99 (2016)
9. Al-Jaroodi, J., Mohamed, N.: Blockchain in industries: a survey. IEEE Access **7**, 36500–36515 (2019)
10. Wang, Y.: Making sense of blockchain technology: how will it transform supply chains? Int. J. Prod. Econ. **211**, 221–236 (2019). https://www.sohu.com/a/295644821_742092
11. Huang, J., Gao, L., Xu, Y.: The design of smart contracts on crowd funding private blockchain. J. Inf. Secur. Res. **3**, 211–219 (2017)
12. Tian, Z., Xiaoyue, S., Pingwen, Q., Dongge, L.: Research on optimization of intelligent logistics model based on blockchain. Cyberspace Secur. **11**(09), 78–83 (2020)
13. Wang, H., Chen, K., Xu, D.: A maturity model for blockchain adoption. Financ. Innov. **1**, 17 (2016)
14. Jia, J., Xuguang, C.: Application status and future prospect of blockchain technology in logistics. Sci. Technol. Logist. **42**(08), 40–42 (2019)
15. Bintao, L.: Study on emergency supplies reserve supply model in big data environment. Value Eng. **35**, 39–41 (2019)
16. Litke, A.: Blockchains for supply chain management: architectural elements and challenges towards a global scale deployment. Logistics **3**(5) (2019)

Drawing Fairs and Green Development

Yourglich Rachel(✉)

Department of Fine Arts, Paris College of Art, 15 rue Fenelon, Paris, France
rachel.yourglich@paris.edu

Abstract. This conference paper examines the recent shift in today's art market dedicated specifically to drawing. The main scope of this paper intends to answer the question of how the market of drawing fits into the global art market as a whole and what the effects art and drawing fairs have on the environment. Through the calculations of environmental issues, such as carbon footprint and the offset of carbon emissions, one is able to understand the impact that these fairs have on green development. Qualitative research has been conducted in the form of gathering information through books and articles along with a series of interviews. Those interviewed are closely involved in various institutions public and private, such as The Drawing Now Art Fair in Paris and Art on Paper, Brussels; these institutions all focus on the practice of drawing within the contemporary art world. The implications of this paper are to understand the impact that drawing fairs have on the environment in contrast to art fairs in general.

Keywords: Contemporary drawing · Drawing fair · Art fair · Art market · Environment · Carbon footprint · Offset carbon footprint · Green development

1 Introduction

One of the biggest players in the economic mechanics of the art market are the fairs [1]. Whether they are art fairs showcasing a variety of artworks and artists, or more niche markets, which focus on drawing, photography or video art [2], these fairs greatly impact the art world as a whole [3].

The top art fairs that occur worldwide, listed in order of their prestige are: (1) Art Basel, Basel, (2) Frieze Art Fair, London, (3) TEFAF, Maastricht, (4) The Armory Show, New York, (5) FIAC, Paris, (6) Art Dubai, Dubai, and (7) Scope Art Show, New York, Basel, Hamptons, London, Miami [4].

In the 2020 Art Basel report, dealers reported that \$2.5 billion or 15% of these sales were made before the fair began, while \$10.6 billion or 64% of sales were made during the fair, and \$3.5 billion or 21% of sales were made after the fair had ended [5].

2 Drawing Fairs

A drawing fair is similar to an art fair in that it is also a marketplace for selling and exhibiting works of art, but those classified as drawings, not in any other mediums. It

C. Yuan et al. (Eds.): *Proceedings of the 4th International Conference on Economic Management and Green Development*, AEPS, pp. 273–285, 2021.
https://doi.org/10.1007/978-981-16-5359-9_32

should be noted that the clientele of art fairs versus those of drawing fairs are completely different. While most of those who attend the major art fairs of the world are the global millionaire and billionaire class, or the one percent of the wealthy class in the world [6], drawing fairs have a much more local clientele, with most of their attendees being from central Europe [7].

The largest, most respected drawing fairs occur mostly in Europe, with the following list being provided as an example of the contemporary fairs that take place: (1) Drawing Now Art Fair, Paris, France, (2) DDESSIN, Paris, France, (3) Art on Paper Brussels, Brussels, Belgium, (4) Drawing Room Fair, Madrid and Lisbon, (5) Draw Art Fair, London.

Among the popular art fairs were drawing fairs and other niche medium-based fairs occurring in Europe. The Fair Director of Art on Paper, Brussels, Gilles Parmentier was interviewed for this research. He has been managing the drawing fair for two years and his role is based around developing and organizing the quality of the fair and working with galleries to maintain sustainable relationships [8]. Parmentier states that within the sector of the art market, drawing is particularly easy to show and to sell, as it generally encompasses small format works. This enables easier transportation, exhibition, and storage of the works of the artists, galleries and collectors involved [8]. He affirms that drawings are, in general, less expensive than paintings and this difference in price point has allowed collectors, even young collectors, to acquire works. The collectors coming from major art fairs find the artists at drawing fairs, like Art on Paper, Brussels more easily accessible. Art on Paper, Brussels is particularly sought after for collectors who are less well off, in that they can both collect drawings at reasonable prices and dine with the artists [8].

Parmentier says drawing fairs are putting drawing in the spotlight and organizing fairs in a way that brings together all the sector's actors. This creates a goodwill for drawing [8]. This goodwill for drawing has been increased further, as Elizabeth Tenenbaum, a private curator, advisor and appraiser, as well as the manager of private collections, such as for the JoAnn Gonzalez Hickey (a collection of drawings), states that "auction houses began adding drawing by contemporary artists to day and evening contemporary sales rather than segregating drawing into prints and drawings sales," which has increased the value, prestige and consideration of drawings [9].

Drawing fairs have been deemed particularly important for the future. Attracting more local clientele makes the fairs more environmentally friendly. Research shows that fairs like Art Basel impact the environment more heavily than niche fairs in this respect [8].

Don Thompson, author of *The $12 Million Dollar Stuffed Shark: The Curious Economics of Contemporary Art* affirms this: there are some large art fairs who chase "jet-setters" when instead they should "concentrate on the pool of collectors close to home, those who spend $5,000 to $50,000 at a time…every fair needs to know what its niche is [6]." That being said, Christine Phal, the President of Drawing Now Art Fair and creator of The Drawing Lab in Paris states that "we will have completely won when I will not be told that I have a «niche» art fair. I want to leave my «niche» (kennel in French) and leave my leash on the sidewalk [10]!".

3 Carbon Emissions Offsets

One can calculate their carbon footprint in order to determine the environmental impact of attending or shipping works to art fairs, as well as the carbon offsets required to "balance out" their carbon footprints [11]. According to Naomi Rea of Artnet, "Money can be invested in projects to plant trees and protect forests, or to fund renewable-energy programs," which often "take place in developing nations [11]."

Rea calculates that a team of five individuals traveling a round-trip economy flight from New York to Art Basel Miami on a commercial airline would emit between 2.43 and 3.3 metric tons of carbon dioxide. A business or first-class flight of the same destination would emit between the equivalent of 4.2 and 7.06 metrics tons of carbon dioxide. In order to offset their carbon footprints, those individuals would have to spend a total of between $24.60 and $71.48 depending on their seat class [11]. It should also be kept in mind that these figures do not include the carbon emissions of those who fly on private jets [11]; as the majority of those who attend art fairs are of the one percent of the wealthiest class in the world [6], it is suspected that the number of those who fly on private jets is quite high.

The same formula can be used to calculate travel from London to Paris on a commercial airline [12]. If the same team of five were instead travel more locally by staying within Europe, the carbon emissions would be on 0.54 metric tons of CO_2 for the 432.15 miles traveled by flying economy. Those destinations traveled on business or first class would create an equivalent of 0.81 metric tons of CO_2 [12]. In comparison, the offset of carbon emissions would cost between $4.06 to $8.22 for a combined price of all five travels flying roundtrip. That creates a difference of between $20.54 and $63.26 saved carbon emissions offset dollars by attending more local drawing fairs such as Drawing Now Art Fair, rather than larger art fairs, such as Art Basel.

It is also important to consider the artwork that is shipped from the artist or gallery's location to the art fair itself. Rea uses the example of shipping the "eight 40-ton steel blocks that make up Richard Serra's *Equal* (2015) from his fabricators in Wetzlar, Germany to the new MoMA in New York" by cargo ship, which would equate to the emissions equivalent of about 73.5 tons of CO_2 [11]. But the work must also travel to and from the ports in Hamburg and New York, equating to an additional 20 tons of CO_2 emissions. In order to offset the emissions of transporting a work of this scale, it would cost $1,012.50 [11].

While the research of drawing fairs' carbon footprint and emissions are rather new, requiring further investigation, drawings are much more accessible and easier to transport than paintings and sculptures, as the scale is normally comparably smaller and lighter than the latter [8]. In addition, the location of shipment from gallery or artist's location to the location of the drawing fair is considerably less distant than art fairs as a whole. One can therefore conclude that the offset of carbon emissions would be much lower in comparison for shipment, although further research is still required to confirm exact numbers in this area of research.

As the pandemic of COVID-19 has had a dire impact on the art market at large, according to Patrick Heide, owner of the Patrick Heide Gallery in London and participator in drawing fairs in Europe, states that "what is apparent is that fairs will probably have a tough time next year" as less galleries will likely be able to attend [13]. Yet, while large fairs such as Art Basel will quite likely stay around, the number of art fairs who have a large budget without a comparable quality of art presented at the fair may see a reduction in participation and attendance or might even disappear. In contrast, drawing fairs, whose price points are typically much lower and have a very loyal clientele may be able to benefit from the financial crisis through accruing or maintaining participants and attendees from central Europe [13].

4 Conclusions

Rather than spending money to offset the carbon footprint of art fairs, the focus could, and arguably should, be placed on drawing fairs, which have a much smaller carbon footprint by nature. This alternative model has a more local clientele as opposed the much larger model of art fairs, such as Art Basel. It should be noted that this research is part of a larger body of work, which also worked to study the hypothesized effects that the pandemic of COVID-19 will have on the art market. The effects of the pandemic on larger art fairs expects to have a reduction in the number of fairs, as travel will be limited. This allows for drawing fairs to take a larger role in the art market, through the use of less travel and less shipment of artwork.

Appendix

Gilles Parmentier, Personal Interview, Conducted by phone on July 9, 2020.*
RY: So my first question is, can you provide a brief description of who you are, your relationship to the drawing market and your relationship to drawing?
GP: Okay, so my name is, uh, Gilles Parmentier and, I'm a fair director of Art on Paper Brussels. So I'm now managing the fair for two years, so I managed the two last editions, that means that we don't have a complete period, we don't have an edition this year but I've managed the two last editions, 2018 and 2019. Uh my background is more communication and management oriented so I'm not, uh, an expert in the art sector and in the drawing sector, but I'm, uh, generally advised by more of drawing specialist and also advised by an art director. So my role consists more in developing, uh, the fair of controlling the quality of the fair, the organization of the fair, in the communication of the fair in elements of the fair and in relationships with galleries, collectors and partners and institutions. So that consists of mainly my part of the job and it's really, really a wonderful job. And I manage in a team of two persons so I'm working more with the art director but also with a fair coordinator and someone who is taking care of the gallery section. Galleries are really important for us because they are our clients and partners. As organizators of the fair we are working with the galleries and the galleries are selected by a selection community each year, so they have to apply to take part in the fair. And when they are selected they can start to communicate with the fair and have contribution with the fair in Brussels with us. And every year we, I mean for the two last editions, we selected around 40 galleries because we have the chance to explore new locations in Bozare, so we have the bigger place. And that means we go up and that means we continue to improve the quality of the fair and also because edition, because compared to the first editions of Art on Paper, we are now doubling the participation of galleries.

RY: Yes, so what is your definition of a drawing, or when does a drawing stop being a drawing, in your mind?
GP: It's a really hard definition. I think that, and it's really interesting to discuss about that because in the whole drawing, not the drawing world, but I mean the drawing sector, artists, gallerists, art directors, fair directors, institutions, curators, everybody has their own definition and it's always difficult to find the word of drawing. My personal definition is that drawing starts with gesture and as Art on Paper, we're linking generally drawing to the paper. But it's really hard to find the start of the drawing because, for me, drawing is a gesture and it can start also in your mind. You can start drawing in your mind for such you are living your art piece and then realizing the art piece on paper for example of some piece of support. So drawing is really a world in itself.
RY: Okay, thank you. And what is the role of drawing in the art market?
GP: I think that the world of drawing is really evolving and evolving in the sense that 50 years ago, I cannot say that no one was interested in drawing. That isn't true. But we see that more and more people are starting to collect, to show, to propose, to realize also, drawings. And I feel that this is starting to increase from the year 2000 and there is a tendency, so we see now that there are more art fairs, but also more galleries, more artists starting to focus on drawing as a main practice. And I think it's a positive, it's a positive evolution. You will have to ask me why is it like that, why that situation. It's hard to analyze. But I think there are some facts. The first fact is that for artists but also for galleries, for the market sector these are pieces that are generally from small format, generally easy to take with you, to show, easy to, to, also sometimes to make in storage. And these are pieces where in the past decades were not that much showing. So I think it's interesting to show that again because its a way to a way to reveal starting the creation process of artists. Another thing is that drawings in general, comparing to paintings for example, in general are less expensive generally. And so that's also an opportunity for collectors but also for young collectors to start a collection. To start to discuss and enter and dine with artists. Another thing is that a lot of actors, like institutions also are starting to make drawing centers, to make drawing or institutions that focus on drawings. So you have a lot of exhibitions specially that focus on artists who are maybe practicing drawing but also drawing as a teen, so this is a contribution. It's a contribution that puts drawing on the spotlight. And also the fact that more and more fairs are organizing fairs that are specifically organizing drawings, fairs that bring together all the actors in the sector. And it creates also a simulation, it creates also a good will for drawing that we see now.
RY: And you mentioned the price point of drawing being less than like, say, a painting. Why do you think that is?
GP: I don't have a rational explanation. I'm not sure about that. I think it's historical, but I don't have a precise explanation about that because for me, drawings are, as a sort of gesture, are a way to explore an art piece with a lot of intention because you are really close to the creation, really close to the artist's, to what you want to express. And, for me, it has a lot of values because it's really beauty. Drawing doesn't lie. I mean, you cannot lie with drawings. You cannot use a lot of techniques to hide something. If you're looking at the drawings you are completely and directly consumed with the artist's intention. So for me it has a lot of value, but maybe it's undervalued.

RY: So, why is the economy of drawing a new and somewhat under researched field? Does it have to do with, what does it have to do with, or is it?
GP: I think, I think it's really totally an undersized and drawing will get higher and higher from the communities, the drawing fairs I mean. The role of drawing is evolving I mean taking more and more places. It's not, it's still I mean, in the like the southern market comparing it with paintings. It's a little bit like if you compare with photography against I feel that drawing is still finding like photography, it's totally different mediums, totally different approach but it's helpful to tie them together as recognition that it's normally not had.
RY: And why in your opinion were drawing fairs created?
GP: I mean, I think that drawing fairs are responding to a need. A lot of galleries, a lot of artists, as I said, are focusing on drawing. And there is a space in the art market because if you are comparing like Art Basel and niche and contemporary drawing fairs, niche fairs, so fairs that specialize in a dedicated medium has an increasing opportunity to propose another way to discover a practice. And I think that this kind of fair, like us, has a role to play to propose to discover to attract the attention to the public to new proposals, to new artists and new practice. And I think it's a good position, like niche fairs, are generally located on the first border market, but on the local market like we are focused on the European market mainly. And this fair has a like a human level, has a human, its not a transnational, it's not working worldwide. We have, we have a local level, which is really interesting because you can meet the artists, you can go to their workplace, you can discuss with them, you can follow them during the years. And we, I think our work proposal is really I think it will go further in the coming decades. And comparing with the huge worldwide fairs like Basel who will have, I think, some problems regarding the environment, regarding the need of people to go – the rush to go everyday or every week to another place around the world, just to buy art. I think it's a total different corporation…
RY: What type of success have you seen with drawing fairs?
GP: Lots of success. Just have a look at the importance that drawing is now taking in the main galleries around the world. Just look at the exhibitions specially dedicated on drawing. Just look at the artists deciding to draw as a main practice. I think drawing is really evolving and we have lots of, all sorts of transactions, lots of sellings of drawing everyday. I mean clearly, if you look at the auction houses, drawing is taking now a place more and more importance. But, it's still difficult. It's more drawing well-known international drawing pieces who can be sold with the internet. If you don't know the person, their piece, you have to feel it, to discover it, by your hands or by your eyes. And so it's all part of the business for the piece like the drawing of Leonardo DiVinci, then you can sell it around the world by the internet with auction houses. But for new artists, for emerging artists, you have to feel it and you have to [untranscribable word]…
RY: Okay. And what influence do auction houses play in the creation of drawing fairs?
GP: I think we have to work closely with them. It's difficult to find the balance, the good balance because at the fair we are mainly working with galleries and the artists, emerging artists or established artists. And the auctions are a little bit difficult because we see them as a competitor for galleries or for the fair. But I think there is a good complimentor we can feel and I think, for example, we are beginning to work with auction houses in the

coming years. As a partner, we can focus together on drawing. And I think if we have, the more actors you have working in the same direction, the more it will be positive for the sector…

RY: Okay, yeah. That makes total sense. Is there more or less respect towards drawing in the expansion of the drawing market?

GP: I think there is more. Definitely more respect towards drawings. And, as I said, it's a movement with more involved everyday or every year. So there is more and more respect for drawing, that's totally clear.

RY: Okay, yeah. Does this, does this rise in drawing fairs reflect a shift in the value of drawing?

GP: It's evolving, yeah. I think it's evolving. For emerging artists, it's still difficult to, I mean, to go upper quickly. You have like a window, a glass window which reach you to go a high level of price quickly. But for established artists, yeah, we've seen that the price of art pieces are really evolving, and we've seen that it's evolving on the high tradition. So yes, the rise of drawing fairs reflects also in the value of drawing. We've seen it. Every year we are selling, and the galleries are selling more and more drawings every year, more expensive.

RY: Okay. And where do you see drawing moving towards in the future artistically?

GP: That's really interesting and that's really more for the part of the art director. The practice of drawing is really evolving and every artist, a lot of artists are making evolving the [untranscribable word] of drawing. Drawing isn't only on paper, but it can be also on the wall, but it also can be on the sculpture, and the sculpture can be a drawing as well. So, I mean there is a recycle of resources which are not finished, and I think this is really, really exciting to see how drawing is evolving and how far it can go. Go farther and go back.

RY: Okay. And my last question is, how do you differentia–differentiate, I'm sorry, between the value of art from an artistic point of view and a financial point of view?

GP: This is a discussion we can have with a specially talking about the value of a drawing, I think the value of an art piece is really, is really changing right now. But the obvious point of view is, for me and a personal point of view, is you don't have works of, I don't know, a Jeff Koons, you know, people are paying millions to buy one. So, it's totally personal for me. And so, there is a real personal value based on your expertise, on your preferences, but also a more objective value on the work of the artist, his career, his traject of work, but also, the insurance that curators, galleries and institutions put on you. But, on the final standpoint of view, you know, today, the value of an artwork is really, is really difficult to define besides you have platforms that can give you a lot of information about the auctions of the selling price of an auction or an artist. But it's not objective, it's more subjective for me.

***full interview available by request.**

Elizabeth Tenenbaum, Personal Interview, Conducted by email exchange on August 5, 2020.

Artwiseinc.com

RY: Can you provide a brief description of who you are, your relationship to drawing and your relationship to the drawing market?

ET: I was an artist and I have worked in the contemporary art field for 20+ years. I ended up starting my own business managing private collections and eventually became a private curator, advisor and appraiser. I have managed the JoAnn Gonzalez Hickey collection which is a collection of drawings for the past 12 years. I also became the director of Syzygy-nyc.org, an academic platform through which the JoAnn Gonzalez Hickey collection is lent to colleges and universities for study.
RY: Can you provide a notion of what a drawing is? What constitutes a drawing from another medium for example? Or when does a drawing stop being a drawing?
ET: Drawing is a line in space or on a surface. Line determines the form. Most commonly drawing is defined by marks on paper but not limited to this definition. We often debate the difference between drawing and painting. The medium is less important than the mark making. Drawing can have dimension but the overall feeling would be of lines or marks made. I think of Fred Sandback yarn works as an example of this. The yarn delineates planes and architectural volumes, while others may say this is an installation or a sculpture, I still read them as drawings… No paper or pencils involved. There are works on paper which are more painterly and layered with paint. This I perceive as a painting on paper, rather than a drawing. Negative space plays a role in how a work on paper feels more or less like a drawing.
RY: Why is drawing so hard to define?
ET: As contemporary artists innovate, and push boundaries drawing is a prime playground for this sort of experimentation due to the accessibility and immediacy of drawing. There has been a major blur in artists identifying themselves by medium. Instead of identifying as a "sculptor", "painter", or "photographer" contemporary artists call themselves artists or interdisciplinary artists so as not to be pigeonholed into one way of making art. This shift has contributed to the difficulty we now have in defining drawing. Often contemporary artists seek new means of expression allowing for a great expansion in the definition of drawing. The avant-garde played a key role in this as progress is made in adapting to modern mainstream thinking… see next question.
RY: In your opinion, how has avant-garde influenced the definition of what a drawing is?
ET: The avant-garde, specifically dadaism used paper as a medium and a place to experiment. The key concept of the Dada movement was that art could be made of anything (such as the found object readymades Duchamp presented as art). World War I created some scarcity of materials. Artists still had access to paper and printed materials which naturally led to collages and drawings. The avant-garde presented the idea that art need not be beautiful rather more important to explore, innovate and rebel. Drawing isn't elitist or precious.
RY: Why is the economy of drawing a somewhat new and under researched field?
ET: Drawing was often considered preparatory for painting or sculpture. It wasn't show or shared in exhibition and was rarely considered a principal practice by artists until Dadaism. It has taken a bit less than a century to build an economy around drawing. It runs parallel to photography in this way, even though the technology of photography is much newer than cave drawings.
RY: What is the role of drawing in the art market?

ET: Drawing has become an accepted contemporary medium. The marketplace recognizes artists who draw as their only medium of artmaking.
RY: Is it true that drawings typically sell for less than other art mediums?
ET: This depends entirely on the artist. It is true that drawing is often priced less than painting. I think that people still consider the amount of time or labor to create the artwork and scale when pricing art, which is a very commercialized way of assessing retail prices.
RY: How have artists who draw contributed to the respect of drawing as an artistic medium?
ET: Artists who draw have stuck with drawing. The increase in graduate art programs has also lead to artists who draw teaching and having notoriety as mentors to younger artists. This has helped to build the respect of drawing as a medium as well. More museums have had exhibitions featuring artists who draw. Curators and publishers recognizing artist who draw with exhibitions at MoMA. Vitamin D: New Perspectives in Drawing book published by Phaidon in 2005.
RY: How, in your opinion, did contemporary drawing influence the creation of drawing fairs?
ET: So many more artists are making drawings and exhibiting than ever before. Contemporary drawing fairs exist because drawing is an accessible collecting price point for artists whose paintings and sculptures sell for far more.
RY: How, in your view, did the 2008 economic crisis play a role in the creation of drawing fairs?
ET: This was a big turning point for drawing. Some collectors suddenly did not have the budget or confidence to purchase big ticket items, works on paper came to fill a void at a lower price point.
RY: How have auction houses played a role in the creation of drawing fairs?
ET: Auction houses began adding drawing by contemporary artists to day and evening contemporary sales rather than segregating drawing into prints and drawings sales, thus driving up the prestige, value and consideration in the marketplace.
RY: Why are there not, in your opinion, high end drawing fairs in New York, like there are in Europe?
ET: I think that Americans are not focused on drawing alone. Sales at NY Sothebys and Christies lump major high end drawings into the bigger contemporary and post-war sales. We are newer as a nation to the idea of collecting and patronage which is has a very long history in Europe. We also do not have the same level of art history education as Europe which creates a deeper understanding of the arts by the general population.
RY: Does the increase in drawing fairs in Europe reflect a shift in the value of drawing?
ET: I do think that more and more collectors are examining and considering drawing as an important medium hence more drawing art fairs.
RY: How do European drawing fairs impact the American drawing market?
ET: I think that the European drawing fairs have helped move the needle higher on the price points in America as many galleries are expected to price match an artist that is represented in both Europe and the US.
RY: How do you see drawing fairs being changed by COVID-19?

ET: Absolutely all art fairs are changed by COVID-19 as people will not feel safe to gather in larger groups to view art. The online viewing rooms and virtual are fairs have now become the marketplace for art sold at artfairs. I don't know that galleries will continue to pay booth rent if they can have some successful online sales instead.
RY: How do you see drawing as a medium being changed by COVID-19?
ET: Due to the necessity of online viewing I see that artists whose drawings translates well in digital images will have better success at sales. This may lead to artists making drawing with online viewing in mind (possibly bolder, brighter and more graphic).
RY: How do you differentiate between the value of art from an artistic point of view and a financial point of view?
ET: Works of art can be rich with information, style, beauty etc. and not have much value in the marketplace. So much of the value comes from the artist's pedigree/education, name recognition, curatorial support, institutional exhibitions, marketing and other factors. For me it depends on what is the purpose of making such evaluations, as I am a qualified appraiser. Art appreciation is not limited to value. My 5 year-old son has made some wonderful art from an artistic point of view but is not part of the marketplace at large.

Christine Phal, Personal Interview, Conducted by email exchange on August 16, 2020.*
RY: Can you provide a brief description of who you are, your relationship to drawing and your relationship to the drawing market?
CP: I have wanted to work in the artistic field since my adolescence, and, more specifically in cinema as film director. But I had to go towards another field, more scientific and medical. I worked for about twenty years as a pharmacist, and then I turned the page after meeting visual artists and a formation in the art market. I opened a gallery in 1990 to show young artist's work and, even if painting was quite represented, drawing exhibitions quickly took center stage there. But what interested me more was to allow new audiences to discover contemporary art and have access to it. Drawing proved to be the best vector allowing this access to art and to the artists. Since drawing is cheaper than painting or sculpture, the idea of an art fair dedicated to contemporary drawing sprang in my mind. Thus, the 'first salon of contemporary drawing' was born in April 2007. The market at the time was underdeveloped but the enthusiasm of visitors since the first edition validated my feeling. In 2010, we took the name Drawing Now art fair to better show the international profile of the fair. **RY:** Can you provide a notion of what a drawing is? What constitutes a drawing from another medium, for example? Or when does a drawing stop being a drawing?
CP: It's always difficult to give a definition of contemporary drawing because it can take so many different shapes! At Drawing Now art fair we state that drawing can leave the paper and take on extremely varied substrata, and that the tools for making it are as varied. Contemporary drawing can take over walls, ceilings, space, digital screens et the most varied of substrata. It can be made with the finger, with all sorts of pencils, pens, etc. Maybe the best way of defining it is the presence of tracing! ('le trait', i.e. the line). When painting supplants it, and particularly oil painting, and when the whole surface is covered, then drawing disappears in favor of painting.
RY: What is the role of drawing in the art market?

CP: Drawing has taken a much more important place in the international art market. When we talk about the art market the only sources we have are sales and auction sales. The numbers I quote are communicated by Artprice. In 2019 the sales of drawings represented 23% of worldwide sales. Unfortunately in France they only represent 4% of those and China leads with 54% of its global volume! The volume of sales in France was multiplied by 3 in France. I insist that the only numbers we have are those of public sales and that we do not have the numbers of gallery sales or artist ones!
RY: Why is the economy of drawing a new and somewhat unresearched field?
CP: Maybe because the sums in question do not spike the same interest as the record sales of Banky or Damien Hirst?
RY: What role does the history of drawing play in the market of drawing?
CP: In times of crisis, buyers reassure themselves with artists with a well-established price level. When things calm down, the prices attained by the artists in the progression of their notoriety reassure buyers on the fact that the drawing market is in constant progression and that drawing became a medium in itself, with its place alongside photography, painting and sculpture.
RY: In your perspective, how has *avant-garde* influenced the drawing market?
CP: I don't think that the role of drawing in the avant-garde has been well identified and therefore it doesn't, I think, have a real influence in the market. On the other hand, and echoing my previous answer, we note in fairs that there is an attraction to that avant-garde!
RY: In your perspective, what role does Outsider Art play in the market of drawing?
CP: The art named Outsider in Anglo-Saxon countries and Art Brut in Europe has, in 10 years, had a progression in parallel with that of contemporary drawing, by being integrated simultaneously in famous private collections and institutional collections like the Centre Pompidou. An important amount of work is the responsibility of some art dealers who very seriously took Art brut at the heart of contemporary drawing by informing institutions and creating a faithful following. Drawing is particularly present in Art Brut for economic reasons. Certain famous examples showed how these artists use all the surfaces they can find to draw!
RY: Why, in your opinion, were drawing fairs created?
CP: We mustn't forget that the Salon du Dessin ancien (*meaning Ancient Masters up to the beginnings of the xxth century*) has existed for more than 30 years and that it draws attentive and passionate collectors. It was created by dealers that wanted to establish a given and fixed moment in the year in Paris to present their best works. To my knowledge, it's the oldest drawing fair. We followed suit in 2007 by taking on the contemporary field of drawing and focusing on the last 50 years of contemporary drawing in its diversity. I am convinced that the particularities of drawing allow for a real meeting with the work of the artists. The other fairs that followed us profited from this interest of collectors for the medium.
RY: What type of success have you seen with drawing fairs?
CP: Since the first Salon in 2007 the public was present in great number, attentive and enthusiastic and the galleries participating had an never seen before commercial success. With time, success was established and sensed by collectors, artists, galleries and institutions who never miss an edition of the fair. Personally, what moves me is that

a great majority of the collectors I meet say that Drawing Now is their favorite fair! This is rather difficult to say after the cancelling of the 14th edition and it is difficult to show my enthusiasm!

RY: How, in your view, did the 2008 economic crisis play a role in the creation of drawing fairs?

CP: In 2008 there was only us and the Salon du Dessin Ancien so I don't feel that there was a specific role on the fair played by the crisis.

RY: What influence do auction houses play in the creation of drawing fairs?

CP: None!

RY: Why are there not, in your opinion, drawing fairs in New York?

CP: I think that there are a lot of fairs in The United Sates and New York! Maybe the fact that drawing is now very present in general big art fairs is enough for the market!

RY: Is there more or less respect towards drawing with the expansion of the drawing market?

CP: Drawing has really taken on a place for itself. We will have completely won when I will not be told that I have a «niche» art fair. I want to leave my « niche» (kennel in French) and leave my leash on the sidewalk!

RY: Does this rise in drawing fairs reflect a shift in the value of drawing?

CP: I think that mostly it is a way to stimulate new collectors and amateurs. There is a new reconfiguration of drawing by young artists who cannot aspire to exhibit at Art Basel or FIAC! The more drawing is shown, the more value it takes on.

RY: Where do you see drawing moving towards in the future artistically?

CP: Surely by taking on all shapes of drawing, from performance to Comics, from Art Brut to animation! The borders are more and more permeable and artists play with that really well.

RY: Where do you see drawing as a value in the art market?

CP: When I see the record prizes of drawing Chinese artists, I think that there is still a long way to go. But little by little drawing takes its own place!

RY: How do you differentiate between the value of art from an artistic point of view and a financial point of view?

CP: For me there are several markets: the art works one buys according to their value with a notion of financial gain from the get go and the buyers who have a real interest for art and artists. In France, we say « those who buy with their ears and those who buy with their eyes and heart». Drawing belongs to the latter category!

***Translated from original French response by Joana P.R. Neves and available upon request.**

References

1. Adam, G. (2014) Big Bucks: the Explosion of the Art Market in the Twenty-First Century. 10.
2. Neves, J. P. R. Personal Correspondence. April 3, 2020.
3. Beuendorf, H. (2016) "Art Demystified: How Are Wealthy Busy People Changing the Art Market?." Artnet News. Retrieved from https://news.artnet.com/market/art-demystified-art-fairs-market-698750.

4. Kelly Richman-Abdou. (2018) "7 Most Important Art Fairs People Travel Across the World to See." My Modern Met. Retrieved from https://mymodernmet.com/art-fairs/.
5. "The Art Market Highlights Art Basel - Art Fairs." (2020) The Art Market Highlights Art Basel - Art Fairs.. Retrieved from https://theartmarket.foleon.com/2020/artbasel/art-fairs/.
6. Thompson, D. (2008) The $12 Million Stuffed Shark: The Curious Economics of Contemporary Art. 171.
7. Neves, J. P. R. (2020) Personal Correspondence.
8. Parmentier, G. (2020) Personal Interview. Conducted by phone.
9. Tenenbaum, E. (2020) Personal Interview. Conducted by email exchange.
10. Phal,C. (2020) Personal Interview. Conducted by email exchange.
11. Rea, N. (2019) "What Would It Cost for the Art World to Offset Its Enormous Carbon Footprint? We've Compiled a Helpful Menu of Prices." Artnet News. Retrieved from https://news.artnet.com/art-world/carbon-offset-art-world-1720782.
12. "Carbon Footprint Calculator." (2020) ClimateCare. Retrieved from https://climatecare.org/calculator/.
13. Heide, P. (2020) Personal Interview. Conducted by phone.

Experimental Study on the Effect of Thermal Comfort on Parameters

Junyi Zhao(✉)

Country College of Arts, Humanities and Social Sciences, The University of Edinburgh, Edinburgh, UK

Abstract. The thermal comfort of buildings affects people's lives from a variety of aspects. Thermal Comfort is the degree of people's satisfaction with the current thermal environment through their subjective thinking. This paper introduces PMV model, Berkeley model and the results of local thermal sensation experiment, to further understand the relationship between human physiological measurement and indoor and outdoor environment, it is concluded that in the experiment of Thermal Comfort, the experimental variables will be affected by many factors, and the future research should further consider various experimental factors.

Keywords: Thermal comfort · Individual differences · Experimental research

1 Introduction

In modern architecture, thermal discomfort is still prevalent, and this phenomenon will significantly reduce the building quality of life, so people need to create a thermal comfortable living environment. How to create a loose and comfortable indoor thermal environment for people to live in has been a popular issue in the study of building structure. With the development of building technology and code, the thermal comfort of facilities needs to meet future growth and demand. According to the experimental results, it is found that the research on the thermal reaction of the human body is not mature at home and abroad. Based on the above background, this paper makes further theoretical and experimental research on the individual difference of human thermal response, physiological measurement of the human body and related indoor and outdoor environmental factors.

1.1 Definition of Thermal Comfort

ASHRAE 55 defines thermal comfort as the human psychological state of being satisfied with the thermal environment and evaluates it through subjective thinking [1]. This suggests that the heat balance of the body and its surroundings is related to the sensation of heat [2]. And (ISO 7730, ASHRAE 55) specifies the exact physical criteria for creating an acceptable thermal environment, such as temperature, airflow and humidity limits [1]. Generally, Thermal Comfort is classified according to the type of environment: outdoor,

C. Yuan et al. (Eds.): *Proceedings of the 4th International Conference on Economic Management and Green Development*, AEPS, pp. 286–301, 2021.
https://doi.org/10.1007/978-981-16-5359-9_33

semi-outdoor or indoor. At present, there are two main research methods for indoor comfort: one is the classical steady-state model based on human body heat balance model and developed for air-conditioning space, the other is an adaptive model [1]. Thermal Comfort related to thermal sense is one of the most critical performance indexes of ventilation, heating and air conditioning systems, including physiological models (thermal regulation in the human body and psychological models [3].

1.2 Individual Differences

When setting up a research model, the experimental process needs to take into account individual characteristics that are independent and accessible. Individual differences in thermal comfort can be divided into individual differences and intra-individual differences. Among them, the unique differences refer to the differences of heat sensation between people in the same environment, and the intra-individual difference refers to the difference of heat sensation on different occasions of the same domain. There are two ways to address individual differences in heat regulation models. One is the passive system of the human body; the other is the active control system of the human body [3]. In the present research of individual difference thermal comfort, the subjects are usually divided according to certain conditions. After controlling the quantity and the variable, the experiment is carried out to collect the thermal response data of different human bodies. Finally, the collected data are analysed. Subjects were classified by age, sex, country, climate, level of exercise mood and so on. Gender and age are considered to be the two primary sources of individual differences. At present, many studies still cannot draw a relatively consistent conclusion. That is the reason for individual differences and the factors that exist differences are still unclear. A person's thermal comfort depends in part on subjects' activities and clothing, and in part on four environmental variables: Air Temperature, average radiant temperature, air velocity, and humidity [4]. Therefore, some studies show that if clothing and anthropometric parameters can be effectively controlled by variables, then the differences in thermal comfort will be more accurate. It is concluded that the difference between clothing and physiological response of the human body is probably the most significant parameter for the thermal comfort differences between groups and individuals.

1.3 Indoor Environment Control Mode

The interaction between the indoor background and the human body is also one of the factors involved in thermal comfort. The architecture account for about 30% of global energy consumption and contribute significantly to climate change [5]. With the PMV model as the core, the difference of indoor environment perception will become the main factor that affects people's work. The study found that the thermal environment is the main factor that affects the comfort of a particular office worker. The difference between indoor environment perception will have a negative influence on their working environment rating [6]. Rapid advances in technology provide opportunities to address individual differences in the built environment from centralized fixed set points to personalized air conditioning. Personalized building air conditioning strategies work best for people, including women and the elderly Air-conditioning strategies in personalized buildings are most beneficial to people who include women and the elderly [7]. Occupants of modern buildings are often affected by the climate in which they work, and adaptive models of thermal comfort predict that humans will become more comfortable indoors rather than outdoors [8].

1.4 Thermophysical Parameter

The acquisition of a human physiological signal is an essential step in the experiment. It has long been thought that skin temperature is a significant factor in judging the temperature perception caused by the environment. And then it also affects thermal comfort in cold climates. The distribution of skin temperature on the body surface is also a significant factor in causing cold. The higher the uniformity, the higher the comfort. There are three main ways to collect physiological signals: 1. In Vitro measurement and in Vivo measurement are divided according to the objects of size. 2. According to the measurement conditions, it is divided into non-invasive measurement and invasive measurement. 3. According to the measurement results, it can be divided into one-dimensional information measurement and multi-dimensional information measurement. In the study of thermal response of the human body, the measure of thermal physiological parameters belongs to in Vivo measurement, and most of them are non-invasive contact measurement.

Mathematical modelling of the thermal response of the human body is a useful tool, which can understand human activity at different environmental conditions and levels. Benefit from the development of modern bio-thermal models, the transient or steady-state non-uniform thermal response of human body can be predicted [9]. From a mathematical point of view, the human body can be divided into two mutually regulated temperature control systems: Active Control System and Passive Control System. Cybernetics model simulates the active system. The passive control system is modelled by simulating the heat transfer phenomena occurring in the physical human body and its interior and surface. Finally, a comparison is made between the predictions of the model and the known solutions [10].

1.5 Existing Problem

(1) Current studies tend to suggest that the differences between individuals are primarily due to clothing and physiological responses (such as metabolic rate, peripheral vasoconstriction/vasodilation, etc.). Still, there is no substantial evidence to support this view, therefore, more in-depth research should be carried out from these two aspects.
(2) When measuring the physiological parameters of human skin, it is necessary to select the local-part to be calculated, and the usual approach is to use the most common parts (DUBOISE's 7-point method) directly from the formula or to use the factors that can more accurately and sensitively calculate the average skin temperature. In fact, with the proliferation of wearable products, the elements that are measured should be chosen as much as people want and are easier to measure, rather than based on formulas. This requires further research on the relationship between local physiological/subjective parameters and global physiological/personal parameters.

(3) Although the control mode of air conditioning has changed, there are still some problems. For example, there is a lack of consideration of human body parameters, or the skin temperature is only represented by the skin temperature of the wrist, or SSDI model replaces the PMV model because of its complexity and slow calculation speed. Whether other parts and other thermal physiological parameters can be used as a better regulation basis, and whether the PMV model needs to be simplified, these questions need to be further studied.

1.6 Research Method

(1) Individual clothing differences can be reflected by clothing thermal resistance, and other parameters can be directly obtained by using a questionnaire, that is, by asking participants to fill in the current clothing situation. Also, it can be obtained indirectly by measuring the thermal parameters of the human body surface and clothing surface. An individual difference prediction model can get individual Metabolic rate differences. Individual differences in the contraction/relaxation of blood vessels can be expressed by parameters such as heart rate, blood pressure, skin temperature, skin heat flow, etc. To sum up, the measurement of physiological signals is fundamental in this step, and that should first determine the appropriate experimental equipment.
(2) The research chooses the measuring position considering the conditions of use of the non-invasive contact measuring instrument and people's will. Through multi-statistical analysis, regression analysis and other data analysis methods, the laws of these experiments are summarized.
(3) Based on the PMV model, the strategy of environmental regulation is put forward according to the collected physiological parameters, subjective feeling and ecological parameters. The system can be programmed into the receiving terminal, and an infrared transmitter can be installed on the receiving terminal so that when the human body is in the indoor environment, the receiving airport can collect data in real-time and control the indoor air conditioning by adjusting the strategy.

2 Investigation of Experimental Model

In this section, the research background and the purpose of the PMV model, the Berkeley model and local thermal sensing are introduced.

2.1 The PMV Model

The PMV model, which means the predicted average vote, is the most widely used measure of thermal comfort for medium indoor buildings in the international standard ISO 7730. Based on the steady state heat transfer theory, the PMV model was calibrated with data from Nevins and McNall, and a series of climate chamber studies that measured the thermal experience of large numbers of American students. It predicts ASHRAE's subjective warmth ((cold (3), cool (−2), slightly cool (−1), neutral (0), slightly warm

(1), warm (2), hot (3)). The called for input variables are air temperature, average radiant temperature, air humidity, speed, metabolic rate, and clothing insulation [11].

The ASHRAE database is a combination of comprehensive measurements of a wide range of climates, the number and types of buildings, the diversity of cultural backgrounds, and the thermal environment, also it is very suitable for PMV model validation. The results of Humphreys and Nicol show that [11] the relationship between PMV model and comfort is not as close as that between air temperature and earth temperature. In subsequent studies, Humphreys found that average indoor temperature may be associated with differences in PMV and average comfort. In their paper "The validity of ISO-PMV for Predicting Comfort votes in every-day thermal environments", Humphreys and Nicol discuss how the difference between PMV, and actual voting depends on the variables used to calculate PMV, and further discuss the effects of climate. Finally, they discuss the scope of improving PMV and prove that it is feasible to improve PMV to improve its performance.

Because people's perception of temperature is different, PMV models can not accurately predict a person's comfort level in some cases. Instead, it predicts the average comfort level of the human body in the same level of physical activity, wearing the same clothing with the same thermal insulation, and in the same thermal environment. Although the PMV data in the database represent only a sufficiently large population, they differ in terms of thermal clothing insulation, thermal environment and metabolic rate, so this rarely happens in practice. In the experiments in this paper, to avoid this situation, they calculated the PMV for each particular occasion and subtracted the corresponding actual comfort vote. The individual differences available in this process are unbiased but low-precision estimates of the real differences between the PMV model and the actual voting. If there are no deviations during the experiment, the distribution of any batches extracted from these deviations has an average value that is not significantly contrasting from zero. Its standard deviation results are based on individual differences between people and any errors related to the development and evaluation of PMV models.

There are individual differences in the factors that lead to the differences in the experiment. The differences in body temperature are considered to be neutral, and personal judgments vary over time. This dispersion from individual differences cannot be ignored, and typical thermal comfort climate studies have found that it has a scale unit, which standard deviation of less than one. But in the practical experiment application, the data used to calculate PMV will inevitably be affected by the random error of measurement and estimation. At the same time, the use of the equation itself may also lead to unexpected errors. PMV is an approximation of a hugely complicated physical, biological, and cognitive system that includes the thermal surroundings, the human body, and the mind, it heralds comfortable voting, a definition of a state of mind. It cannot be ruled out that some of the factors that add to comfortable voting have been absent or included in a less precise manner. The PMV equation is based on the theory of steady-state heat transfer, a state that has never been accurately described in everyday life, which means it can be better described as a dynamic heat balance. In a word, the standard deviation of difference may come from three aspects: measurement error, individual difference and equation error. The direction of the overall PMV bias was an overestimation of the average subjective temperature of a population in a warm environment, and this difference is of technical significance because it has a practical impact on the operation of the building and may lead to the facility providing avoidable cooling. It can even be used as a thermal simulation in the design process to influence design decisions can be misleading in suggesting that the building needs to be cooled to continue a comfortable indoor environment in the summer.

The analysis shows that there is an obvious nonlinear relationship between the mean indoor temperature and the PMV model deviation. The psychology may be influenced by the climate's perception of indoor comfort, but the possibility of ignoring physiological effects is not in line with experimental principles. For instance, metabolic rates are usually appraise using standard activity lists, and because people used to warm meteorological character have decreased metabolic rates for the same task lists. Therefore, with the same activity list, the results of possibility hide metabolic differences. Physiological adaptations do affect the body's thermal comfort preferences for different body states of importance, in which case the transformation to heat is well cite. Still, the adaptation to cold is less individual. However, whether physical, physiological or psychological factors, the outdoor temperature is an essential factor that causes the deviation of PMV measurement [11].

The predictions generated by the PMV model were biased towards operating temperature, humidity, airflow, outdoor temperature, metabolic rate and clothing insulation. When PMV is used to predict the average comfort degree of people in the daily environment of buildings, especially in a warm climate, it will produce severe error. At present, the revision of ISO7730 should pay attention to the limitation of PMV in buildings and give the appropriate range of use according to the empirical results.

2.2 The Berkley Comfort Model

The Berkley comfort model is a modified version of the STOLWIJK model that measures the body's heat regulation. STOLWIJK's 25-node model of thermogenesis presents the basic concepts, algorithms, physical constants, and physiological control subsystems for many modern multi-node models. The STOLWIJK model is established on six parts of the body: the head, torso, arms, hands, legs and feet. The Berkley Comfort model in this paper can simulate any number of systems with significant improvements. In most practical applications, the Berkley Comfort model uses 16 body parts to segment the thermal mannequin. Each part consists of four layers of body tissue, core, muscle, fat and skin, and a layer of clothing. The model uses the standard finite difference Algorithm to calculate the heat transfer between each node. The Algorithm can ensure the optimization of the computational resources while keep the numerical stability. Berkley's comfort model uses a series of discrete stages of fluctuating length to resemble a continuous combination of almost any environment, clothing, and metabolic condition. The model can predict the developments of transient and spatial asymmetry conditions that are entirely lost in global models just as the two-node PMV model [12].

The Berkeley model is mainly improved in the following aspects:

1. The measured body nodes are changed from six to unlimited nodes. In this experiment, 16 nodes are used to directly correspond to the human body model, which

can accurately measure the heat transfer system of each part of the human body and the thermal insulation value of the clothing and can be directly substituted into the thermal comfort model. The anthropomorphic model provides valuable data for each node while measuring the environmental heat flux.

2. Human body heat regulation mainly through regulating blood flow to achieve, through the contraction and expansion of blood vessels, the body to adjust blood circulation to manage skincondition, upturn or reduce heat fall to the environment. The original Stolwijk model simulated that the arterial blood temperature of the whole body was the same. Still, the climate of the limb would be unstable in a relaxed environment, so the central Artery/vein counter current heat exchange was decided after adjusting the blood flow model and improve the hemoperfusion model to estimate the local tissue blood flow.
3. Added clothing nodes to simulate heat and humidity capacitance. Stolwijk model believes that clothing is an insulating material without mass, but transient effect experiments have proved that the heat capacity of clothing will affect the results. Wet Capacity is essential to accurately simulate the evaporation heat loss of the human body through clothing, and a regain approach is used to calculate the moisture absorption of a given fabric at relative humidity.
4. An addition to the conduction of heat through a surface in contact with an object. In practically all environment, the body is in connection with a solid surface by conducting heat, also it can help to increase the insulation treatment of the clothing model under the steady-state condition. Still, under the transient state condition, the concrete surface may have a vast thermal mass to affect the experiment result. In subsequent improvements to the model, the Berkeley model included contact surfaces for each part. The contact surface uses the far-field temperature and heat transfer coefficient as the boundary conditions, which also include thermal conductivity, specific heat, initial temperature and thickness.
5. The convective and radiative heat transfer coefficients are developed based on the existing ones. The average radiative temperature of each body segment can be calculated as a linear model of radiative heat transfer and as an explicit model of Stefan Boltzmann's law. The Stolwijk model uses a mixture of convective and radiative heat transfer coefficients. The Berkeley model separates convective heat transfer from radiative heat transfer. The advantage of this approach is that in non-uniform environments, the results are more accurate using real three-dimensional models of the human body and patten factors between the body and any set of environmental surfaces.
6. In addition to heat transfer with the surrounding surface of the long-wave radiation, the human body is generally solved to solar radiation, heat lamps or other sources of radiation. The Berkeley model already includes a two-band radiative heat flux model that separates the heat flux into short-and long-wave components. The absorption of short-wave radiation through the skin or clothing; the absorption of long-wave radiation through the skin or clothing.
7. Use angle factor to calculation the explicit radiation heat transfer. Each person's Human physiology is very different, and these distinctions affect the perception of thermal comfort. To do this, the Berkeley model used a pre-processor called

bodybuilder to test the degree to which physiological variation affects the body's perception of thermal comfort.

2.3 Local Thermal Sensation

The thermal environment is usually asymmetric, which means that the space is uneven or transitional over time. Environmental conditioning systems can take advantage of nonuniformity and transient operation to curtail the amount of energy required to present a suitable environment. The human body's response to an asymmetrical climate depends on its local thermal perception, but there is no sensory model that can predict these local effects. PMV models are the most commonly used heat sensing models that have evolved under uniform steady-state conditions. Although they cannot predict short-term responses, they can treat the body as an aggregate based on physiological averages and sensations. Although models grow in recent years can forecast rude behaviour, the project can only predict sense at the whole-body level, so they are of limited value for the assessment of heterogeneous environments. This article interprets the development of 19-site local sensory models adapted to non-uniform and transient conditions. Humans can only feel the warming of the environment through the thermoreceptor, located on the skin and at the core. The body's heat receptors sense the temperature of the surrounding tissue and express signals to the brain that interpret the environment as heat. The properties of heat receptors determine the thermal response, with Thermoreceptor being able to adapt to changes in temperature but less efficient in responding to relatively stable temperature states [13].

The local thermal sensation model was developed based on human experiments in which air was used to cool or heat various parts of the body. However, since the model is concerned with sensing skin and core temperature, rather than the surrounding environment, the local thermal sensing model can be applied to other heat transfer models. The body's thermoreceptor can sense tissue temperature, not ambient temperature, so as a result, the local thermal model is appropriate for this experiment. It requires extensive testing to overlays the full area of test conditions in all parts of the body. In this project, only a subset of viable test conditions can be performed, concentrate more on cooling a part of the body in a warm environment than heating it in a cool one. These experiments were carried out under the condition that the temperature of the cooling air produced a reasonably rapid rate of change in skin temperature.

The local thermal sensing model was exploited and validated using information from sedentary subjects. During the course of the study, it was impossible to determine its applicability at a higher level of activity. The experimental results show that the coefficients of the model can be applied to the case of sweating after considering the influence factors of sweating. If this combines a sensory model with a computer model of thermal physiology, it can then calculate the skin and core temperatures and their settings. The structure of the model is reasonable because its factor justifies the processes involved. The coefficients of the model are easy to be tested and modified, which means that there is an extensive area of environmental conditions during the experiment without changing the form of the model.

3 Experimental Data and Analysis of Experimental Model

This section mainly introduces the experiment concrete research data and in the actual application.

3.1 The PMV Model

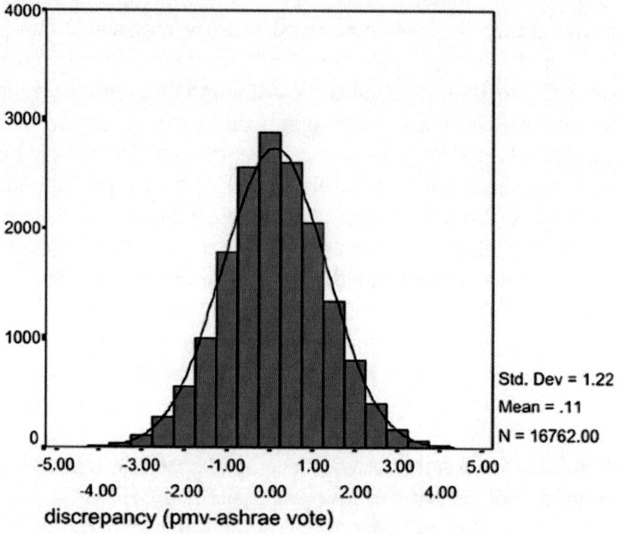

Fig. 1. Difference histogram.

To evaluate the overall accuracy of PMV, the experimental model aggregates all available devices into a unitary distribution. Figure 1 shows a histogram of all the differences, representing the overall picture of the data available worldwide. The average deviation of the chart indicates that the calculated PMV values are as a whole and are higher than the actual ASHRAE voting scale unit of 0.11 ± 0.01. The data, which combined results from different buildings, climates and seasons around the world, showed no significant bias in the PMV model data.

Since the equation error comes from the approximate value in the exponential formula, the equation error can usually be uncovering by testing the different amount of all suspicious variable. In the absence of an equation misstep affecting the variable, there is no compelling deviation from the variable unless the effect is due to a change in the relevant variable, which itself is affected by the equation error. In the database, the air temperature and the mean radiant temperature change together ($r_2 = 0.96$). The difference between them is minimal, with an average absolute difference of 0.5 k. Therefore, these data are not suitable for measuring the relative weights of air temperature and mean radiant temperature in PMV. During the experiment, it is better to use "operating temperature" to indicate their combined effect. The operating temperature in the table is a simple average of the air temperature and the bureau radiation temperature, which is a sufficient approximation to be obtained, given the small differences between the two. As can be seen from the data in Fig. 2, the results of the univariate analysis show that the differences in the experimental results are very significant. The PMV overestimated how warm it would feel at room temperatures above 27°, and at higher temperatures, the bias became severe. Figure 3 shows the same process that applies to airflow. Variance analysis studies show that PMV is not significantly biased at low speeds, but if the square root of air velocity exceeds 0.45, or 0.2 m/s, it exaggerates the sense of warmth. The PMV model thus underestimates the cooling achieve of elevated airflow.

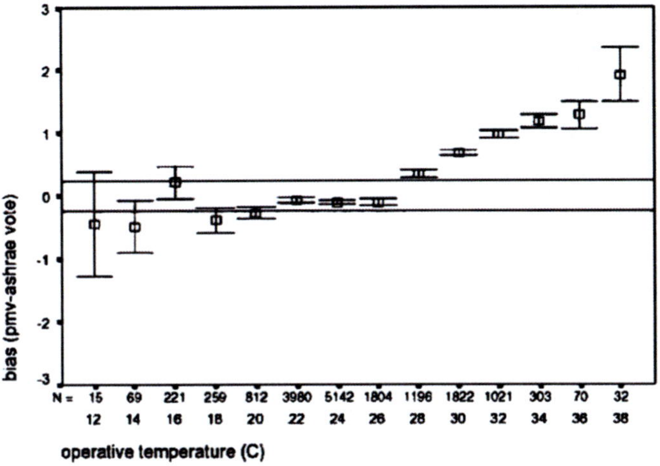

Fig. 2. PMV bias against variables.

Since the equation error comes from the approximate value in the exponential formula, the equation error can usually be detected by testing the different amount of any suspicious variable. Unless the effect of the equation error is produced by a change in a related variable, which itself is affected by the equation error, these effects will not have a significant deviation from the variable. It is, therefore, necessary to examine all the variables that may give rise to errors in the equation and to consider them in their joint action.

The analysis shows that there is a considerable deviation for all the variables that cause PMV. Any variable is more limited in scope than is given in ISO 7730. But these effects are not necessarily independent, and the apparent deviation of any appropriate variable rest with the values of all other biased variables. The results depend on the structure of the database so that differences may vary between buildings and between different types of indoor environments. When look at additional data, the relationships between the variables are other, and the results are different.

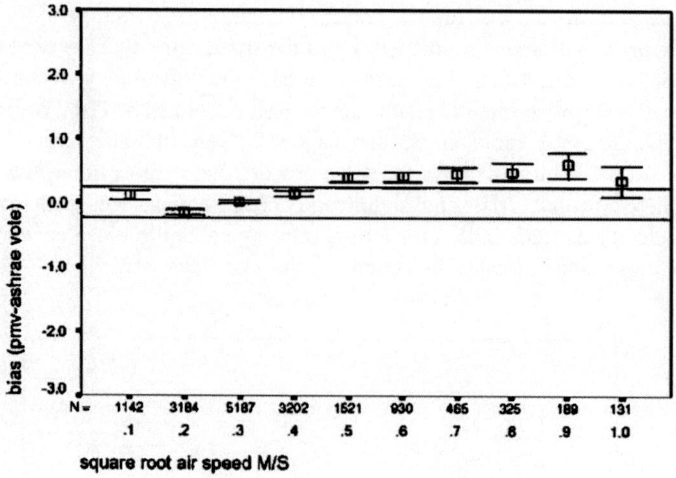

Fig. 3. Bias in PMV against the variables.

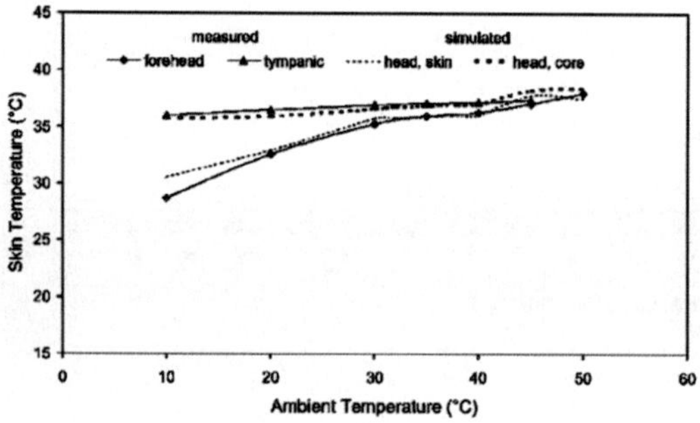

Fig. 4. A analysing of measured and fictitious temperatures as steady-state conditions ranging from 10 to 50 °C.

3.2 The Berkley Comfort Model

As a preliminary validation of the model, the simulated skin temperature was compared in the literature with some physiological studies by other researchers. These studies include several steady-state conditions and three transient environmental conditions.

The flexibility of the internal model structure is maintained in the application of the Berkeley model, which ensures that the model can be changed without recompiling the code. Parts of the body are connected by blood, and nodes conduct and exchange heat with each other. Each piece of the node forms a link tree, and the structure of each segment does not need to be precisely the same. For example, if the subjects were wearing shorts, the model would generate a path for the thighs to wear clothing and bare skin. If the issue is wearing trousers, only the direction of wearing will be created.

Under steady-state conditions, the predicted core temperature is very close to the measured value (within 0.5 °C) (Fig. 4). In the transient process, the actual arm temperature is higher than the simulated arm temperature, but the final results are very similar.

The original STOLWIJK model tended to predict higher skin temperatures from colder environments. The reason is that the STOLWIJK model assumes that in the central blood model, the node receives the same blood temperature as the core of the body. At the same time, counter current heat conduction reduces the skin temperature of the limb, resulting in a decrease in arterial blood temperature. Increasing the counter current blood model can improve the consistency between the model and limb experiment data. It is concluded that the model can accurately predict the core and limb skin temperature in a specific temperature environment.

The adjustable input structure of the Berkley Comfort Model, as well as its characteristics for evaluating transient non uniform thermal environments, has been widely used in experimental applications. For example, a non-uniform or hierarchical environment is provided for a mission/environment system or ventilation displacement device in evaluating innovative HVAC methods. On the other hand, the internal test environment is very short and has a highly uneven radiation load, which will have an impact on the evaluation of occupant comfort. The input structure of the Berkeley Comfort model allows it to evaluate thermal comfort in a more rigorous manner and connect to the building simulation program.

3.3 Local Thermal Sensation

In this paper, the local perception model is a function of the local skin, the average skin temperature and their change rate (Fig. 5), local and mean surface temperatures show responses to safe conditions, and the derivatives of surface and central temperatures perform responses to transient conditions. The local skin temperature represents the thermal state of the local skin in the static part of the model. Although core temperature is difficult to use to determine the value of the body's neutral set point, it may be another way to represent the overall state of the body because it's more variable in the body and the body than it is in the temperature response. But when the goal is to describe a short-term change in body temperature, the deficit disappears. Because the centre temperature has a higher resolution than the average surface temperature, the centre temperature is used in the active part of the model. During the experiment, the prediction of local sensation was positioned filly on physiological data, not environmental parameters, so each body part had a unique model that could capture asymmetric features in any environment.

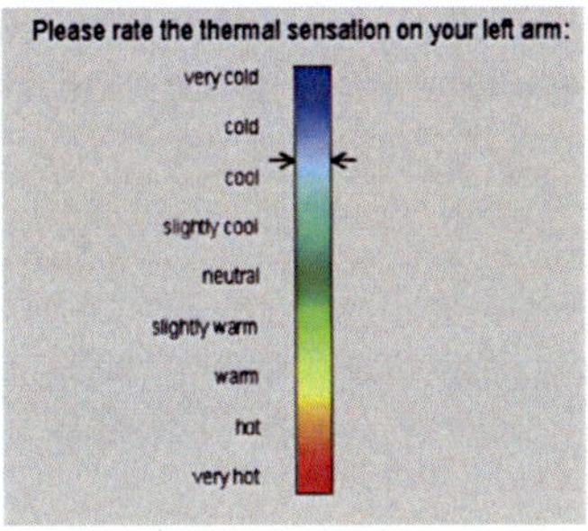

Fig. 5. Thermal sensation scale.

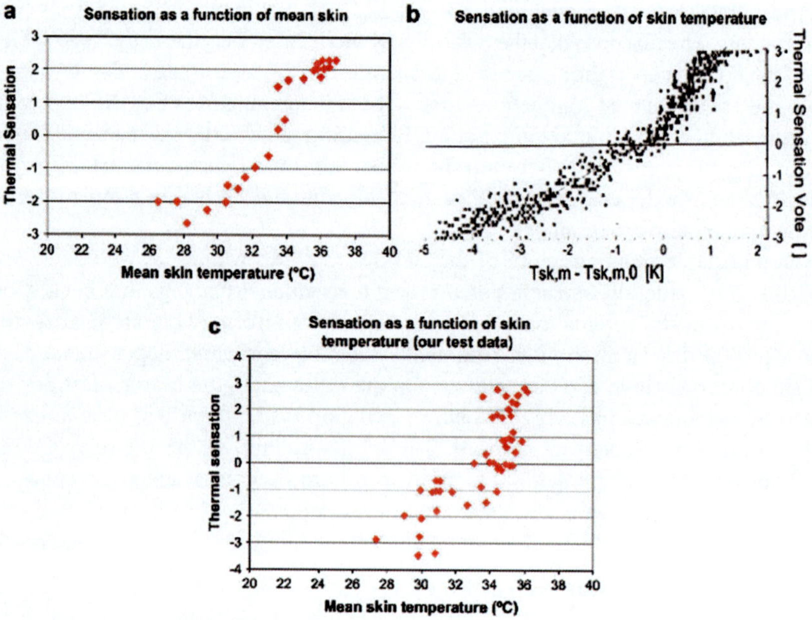

Fig. 6. The relationship from the average skin temperature to the whole-body thermal sensation (a) sensation take for a function of the mean skin temperature (b) feeling as an action of the skin temperature difference (c) sensation as a function of the skin temperature.

When skin temperature was between greatly low and high, the correlation between skin temperature and the thermal sensation was nearly linear. To include the whole range of skin temperature, the logistic function of local skin temperature is needed to express the difference between the local skin temperature and its set value. When the sensation of a body part is uninvolved, the setting point of the body part is the local skin temperature of the position. The trend is approximately a linear function of the average skin temperature from 29 to 34 °C (Fig. 6), or a linear function of the difference between the average skin temperature and the setpoint (from −3 to 1 K) (Fig. 6). The linear relationship disappeared, and the thermal sensation level decreased when the skin surface temperature was above or below the intermediate range. It is suggested that local sensation is the logical function of local skin sensation, and the difference from local skin temperature to set point is used to analysis the whole range of skin temperature variation. The set point for the body part is the limited skin temperature where the body part is neutral (zero). The Logistic function is linear in the middle, as shown in Fig. 3, and tends to level off as skin temperature increases or decreases. The local sensation is affected by both local temperature and global thermal state.

Figure 7 shows an example of this phenomenon. The solid circle is obtained by cold test at room temperature of 16~20 °C and whole body at low temperature. These open triangles are temperature measurements taken at room temperatures from 28 to 32 °C, with the entire body in a warm state. Each data point is from a specific local cooling or heating test at the end of the body's thermal state has been stabilized.

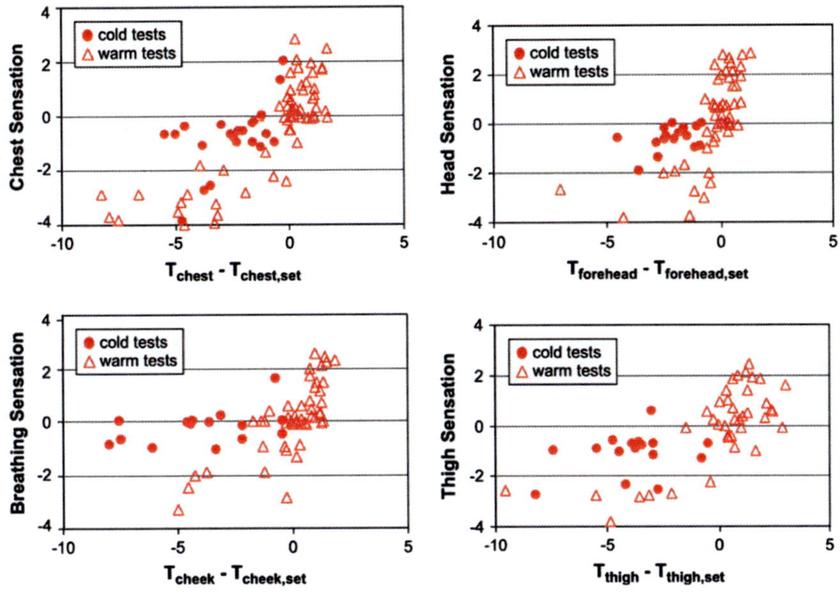

Fig. 7. Local sensation, local skin temperature, and cold or warm whole-body state.

The experimental results show a clear separation of the local sensation relative to the warm or cold overall hot state. Parts of the body that have the same skin temperature feel more generous when the whole body is cold and colder when the entire body is warm.

4 Conclusion

This paper studies the influence of thermal comfort on the human body and its development requirement. The thermal sensation in thermal comfort is linked to the thermal balance of the human body and its surroundings, and in experiments on thermal comfort, priority is given to the social psychology of evaluating environmental conditions through their subjective thoughts. In the course of the investigation, the test of the human body's physiological instinct is to help the interior design of the building to understand the structure of the building, and to build a more extensive indoor thermal environment that covers people. In the course of the experiment, it is necessary to take into account not only the differences between different groups of people and the additional requirements of the thermal environment. It is essential to know how to control the indoor thermal setting in the building.

In this paper, from the individual differences, indoor environmental control and thermal physiological parameters of these three different factors to carry out various experimental studies, the conclusions are as follows:

1. From the PMV model experiment, it is concluded that the difference between different indoor types between other buildings may lead to the change of human physiological feeling. Individual differences in the control of the variables of the data

will also affect the results of the experiment. The power between the variables will affect each other, and the experimental results will change because of the different relations among the variables. Some experimental results show that the PMV model overestimates people's subjective temperature assessment in a warm environment, which results in large differences between the experimental results and the predicted results.

2. The Berkley comfort model is an improvement on the Stolwijk model, which not only expands the experimental node range but also ensures data stability, Berkley's comfort model captures physiological information while maintaining the flexibility of the internal model structure, resulting in more accurate data and a more rigorous way to assess thermal comfort. At the same time, Berkley comfort model can simulate almost any environmental conditions. In the transient and non-uniform thermal environment, the improved model provides a more abundant possibility for the experiment variable deployment.
3. Local thermal sensation is developed based on human experiment, which is based on the development of 19 local sensation models of the human body under non-uniform and transient conditions, by measuring the local temperature and average surface temperature. Comparing the two, the conclusion is that the whole-body temperature feels cold, and the local weather feels warm under the premise of having the same skin temperature. Conversely, when the entire body is in a warm condition, the local feeling will be more inclined to cold.

The results thus far have been skewed by variables and other factors beyond our control because the subjective aspects of human perception cannot be controlled. After the experiment of measuring the physiological parameters of the human body, the local thermal sensation experiment can get relatively accurate results. The subjective, physiological parameters of the human body will change with the experimental variables. Simplification and optimization of the empirical model can effectively improve the accuracy of the experimental results in a specific region. However, in the current experimental research, there is no very perfect experimental setup, and it is necessary to consider further various experimental factors related to thermal comfort research.

References

1. Rupp, R.F., Vásquez, N.G., Lamberts, R.: A review of human thermal comfort in the built environment. Energy Build. **105**, 178–205 (2015)
2. Choi, J.H., Loftness, V.: Investigation of human body skin temperatures as a bio-signal to indicate overall thermal sensations. Build. Environ. **58**, 258–269 (2012)
3. Zhou, X., Lian, Z., Lan, L.: An individualized human thermoregulation model for Chinese adults. Build. Environ. **70**, 257–265 (2013)
4. Fanger, P.O., Højbjerre, J., Thomsen, J.O.B.: Thermal comfort conditions in the morning and in the evening. Int. J. Biometeorol. **18**(1), 16–22 (1974)
5. Ghahramani, A., et al.: Towards unsupervised learning of thermal comfort using infrared thermography. Appl. Energy **211**(2018), 41–49 (2018)
6. Fox, R.H., Lofstedt, B.E., Woodward, P.M., et al.: Comparison of thermoregulatory function in men and women. J. Appl. Physiol. **26**, 444–53 (1969)

7. Wang, Z., et al.: Individual difference in thermal comfort: a literature review. Build. Environ. (2018). https://doi.org/10.1016/j.buildenv.2018.04.040
8. Erlandson, T.M., Cena, K., de Dear, R.: Gender differences and non-thermal factors in thermal comfort of office occupants in a hot-arid climate. Elsevier Ergonomics Book Series, vol. 3, pp. 263–268 (2005)
9. Salloum, M., Ghaddar, N., Ghali, K.: A new transient bioheat model of the human body and its integration to clothing models. Int. J. Therm. Sci. **46**(4), 371–384 (2007)
10. Fiala, D., Lomas, K.J., Strohrer, M.: A computer model of human thermoregulation for a wide range of environmental conditions: the passive system. J. Appl. Physiol. **87**, 1957–1972 (1999)
11. Humphreys, M.A., Nicol, J.F.: The validity of ISO-PMV for predicting comfort votes in every-day thermal environments. Energy Build. **34**(6), 667–684 (2002)
12. Huizenga, C., Hui, Z.: A model of human physiology and comfort for assessing complex thermal environments. Build. Environ. **36**(6), 691–699 (2001)
13. Zhang, H., Arens, E., Huizenga, C., et al.: Thermal sensation and comfort models for non-uniform and transient environments: part I: local sensation of individual body parts. Build. Environ. **45**, 380–388 (2010)

Analyse Different Angles of Editing Customer Information Data in Customer Relationship Management

Qiao Rong(✉)

University of Science and Technology of China, Huangshan Road 443, Shushan District, Hefei, Anhui, China

Abstract. As the market gradually plays a decisive role in social and economic development, seizing customer resources has become the top priority for enterprise development, and customer data is the information foundation for seizing customer resources, retaining customers, and improving customer satisfaction. We are now in the era of big data, and the role of data in the development of enterprises has become increasingly prominent. Mastering data has become a symbol of mastering wealth, and enterprises' data quality construction has become an important way for sustainable development. This article analyses and summarizes several practical methods of editing customer data, hoping to provide some help to Chinese enterprises and governments in maintaining customer relationships.

Keywords: Customer relationship management · Edit customer information data

1 Introduction

The theory of customer relationship management is developed based on the theory of relationship marketing [1]. Its core lies in taking the customer as the centre, fully doing a good job in the collection, exploration, utilization of customer information, implementing customer service and obtaining customer satisfaction [2]. In the actual application process, the most important link is to collect customer information and perform data processing [3]. When performing data processing, customer information data can be classified from different perspectives [4].

2 Edit Data According to Customer Value

Customer value refers to the total benefits that the customer's entire life cycle can bring to the company. It includes two major parts: potential value and current value. On the basis of fully collecting customer information, companies can conduct customer development analysis based on customer business indicators and then determine Value customers. Potential value customers and other customer types provide corresponding levels of

C. Yuan et al. (Eds.): *Proceedings of the 4th International Conference on Economic Management and Green Development*, AEPS, pp. 302–306, 2021.
https://doi.org/10.1007/978-981-16-5359-9_34

service based on value to achieve customer satisfaction, customer loyalty and customer retention, and optimize the input and output efficiency [5, 6].

The evaluation of customer value can be analysed and considered through current value and potential value, and the current value evaluation and potential value evaluation of customers can be decomposed into corresponding indicators, and the category of customer value can be obtained by coordinated analysis [7]. This model is shown as Fig. 1:

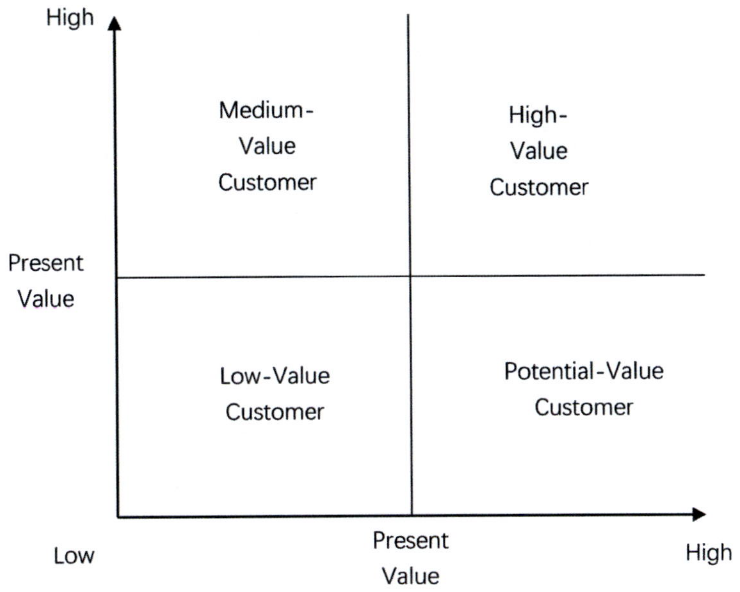

Fig. 1. Customer value evaluation mode diagram 1.

Generally speaking, customers with low contribution to the company also have less expectation of service demand. The higher the value, the higher the demand for service. Therefore, the company can combine customer value evaluation and provide corresponding services referring to high, medium, low and potential distribution, so as to improve the comprehensive satisfaction of all levels of value customers and achieve continuous optimization of customer relationships with limited resources.

3 Edit Data According to Customer Satisfaction Evaluation

Customer satisfaction is the basis of keeping customers and realizing customer loyalty, and it is also one of the purposes of customer relationship management and customer service. Grasping customer satisfaction, improving customer satisfaction in a targeted manner, and developing customer service are key tasks for customer relationship maintenance. Customer satisfaction evaluation model is a more effective satisfaction evaluation tool [8]. This model is shown as Fig. 2:

The customer satisfaction evaluation model is developed from the five aspects of reliability, security, tangibility, responsiveness, and personalised care. Based on the actual situation, corresponding indicators are designed and scored, to evaluate the customer's satisfaction with the service and follow the satisfaction degree to edit customer information.

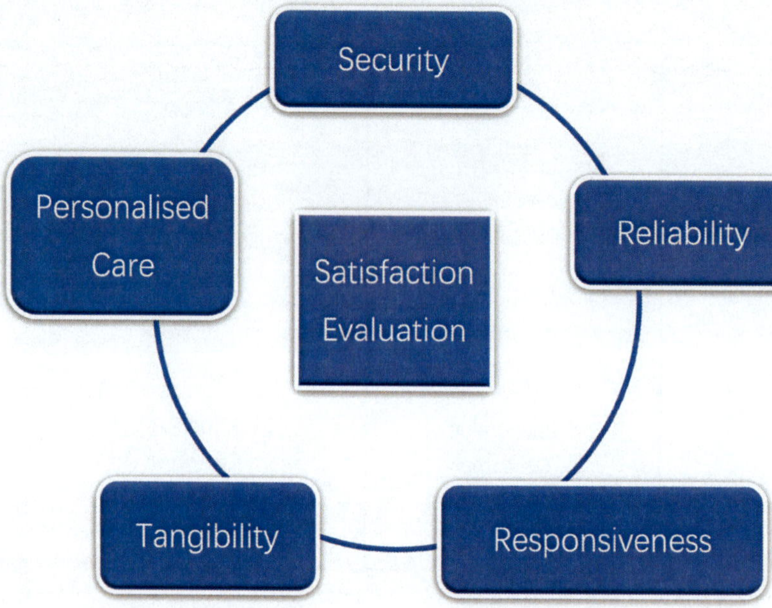

Fig. 2. Customer value evaluation mode diagram 2.

4 Edit Data Information According to the Service Blueprint

Kahneman [9], winner of the Nobel Prize in Economics in 2002, after a long period of in-depth research, he proposed the famous peak-end law, that is, people's evaluation of experience mainly depends on the memory of the peak and the end. Comparing with the peak-end law, it can be concluded that doing a good service must pay attention to the customer needs at the peak moment and the end-point experience. If we want to use this theory to build a service system, it is required to describe the whole processes of the horizontal service, find the peak-end moment, explore the customer response demand and do a good job of service design.

The service blueprint is a deepening of the peak-end law, which includes not only the horizontal external customer service processes, but also the vertical internal company service activities. When an enterprise is editing customer information, it can record customer demand information from multiple contents such as customer behaviour, employee service behaviour, tangible display, background service and supportive activities.

5 Edit Data from the Perspective of Data Quality Evaluation Dimensions

According to specific actual conditions, various international organizations have proposed their own data quality evaluation dimensions [10]. The requirements of US national accounting for data quality are from the three dimensions of accuracy, comparability, and applicability. The British statistics department conducts quality assessment in terms of timeliness, effectiveness, accuracy, and objectivity. The statistical departments of Europe, Canada, Australia, and other countries, as well as international organizations such as the IMF, have also proposed their own corresponding data quality analysis and evaluation dimensions. In general, the analysis dimensions proposed by various scholars and countries are basically similar. Chinese companies can also learn from three dimensions of accuracy, comparability and applicability which are used for data recording.

6 Optimize Data in Terms of Data Quality Factors

When companies edit customer information data, they should also pay attention to some factors affecting data quality, including inherent data quality problems, quality problems during data acquisition, quality problems during data analysis and processing, errors caused by insufficient preparation, errors during data collection, data processing errors and errors caused by personal subjective factors [11, 12].

7 Conclusion

Through the above explanation, I believe that companies can edit customer information data from the aspects of customer value, customer satisfaction evaluation, service blueprint, data quality evaluation dimensions and optimize data from the aspects of data quality influencing factors. When faced with so many angles of data classification, it is often difficult to choose. To solve such problems, the American mathematician Sadie proposed the famous analytical hierarchy process (AHP) [13], which can decompose the core problem into several impact factors and factor supporting indicators. In AHP, the upper and lower logical hierarchical structure between the impact factors and its indicators is constructed and the judgment matrix is established. Through the expert evaluation method, the importance of each factor indicator is determined in pairs comparative judgment. Based on the judgment matrix, the feature vector is calculated, and the corresponding weight of each indicator is obtained, and then the influence degree of the lowest level indicator on the overall goal is obtained. This method can provide a good way for how to select data [14, 15].

References

1. Berry, L.L.: Relationship Marketing, p. 25. AMA, Chicago (1995)

2. Finnie, W., Randall, R.M.: Loyalty as a philosophy and strategy: an interview with Frederick F. Reichheld. Strateg. Leadersh. **30**(2) (2002)
3. Wang, R.Y., Strong, D.M.: Beyond accuracy: what data quality means to data consumers. J. Manag. Inf. Syst. **39**(11), 86–95 (1996)
4. Pipino, L.L., Lee, Y.W., Wang, R.Y.: Data quality assessment. Commun. ACM **45**(4), 211–218 (2002)
5. Lambert, D.M.: Customer relationship management as a business process. J. Bus. Industr. Mark. **25**(1), 4–17 (2001)
6. Berry, L.L.: Relationship marketing of services—growing interest. J. Acad. Mark. Sci. **23**(2), 236–245 (1995)
7. Wang, Y.: Research on the factors influencing the quality of enterprise statistical data. Master's degree thesis of Zhejiang University (2006)
8. Jieru, Z.: Research on the influencing factors of the successful implementation of customer relationship management in enterprises. Mark. Guide **06**, 32–35 (2008)
9. Hongbin, N.: Cigarette Service Marketing. China Finance Press, Beijing (2011)
10. Ye, S.: Research on the evaluation method and application of government statistics data. Ph.D. thesis of Hunan University (2011)
11. Zhu, Q.: Brief analysis of the causes and countermeasures of statistical errors. J. Wuhu Vocat. Tech. Coll. **2003**(01), 55–57+64 (2003)
12. Jincan, L.: Some understandings of statistical errors. Stat. Forecast **02**, 32–34 (2002)
13. Pu, C.: Research and application of analytic hierarchy process. Master's thesis of North China Electric Power University (2008)
14. Landlord: Tobacco enterprise job value evaluation based on analytic hierarchy process. Master's degree thesis of Guizhou University of Finance and Economics (2012)
15. Wang, H.: Research on the suitability of external public space in Chengdu senior-care community based on AHP analytic hierarchy process. Master's degree thesis of Southwest Jiaotong University (2016)

Research on the Design and Guarantee Measures of Enterprise Digital Platform Function Module

Qiao Rong(✉)

University of Science and Technology of China, Shushan District, Huangshan Road 443, Anhui, China

Abstract. The construction of digital platform refers to the realization of major transformation and optimization and innovation of management and operation by enterprises combining with new Internet technologies, bringing new value and transparency to customers and employees in digital business, forming an interconnected company, so as to enhance their core competitiveness. This paper firstly studies the module design of the company's digital platform construction, and then puts forward some measures to ensure the construction of the module project, hoping to provide theoretical help for some enterprises towards intelligent.

Keywords: Design enterprise module · Guarantee function platform

1 Introduction

If an enterprise wants to achieve intelligent manufacturing, it must carry out digitization firstly and then intellectualization, because digitization is the basis of business modeling, as well as the basis of network information flow, business system and intelligent processing equipment integration and big data. Digital platform is a deep integration of the theoretical basis of enterprise information construction, combined with new Internet technology, to realize major transformation and optimization innovation of management and operation, and bring new value to enterprises, customers and employees. It will build a transparent and interconnected company from five aspects of digital marketing, digital R & D, digital procurement, digital warehouse logistics and digital service, Thus, it can form an absolute advantage in a constantly changing digital economic environment [2].

2 Function Module Design of Digital Platform

2.1 Digital Supply Chain Module

The digitalization of supply chain module cannot be separated from the construction of the following functions: the standardization and electronization of sales business process and processing information management [3]; the use of information management system from the sales contract, customer orders, customer delivery, inventory and accounts

C. Yuan et al. (Eds.): *Proceedings of the 4th International Conference on Economic Management and Green Development*, AEPS, pp. 307–311, 2021.
https://doi.org/10.1007/978-981-16-5359-9_35

receivable comprehensive management, and customer demand system docking, accurate acquisition of customer order demand; optimization of sales process, Connect a large number of breakpoints in the previous process [4]; provide the actual warehouse business management of the finished product warehouse of the sales department; provide the management, evaluation and maintenance of customer information; provide the tracking of sales orders; provide the delivery management of goods; provide the statistics and query of sales information to improve the accuracy of sales forecast, so as to reduce inventory and improve the response ability to the market; Establish flexible pricing system and order management system matching with customization; strengthen the closed-loop feedback between after-sales problems and R & D system, quickly classify and handle customer problems according to customer problems, and form a knowledge base. Once similar problems occur, solutions can be given in time.

2.2 Digital Manufacturing Module

In the construction of digital production, it is necessary to establish a flexible multi-organization application architecture, quickly set up multi-factories, multi-organizations and business relationships among them, and help enterprises optimize the business cooperation process among multiple factories and organizations [5]. The production execution layer is mainly divided into workshop, production execution, production control, equipment management and quality management. Among them, the production execution system includes production planning, real-time monitoring, data acquisition, field monitoring and other functions. It mainly provides real-time data and processing functions for specific production scheduling and analysis within the enterprise. The production control system includes processing equipment, control system, detection system, etc., to detect the finished products. Production control level, such as production progress, workshop site records, equipment efficiency analysis, personnel efficiency analysis, etc.; quality control level, such as pass rate, scrap rate, cause analysis of defects, etc.; material management level, such as stock age analysis, parts management, material shelf life management, etc. [6].

2.3 Digital Financial Budget Module

The construction of digital financial budget module should provide the following functions: procurement and payment business: assets, equipment, etc. are included in the procurement asset management and expenses are included in the procurement expense management [7]; prepayment and a/P ratio are determined in the contract link and funds are arranged as a whole; sales collection business: receivables management of enterprises in the digital A/R management platform, which is mainly completed through the circulation of a/R documents, collection documents and sales invoice documents; expense budget management and control business: online application of expenses, reimbursement and payment application, mobile application of expense reimbursement, budget control in advance, and post analysis; cost management and control business: the concept of cost center is established to make the expense distribution more accurate through quota factor and to realize the multi-dimensional cost calculation and analysis table of cost center + product + production order; budget business: establish a sound budget management system of the company as the starting point of internal control of the group and effectively carry out pre-construction, in-process control and post analysis of various expenses.

2.4 Other Business Digital Modules

The digital construction of other business mainly aims at the construction of human resources and mobile platform mentioned in the analysis of other business needs.

Human Resource Management. Organization construction is to realize the organizational structure, position, position and professional title system needed by enterprise strategic development. Basic module, the data in this module is used for organizational performance management. Department salary accounting and human resource construction. Therefore, in all industries and enterprises, the organization construction is the structure. Build other business platforms.

Mobile Office Management Mobile office collaboration can make data managers get rid of the shackles of time and space. By using smart mobile terminals such as mobile phones or tablets, a real-time data management mechanism of enterprises interconnected with computers can be established. Data management and communication can be carried out at any time to improve the efficiency of data management and promote the growth of enterprise benefits. Mobile office supports intelligent mobile terminals based on Android and IOS [9, 10]. Therefore, it is necessary to strengthen the ability of enterprise mobile office management.

3 Safeguard Measures for Establishing Digital Platform Module

3.1 Organization Guarantee

In order to strengthen the management and control in the implementation process of the digital platform module, coordinate and mobilize the strength of all aspects, foresee the possible problems and solve them in time, enterprises should establish a digital platform strategy committee to strengthen the organizational guarantee.

3.2 Method and System Guarantee

As the implementer of the digital platform, the enterprise will strictly ensure that all systems comply with relevant specifications and standards, and comply with the provisions of relevant national quality standards, and also adhere to the management and operation policy of "careful design, standardized construction, quality-oriented, user first" management and operation policy [11].

3.3 Project Implementation Schedule Management

After the establishment of the digital platform module, it is necessary to strengthen the project implementation schedule management. All the specific work is carried out according to the project work schedule. In the process of project implementation, the project implementation progress is controlled. In case of deviation, the project work plan needs to be adjusted accordingly [12].

3.4 Project Progress Monitoring

In the process of digital platform project construction, the enterprise will monitor the overall progress in the following ways: according to the "project schedule" jointly confirmed by both parties, the project manager will work with all relevant teams to work out a more detailed "time schedule" according to the overall project progress, Ensure that the implementation can be carried out in strict accordance with the time schedule, so as to monitor the progress of the project [13].

3.5 Risk Response Guarantee

The risk management of digital platform module construction is one of the important aspects of project management. By identifying risks, actions can be taken before problems occur. Risk management provides necessary input for the planning process and helps to make scheme selection and decision-making. In order to reduce and avoid the project risk caused by the consequences of decision-making mistakes, enterprises should make a project risk management plan in advance, analyze and identify project risks at the beginning of the project [14].

4 Conclusion

This paper describes the construction of digital platform for intelligent manufacturing, and studies the construction of digital platform and the safeguard measures of platform construction, which meets the needs of current information management, and hopes to provide some theoretical help for enterprises in China and other countries in the world to carry out informatization construction.

References

1. Chen, Y., Wu, Y.: Industry 40: intelligent manufacturing wins the future. Shanghai Inf. **2014**(12), 34–36 (2014)
2. Yin, F.: Intelligent manufacturing industry ushers in development opportunities. Internet Econ. **2017**(1–2), 34–39 (2017)
3. Qiu, L.: Planning and Design of Integrated Supply Chain Management System. Chongqing University, Chongqing (2018)
4. Wang, B., Zhang, S.: Construction ideas and key elements analysis of discrete manufacturing intelligent factory. Natl. Defense Manuf. Technol. **2016**(01), 26–29 (2016)
5. Wang, Z.: Building an Intelligent Manufacturing Platform in Nanjing

6. Xiong, X.: Exploration on Intelligent factory construction of large process enterprises. Contempor. Petrol. Petrochem. **24**(07), 9–12 (2016)
7. Wang, G.: Research on informatization planning for intelligent manufacturing of HF Company. Guangxi University, Guangxi (2017)
8. Fan, X.: Analysis of warehouse management system selection, enterprise guide (2012)
9. Wang, Y., Chen, T.: Review of key technologies of machine vision perception and control in smart factory. ZTE Technol. **2016**(22–5), 26–30 (2016)
10. Zhang, C., Yao, X., Zhang, Y., Zhou, J., Yi, A.: From "CNC generation" to "intelligent generation." Comput. Integr. Manuf. Syst. **2015**(7), 1735–1743 (2015)
11. Fu, S.: Application of artificial intelligence in national container transportation business information processing of shipping companies. Containerizatio. **24**(07), 9–12 (2018)
12. Xu, S.: Design and implementation of China Tobacco safety production management information system. Yunnan University, Kunming (2010)
13. Wang, B.: Research on Life Cycle Project Cost Management of Self-use Projects. Shandong University, Jinan (2016)
14. Li, X.: Application of information management in Karamay Petrochemical Company. Xinjiang. Petrol. Technol. **2014**, 34–36 (2009)

Environmental Operations in the Airline Industry: Comparison Between Chinese and Overseas Airline Groups

Minyou Qing(✉)

Management School, University of Liverpool, Brownlow Hill, Liverpool, UK

Abstract. Air transportation, with the fastest speed, the highest reliability, and the best level of safety, is still one of the main methods of long-distance passenger transportation and time-constrained transportation. Due to the high costs in fuel and maintenance, and limited transportation capacities, it becomes the most expensive type of transportation. However, the operational support for air transportation inevitable brings a lot of energy consumption and greenhouse gas (GHG) emissions. Both the in sky emissions and on-ground emissions would together generate great air pollution issues for airlines. As the core parts of airline operations, both fleet management and energy management are important for airline groups to take continual sustainable optimizations. This paper would adopt some Key Performance Indicators (KPIs) to benchmark some Chinese and overseas airline groups' environmental-related figures and to evaluate their environmental performance situation.

Keywords: Environmental operation · Fleet management · Energy management · KPIs

1 Introduction

In the context the Reformation and Opening, and trade globalization, China has become increasingly connected with the world. Whether the official exchanges, business exchanges, cargo transportation, or vacation trips, the demand for air transportation are constantly increasing. According to the Ministry of Transport of China, in 2019, China has 62 air transportation companies, 238 airports engaged in transportation, 3,818 registered aircraft, 5521 operating routes, and the total civil aviation transportation turnover of 129.33 billion ton-kilometres [1, 2]. Those airlines consume plenty of resources and emit numerous GHG, which will have a severe impact on the ecological environment. In 2019, the four major aviation group companies accounted for 84.9% of Chinese total air transport turnover, China Southern Airlines (25.2%), Air China (24.6%), China Eastern Airlines (19.5%), and Hainan Airlines (15.6%) [1, 2]. Taking the fourth-ranked Hainan Airlines as an example, it consumes a total of 5,360,917 tons of energy and causes 11,476,749 tons of carbon dioxide emissions [3].

C. Yuan et al. (Eds.): *Proceedings of the 4th International Conference on Economic Management and Green Development*, AEPS, pp. 312–321, 2021.
https://doi.org/10.1007/978-981-16-5359-9_36

Committing to the environment issue, multiple airline groups around the world already launched a series of special initiatives. Emirates Airlines through establishing the "Amarula Trust" program to supports elephant conservation projects and the development of eco-tourism skills [4]; All Nippon Airways Group developing the next-generation volatile organic compounds free paint system to provide better material weather resistance [5]; Air France launched optional carbon footprint reduction service "Trip and Tree", through passengers' donation supporting afforestation projects [6]; Qantas Airways through the "Reconciliation Action Plan" inspiring Indigenous rangers to work for the regeneration of native vegetation [7]; Air New Zealand cooperate with the Department of Conservation's to protect the biodiversity of New Zealand [8].

At the meantime, the Chinese civil aviation industry also gives serious concern about this topic. China Eastern Airlines promotes three-colour garbage bags help complete nearly 30 tons of cabin garbage classification [9]; Hainan Airlines insists on "Fly safe, Fly green" operation strategy [3]; Xiamen Airlines replaces and upgrades cabin disposable consumables with bamboo products [10]; Air China established Phoenix Cultural and Creative Studio to make aviation-themed products by recycling scrap materials from retired aircraft [11]; China Southern Airlines developed "Aviation Fuel 'E' Cloud" real-time big data sharing platform [12]. Overviewing these airline groups' practices, including direct green operation measures and indirect green business cooperation. We could affirm their efforts to be greener, but due to the mode of action and the effective period duration are various, the environmental influence is hard to evaluate.

To identify the level of the environmental performance of Chinese domestic airline groups, this paper tends to conduct business analysis in operation field for them and take some overseas airline groups as its contrasts. Benchmarking the acquired data will help airlines clarify the advantages and disadvantages of green operations to affect the future resource allocation for environmental protection investment.

2 The Environmental Operations of Airlines

2.1 Method and Data

This paper tends to select two each for Chinese domestic airline groups and overseas airline groups as research samples. According to the "Best Airlines 2020 Award" ranking data from the Airline Rating website [13], the best two airlines around the world are Air New Zealand (ANZ) and Singapore Airlines (SIA). Due to no Chinese airlines show on this ranking list, we take another rating data from this website, "Compare Airline Safety Ratings" [14], as the reference. In which, the Air China (CCA) and China Southern Airlines (CSN) achieve the best performance in Chinese Airline Groups. Those four corporations in Table 1 are the samples of the comparison analysis.

CCA is the only civil airline flying the national flag in China. In addition to its normal passenger and cargo business, it also undertakes important domestic and foreign flight support tasks such as travel by state leaders, charter flights, and emergency flights. It built a green and low-carbon circular operation model to achieve its "Become a world-class air transport company with global competitiveness" vision.

CSN is the airline with the most transport aircraft, the most developed route network and the largest annual passenger volume in China. As a huge level of carbon emitter, it has formulated the "Energy Conservation and Emission Reduction Management Manual", "Energy and Environmental Protection Management Business Process" and other systems to achieve green flight.

Table 1. ICAO code chart.

Airlines	ICAO code
Air New Zealand	ANZ
Singapore Airlines	SIA
Air China	CCA
China Southern Airlines	CSN

ANZ as the largest airline in New Zealand and the main air carrier in the South Pacific, it pursuits to enriching New Zealand by connecting New Zealand to the world. By following the United Nations Sustainable Development Goals (SDGs), it established a sustainable framework to monitor its actions, optimizing its operational efficiency.

SIA might be the most special one, which pays attention to the future of travel, by reducing the current environmental footprint. For achieving the ambition of long-term sustainable development, it proposes several environmental targets, such as eliminating plastic straws and reducing 30% waste from buildings.

The source of collecting data on these corporates' environmental operation measures has two main types. Both ANZ and SIA demonstrate their measures in Sustainability Reports (SD report), while CCA and CSN publish this information through their Corporate Social Responsibility Report (CSR report). Besides, other relevant activities are shared on their official websites also become a significant data source. To process information, this paper would apply the Benchmarking Analysis Method, which could compare corporates' internal data as well as conducting data analytics with external competitors [15–17]. And the KPIs figures during the analysis would contribute to measuring and monitoring, which might be help identify performance weakness [18, 19].

2.2 Environmental Operations

All of the chosen airlines treat environmental sustainable development as an important part of their social responsibility. As shown on their reports, they devote themselves to conducting a series of operation actions to improve its environmental performance. In this part, we would benchmark their performance from two core dimensions by using some KPI's to display, fleet operation management, and energy operation management.

Fleet Operation. In the scope of fleet operation management, these corporates adopt various measures to make themselves more sustainable and environment-friendly. In FY19/20, CSN promoted the single-engine taxi-in project, in which one engine will be turned off when the aircraft is taxiing to reduce taxiing fuel consumption. It implemented a total of 160,882 single-engine slip-in flights, saving 3,042 tons of fuel [12]. SIA develops digital green operating procedures to help pilots make more informed decisions, as well as adopting a similar single-engine taxi-in project, which reducing 10,000 tonnes fuel consumption [20]. Besides these specific approaches, the most typical actions are fleet structure improvement. Along with the increase in the service life, the operating efficiency of old aircraft will inevitably show a downward trend as time goes by. All of them continually purchasing higher efficiency aircraft, which used to replace outdated aircraft. For example, in FY19/20 the SIA order nineteen Airbus A350–900 and twenty-nine Boeing 787–10, which would bring 25% more fuel-efficient than previous-generation aircraft [20], while ANZ through technology innovation and new aircraft procurement achieving 1.8% average annual fuel efficiency improvement compared to 2009 baseline [21]. Compared with ANZ and SIA, as huge scale fleet owners, both CCA and CSN also committed to fleet structure improvement in recent years (Table 2).

Table 2. Aircraft introduce.

Airline	FY15/16	FY16/17	FY17/18	FY18/19	FY19/20
CCA	66	54	56	50	48
CSN	58	53	79	104	54

Although the overall performance of ANZ and SIA are the top around the world, in the perspective of fleet size, they far lower than CCA and CSN, especially ANZ just has 112 aircraft. In contrast, both CCA and CSN had owned more than 500 aircraft fleet five years ago (Table 3).

Table 3. Fleet size.

	FY15/16	FY16/17	FY17/18	FY18/19	FY19/20
ANZ	104	103	114	113	112
SIA	173	178	186	202	203
CCA	590	623	655	669	699
CSN	667	702	754	840	862

Usually, operating a larger scale fleet would cause more energy consumption and green gas emissions. However, the positive correlation growth of emission volume that accompanies the fleet scale increase, does not reflect the efficiency of green operations. This part attempt to adopt the Average fleet age and CO_2 emission (CO_2-e) per tonne-kilometre as the KPI to benchmark their performance.

As shown in Table 4, during the period between FY15/16 and FY19/20, although ANZ shows a decreasing trend for the average fleet age in recent two years, it has the worst overall performance, in each year it ranks the bottom. The performance of SIA is the opposite of ANZ, it has the best average performance during the past five years,

Table 4. Average fleet age.

	FY15/16	FY16/17	FY17/18	FY18/19	FY19/20	Average
ANZ	7.50	7.00	7.50	7.10	7.10	7.24
SIA	6.44	6.68	6.55	6.42	5.92	6.40
CCA	6.30	6.36	6.53	6.62	6.96	6.55
CSN	6.30	6.60	6.67	6.50	6.70	6.55

even keep the youngest fleet in recent two years, especially decrease to 5.92 years in FY19/20. As for two Chinese airlines, both of them place at the middle level with similar performance.

The performance of each airline in Table 5 demonstrate a different situation than Table 4. In the field of CO_2-e per tonne-kilometre, ANZ achieves the best overall performance during the period and controls its figures remaining at a stable level (between 0.72 kg and 0.75 kg). After undergoing an improvement from 0.89 kg to 0.74 kg, SIA data face a significant rebound to 0.81 kg. And then, even not be the worst one, the CCA still demonstrates a negative decrease data trend, which got worse to 0.93 kg. Lastly, although the five-year performance far away from other airlines, the figures of CSN has a continual decrease trend, which already down below 0.90 kg (0.87 kg in FY19/20). In general, in the field of fleet operation management, Chinese airline groups have an obvious advantage in their fleet scale, which might create more revenue, but from the perspective of sustainable green operation, they do not as good as overseas airline groups.

Table 5. CO_2-e per Tonne-kilometer (kg)

	FY15/16	FY16/17	FY17/18	FY18/19	FY19/20
ANZ	0.74	0.73	0.72	0.73	0.75
SIA	0.89	0.88	0.84	0.74	0.81
CCA	0.87	0.86	0.87	0.87	0.93
CSN	0.95	0.95	0.92	0.89	0.87

Energy Operation. GHG emissions are one of the important influencing factors leading to global climate change [22], and extreme weather caused by climate change will also have a huge impact on air transportation, involving various aspects including cost, safety, and the environment [23]. As the airline corporate, aviation fuel consumption is one of the main sources of GHG emissions. Therefore, how to achieve more efficient energy operation management has become the focus of these companies in the aspect of environmental sustainability. In FY19/20, CCA further expands its energy management system, incorporating its southwest branch into the framework of the energy management system to enhance entire energy consumption efficiency. Besides, it continually develops its energy structure optimization, using bridge-mounted equipment power supply mode to replace the aircraft auxiliary power unit, and applying "ground support vehicle 'oil-to-electricity' project". CSN adopts new biofuel, 1 megajoule of heat produces 24 g of carbon dioxide equivalent, to perform intercontinental flight missions, which would reduce 73% CO_2-e. SIA launched a pilot project on the Data-driven Statistical Contingency Fuel approach, which estimates saving aviation fuel 4,000 tonnes each year. And ANZ continually implements its "Carbon Reduction Programme", which saved 3,341 tonnes of fuel or 10,557 tonnes of CO_2-e. To benchmark their energy operation performance, we would take the other KPIs as the reference, and the GHG emission figures would use the sum of Scope1 and Scope2 [24].

Firstly, considering the different aircraft numbers in the fleet would affect the total CO_2-e volume, as a result, we tend to calculate the GHG emission (tonnes CO_2) per hundred aircraft for each airline.

Table 6. GHG emission (tonnes CO_2) per hundred aircraft

	FY15/16	FY16/17	FY17/18	FY18/19	FY19/20
ANZ	3,313,419	3,515,342	3,274,816	3,476,768	2,838,809
SIA	8,055,008	7,910,699	7,620,742	8,168,415	8,035,856
CCA	3,273,288	3,349,053	3,272,153	3,519,133	3,325,894
CSN	3,196,042	3,280,385	3,352,706	3,202,536	3,320,916

Observing Table 6, two Chinese airlines corporate achieve ideal performance, they keep a stable energy efficiency even operating huge scale fleet with continual expansion. After four years of steady, ANZ's figures achieved a significant decrease, down to 2,838,809 in FY19/20. As for the performance of SIA, which is unexpected. Its figures are two times more than other airlines, even close to three times than ANZ in FY19/20. The result of SIA might have its special factors, but to a certain extent, its energy operation efficiency does need further review and enhancement. However, due to each market and route has specific features, there are obvious differences for these airlines in matching aircraft types and fleet sizes, which might influent their CO_2-e. Therefore, this paper used another KPI for energy operation performance benchmarking.

The second KPI is the Scale-CO_2-e Ratio (SCR), which reference to enterprise ecological efficiency because the fleet scale can be seen as the value of a product or service. Based on this KPI, this study could measure the comparison of growth speed between carbon dioxide emissions and fleet scale. It involves the annual fleet scale growth rate (FSGR) and annual CO_2-e growth rate (CEGR). Due to the figures of FSGR and CEGR are in percentage, as a result, the final formula would do subtraction directly, which is "SCR = FSGR-CEGR", the larger the value, the higher the efficiency. And if the SCR value is less than 0, which means the speed of CEGR faster than FSGR. Through comparing the value size, we could figure out which airlines', fleet size growth with CO_2-e as opportunity cost, has higher efficiency (Table 7).

Table 9 demonstrate the SCR figures in the past four years, if judge from the size of the data discreteness, CSN has the smallest standard deviation, CCA and SIA have close figures, while ANZ shows a much higher figure. The size of the standard deviation could reflect the steady of operation efficiency under Scale-CO_2-e dimension. Because, when establishing a mature operation management system, the CO_2-e would increase in a certain ratio related to the fleet scale growth. However, the data discreteness could just represent the data stability, as for the real performance, this study needs to observe its SCR figures in each year (Table 8).

Table 7. Four airlines FSGR Data

	FY15/16	FY16/17	FY17/18	FY18/19	FY19/20
ANZ	104	103	114	113	112
ANZ FSGR	-	-0.96%	10.68%	-0.88%	-0.88%
SIA	173	178	186	202	203
SIA fsgr	-	2.89%	4.49%	8.60%	0.50%
CCA	590	623	655	669	699
CCA FSGR	-	5.59%	5.14%	2.14%	4.48%
CSN	667	702	754	840	862
CSN FSGR	-	5.25%	7.41%	11.41%	2.62%

Table 8. Four Airlines CEGR Data

	FY15/16	FY16/17	FY17/18	FY18/19	FY19/20
ANZ	3,445,956	3,620,802	3,733,290	3,928,748	3,179,466
ANZ CEGR	-	5.07%	3.11%	5.24%	-19.07%
SIA	13,935,164	14,081,045	14,174,581	16,500,198	16,312,787
SIA CEGR	-	1.05%	0.66%	16.41%	-1.14%
CCA	19,312,400	20,864,600	21,432,600	23,543,000	23,248,000
CCA CEGR	-	8.04%	2.72%	9.85%	-1.25%
CSN	21,317,600	23,028,300	25,279,400	26,901,300	28,626,300
CSN CEGR	-	8.02%	9.78%	6.42%	6.41%

As shown in Fig. 1, ANZ undergoing a violent fluctuation, it successfully reaches to around 20% in FY19/20, but once down to less than –5% twice. In most years, the data of SIA is positive, besides a dramatic decline in FY18/19. It is worth noted that, in that

Table 9. SCR Data

	FY16/17	FY17/18	FY18/19	FY19/20	Standard Deviation
ANZ	-6.04%	7.57%	-6.11%	18.19%	0.118
SIA	1.84%	3.83%	-7.80%	1.63%	0.052
CCA	-2.44%	2.41%	-7.71%	5.74%	0.059
CSN	-2.78%	-2.37%	4.99%	-3.79%	0.040

year, both ANZ and CCA meet the same problem and demonstrate similar data. In fact, during this period, CCA also faces with the data fluctuating like ANZ, but in most years it not as good as ANZ. And CSN, which has the smallest standard deviation, actually SCR value performed the worst. As a result, the status of the ideal figures should be as achieve SCR value as higher as possible and avoid dramatic data fluctuation.

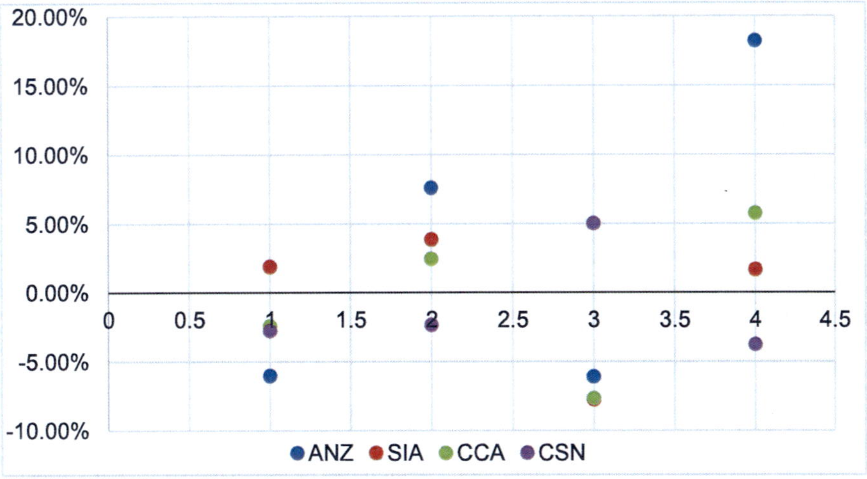

Fig. 1. Four airlines SCR data scatter plot.

3 Discussion

This paper has conducted a series of benchmarking for environmental operations between Chinese and overseas airline groups. During the future researches, not only fleet operation management but also energy operation management could reference more aspects of data.

As mentioned before in the chapter about fleet operation management, this paper pays attention to Average fleet age and CO_2-e per Tonne-kilometer. However, all of these figures are whole company dimension data, if this study attempt to conduct in-depth research on fleet operation management, airline corporates could grasp these two types of data by branch. Firstly, adopting internal benchmarking to compare the performance of each branch in these two aspects, by which they could figure out which branch has the best performance and which should get enhancement immediately. Secondly, taking external benchmarking to compare branches in the same market or route with other airlines. Through the external benchmarking they could identify in which market has more competitiveness. Based on considering these two benchmarking methods, it would be meaningful for helping these airline corporates to improve resource allocation in the future.

Energy operation management also has some potential research aspects. For example, the analysis of fuel consumption and CO_2-e for each aircraft type. Whether in CSR reports or SD reports, airlines just disclosure their total figures, but the data of each fleet, consist of the same aircraft type, did not be published. We all know, most airlines are researching and developing new biofuel. However, considering many factors such as aircraft engine, flight load, fuel tank volume, and fuel characteristics, various types of biofuel with different compositions may only apply to certain aircraft models. From the perspective of energy utilization, formulate specific energy operation management system by aircraft types could enhance operational efficiency and decrease unnecessary costs.

4 Conclusion

This paper found that based on the analysis of multiple dimensions of data, the overall environmental operations performance of the two Chinese airlines is not as good as that two overseas airlines, especially for data involving carbon emissions such as CO_2-e per tonne-kilometre. However, both Air China and CSN are actively seeking new and more environmentally friendly operations, including management systems, big data services, flight technology innovation, and infrastructure construction. Overall, the analysis of environmental operations shows that various airlines have paid enough attention. Whether it is China Southern Airlines with a large fleet or Air New Zealand with a small fleet but a good evaluation, they all agree that green operations can help companies achieve their goal of reducing carbon emissions and becoming environmentally friendly. It is undeniable that the continuous expansion of the scale of the company will indeed increase the difficulty of management and reduce the efficiency of operations. However, the environment is an important part of sustainable development, airlines as one of the numerous CO_2 emitters, how to achieve green operations and green flights might be a long-term topic for them that will never be able to ignore.

References

1. The Ministry of Transport of the People's Republic of China: Statistical Bulletin of Civil Aviation Industry Development (2019). http://www.mot.gov.cn/tongjishuju/minhang/
2. The Ministry of Transport of the People's Republic of China: Statistical Bulletin of Civil Aviation Airport Production (2019). http://www.mot.gov.cn/tongjishuju/minhang/

3. Hainan Airlines Holding Co. Ltd.: 2019 Corporate Social Responsibility Report. HNA, Hainan (2019)
4. The Emirates Group: Food and product sustainability (2019). https://www.emirates.com/uk/english/about-us/our-planet/sustainability-in-operations/sustainable-onboard-products/
5. All Nippon Airways Co., Ltd: Flight Operations Initiatives (2019). https://www.ana.co.jp/group/en/csr/environment/operating/
6. Air France Limited: Acting Together for Responsible Travel (2019). https://www.airfrance.fr/FR/en/common/page_flottante/information/developpement-durable.htm
7. Qantas Airways Limited: Carbon offsetting (2019). https://www.qantas.com/us/en/qantas-group/acting-responsibly/our-planet/carbon-offsetting.html
8. Air New Zealand Limited: Air New Zealand and the Department of Conservation (2019). https://www.airnewzealand.com/doc
9. China Eastern Airlines Co., Ltd.: 2019 Corporate Social Responsibility Report. China Eastern Airlines, Shanghai (2019)
10. Xiamen Airlines Co., Ltd.: 2019 Corporate Social Responsibility Report. Xiamen Airlines, Xiamen (2019)
11. Air China Co., Ltd.: 2019 Corporate Social Responsibility Report. Air China, Beijing (2019)
12. China Southern Airlines Co., Ltd.: 2019 Corporate Social Responsibility Report. China Southern Airlines, Guangzhou (2019)
13. AirlineRating: Best Airlines 2020: Airline Ratings Names Air New Zealand Top Carrier (2020). https://www.airlineratings.com/news/airline-ratings-names-best-airlines-2020/
14. AirlineRating: Compare Airline Safety Ratings (2020). https://www.airlineratings.com/safety-rating-tool/
15. Puig, M., et al.: Benchmark dynamics in the environmental performance of ports. Mar. Pollut. Bull. **121**(1–2), 111–119 (2017)
16. Galingdro, B.M., Zanghelini, G.M., Soares, S.R.: Use of benchmarking techniques to improve communication in life cycle assessment: a general review. J. Clean. Prod. **213**, 143–157 (2019)
17. Park, Y.S., et al.: Benchmarking environmental efficiency of ports using data mining and RDEA: the case of a U.S. container ports. Int. J. Logist. Res. App. **22**(2), 172–187 (2019). https://doi.org/10.1080/13675567.2018.1504903
18. Lindberg, C.-F., et al.: Key performance indicators improve industrial performance. Energy Proc. **75**, 1785–1790 (2015). https://doi.org/10.1016/j.egypro.2015.07.474
19. Kylili, A., Fokaides, P.A., Jimenez, P.A.L.: Key Performance Indicators (KPIs) approach in buildings renovation for the sustainability of the built environment: a review. Renew. Sustain. Energy Rev. **56**, 906–915 (2016). https://doi.org/10.1016/j.rser.2015.11.096
20. Singapore Airlines Limited.: Sustainability Report FY2019/20. Singapore Airlines, Singapore (2019)
21. Air New Zealand Limited.: Sustainability Report 2020. Air New Zealand, New Zealand (2020)
22. Yu, J.L.: China's aircraft-related CO2 emissions: Decomposition analysis, decoupling status, and future trends. Energy Policy **138**, 111215 (2020). https://doi.org/10.1016/j.enpol.2019.111215
23. Ritchie, B.W., et al.: Effects of climate change policies on aviation carbon offsetting: a three-year panel study. J. Sustain. Tour. **28**(2), 337–360 (2020)
24. United States Department of Transportation: Airport Carbon Emissions Reduction (2020). https://www.faa.gov/airports/environmental/air_quality/carbon_emissions_reduction/

Ethical or Abusive? A Review of Two Leadership Influencing Mechanisms

Dan Qin(✉)

School of Foreign Languages, University of Electronic Science and Technology of China, Xiyuan Avenue, West Hi-Tech District, Chengdu, China

Abstract. Few studies have reviewed the mechanisms of ethical leadership and abusive supervision. Based on the review of the leadership literature of top journals in the past two decades, this paper summarizes the influencing mechanism and research perspectives of these two kinds of leadership. Previous studies have shown that ethical leadership has positive effects on employees and teams, while abusive supervision has negative effects on employees and team performance. However, existing literature mainly uses employees' self-concept, motivation, emotion, and other mediations to explain the influence of leadership on subordinates. This paper summarizes the main research perspectives of ethical leadership and abusive supervision. It sorts out the development background of two kinds of leadership research, which has a guiding role for follow-up research.

Keywords: Ethical leadership · Abusive supervision · Influencing mechanisms

1 Introduction

Leadership is a key resource for employees and teams. In leadership research, scholars have found that both positive leadership (e.g., ethical leadership) and negative leadership (e.g., abusive supervision) are common in organizations and affect employee behavior and team performance. At present, ethical leadership and abusive supervision have been widely concerned by scholars, but few articles have compared them together. There is a lack of literature to systematically review their influencing mechanism and specific context. Therefore, to fill the research gap, this paper combines the two for comparative analysis.

Ethical leadership and abusive supervision are two contrasting styles of leadership. Ethical leadership has been defined as "the manifestation of normative and appropriate conduct in individual behaviors and interpersonal relationships, and the enhancement of subordinates' moral behaviors through reciprocal communication, reinforcement, and decision-making" [1]. Abusive supervision refers to "the extent to which followers perceive chronic supervisor hostility through verbal and nonverbal behaviors to the exclusive of physical contact." [2]. They have different styles and different functional boundaries, so they have different effects on employees and teams. This paper will explore the influence of ethical leadership and persistent supervisor hostility in different

C. Yuan et al. (Eds.): *Proceedings of the 4th International Conference on Economic Management and Green Development*, AEPS, pp. 322–330, 2021.
https://doi.org/10.1007/978-981-16-5359-9_37

specific situations. It focuses on the influencing mechanism of moral leadership and supervisory abuse. Based on a review of 122 relevant works on leadership research in top journals in the past two decades, this paper summarizes their mediations and boundary conditions for employee work results and team results, which is of certain guiding significance for future leadership research.

2 Research Mechanism

2.1 Mediators

This section reviews the leadership research literature from top management journals (ABS rated 4 or above) over the past two decades, as shown in Fig. 1. Research trends have changed over the past two decades. First of all, leadership research has changed from performance-oriented to employee-oriented, and the research results have changed from simple performance to diversified employee behaviors. Secondly, the research perspective is transformed from static to dynamic, and scholars gradually realize the dynamic changes of key situations such as attention, emotion, and pressure between leaders and subordinates. Finally, the understanding of leadership has become more comprehensive. Instead of simply believing that positive leadership will bring positive effects, more emphasis has been placed on the role boundary of leadership.

Mediator A. Employees are influenced through five mediators – self-construal, efficacy, cognition, collective outcome, and organizational support – by ethical leaders, which are expounded as follows.

Firstly, as a social referent, ethical leaders can cultivate employees' moral traits thus altering or shaping their self-construal both vicariously and directly. Previous studies have revealed that followers are inclined to learn moral values and politics from their ethical leader by regarding him as a role model who also contributes to the rise of employees' self-esteem and feelings of organizational respect. Hence, organizational identification is enhanced, through which ethical leaders can affect followers' job performance. Zhu et al. added relational identification into focus and deemed that employees may identify with the core member to meet their fundamental needs like affiliation [3].

Secondly, ethical leadership relates to efficacy, especially resolution efficacy and group ethical voice efficacy. By defining success not only in the attainment of goals but also in social processes employees learn, ethical leaders enhance their subordinates' resolution efficacy, which is a specific form targeted at conflict resolution [4]. With the help of the social learning process which is especially important to their leaders' ethical behaviors, employees' group ethical voice is activated, thus forming control beliefs, conceptually related to the concept of self-efficacy [5].

Thirdly, when ethical leaders offer narratives expressing proactive organizational expectations, subordinates garner a cognitive anchor which guides moral equity judgments. If retrospective evaluations of ethically-related situations are developed, followers will form judgments about workplace deviance as being ethically inappropriate while they gain an understanding of the importance of cooperative behaviors when supervisors account for the outlook about how to interact with co-workers. Both are fulfilled through sense-making and social learning [6].

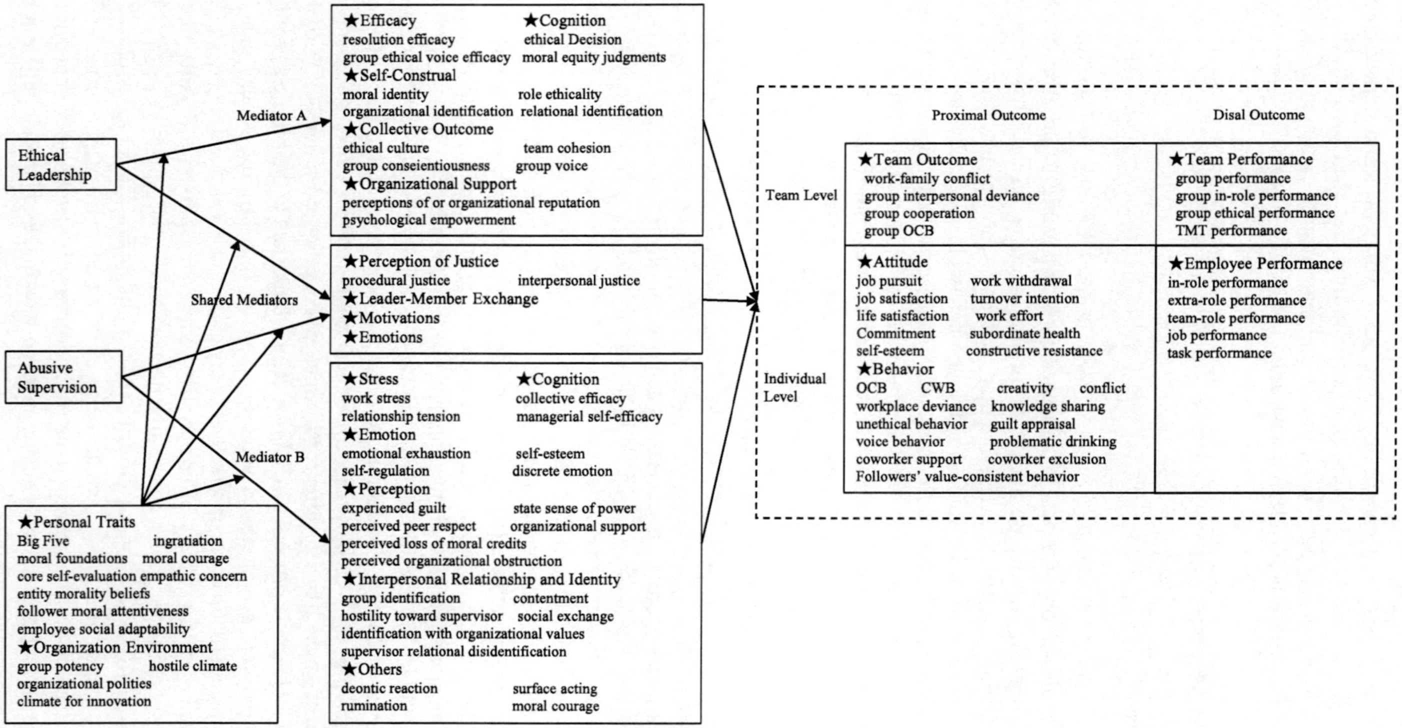

Fig. 1. Summary of research mechanisms

Fourthly, research focuses on collective outcome as a mediator, through which ethical guidance affects employees. By meeting the needs of team members, decreasing the occurrence of interpersonal disputes among subordinates, and enhancing reward expectancy for diligent work, ethical leaders yield team cohesion effectively [7]. Walumbwa et al. revealed a link between ethical leadership and workgroup conscientiousness [8]. Moral leaders display behaviors signaling honesty, self-discipline, reliability, responsibility, high standards, and truthfulness which are consistent with hallmarks of conscientiousness and therefore shape the norms the workgroup is likely to formulate.

Finally, organizational support links moral guidance and employees. As a comprehensive motivational mechanism, psychological empowerment operates through four cognitions: meaning, impact, self-determination, and competence, two of which are influenced by ethical leadership. To be more specific, through developmental interactions, such as reciprocal dialogue, moral leaders perceive subordinates' concerns and provide constructive and positive feedback accordingly, thus enhancing confidence in their followers' competence. Also, the model for employees is critical to understanding the full scope of one's decisions, which is beneficial to foster a sense of self-determination [8].

Mediator B. Figure 1 displays stress, emotions, perceptions, efficacy, identifications, and other factors mediating the relationship between abusive supervisors and subordinates, which are delineated as follows.

Firstly, abusive supervision influences employees mediated by stress. Supervisory abuse—such as deriding, undermining, and censuring subordinates—signals a chronic stressor that is in connection with psychosomatic complaints [9]. Prior study unfolds that to reconstruct fairness and relieve frustration got from abusive leaders, followers are inclined to transmit the displaced aggression by annoying and irritable interactions with a partner instead of striking back to their leaders on account of concerning the fallout related to their career lives, which reinforces the tension between employees and their partners [11].

Secondly, abusive supervisors lead to destructive behaviors of followers through negative impacts on their emotions. From a psychological perspective, receiving negative treatment in the social group, such as criticism, ridicule, and abuse, a person will be perceived as not valued and the supervisory hostility represents a continual deterioration in relationships. Employees are in consequence marginalized; thus their self-esteem is diminished [12].

Thirdly, abusive supervision associates with perceptions. For instance, perceived organizational obstruction is positively related to abusive supervisors on account of the influence that the negative social exchange between leaders and employees has on employees' negative social exchange perceptions of their organizations [13]. Schaubroeck et al. have revealed that when receiving more supervisory abuse than group members who report to the same leader, the person perceives less peer respect because subordinates show great deference to supervisors, viewed as experienced evaluators and the authority in organizations, which adversely impact employees [14].

Fourthly, some research discusses efficacy as a mediator of the relationship between supervisory abuse and subordinates. Abusive supervision relates to low psychological safety which conduces to the development of collective efficacy. When receiving persistent supervisor hostility, employees are inclined to be wary of putting forward ideas and even deem themselves incompetent, which is harmful to task-related teamwork, thus lessening collective efficacy [15].

Finally, identification is an important explanatory mechanism. Through persuasions and serving themselves as exemplars, supervisors influence employees in identifying with organizational values. Abusive supervision actions, such as belittlement and ridicule, don't abide by core values which are crucial components of organizations, hence a negative impact on followers' identification with those values [16]. Besides, there are some other types of mediators – deontic reaction, surface acting, rumination, and moral courage. Targets of supervisory abuse are inclined to use surface acting to comply with social norms of emotional displays, such as fake positive emotions to avoid further abuse and suppressions of negative emotions to avert coworker derision [17].

Shared Mediators. There are some shared factors mediating the relationship between ethical leaders and employees and between abusive supervisors and their followers, among which motivations are the most common one. To motivate the work unit's behavior to meet organizational requirements, supervisors implement policy control systems aligning with their moral principles. Extrinsic motivation, controlled motivation in particular, such as monetary benefits, conduces to employees' behavioral regulation [18]. H. Zhang et al. Asserted that continual abusive supervision was liable to form a negative organizational climate which may undermine intrinsic motivation, hence a decrease in creativity and work enjoyment [19].

2.2 Moderators

Moderators are mainly divided into two parts – individual characteristics and organizational environment. In terms of the former part, the trait of moral judgment is the leading factor, such as moral foundations, moral courage, follower's moral attentiveness, and so forth. Employees with high moral judgment are more likely to be influenced by moral leadership, and more averse to abusive leadership in contrast.

2.3 Results

Figure 1 classifies results into proximal and distal from both individual and organizational level, among which, proximal results are mainly divided into employee attitudes, employee behaviors, and team results, while employee performance and organizational performance are included in distal results.

In terms of employee attitudes, there exist general work attitudes, such as job pursuit, turnover intention, and job satisfaction, and specific attitudes, such as work withdrawal, constructive resistance, and organizational identification. For instance, when abused by supervisors, subordinates perceive lower peer respect in social groups, which undermines the potential for them to foster a sense of belonging, a crucial part in connecting an individual to teams. Hence, employees' turnover intentions increase [20]. Also, if followers of abusive supervisors don't take measures (e.g., suppression) to manage their negative emotions, they will develop high levels of emotional exhaustion, thus increasing work withdrawal [21].

General behaviors like organizational citizenship behavior, workplace deviance, and voice behavior, and specific behaviors (e.g., problematic drinking) are included in employee behaviors. Prior study reveals that leader-member exchange is generally of high quality in the context of ethical leadership, which compels employees to foster a sense of obligation to engage in the workplace in a reciprocate way [22]. More commonly discussed in the area of team results is the engendered impact on the working groups and families, such as group cooperation and work-to-family conflict.

In terms of performance, followers of ethical leaders perform better on account of high organizational identification with which individuals are inclined to align their interests with those of their teams [3]. Besides, Walumbwa et al. asserted that group members' high performance related to ethical leaders' role-modeling process, high LMX, and behaviors of interacting directly with team members' performance management [8].

3 Theoretical Perspective

When summarizing and refining the existed literature, the author finds that the influencing mechanism of the two kinds of leadership is analyzed generally from three various perspectives – the resource, interpersonal and emotional perspective, which will be expounded as follows.

3.1 The Resource Perspective

From the perspective of conservation of resources, leadership is regarded as a kind of situation of increasing and losing employee resources. Ethical leadership conduces to resource increment. Its presence supplements the consumption of employees' resources, through which employees are encouraged to actively respond to the changes in the organizational environment and devote themselves to the work processes. For example, moral leadership can improve employees' self-efficacy, and this mechanism plays a more significant role in proactive personality. However, abusive supervision consumes resources. Supervisory abuse, such as gruff manners and harsh behaviors toward subordinates, results in resource depletion, emotional exhaustion, and a sense of pressure. Followers' resources are used to deal with such intensive circumstances. Hence, it is difficult for employees to concentrate their energy on the work processes and actively contribute to the team. Some research reveals that out of the fear of their resource loss, employees are inclined to take reflexive behavior and reduce the initiative behavior and voice behavior, thus leading to negative results (e.g., work-to-family conflict, work withdrawal, and lower performance).

3.2 Interpersonal Perspective

The interpersonal perspective focuses on the influence of leadership style on subordinates' interpersonal relationships, the relationship between the two, and the impact of the relationship quality on employees' behavior and performance. Generally, the relationship between moral leadership and subordinates is better. Ethical leaders tend to be more friendly and they would like to admit their own mistakes frankly. Equal dialogues with employees and attention to the development of subordinates also conduce to higher exchange quality. In this case, followers are more willing to contribute to the organization in return for the leader's efforts. For example, ethical leadership increases the organizational identity of employees, thereby reducing turnover intention and improving job satisfaction. On the contrary, abusive supervision, such as ridicule, negligence in emotional care, and lack of empathy towards employees, curtails the relationship between supervisors and their followers. Subordinates perceive irrationality and have a lower psychological sense of belonging to social groups. Therefore, unwillingness for them to pay for the organization and repay leaders is fostered. Supervisory abuse reduces individuals' perception of justice who feel the interpersonal justice is much unfair because leaders scold without any consideration, which results in turnover intention, reduction in employees' sense of happiness, and even counter-productive, deviant, and revengeful behaviors.

3.3 Emotional Perspective

From the perspective of emotion, leadership is regarded as an important factor in influencing employees' emotions. Ethical leadership can generate good emotions for employees. Ethical leaders elicit a moral atmosphere to the team. As an employee, affected by the positive atmosphere within the team, he tends to be in a good mood and feels that his co-workers are amicable and reasonable, hence more willing to contribute to teams (i.e., a positive emotion). At the team level, due to the positive ethical climate brought by the exemplary role of moral leaders and their friendly behaviors, team members are in a positive attitude and become more united, thereby increasing organizational citizenship behavior of the team. Abusive supervisors engender negative emotions. Intuitively, when employees encounter injustice and unreasonable abuse and criticism from leaders, they perceive inequity in the team. Therefore, they not only produce negative emotions but also fear leaders for punishment and abuse. Such emotions all result in subordinates' inability to concentrate on their work, thus reducing job satisfaction.

4 Conclusion and Discussion

Through reviewing existing research, this paper summarizes the main influencing mechanisms of ethical leadership and abusive supervision. The main findings are as follows: First, ethical leadership and abusive supervision, as two distinct leadership styles, have different impacts on employees and teams. In general, ethical leadership can improve employees' initiative, employee performance, and team performance. Abusive supervision undermines motivation and harms employees and teams. Secondly, existing studies have established a rich explanation mechanism between the two kinds of leadership and work results, including common intermediary mechanisms (such as motivation and emotion) and their unique intermediary mechanisms. Finally, the existing literature generally analyzes the influencing mechanism of two kinds of leadership from three perspectives: the resource perspective, interpersonal perspective, and emotional perspective.

This research has the following three contributions. First, by summarizing the research results of top journals, this paper sorts out the research mechanisms of the two kinds of leadership respectively, which lays a foundation for follow-up research to some extent. Secondly, by further refining the literature, this paper summarizes three common theoretical perspectives and clarifies the characteristics of each. Finally, this paper makes a comparative study of ethical leadership and abusive supervision to clarify their results for employees and teams. Of course, this study inevitably has some shortcomings, such as the absence of further discussion on the situational nature of ethical leadership and abusive supervision. Recent studies have shown that abusive supervision has positive effects under some conditions, and ethical leadership may also bring negative effects such as the leader's moral superiority. On this basis, the following research can further discuss the functional boundary of the two kinds of leadership.

References

1. Brown, M.E., Treviño, L.K., Harrison, D.A.: Ethical leadership: a social learning perspective for construct development and testing. Organ. Behav. Hum. Decis. Process. **97**, 117–134 (2005)
2. Tepper, B.J.: Consequences of abusive supervision. Acad. Manage. J. **43**, 178–190 (2000)
3. Zhu, W., He, H., Treviño, L.K., Chao, M.M., Wang, W.: Ethical leadership and follower voice and performance: the role of follower identifications and entity morality beliefs. Leadersh. Q. **26**(5), 702–718 (2015)
4. Babalola, M.T., Stouten, J., Euwema, M.C., Ovadje, F.: The relation between ethical leadership and workplace conflicts. J. Manage. **44**(5), 2037–2063 (2016)
5. Huang, L., Paterson, T.A.: Group ethical voice: influence of ethical leadership and impact on ethical performance. J. Manage. **43**(4), 1157–1184 (2017)
6. Resick, C.J., Hargis, M.B., Shao, P., Dust, S.B.: Ethical leadership, moral equity judgments, and discretionary workplace behavior. Human Relat. **66**(7), 951–972 (2013)
7. Zheng, D., et al.: Effects of ethical leadership on emotional exhaustion in high moral intensity situations. Leadersh. Q. **26**(5), 732–748 (2015)
8. Walumbwa, F.O., Morrison, E.W., Christensen, A.L.: Ethical leadership and group in-role performance: the mediating roles of group conscientiousness and group voice. Leadersh. Q. **23**(5), 953–964 (2012)
9. Dust, S.B., Resick, C.J., Margolis, J.A., Mawritz, M.B., Greenbaum, R.L.: Ethical leadership and employee success: examining the roles of psychological empowerment and emotional exhaustion. Leadersh. Q. **29**(5), 570–583 (2018)
10. Zhang, Y., Liu, X., Xu, S., Yang, L.Q., Bednall, T.C.: Why abusive supervision impacts employee OCB and CWB: a meta-analytic review of competing mediating mechanisms. J. Manage. **45**(6), 2474–2497 (2019)
11. Carlson, D.S., Ferguson, M., Perrewé, P.L., Whitten, D.: The fallout from abusive supervision: an examination of subordinates and their partners. Pers. Psychol. **64**(4), 937–961 (2011)
12. Vogel, R.M., Mitchell, M.S.: The motivational effects of diminished self-esteem for employees who experience abusive supervision. J. Manage. **43**(7), 2218–2251 (2017)
13. Mackey, J.D., McAllister, C.P., Brees, J.R., Huang, L., Carson, J.E.: Perceived organizational obstruction: a mediator that addresses source–target misalignment between abusive supervision and OCBs. J. Organ. Behav. **39**(10), 1283–1295 (2018)
14. Schaubroeck, J.M., Lam, S.S.K., Peng, A.C.: Can peers' ethical and transformational leadership improve coworkers' service quality? A latent growth analysis. Organ. Behav. Hum. Decis. Process. **133**, 45–58 (2016)

15. Priesemuth, M., Schminke, M., Ambrose, M.L., Folger, R.: Abusive supervision climate: a multiple-mediation. Acad. Manage. J. **57**(5), 1513–1534 (2014)
16. Hannah, S.T., et al.: Joint influences of individual and work unit abusive supervision on ethical intentions and behaviors: a moderated mediation model. J. Appl. Psychol. **98**(4), 579–592 (2013)
17. Carlson, D., Ferguson, M., Hunter, E., Whitten, D.: Abusive supervision and work-family conflict: the path through emotional labor and burnout. Leadersh. Q. **23**(5), 849–859 (2012)
18. Bavik, Y.L., Tang, P.M., Shao, R., Lam, L.W.: Ethical leadership and employee knowledge sharing: exploring dual-mediation paths. Leadersh. Q. **29**(2), 322–332 (2018)
19. Zhang, H., Kwan, H.K., Zhang, X., Wu, L.Z.: High core self-evaluators maintain creativity: a motivational model of abusive supervision. J. Manage. **40**(4), 1151–1174 (2014)
20. Schaubroeck, J.M., Peng, A.C., Hannah, S.T.: The role of peer respect in linking abusive supervision to follower outcomes: dual moderation of group potency. J. Appl. Psychol. **101**(2), 267–278 (2016)
21. Chi, S.C.S., Liang, S.G.: When do subordinates' emotion-regulation strategies matter? Abusive supervision, subordinates' emotional exhaustion, and work withdrawal. Leadersh. Q. **24**(1), 125–137 (2013)
22. Thiel, C.E., Hardy, J.H., Peterson, D.R., Welsh, D.T., Bonner, J.M.: Too many sheep in the flock? Span of control attenuates the influence of ethical leadership. J. Appl. Psychol. **103**(12), 1324–1334 (2018)

Does Emotion Affect People's Decision-Making: Evidence from an Experimental Study

Xinyu Zhu(✉)

WHBC Wuhan Foreign Languages School, Wuhan, Hubei, China

Abstract. Normally, economists believe that predictions are purely based on relevant information and collective understanding acquired from well-researched surroundings. Also, many works in classical economics suggest that emotions– both positively active and negatively calming – could widen gaps between rational human and real people. in this study, this author conducts an experiment seeking to understand the impact of emotion intervention on performance using 90 high school students to test how different emotions affect their predictions towards the traditional Chinese game named pitch-pot which requires players to throw sticks from a set distance into a large canister. The main finding of this work is that people's predictions do have a relationship with their emotions and this connection is especially strong when the emotions are in positive states.

Keywords: Emotional effect · Decision making · Experimental study · Determinants of rationality

1 Introduction

Classical works in economics tend to assume the rational economic man as the guiding yardstick for understanding human behaviour [1]. However, this assumption has recently been challenged by behavioural economists with experimental and observational evidence [2]. One of the main claims behavioural economists made is that emotions which are considered as variables that play a very significant role in our daily life and social connection, have the potential to affect the real behaviours of people in everyday life [3] on the one hand, but also make people commit some simple heuristics mistakes [4] on the other hand. This study builds on the research by looking at the impact of emotion on decisions making on a group of high school students using experimental design.

2 Manuscript Preparation

2.1 Experimental Design

To assess the impact of emotions on people's decisions, the author applied three treatments to the randomized-allocated subjects [5]. To apply the interventions on emotions using random assignment techniques [6], 90 students were randomly divided into three

C. Yuan et al. (Eds.): *Proceedings of the 4th International Conference on Economic Management and Green Development*, AEPS, pp. 331–337, 2021.
https://doi.org/10.1007/978-981-16-5359-9_38

groups and one student (who is not familiar with the test group) was randomly picked. He or she was asked to throw a stick to a pitch pot that is put a mile away. The selected member had 5 chances to throw the sticks overall. All tested members needed to predict the result of each throw, as shown in Table 2 below. They did one prediction every time before the person is throwing, which means they were perfectly able to learn from the previous result and remake the prediction for the next throwing [7].

Before the experiment started, all students were asked to finish questionnaires to collect basic information of tested members, including age, gender, class, GPA, nativeness. Then, students that were in group one were asked to predict the result without giving any affections. And then the conductor of the experiment told them people who get the right prediction in the following five turns would get a small award, so students were assumed to be more motivated than they were in the first round.

However, the rest two groups of people's emotions were given controlled treatment [6]. People from the second group were asked to watch a clip from a comedy after answering the basic questions on the questionnaires and before making predictions. Then they were asked to write down what they saw in the film how they feel about it on the questionnaire. By contrast, The third group watched a horror movie before making predictions, as shown in Fig. 1. So basically, each participant in these two groups experienced typically followed three steps: (1) completed a survey about personal information, (2) watched movies, and wrote down feelings, (3) made predictions. In Step 1, the conductor approached potential subjects by entering each classroom during the break time and inquired about their interest in filling out a survey that would take about ten minutes. Most of them just agreed without hesitation, as they were told that everyone who participated in this experiment would receive a small gift as a reward. Also, it seemed quite interesting for a high school student who always struggled with schoolwork to join in unusual activity. I briefly explained the whole procedure and then I handed out the questionnaires. After inviting a student to stand in front of the class to throw the stick, I gave time for all students to inspect the pitch pot, the distance, and the stick. For all three groups of students, the physical products which were used in the experiment are the same.

A few noteworthy aspects of the experimental design merit further consideration. Asking students to write their feelings down helps me to figure out if their emotions have been affected successfully and what are their current emotion [8]. The result shows that most students' emotions have been successfully transformed into the right state. For example, students who watch horror movies always use words like disgusting, dark, nervous, scared to describe their feelings.

Given the research design, I have the following three predictions:

1. People are more likely to make positive predictions when given positive interventions.
2. The predictions people made are more negative under negative interventions.
3. The effect of the intervention on emotion would diminish as the times of repetition increases. This might be a result of the learning/adaptational process [2] where people are trying to adjust themselves.

2.2 Experimental Results

Tables

Table 1 presents the summary statistics of the variables collected from the participants in the experiment by groups. The following variables are collected from the participants. The actual age of the students is represented by the variable, age. Using male as a benchmark, the female is coded as 1 under the variable, gender. Grade denotes the actual year that the person spent in school and GPA measures the average score the person has at school: 0 for high, 1 for midium level, 2 for low level. Native-ness characterizes the fact that whether the person is born and lived in the city where the experiment took place: 0 if native, 1 if non-native and emotions denote their self-assessment to their emotional state: 0 if emotional, 1 if non-emotional. Table 1 presents the summary statistics of the participants in the experiment by groups.

Table 1. Selected characteristics of participants

	Group 0 mean (std. dev.)	Group 1 mean (std. dev.)	Group 2 mean (std. dev.)
Age	16 (0)	15.14 (0.73)	15.95 (0.83)
Gender	0.21 (0.42)	0.38 (0.50)	0.65 (0.49)
Grade	11 (0)	10 (0)	11.05 (0.51)
GPA	0.5 (0.62)	0.55 (0.51)	1 (0)
Native	0.21 (0.42)	0.05 (0.22)	1 (0.31)
Emotions	0.58 (0.51)	0.70 (0.47)	0.70 (0.47)

Notes: See text descriptions

Table 1 provides a summary of the basic information of all subjects. Looking across all these variables, the variable age and grade share similar characteristics across the groups. Disparities across groups for variables Native could be a concern, which would be controlled in the regression specifications. Group 0 and Group 1 share a similar average age and average GPA but they a big disparity with those of Group 2. By analyzing the numbers of standard deviation, it is easy to tell that many subjects in one group have the same characteristics. For example, Group 0 includes students that have the same age and are in the same grade. Students in Group 1 are also in the same grade level. And in Group 2, the average GPA (stated by themselves) is at the middle level for all students. By analyzing the numbers of standard deviation, it is easy to tell that the subjects in one group are relatively similar in many aspects like Group 0 has the same age and Group 1

Table 2. Summary of the predicting results from the experiment

	Group 0	Group 1	Group 2
Panel A: mean of different predictions			
1st	0.05 (0.229)	0.52 (0.512)	0.15 (0.366)
2nd	0 (0)	0.38 (0.498)	0 (0)
3rd	0.10 (0.315)	0.42 (0.507)	0 (0)
4th	0.16 (0.375)	0.24 (0.436)	0.15 (0.366)
5th	0.74 (0.452)	0.29 (0.463)	0 (0.308)
Total	0.95 (0.621)	1.86 (1.014)	1.40 (0.754)
Panel B: Percentage of Making positive predictions			
0	19.05	9.52	71.43
1	60.00	30.00	10.00
2	25,00	50.00	25.00
3	0	100	0
Panel C: Percentage of making positive predictions in the different turns			
1st turn	6.67	73.33	20.00
2nd turn	0	100	0
3rd turn	18.18	81.82	0
4th turn	27.27	45.45	27.27
5th turn	63.64	27.27	9.09

includes students who are in the same grade level. Students in group 0 are all the same age, and students in group 2 have similar academic performances.

Table 2 offers the average predicting results people made in the experiments. Throughout the experiment, numbers are used to denote the predicting results: 1 demotes "hit" and 0 denotes "miss". From Panel A, we can tell the data of Group 0, has the tendency of converging to 1 which means the tested members are more likely to predict "hit" as the repetition of the game increases. And the probability turns to be the highest in the last round.However, for Group 2, the average predicting result is the closest to 1 in the first round. Results of Group 2 consist of the lowest number at any term and the lowest total prediction number. From Panel B, it is obvious to tell that people without intervention are more likely to make 1 positive predictions throughout the game, while those who are under positive intervention generally tend to make 3 positive predictions out of 5 trials. In opposite, those who are controlled to have negative emotions have the highest possibility to make 0 positive prediction. Panel C indicates a similar pattern as Panel A. People in Group 0 are more likely to predict "hit" for the last trial while Group 1 shows a higher possibility of predicting "hit" at the first three trials. Group 2 shows a slightly higher average number both for the first and for the last trials.

Based on the assumption of equal variance, to conduct a formal statistical test, this author performs a t-test for equalization of means and reports the associated p-values below for each of the interesting outcomes.

Although the analysis above has already provided conclusions that are consistent with the main predictions of the study, other factors that may affect the decision-making process still need to be controlled. These other subject-specific variables [9] can be related to the outcome by using a typical econometric model:

Y_{i,t} = constant + \β * D_i + \γ* control variables + error term.

X includes subject-specific variables that may affect the prediction results. Variables in X are listed in Table 1 and include age, gender, class, GPA, native-ness, and a variable indicating whether the subjects consider themselves as emotional people in their daily lives. While Y is determined by all variables。Using group 0 as baseline control. Selection bias is eliminated via experimental design (Table 3).

Table 3. Summary of P-values from pairwise T-test for equalization of mean

	Group 0 vs 1	Group 0 vs 2	Group 1 vs 2
1st predict	0.0007	0.3293	0.0108
2nd predict	0.0019	0.1436	0.0015
3rd predict	0.0019	0.9473	0.0005
4th predict	0.5387	0.0000	0.4892
5th predict	0.0035	0.0184	0.1405
Total predictions	0.0017	0.0000	0.0000
Times of correct prediction	0.0002	0.3298	0.0000

Table 4. Summary regression

	– 1	– 2	– 3	– 4	– 5
	Total Predict	Total Predict	Total Predict	Total Predict	Total Predict
0.Group	0	0	0	0	0
	(.)	(.)	(.)	(.)	(.)
1st prediction					
1.Group	0.471***	0.489***	0.434***	0.470***	0.442**
	– 3.8	– 3.9	– 3.51	– 3.68	– 3.41
2.Group	0.0974	0.144	0.00611	0.0967	0.037
	– 0.78	– 1.07	– 0.04	– 0.76	– 0.25
2nd prediction					
1.Group	0.381***	0.368***	0.349***	0.398***	0.350***
	– 4.08	– 3.89	– 3.78	– 4.16	– 3.67
2.Group	0	– 0.0331	– 0.00676	0.0116	– 0.031
	0	(–0.33)	(–0.07)	– 0.12	(–0.28)
3rd prediction					
1.Group	0.323**	0.348**	0.364**	0.300*	0.379***
	– 2.93	– 3.15	– 3.42	– 2.66	– 3.5
2.Group	– 0.105	– 0.0411	0.0252	– 0.21	0.0848
	(–0.94)	(–0.35)	– 0.21	(–1.07)	– 0.68
4th prediction					
1.Group	0.0802	0.0464	0.0998	0.048	0.0365
	– 0.64	– 0.37	– 0.78	– 0.38	– 0.28
2.Group	– 0.00789	– 0.0952	0.0448	– 0.0298	– 0.0682
	(–0.06)	(–0.72)	– 0.32	(–0.23)	(–0.45)
5th prediction					
1.Group	– 0.451**	– 0.444**	– 0.411**	– 0.491***	– 0.427**
	(–3.44)	(–3.32)	(–3.28)	(–3.70)	(–3.27)
2.Group	– 0.637***	– 0.618***	– 0.460**	– 0.664***	– 0.460**
	(–4.80)	(–4.32)	(–3.32)	(–5.00)	(–3.05)
Total prediction					
1.Group	0.910***	0.908**	0.934***	0.842**	0.882**
	– 3.5	– 3.43	– 3.58	– 3.19	– 3.23
2.Group	– 0.547*	– 0.553	– 0.328	– 0.593*	– 0.382
	(–2.08)	(–1.95)	(–1.14)	(–2.25)	(–1.21)

t statistics in parentheses.
* $p < 0.05$, ** $p < 0.01$, *** $p < 0.001$.

Finally, Table 4 reports regression estimates. Dependent variables equal 1 if the subject chose to predict "in", 0 otherwise. Gender = 0 if male, 1 otherwise. GPA = 0 if the subject said his GPA is at a high level, GPA = 1 if it is at the middle level, 0 denotes for low level. Native = 0 if the subject is native-born, 1 otherwise. ***Denotes coefficient estimate is significant at the P <. 05 lev. Overall, compares the results with Group 0, the significant level of Group 1 is very significant for all five trials while that of Group 2 is only significant at the first and the fourth trial. Also, according to the regression estimates table of the first prediction, by comparing the results with Group 0, the significant level of Group 1 is very significant for the first four trials and relatively less significant for the last trial. However, none of the trials done with Group 2 shows a significant result.

3 Conclusions

In this study, several unique insights are yielded through examining the behavior in three groups with positive or negative affect, allowing me to consider whether optimization as a modeling tool in neo-classical predictions is reasonable.

The first insight is emotions in all kinds of structures would interrupt our predictions. Secondly, positive affect is more obvious than negative affect when it comes to predicting.

To find out whether humans differ from people that the set of assumptions defined [9] in other ways, further investigations and more experiments are needed. If stronger evidences can prove that the divergence is large, then economic models may need to stick to some new things. Making central concepts unchanged, economists will become more sophisticated in their ability to characterize the optimal solutions to problems as well as to make predictions about how people would react to new environments.

References

1. Geanakoplos, J.: Arrow-Debreu model of general equilibrium. Palgrave Macmillan, UK (2008)
2. Thaler, R.H., Sunstein, C.R.: Nudge: Improving Decisions About Health, Wealth, and Happiness. Penguin, New York (2009)
3. Grewal, D., Salovey, P.: Feeling smart: the science of emotional intelligence: a new idea in psychology has matured and shows promise of explaining how attending to emotions can help us in everyday life. Am. Sci. **93**(4), 330–339 (2005)
4. Kahneman, D. (2011). Thinking, fast and slow (Kindle Edition). [8] List, J.A.: Does market experience eliminate market anomalies? Q. J. Econ. **118**(1), 41–71 (2003)
5. Harrison, G.W., List, J.A.: Field experiments. J. Econ. Literat. **42**(4), 1009–1055 (2004)
6. List, J.A.: Does market experience eliminate market anomalies? Q. J. Econ. **118**(1), 41–71 (2003)
7. Ariely, D.: Predictably irrational (p. E2). Harper Audio (2008)
8. Tamir, M.: What do people want to feel and why? Pleasure and utility in emotion regulation. Curr. Dir. Psychol. Sci. **18**(2), 101–105 (2009)
9. Wilson, T.D.: Strangers to Ourselves. Harvard University Press, Cambridge (2004)

A Review of Constructive Deviation

Wan Wei and Zhang Meiyu(✉)

School of Management, Wuhan University of Technology, Luoshi Road, Wuhan, China
wanwei_iris@whut.edu.cn

Abstract. Constructive deviation, as a new type of behavior, has attracted wide attention from scholars. Previous studies have explored the influence and mechanisms through which constructive deviation operates as a dependent, intermediate, and moderating variable. This article reviews the research on the concept of constructive deviation and its related theories. It also discusses the limitations of constructive deviation and future research directions.

Keywords: Constructive deviation · Conceptual connotation · Related variables · Future research directions

1 Introduction

As organizations become more flexible, performance-oriented and globalized, it is necessary for employees to be more creative and innovative at work. If the employees want to innovate in their workplace, they may deviate from the institutional norms and procedures. While some studies have shown that deviation from the norm may be harmful [1], others found that it can be beneficial to organizations and significantly improve their effectiveness [2, 3]. In addition, although employee compliance with norms is important to the operation and development of an organization, strict compliance with all rules and regulations may inhibit employees from seeking innovative ways to solve problems at work [2]. With the ever-changing and developing environments of organizations, the rules need to be adapted to the environment and the development. Sometimes the rigid and unchanging rules not only harm the interests of the organization, but also prevent its development. It is common to find that the employees violate the institutional rules due to pro-social intentions, and such behavior can promote innovation and development. This study will systematically revise and discuss the existing research on constructive deviation, and then make suggestions for future research directions.

2 The Concept of Constructive Deviation

The scholars developed the concept of constructive deviation, advancing their own understanding and definition of the term. According to Galperin (2003), constructive deviation is a behavior that violates the rules of an organization that is usually voluntarily conducted by employees to increase the well-being of the organization or its stakeholders

C. Yuan et al. (Eds.): *Proceedings of the 4th International Conference on Economic Management and Green Development*, AEPS, pp. 338–344, 2021.
https://doi.org/10.1007/978-981-16-5359-9_39

[2]. Warren (2003) defined constructive deviance as a behavior that violates the formal rules of the group but that is beneficial to the group [3]. Vadera et al. (2013) proposed that, in an unpredictable and suddenly changing world, employees may deviate from the established path and become more creative, more efficient, and more relevant [4].

In summary, although scholars have different definitions and explanations of constructive deviation, they all agree that it ultimately benefits an organization. While the various definitions contain similar underlying dimensions, they vary slightly in their description of specific behaviors (see Table 1).

Table 1. Definition of constructive deviation

Author	Definition	Similarities
Galperin (2003) [2]	Voluntary behavior that violates the organization's core norms in order to increase the well-being of the organization or its stakeholders	Ultimately beneficial for the organization
Warren (2003) [3]	Behavior deviates from the norms of the reference group but conforms to the super-norms and benefits the reference group	
Vadera et al. (2013) [4]	Employees may deviate from the established path and become more creative, and this beneficial deviation is called constructive deviation	

3 Constructive Deviation and Other Similar Concepts

Warren (2003) outlined that constructive deviation is a general term that includes many types of behaviors, such as principled organizational dissent, anti-role behavior, moderate activism, whistleblowing, exercising the right to speak, and pro-social behavior [3]. According to the theory of organizational citizenship behavior (OCB), there are some differences between constructive deviation and these different types of behaviors. Constructive deviation, pro-social rule breaking (PSRB), and OCB have both similarities and differences. In order to further define the concept of constructive deviation, this article summarizes and compares the views of different scholars.

3.1 Constructive Deviation and PSRB

The concept of PSRB was first introduced by Morrison [5], who defined it as the process through which employees deliberately violate an organization's formal policies, regulations, or prohibitions for the benefit of the organization or its stakeholders. Compared with the traditional view that rule violation is a form of opportunism or laziness from the part of self-serving employees, some types of violations of organizational policies may be due to good intentions. In particular, individuals that have a strong tendency to take risks can significantly contribute to constructive deviance [6] and PSRB [5]. Different leadership styles also have an influence on these two behaviors. For example, authoritarian leadership was found to have a significant negative effect on constructive deviation [7] and PSRB [8]. As we can see, both constructive deviation and PSRB are spontaneous acts carried out by employees for the purpose of improving the interests of the organization and violating the organizational rules. The rules violated by constructive deviation include formal and informal norms, while the notion PSRB emphasizes that the violations are formal rules and regulations.

3.2 Constructive Deviation and OCB

Galperin proposed that constructive deviance is related to OCB because the various aspects of employee performance involved in these behaviors exceed their formal work responsibilities. However, these two concepts are very different. OCB is a discretionary behavior that is not recognized by the formal reward system but that can promote the effective operation of the organization. Unlike organizational citizenship behavior, which is more subordinate and passive in nature, constructive deviations include intentional measures (Table 2).

Table 2. Comparison between constructive deviation and other similar behavior

	Constructive Deviation and PSRB	Constructive Deviation and OCB
Similarities	Both are non-mandatory, voluntary actions conducted by employees; individual behaviors outside the scope of official work responsibilities; behaviors are beneficial to the relevant interests of the organization; behaviors are deviant and violate organizational norms or regulations	Both are discretionary behaviors that are not recognized by the formal reward system but can promote the effective operation of the organization
Differences	The organizational norms violated by constructive deviations include both formal and informal ones (Bennett & Robinson, 2000); the rules violated by PSRB are the official ones set by the management of the organization [5]	Unlike OCB, which is more subordinate and passive in nature, constructive deviation includes intentional and significant violations of organizational norms [9]

4 Theory

Until now, different scholars illustrate constructive deviation from different perspectives. The main theories go as follows:

4.1 Social Exchange Theory

Social exchange theory mainly refers to the phenomenon where individuals create and maintain exchange relationships with others under the assumption that they will get a reward [10]. If employees perceive lower organizational support, they will build some support networks to reduce avoid any resistance in their violation of rules. As such, there is an interaction between attitude, perception, and behavior. According to social exchange theory, constructive deviation is a way for employees to contribute to the positive exchange relationship they experience within the organization.

4.2 Social Identity Theory

Social exchange theory refers to an individual's knowledge about belonging to a certain social group and the emotional and value meaning that arises from identifying with a particular group. Individuals realize that they belong to certain social groups, and at the same time realize the emotional and value meaning that this identity brings to them [11]. Social identity theory illustrates that people's efforts to achieve and maintain a positive social identity may cause deviations from the norms.

4.3 Social Anomie Theory

Social anomie theory explains deviant behavior by emphasizing that social imbalances and changes lead to a decline in the acceptance of social rules and norms [12]. Anomie theory assumes that specific conditions within the social structure produce a certain environment in which deviation from convention and norms becomes an expected response. Therefore, social anomalies are the erosion of values, norms, and rules. Furthermore, certain institutional environments put pressure on individuals, prompting them to act unconventionally. That is, they push past the rules and their restrictions of rules and engage in deviant behavior, including constructive deviation behavior.

5 The Causes and Consequences of Constructive Deviation

Existing studies have explored the antecedent variables of constructive deviant behavior from multiple levels. The individual level includes active personality [13], psychological empowerment, and risk tolerance [6, 14] is positively related to constructive deviant behavior. The team level includes empowered leaders [7, 14–16] authoritarian leadership [7], inclusive leadership [17].The results show that empowered leaders and inclusive leadership has positive impact on constructive deviant behavior [7, 14–16],but authoritarian leadership is negatively related to constructive deviant behavior [7].The organizational level includes organizational justice [18], moral identity [18], and organizational culture [19] is positively related to constructive deviant.

After reviewing the relevant literature, we noted that most of the existing studies pay more attention to the positive impact of constructive deviation. This study describes the effects of constructive deviation at the three levels (individual, team, and organization). At the individual level, constructive deviant behaviors can make employees feel marginalized or socially isolated [20]. Conversely, employees who adopt constructive deviant behaviors can improve work efficiency and services [14]. In addition, constructive deviant behaviors can also increase employee performance, promote self-improvement, and help others gain more knowledge [21]. At the organizational level, constructive deviation can improve the organization's environmental adaptability. That is, employees who participate in constructive deviant behaviors can be described as active change agents, who help the organization adapt to changes and to new environmental constraints in a dynamic world market [4]. In short, the existing research on constructive deviation mostly focuses on the organizational level and the individual level. There is a lack of research at the team level, and most of the research focuses on the impact of constructive deviation.

6 Discussion of Constructive Deviant Behavior and Future Research Prospects

For an organization, constructive deviation has both advantages and disadvantages. On the one hand, it can increase the management costs and risks and, on the other hand, it can improve the creativity of the organization. Managers must learn how to motivate constructive deviation among their employees in order to promote their organization's development. While exploring the theoretical roots of constructive deviations, scholars have made several constructive suggestions based on the management style of the organization.

6.1 Managerial Strategies for Constructive Deviation

According to the literature, constructive deviation is usually a bottom-up, unplanned activity that occurs without the formal authorization of the relevant management department. In general, constructive deviation is not part of the organization's formal planning, so there is a lack of corresponding resources to encourage and promote this behavior among employees. Although constructive deviation ultimately benefits the organization and its stakeholders, we do not fully know how to effectively identify the constructive deviation of employees. Recognizing the constructive deviant behavior of employees brings great challenges to leaders and their organization. Although leadership styles such as authoritarian leadership and inclusive leadership can promote constructive deviant behaviors to a certain extent, scholars also call for the need to explore more types of leadership style and constructive deviant behavior [19],we also need to explore the relationship between other leadership styles and constructive deviation.

6.2 In-Depth Exploration of the Mechanisms Through Which Constructive Deviation Operates

After years of research on constructive deviation,a majority of the results shows that the benefits of constructive deviation for individuals and organizations include improved work efficiency (personal level) and increased ability of the organization to adapt to the environment (organizational level). Future research can further explore the potential negative effects of constructive deviation and its influence mechanisms at both levels.And future research also needs to illustrate the mechanisms of constructive deviation from different theories.

6.3 Research on Constructive Deviant Behavior in Different Regions

More research is needed to understand how constructive deviation operates in different regions (i.e., how it works in Chinese versus Western cultures). For example, the collective tendencies of Chinese culture and the individualism of Western society may lead to large differences in the constructive deviant behavior of Chinese and Western employees. In the context of a Chinese society in favor of institutional norms, it would be interesting to see whether employees' behavior is consistent with the Western notion of creative freedom still needs to be explored. In addition, most of the constructive deviation measurement tools were developed in the Western context. Whether these tools are suitable for employees in Chinese organizations remains to be verified. Therefore, research on constructive deviation in China should also incorporate more regional cultural factors that are characteristic of the Chinese environment.

References

1. Lee, K., Allen, N.J.: Organizational citizenship behavior and workplace deviance: the role of affect and cognitions. J. Appl. Psychol. **87**, 131 (2002)
2. Galperin, B.L.: Determinants of deviance in the workplace: An empirical examination in Canada and Mexico (2003)
3. Warren, D.E.: Constructive and destructive deviance in organizations. Acad. Manage. Rev. **4**, 622–632 (2003)
4. Vadera, A.K., Pratt, M.G., Mishra, P.: Constructive Deviance in Organizations: Integrating and Moving Forward. J. Manag. **39**, 1221–1276 (2013)
5. Morrison, E.W.: Doing the job well: An investigation of pro-social rule breaking. J. Manage. **32**, 5–28 (2006)
6. Yildiz, Ö., Radtke, J.: Energy cooperatives as a form of workplace democracy? A theoretical assessment. Econ. Soc. Eur. Electron. Newslett. **16**, 17–24 (2015)
7. Sun, J.: Constructive Deviance in Chinese Enterprises: Construct, Formation Mechanisms, and Impacts, Soochow University (2016). (in Chinese)
8. Jianqun, S., Xiaoming, T., Rui, L.: Constructive bias in organization: concept definition, formation mechanism and influence. Psychol. Sci. **39**, 1242–1247 (2016)
9. Galperin, B.L.: Determinants of deviance in the workplace: an empirical examination in Canada and Mexico. Unpublished Doctoral Dissertation. Concordia University, Montreal (2002)
10. Blau, P.M.: Social exchange theory. Retrieved September 3, 62 (1964)
11. Tajfel, H.E.: Differentiation Between Social Groups: Studies in the Social Psychology of Intergroup Relations. Academic Press, Cambridge (1978)
12. Lincoln, J.R., Guillot, D.: Social Theory at Work. Oxford. Oxford University Press, UK (2006)
13. Li, C., Sun, L.: Constructive deviance as a planned behavior. Acad. Manage. Proc. **2015**, 13656 (2015)
14. Mertens, W., Recker, J., Kohlborn, T., Kummer, T.F.: A framework for the study of positive deviance in organizations. Deviant Behav. **37**, 1288–1307 (2016)
15. Mertens, W., Recker, J.: Can constructive deviance be empowered? A multi-level field study in Australian supermarkets. J. Retail. Consum. Serv. **54**, 102036 (2020)
16. Mertens, W., Recker, J., Kummer, T., Kohlborn, T., Viaene, S.: Constructive deviance as a driver for performance in retail. J. Retail. Consum. Serv. **30**, 193–203 (2016)
17. Tian, Y., Wang, Y.: Incentive mechanism of inclusive leadership on employees' constructive deviant behavior. Foreign Econ. Manage. **41**, 54–69 (2019). (in Chinese)

18. Kura, K.M., Shamsudin, F.M., Chauhan, A.: Organisational trust as a mediator between perceived organisational support and constructive deviance. Int. J. Bus. Soc. **17**, 1 (2016)
19. Cohen, A., Ehrlich, S.: Exchange variables, organizational culture and their relationship with constructive deviance. Manage. Res. Rev. **42**, 1423–1446 (2019)
20. Kim, M.J., Choi, J.N.: Group identity and positive deviance in work groups. J. Soc. Psychol. **158**, 730–743 (2018)
21. Gatzweiler, A., Blazevic, V., Piller, F.T.: Dark side or bright light: destructive and constructive deviant content in consumer ideation contests. J. Product Innov. Manage. **34**, 772–789 (2017)

Research on the Application of the Sharing Economy Based on Block Chain Technology

Jin Chen(✉)

Business School, Central South University, 932 Lushan South Road, Changsha, Hunan, China

Abstract. This paper introduces the application research of sharing economy based on block chain technology, which is divided into four parts. At the beginning, the author starts with the problems and describes the definition of the sharing economy. Then, the problems existing in the sharing economy are discussed, namely, network effect and unsound credit system. In addition, the sharing economy also enhances the harm caused by monopoly. At the same time, once unqualified resources appear on the shared platform, users' sense of use will be greatly reduced, and platforms are easy to go bankrupt. Next, the paper discusses the possibility of using blockchain to solve problems existing in the sharing economy, and describes the principle of using blockchain to solve network effects, as well as the reasons why blockchain can be used to solve the problem of sharing platform: 1. Trust sharing to match supply and demand. 2. Data sharing, providing credit protection. 3. Smart contracts to provide solutions. 4. Permission sharing, revolutionary change. After that, the block chain technology is formally applied in the field of car sharing, here the author creates a fully functional, convenient and fast car networking economy ecosystem based on shared car functions, integrated taxiing capabilities, car networking systems, and so on, as a case in point, to illustrate the advantages of introducing blockchain into the sharing economy. Finally, the author summarizes the above. In fact, it can be seen from the previous content that it is feasible to introduce the block chain into the sharing economy, which can smoothly solve the problems existing in the sharing economy in the market. But at the same time, the application of blockchain technology in the sharing economy scenario also has some obstacles. Therefore, the application of blockchain technology in the sharing economy is desirable, but still needs to be improved.

Keywords: Sharing economy · Blockchain · Network effect

1 The Problem Analysis of the Sharing Economy

1.1 Concept of the Sharing Economy

The sharing economy is a term used in applied economics, first proposed in 1978 by Marcus Felson, a sociology professor at Texas State University, and Joel Spaeth, a sociology professor at the University of Illinois. The sharing economy can be defined as

C. Yuan et al. (Eds.): *Proceedings of the 4th International Conference on Economic Management and Green Development*, AEPS, pp. 345–357, 2021.
https://doi.org/10.1007/978-981-16-5359-9_40

a new economic model based on strangers and the temporary transfer of the right to use goods for the main purpose of obtaining certain remuneration.

The concept of the sharing economy first emerged in the 1970s by foreign researchers Marcus Felson and Joe Spaeth, whose research focused on individual car rental and sharing, calling it a new cooperative consumption model called "individual-to-individual" that enables effective control of transportation costs. As a result of the reform of information technology, the sharing economy has also been updated and developed rapidly, becoming the most active business model in recent years. Botsman and Rogers [1] came up with three sharing economies by analyzing a large number of real-world cases 0. Those are the paid-for-value product service system for tangible products, the social wealth redistribution system for second-hand trading of salty fish assets, and the cooperative lifestyle system for virtual asset sharing. In this regard, Acquier et al. studied the sharing economy in terms of overcapacity, and he believed that the excess resources occupied by individuals should be shared among other members of the society and that there should be some benefit [2]. The sharing economy is an economy that collaborates more than competes, which is the view of Murillo et al. [3]. He believes that in the digital economy with information technology as its core, social capital is more important than financial capital. Consumers attach great importance to the right to use goods, which is more important than ownership. Nowadays, people pay more attention to the concept of environmental protection and sustainable consumption. Shared value has replaced exchange value, and the sharing economy has brought about radical changes in the original economic model.

However, there are still many problems to be solved in the practice of sharing economy.

1.2 Network Effect

One of the common characteristics of Internet enterprises is the network effect, and the sharing industry is no exception. In economics, network effects are often explained by network externality.

Research on "network effects" or "network externality" dates back to the 1950s, when Leibenstein (1950) analyzed literature on the effects of comparison and vanity [4]. The first economist to analyze the phenomenon of network effect was Rohlfs (1974), who systematically analyzed the interdependence of consumer demand in telecommunications services, and concluded that the utility a user obtained from communication services increased with the number of people joining the system [5]. Rohlfs' research on communication services includes inverted U-shaped demand curve, and start-up and critical capacity problems in network expansion, etc., which play a key role in laying the foundation for subsequent research. In the 1980s, Katz and Shapiro (1985) formally defined the concept of "network externality", i.e., when a consumer uses or consumes a product or service, the utility or value he receives increases as the number of people using that product or service increases [6]. Katz's research on "network externality" has promoted the development of network effect-related research, and has become one of the most widely cited definitions by follow-up researchers, calling the network externality network effect. In the early 1990s, with Liebowitz, the scholars represented by Margolis (1994) questioned the study of the mainstream literature, pointed out that network effect and network externality were different concepts, and used the externality theoretical framework to analyze the two concepts in depth [7]. They promoted the meaning of network effect, and thought that the network effect referring to the value of an action was affected by the number of participants taking the same action, while network externality refers to the unutilized benefits in the transactions of network participants when balance occurs, and the difference is the ability to internalize the influence of other participants in the network. In summary, Liebowitz et al. believed that when the net value of an action is affected by the number of agents using the same action, the network effect exists, and Liebowitz and others defined the concept of "network effect" more broadly, including Katz and Shapiro's definition of "network externality", and better covered the research content of Leibenstein and Rohlfs, etc. Liebowitz and Magrolis' views were later accepted by scholars such as Katz, but in many scholars' articles, network externality and network effects were used as the same concept. Economides (1995) then pointed out that network externalities arose from the "complementarity" of network structure based on the analysis of network types and network externality, and summarized relevant literature research as macro-method and micro-method, in which macro-method assumed the existence of network externality and tried to model its results.

Externality generally includes positive externality and negative externality. Network externality refers to positive externality, i.e., the more users there are, the more utility each user gets, and the value of each person on the network is in direct terms with the number of other people on the network. Network externality is divided into direct externality and indirect externality. Direct network externality is the change of economic benefit caused by the change of the number of users consuming the same product, that is, the increase of the value of goods is directly caused by the increase of the number of users consuming a product. Indirect network externality is the change of value caused by the increase of the number of users of a product and the increase of the number of complementary products and the decrease of price [8].

It is generally believed that with the increase of the number of network users, the utility of total users will increase geometrically. In practice, however, accurate calculation of network effects is more difficult. Shapiro and Walliams (2000) found that consumers can obtain two basic utility and network utility from the network effect product, which refers to the utility that the product itself brings to the consumer, and does not change with the size of the user of the product [9]. However, network utility mainly comes from the demand dependence between user networks, which changes with the expansion of the user scale of products. Network utility comes from the size of the network, which is positively related to the size of the network, and also reflects a kind of economies of scale, but this kind of particular economies of scale does not come from the supply side of the market, but from the demand side of the market, so many scholars refer to the network effect as the demand-side economies of scale. The network effect is based on the user base effect plus the effect increment with the increase in the number of users.

The utility function of the product with network effect is:

$$U = r + v(Xe) \tag{1}$$

where U is the total utility of the product, r is the basic utility of the product, and v(Xe) is the network effect and a function of the expected size of the user, in terms of Xe.

This model based on the size of the user's expected network is adopted by most scholars, such as Katz and Shapiro (1986) [10], and Economides (1996) [11]. We usually refer to the equation of network effects, which change with the expected size of the user, as a network effect function. A typical network effect function is the Metkav Law, which holds that the value of the network is squared by the number of conference users. This is actually assuming that the value of each consumer's network product is directly related to the size of the network, i.e., a linear network effect function.

The network effect function is as follows:

$$v(Xe) = rXe \tag{2}$$

where r is generally referred to as network effect intensity, which indicates whether the network effect is obvious or not. In particular, r = 0 indicates that the network effect is 0. The network effect intensity of the product can be controlled by the manufacturer. For example, the network effect intensity of the online game with the friend function is greater than that of the online game without the friend function.

Many subsequent studies have shown that it is more likely that the network effect increases in square multiple in the early stage, diminishing marginal utility may appear in the later period. In the later period, the network effect evolves into a long tail effect.

We can also draw inspiration from the long tail curve, which is a logarithmic function:

$$y = logax(0 < a < 1) \tag{3}$$

The network effect can be understood as a reverse function of the long tail effect, and the function is as follows:

$$y = logax(a > 1) \tag{4}$$

In reality, this function can also be understood in two ways. For example, the early growth of drivers and users of Uber may be very important to you, but by the end of hundreds of billions of drivers, users to join have little impact. In addition, if the network effect grows by square times, competitors will have no place to live, which cannot explain the rivalry between platforms such as Google and baidu, Eleme and Meituan.

The network effect not only drives the rise of many Internet services, but also prevents users from migrating to competitors. Imagine if you want to leave WeChat and find a competitor. Even if a competitor has better features or better protection of your privacy, or reflects your values, you can't change because your friends are not there. If you're not there, your friends won't change. "Thus, it is difficult for competitors to compete simply by offering better products."

Because of the network effect, it also has a series of problems, such as that the Internet enhances the ability of monopolies to behave badly. Start-ups need to attract users, but it's hard to attract from existing users based on network effects. The wide reach, ubiquitousity and openness of the Internet mean that businesses can explode. And the nature of Internet companies - information - means that there are dangers, and the harm is not just economic. Internet monopolies contribute to the worst aspects of human nature, exacerbating income inequality and even profiting from the use of their platforms to undermine democracy. Under the influence of this effect, the survival space of start-up enterprises is very small. It is difficult to compete with monopoly enterprises, which also hits the enthusiasm of start-ups and further strengthens the degree of monopoly in the industry.

Common solutions are usually to regulate monopolies, but the results are often rough, clumsy and undesirable. It is time for more competition to address these issues more effectively.

1.3 The Credit System is Imperfect

The participants of the generic economy include the Shared platform, the supplier of idle resources, and the consumer who needs idle resources and the Shared platform.

The sharing platform serves both the supplier and the demander. At present, sharing platforms adopt the centralized storage mode, storing the personal information and transaction records of users (suppliers and consumers) into central databases. There are two main problems in this management mode. One is that once the central database is attacked illegally, the information of users and the platform will be leaked. Second, the Shared platform maintains the database unilaterally and has the ownership of the data. Therefore, the rights and interests of users (suppliers and consumers) are not equal to those of the Shared platform, causing certain credit risks between users and the Shared platform. In order to expand the market share, sharing platforms blindly expand the amount of information in the database without paying attention to the credit assessment of relevant users, which will lead to the deterioration of the service quality of sharing platforms and the aggravation of the trust crisis between users and sharing platforms. Since Shared resources are available to everyone, quality assurance of resources is an issue that requires careful consideration. Once unqualified resources appear in the Shared platform, the user experience will be greatly reduced and the platform will easily go out of business.

The crisis of trust between users (suppliers and consumers) also threatens the sustainable development of the sharing economy. There is evidence that in the sharing economy, people choose to trade or partner in all kinds of exclusive behaviors. Peer-to-peer sharing economy transactions, for example, may also increase peer-to-peer discrimination. According to a survey and analysis of Airbnb users in the United States, African American men pay 12% less rent on Airbnb platforms than other hosts for the same location and type of house. A follow-up study found that African Americans were more likely to be rejected by landlords. In addition, experiments have shown that Uber drivers generally discriminate against African Americans, with African American consumers waiting longer on average and having their appointments cancelled more often when using Uber.

2 Appling Blockchain to Solve the Problem

2.1 Blockchain Background Description

The blockchain technology proposed by Ben Cong in 2008 has great potential in the field of information resource sharing by virtue of its characteristics of open consensus, centering, anonymization and traceability [3]. It is generally believed that blockchain is a distributed ledger and consensus algorithm to generate effective blocks through timestamps and non-symmetric encryption technology encryption chain data blocks, to ensure that it is not tamperable and non-repentant, and the use of smart contracts to achieve data verification and circulation [3]. Researchers have made a large number of literature reviews on blockchain technology. Christidis and Devetsikiotis noted that blockchain technology gave us a distributed and equivalent network in which distrustful members can interact with each other verifiably without the need for trusted authority [3]. Ahram et al. believed that blockchain technology promised to increase transparency and accountability in supply chain networks, resulting in more flexible value chains [3]. Yli-Huumo and others found that most of the papers focused on bitcoin systems, less than 20% focused on other blockchain applications, including smart contracts and licensing, and most of the research focused on uncovering and improving blockchain limitations from a privacy and security perspective, but many of the proposed solutions lacked a specific assessment of its effectiveness [3]. Casino and others have thoroughly categorized blockchain applications in different sectors, including supply chain, business, healthcare, Internet of Things, privacy, and data management, identifying key topics, trends, and emerging research areas, noting deficiencies identified in the literature, in particular the limitations of blockchain technology, and how these limitations arise in different sectors and industries [3]. Hughes and others reviewed the literature in detail, combing through the commercial value, security, business models, crypto transactions, limitations, and practicality of blockchain technology, emphasizing that while there are currently few commercial-grade blockchain applications, the technology shows significant potential to benefit from industry-wide use [3].

2.2 Blockchain Solves the Network Effect

Virtual currencies can help start-ups compete with existing ones that benefit from the network effect.

Virtual currencies pave the way for start-ups to compete with Internet monopolies, giving start-ups a chance to kick-start their own network effects. A start-up can create its own cryptocurrency and distribute it to users as an incentive to experiment with the service. Cryptocurrencies are used to encourage users to join the service. This is basically a free trial of a cryptocurrency version of classic marketing technology. Cryptocurrencies can become more valuable as services are available. Basically, early users are tied to the success of startups.

Blockchain has the advantages of distributed data storage, encryption, irreversible, and disintermediated, and it may be available to build a larger network than the Internet with distributed data storage, which can cause greater network effects. We see this possibility in Bitcoin. On the Internet, we have the Alipay system, WeChat payment system, and various commercial banks in various countries' payment systems. Although some of these systems can transfer money to each other, they are affiliated with the central database of different countries, and transaction costs are relatively high. However, the Bitcoin network is a distributed data storage system, and anyone in the world can generate a very large number of addresses in the Bitcoin network, the real-time global transfer, whose transaction costs are lower and the network effect is greater.

The privatization of blockchain data has the external effect of big data. In the current Internet network, data are stored in various central databases, and users do not have data ownership and use rights. In a blockchain network, data are privately owned by the users, who have their own data, which can be authorized for use and sharing. This privatization mechanism can greatly stimulate user participation in blockchain networks, as well as the protection and sharing of data. Data sharing brings the network effect of big data.

2.3 Blockchain Solves the Problem of Shared Platforms

Trust Sharing to Match Supply and Demand. The great thing about blockchain technology is that it is centralized and trusted, enabling transactions through machines and algorithms in a completely unfamiliar network environment. Without the trust endorsement of the central organization, all participating network nodes can determine their identity through blockchain code and encryption technology, resulting in trust sharing, which can solve the problem of mutual trust in the process of anonymous transactions, greatly improving the speed and timeliness of data processing. The transaction process in the traditional Internet involves the supply side, the demand side and the intermediary agencies to achieve a three-way joint realization. Through the use of blockchain technology, the traditional Internet transactions in the "intermediary system" can be completely abandoned, and the supply and demand of the two sides directly connect together to achieve the optimal match between supply and demand. Blockchain technology is an ideal solution for the sharing economy because of the frequent matching processes that occur between products and users in a shared economy scenario.

Data Sharing, Providing Credit Protection. Blockchain is essentially a distributed sharing system, ledger distributed sharing, data distributed storage, transaction distributed records, participants distributed collaboration, system distributed maintenance. "Sharing" is brought about by the revolutionary changes, that is, the system's data storage, transaction verification, and information transmission process are all reassured, distributed, open, and transparent. Shared data are running simultaneously in all participating network nodes, and no one can tamper with or destroy the ledger, because no one can control all the network nodes at the same time. This kind of technical advantage, which is embodied in "fairness", makes blockchain technology have extensive and in-depth application value in finance, insurance, intellectual property rights, charity and public welfare. As for the sharing economy, it also provides a fundamental guarantee for the formation of user experience as the core of the credit system.

Trust sharing to match supply and demand. Smart contracts are contracts that can be executed automatically in network information platforms and systems when certain conditions are met. The smart contract is one of the most popular business models in the development of blockchain technology, and the smart contract system based on blockchain technology has the dual advantages of automatic execution and trustworthy, so that it can help realize the business scenarios of the sharing economy, such as product reservation, default compensation and so on, which involve online trust and make the sharing economy more perfect and reliable. With the continuous improvement of blockchain technology, smart contracts are expected to become a standardized solution for the sharing economy in specific applications in the future.

Permission sharing, revolutionary change. Based on blockchain technical rules, data require the consent of most participants to the validity of the data before it can be confirmed. In the form of rights sharing, each participant acts as a data provider, validation party, and user, working together to maintain the security and validity of blockchain data. As a result, no organization has full control over the data. The biggest thing about the Internet age is that not only data sharing, but also data permission sharing is a revolutionary change in blockchain.

3 Blockchain Application Examples

Shared cars are the most commonly used shared product for today's users, so we're here to create a fully functional, convenient and fast car networking economy ecosystem based on shared car functions, integrated taxiing capabilities, car networking systems, and so on, as a case in point, to illustrate the advantages of introducing blockchain into the sharing economy.

3.1 Analysis of the Existing Internet Ride-Sharing Platform

In addition to the existing problems in the online taxi platform we said above, there are other more specific drawbacks summarized here. It is well known that for Didi, Uber, Lyft and other taxi platforms, the revenue from the calling business is the core. In the "call service", "commission" is an important source of income. When a platform grows up, can it only charge "the channel fee"?

Moreover, the decline in subsidies, long waiting times for orders caused by a decline in word-of-mouth, safe travel issues, and competitive pressure from the industry are mostly the existing dilemma faced by ride-sharing platforms. Passengers, drivers, platforms and others have their own different concerns and an urgent need for solutions.

On the one hand, in the existing shared platforms, there are also problems that cannot be ignored. Under the "black swan" of the outbreak, pessimism and helplessness are spreading to the entire automotive industry. Supply chain disruptions, factory shutdowns, dealer shutdowns, and shrinking market demand are all challenges faced by the automotive industry today. A series of hidden dangers in the automotive industry have been multiplied by the outbreak.

The continued growth of the automotive industry over the years has come from the fact that cars were used in the past as an asset to a family's must and a good life. However, with everyone's understanding of the car changes from "a symbol" to "a tool", the popularity of shared cars and other modes of travel, innovation between new energy vehicles and traditional vehicles, coupled with too much cost of the holding of vehicles, too expensive to keep cars, consumer choice, and serious vehicle homogenization competition, it results in consumers more and more cautious about the consumption of cars.

On the other hand, auto companies do not know what cars to build. If cars are built but cannot sell out, they can only be backlogged. "The current situation is that there is supply-side and consumer demand, but the core problem is that there needs a connecting medium between the supply side and the demand side."

3.2 Car Networking Economic Ecosystem Design

The original intention of this paper is to build the blockchain and taxi module and shared car module as a car networking system, because fast and affordable taxiing and sharing car functions can attract users, and to build a follow-up database that contains many convenient functions for car owners. The solutions are as follows:

Blockchain technology is a centered organization. If each driver can add relevant information, such as driver identification, vehicle and service location, and evaluation, etc. to a specific menu, and then record it on the blockchain, when someone needs to take a taxi, blockchain can filter out the "suitable driver", and then the transaction between the passenger and the driver can be made through a point-to-point mode. It can better improve the quality of the driver and the protection of the legitimate rights and interests of passengers and personal safety.

Among the above user pain points, the most obvious commonality is the balance of interests. So how to achieve blockchain? Or how to do it? We know that blockchain trust is primarily reflected in the fact that users in the blockchain do not need to trust the other side of the transaction, nor do they need to trust a centralized institution. They only need to trust the software system under the blockchain protocol to implement the transaction.

The premise of this self-trust is the consensus mechanism of blockchain, i.e., in a market of mutual distrust, the sufficient condition for the nodes to agree is that each node, out of consideration for maximizing its own interests, will spontaneously and honestly abide by the pre-set rules of the agreement, judge the authenticity of each record, and ultimately record the record as true into the blockchain.

In other words, if nodes have their own independent interests and compete with each other, it is almost impossible for them to conspire to deceive you, especially when the nodes have public credibility in the network.

Blockchain technology is the use of a set of consensus-based mathematical algorithms to establish a network of "trust" between machines, so that to carry out a new credit creation through technical endorsement rather than central credit institutions.

We can change the revenue model of the balance sheet often adopted by traditional corporate organizations through consensus mechanism, which is often based on profit. Although many companies are also trying to apply blockchain technology, it is difficult to really change the original logic if the valuation model does not change, and blockchain can allow the industry chain upstream and downstream to share the benefits. Such as in the platform system, a large number of value circulation can be formed through the consensus mechanism, resulting in a wealth of application scenarios. The resulting data can be used as a basis for building complex platforms with many features.

For the shared car business, there are vehicle suppliers upstream. Vehicle suppliers involve manufacturers, dealers, car rental companies, auto finance companies, and individual car owners, etc. Manufacturers, dealers, renters put more energy directly to the user to sell cars, make car rental, and do shared travel. This matter is not a priority consideration, but now in the transition period, manufacturers, dealers, renters' willingness to participate in is greatly enhanced.

Based on the accumulation of user data, the platform can provide a series of solutions for manufacturers. For example, by sharing test drives to the market, consumers can experience more models, thus affecting consumers' ultimate spending decisions.

For example, the platform can work with manufacturers, using "travel packages" to allow car owners to take out part of their idle time to share with car owners to offset some of the cost of car purchases. While consumers can buy cars at low cost, they can also change the purchase of vehicles from "depreciation" to "value added" and drive car sales, and subscription services allow users to enjoy more flexible car use rights on the platform.

Consumers can also purchase service package from the platform if they agree to participate in shared travel. Service package includes a certain number of door-to-door car wash, door-to-door repair, routine maintenance, door-to-door insurance and emergency rescue, door-to-door annual inspection and other services. The entire service process is transparent and traceable.

Previously difficult to match, owners can enjoy more models, more revenue, and protect assets when they are used efficiently. Through word-of-mouth, the platform will become more and more popular. The car scene can also be extended from time-between daily rental to full rent, and even from passenger cars to mini-passengers and cargo.

For the upstream end, the pressure of overcapacity can be alleviated and the increment of vehicle operations can be generated, while the platform can expand the scope of services, such as: upstream and various technology suppliers, including intelligent drivers, car networking suppliers, AI technology suppliers, and big data technology suppliers, etc., to meet the country's development direction and promote the development of these technologies. For consumers, the pressure to buy a car and keep a car has been greatly eased, which is after the three markets adding up to the market space. Even after the platform can also consider entering the second-hand car market, after all, the platform has mastered the use of vehicles. So, maintenance of the entire process of data can create a relatively fair second-hand car trading market.

Next, carefully describing the specific functional design: the ride blockchain linkage module is made through zero pumping, to quickly accumulate drivers and riders and to achieve a fast-matching supply and demand platform. Both parties will earn "points" after the user calls the car and the driver serves. All the data of the platform is uploaded and stored by the user, with camera real-time scanning traffic information. Then the video content is encoded and translated, graded, weighted and labeled. The users in the process of driving collect travel data through the platform. By contributing bandwidth to store data, they will get points, that is, "drive mining" and "storage mining". This enables the sharing of traffic data resources and provides data for technology providers such as driverless vehicles.

The core problems such as user authentication, credit system, data sharing and data security are solved well by using the decentralized shared car blockchain linkage module developed by blockchain technology. Shared automotive blockchain linkage module can support franchisees and other partners to quickly access the headquarters platform, support the sharing of data, build alliances, reduce data redundancy, and reduce enterprise costs, while the platform can provide a rich DAP API interface, expand the scope of services, upstream and downstream and different service providers to create a huge car networking system. Points can be earned by using a shared car and returning it as required, charging the car and completing various "small tasks" on the app each day. And the promotion of civilized cars can also greatly improve the points. Civilized car points are based on the user's historical car behavior comprehensive score, with regular dynamic updates.

Points can be exchanged directly for "token", and customers can freely use "token" in the car network economy system. This economic system is a manufacturing and travel system after the three markets add up to the market space, covering a variety of car-related consumption, and bringing great convenience to consumers.

4 Conclusions and Prospects

As can be seen from the above cases, it is feasible to introduce blockchain into the sharing economy, and it can smoothly solve the existing problems of the sharing economy in the market that need to be solved urgently. Block chain technology is established between point to point in the network and reliable trust. It removes the interference of value transfer agent, public information and privacy, improve decision-making and protect individual rights and interests, in order to make sharing economy provide a new technical support, and make the supply and the demand for cars provide a connection between the media. To realize the sharing economy and build car networking system is a kind of ideal solution. The Internet of Vehicles provides an interactive platform for intelligent vehicles to share traffic safety-related information, such as road conditions and brake warnings, with surrounding vehicles. These information can help vehicles timely know the traffic environment, so as to improve the safety and efficiency of traffic [19, 20]. However, due to the high mobility of vehicle networking nodes, vehicles often encounter many unfamiliar vehicles in the process of driving, making the credibility of the messages they receive questionable, especially when malicious nodes appear in the network, by deliberately broadcasting false messages to disrupt the normal operation of the network [21]. The trust management mechanism can help the vehicle to judge whether the received message is credible or not, and help the manager of the Internet of vehicles to know the reputation of each vehicle, so as to take reward or punishment measures for specific vehicles [22, 23]. The trust management mechanism usually sets a trust value for a vehicle based on its historical behavior, which is used to comprehensively evaluate the credibility of the messages broadcast by the vehicle [24]. The higher the trust value is, the higher the credibility of the information released by the car will be. The generation of trust value usually requires the receiver of the message to make a posterior score on the credibility of the message based on other information, with direct observation and other methods. The trust value of a car can be formed when all the scores about a car are accumulated by using a specific method. Compared with the traditional wireless network, such as cellular network, the Internet of vehicles has stronger and stricter requirements for trust management. First of all, most of the data transmitted in the traditional wireless network are information and entertainment businesses. Even if there are false messages, it will not seriously affect the personal security of users. The most important function of the Internet of vehicles is the transmission of traffic safety messages. Whether such messages are true or not is directly related to the driving decisions of vehicles and the safety of passengers' lives. Therefore, trust management mechanism is urgently needed to deal with false messages. Secondly, the time delay tolerated by traffic safety services in Internet of vehicles is much lower than that in traditional wireless networks, which puts forward higher requirements for the performance of trust management mechanism. Therefore, the trust management mechanism in the Internet of Vehicles is obviously higher than the traditional wireless network in terms of necessity and performance requirements, which needs to be further studied. The continuous development of blockchain technology brings new ideas to distributed data storage and management, which can be an effective method to solve the above problems. The application of block chain technology in the field of Internet of vehicles trust management can give full play to its advantages. By connecting the base stations into a distributed blockchain network, the problems such as high delay, poor scalability and single point failure in the centralized trust management mechanism can be effectively solved, and the security and consistency of data storage can be guaranteed. Even if a small number of base stations had storage errors or were controlled by an attacker, it would not affect the consensus results of the entire network. Using the trust data stored in the base station, the vehicle can efficiently learn the trust value of the message sender, and then evaluate the trustworthiness of the received message.

Of course, there are some hurdles to applying blockchain technology to the sharing economy scenario. At present, there is still no perfect consensus mechanism that can simultaneously solve security, environmental protection, efficiency and other problems, which poses a certain obstacle to the large-scale application of block chain technology in the sharing economy. In addition, regulators and arbitration methods are not yet clear. Under the existing blockchain technology, users cannot appeal to any institution when they have doubts about the fairness of transactions. At the same time, the blockchain is designed to be linked in a loop that can trace back from any point to the original block, querying all the information on the chain. However, this can only ensure the fairness of blockchain transactions, but cannot guarantee the legitimacy of blockchain transactions, and it will become extremely difficult to track criminal ACTS. In the long run, blockchain technology faces potential security concerns. At present, the algorithm of blockchain technology is relatively safe, but with the development of quantum computer and other new computing technologies, asymmetric encryption algorithm has a certain possibility to crack in the future, which is also a potential security threat that blockchain technology faces.

Therefore, the application of block chain technology to the sharing economy scenario is desirable, but it still needs to be continuously studied and improved.

References

1. Botsman, R., Rogers, R.: What's mine is yours; How collaborative consumption is changing the way we live. Majeure Altern. Manage. **12**(02), 32–36 (2010)
2. Acquier, A., Daudigeos, T., Pinkse, J., et al.: Promises and paradoxes of the sharing economy: an organizing framework. Technol. Forecast. Soc. Change **125**(1), 1–10 (2017)
3. Murillo, D., Buckland, H., Val, E., et al.: When the sharing economy becomes neoliberalism on steroids: unravelling the controversies. Technol. Forecast. Soc. Change **125**(12), 66–76 (2017)
4. Leibenstein, H.: Bandwagon, snob, and veblen effects in the theory of consumers' demand. Quart. J. Econ. **64**(2), 183–207 (1950)
5. Rohlfs, J.: A theory of interdependent demand for a communications service. Bell J. Econ. Manage. Sci. **5**(1), 16–37 (1974)

6. Katz, M.L., Shapiro, C.: Network externalities, competition, and compatibility. Am. Econ. Rev. **75**(3), 424–440 (1985)
7. Liebowitz, S.J., Stephen, M.E.: Network externality: an uncommon tragedy. J. Econ. Perspect. **8**(2), 133–150 (1994)
8. Economides, N.: The economics of networks. Int. J. Ind. Organ. **14**(6), 673–699 (1995)
9. Shapiro, W.: Information Rules: Strategic Guidance for the Network Economy. Chinese University Press, Hong Kong (2000)
10. Katz, M., Shapiro, C.: Technology adoption in the presence of network externalities. J. Polit. Econ. **94**, 822–841 (1986)
11. Economides, N.: Network externalities, complementarities, and invitations to enter. Eur. J. Polit. Econ. **12**, 211–233 (1996)
12. Yang, B., Zhu, W., Zhao, H.: Research on supplier - oriented supply chain finance model. J. Financ. Res. (12), 175–190 (2016)
13. Jin, S., Bo, Y., Jie, M.: Research on competitive intelligence sharing platform of small and medium-sized enterprises based on block chain. Libr. Inf. Serv. 1–9
14. Christidis, K., Devetsikiotis, M.: Blockchains and smart contracts for the internet of things. IEEE Access **4**, 2292–2303 (2016)
15. Ahram, T., Sargolzaei, A., Sargolzaei, S., et al.: Blockchain technology innovations. In: 2017 IEEE Technology & Engineering Management Conference (TEMSCON), IEEE (2017)
16. Yli-Huumo, J., Ko, D., Choi, S., Park, S., Smolander, K.: Where is current research on blockchain technology?—A systematic review. PLOS ONE **11**(10), e0163477 (2016)
17. Casino, F., Dasaklis, T.K., Patsakis, C.: A systematic literature review of blockchain-based applications: current status, classification and open issues. Telematics Inform. **36**, 55–81 (2018)
18. Hughes, L., Dwivedi, Y.K., Misra, S.K., Rana, N.P., Raghavan, V., Akella, V.: Blockchain research, practice and policy: applications, benefits, limitations, emerging research themes and research agenda. Int. J. Inf. Manage. **49**, 114–129 (2019)
19. Zhou, H., Liu, B., Hao, T., et al.: ChainCluster: engineering a cooperative content distribution framework for highways vehicular communications. IEEE Trans. Intell. Transp. Syst. **15**(6), 2644–2657 (2014)
20. He, S., Shin, D., Zhang, J., et al.: Full-view area coverage in camera sensor networks: dimension reduction and near optimal solutions. IEEE Trans. Veh. Technol. **65**(9), 7448–7461 (2016)
21. Zhang, K., Ni, J., Yang, K., et al.: Security and privacy in smart city applications: challenges and solutions. IEEE Commun. Mag. **55**(1), 122–129 (2017)
22. Li, Q., Malip, Q., Martin, K., et al.: A reputation-based announcement scheme for VANETS. IEEE Trans. Veh. Technol. **61**(9), 4095–4108 (2012)
23. Roosta, T., Meingast, M., Sastry, S.: Distributed reputation system for tracking applications in sensor networks. In: Proceedings of 3rd Annual International Conference on Mobile and Ubiquitous Systems, pp. 17–21. IEEE Press, San Jose, (2006)
24. Li, S., Wang, X.: Quickest attack detection in multi-agent reputation systems. IEEE J. Sel. Top. Sign. Process. **8**(4), 653–666 (2014)

Research on Industrial Chain Development of Web Celebrity Economy in China

Siying Liu(✉)

The Attached Middle School to Jiangxi Normal University, Nanchang, China

Abstract. With the Internet wave rushing to the public, since the advent of the self-media era, the rapid development of the Internet celebrity economy industry has gradually become a trend of the times, and it is an important opportunity for China to achieve high-quality development and economic modernization. Based on the analysis perspective of the industrial chain, this article will demonstrate the concepts and functions of the upstream, midstream and downstream of the Internet celebrity economic industry chain, and hope to provide relevant policy recommendations for the development of this emerging industries.

Keywords: Internet celebrity economy · Industrial chain · Content production side · Communication channels · Monetization

1 Introduction

In recent years, the Internet in China is developing rapidly. Figure 1 shows the development trend of Internet users in China from 2000 to 2019, and Fig. 2 shows the development trend of China's Internet penetration rate in the past seven years from 2002 to 2018. As shown in Figs. 1 and 2, both the number of Internet users and the Internet penetration rate in China have been developing rapidly in the past 20 years. With the help of China's Internet development advantages, since the self-media age has quietly arrived, the Internet celebrity economy has begun to develop rapidly as a derivative business economic model. In recent years, China's Internet celebrity economy has shown a development trend of diversified output content, professionalization of online celebrities, diversified forms of realization of online celebrities, continuous derivation of the online celebrity industry chain, and professionalization of online celebrity economic operations. And the branch industries of the Internet celebrity economy have penetrated into every aspect of people's lives.

As of 2018, the total number of audiences in China's Internet celebrity economic system has continued to grow, reaching 588 million people, and the trend of industrialization has become more obvious. The rapid growth of the Internet celebrity economy has not only provided more employment opportunities for young people, but has also been injected with a large amount of capital to continue to help this industry into the next peak. Therefore, the Internet celebrity economy has the potential to make a significant contribution to the economic growth of China, even the whole world. This article will

C. Yuan et al. (Eds.): *Proceedings of the 4th International Conference on Economic Management and Green Development*, AEPS, pp. 358–366, 2021.
https://doi.org/10.1007/978-981-16-5359-9_41

discuss from the perspective of the Internet celebrity economic industry chain, analyze the role and existence value of the upstream, midstream and downstream of the Internet celebrity industry chain, and try to give relevant suggestions.

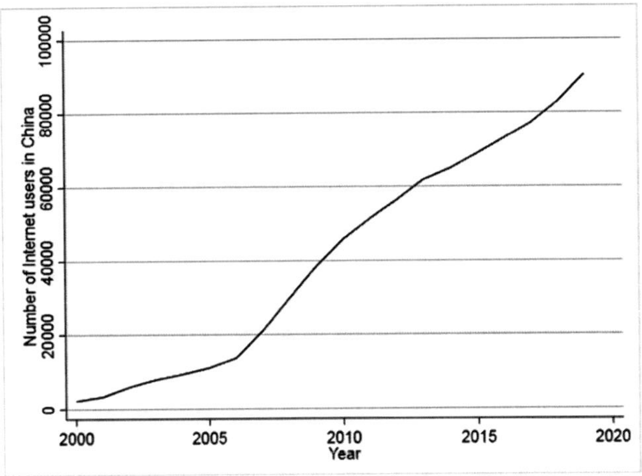

Fig. 1. Number of Internet users in China

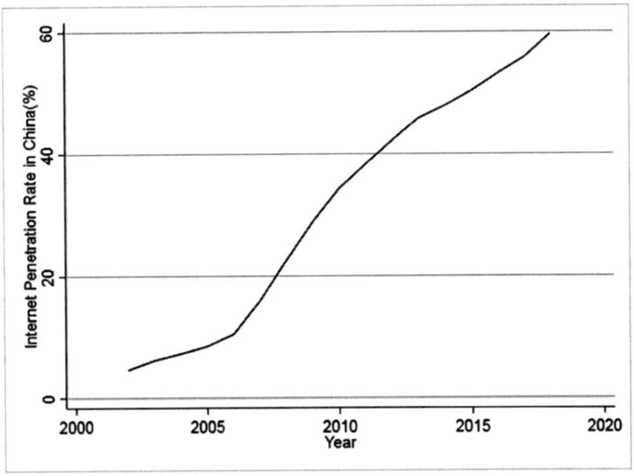

Fig. 2. Internet penetration in China

2 Literature Review

Related to the subject of this article, there are mainly the following two aspects of literature. The first aspect focuses on the research of the Internet celebrity economy and its business model. Shen Xiao and others [1] used the method of content analysis to interpret the incubation platform, identity image, content dissemination, fan characteristics, and evolution mode of Internet celebrities, and finally summarized the four major trends of Internet celebrities' future development, and from the Internet celebrities themselves, live Platform, public, society, government and other perspectives put forward countermeasures for Internet celebrity governance. Sun Jing and Wang Xinxin [2] commented on Internet celebrities and Internet celebrity economy from the perspective of celebrity theory and related domestic and foreign research. Cui Maohui [3] conducted a SWOT analysis of online celebrity live-carrying goods, so as to summarize the development trend of online celebrity live-carrying goods, and proposed how to make the live broadcast of goods go further. Wu Bingjie [4] takes the content-based influencer economy as the research object, and on the basis of clearly combing the concept, background and classification of the content-based influencer economy, from the main body of communication, content production, social asset accumulation and commercial realization In all aspects, combined with specific cases, conduct in-depth interpretation and analysis of the business model of the content-based influencer economy. With the rise of the "Internet celebrity economy" business model, Yu Luying [5] used this to analyze and elaborate on the innovation of business models in the form of new media; Xiao Zanjun and Kang Lijie [6] pointed out that the business model of the Internet celebrity economy mainly includes advertising, e-commerce and Content rewards, etc., and talked about the unique characteristics of online celebrity e-commerce and online celebrity advertising, and also proposed that the business model of the online celebrity economy is facing two hidden concerns.

The second aspect of research mainly focuses on Internet celebrity marketing. Liang Xinmeng [7] discusses why internet celebrity marketing is so popular, how the cooperation between internet celebrities and brands will develop, and the development prospects of internet celebrity economy. Li Yunqing [8] discussed the rise of "Internet celebrity" marketing, the impact of the "Internet celebrity" marketing model on the advertising industry, and solutions. Wang Miao [9] used Taobao Internet celebrity shopkeepers as the research object to analyze the influence factors of their Weibo marketing. Based on the online consumer behavior analysis model "AISAS model" and the theory of ABC consumption attitudes, he constructed a microblog marketing influence factor pair The influence model of the audience's consumption attitude; using SPSS21.0 to analyze the sample data, analyze the empirical results from the perspectives of communication and psychology, and make suggestions for the development of Internet celebrities. Jialing Liu [10] takes papi sauce as an example, analyzes the marketing strategy of Papi sauce based on the development history of mobile internet celebrity short videos, clarifies its successful marketing experience and analyzes its shortcomings, and has a great impact on the development and marketing of mobile internet celebrity short videos in the future Put forward prospects and suggestions. Fang Xin [11] observes the marketing model of "net celebrities" from the perspective of subculture, focusing on the analysis of the characteristics of "net celebrity" marketing from the perspective of style shaping, promotion and dissemination and the process of flow monetization during the development of "net celebrities", sorting out and constructing How the audience's initiative can improve the current "net celebrity" marketing model, and put forward suggestions for its development and optimization. Li Pingping et al. [12] conducted research and analysis on the "Internet celebrity" marketing model and the future development prospects of the "Internet celebrity economy", and explored the pros and cons of the existing marketing model. Yin Hanzhi [13] takes the use of Chinese influencer marketing by Korean companies as the research object to analyze the influence of influencer marketing characteristics on consumers' purchase intention.

Looking at the existing literature, the research of this article may have the following innovations: (1) In terms of research topics, this article selects the newly emerging Chinese Internet celebrity economic industry in the past ten years. At present, domestic research focusing on this aspect is not very sufficient. (2) In terms of research methods, this article uses the analysis perspective of the industrial chain to explore the upstream, midstream and downstream of the Internet celebrity economy. (3) In the research sense, this article attempts to reveal the content and role of the industrial chain under the rapid development of China's Internet celebrity economy industry, and tries to provide policy recommendations on the Internet celebrity economy for achieving high-quality development in China.

3 Industrial Chain

The industrial chain is a concept in industrial economics. This means that due to the same production goals, there is a link between multiple production departments that closely links them. This association line is sorted according to a specific logical relationship and timeline layout. There's just like a bond in them and they are indispensable to each other. Regarding the Internet celebrity economic industry chain, this article believes that there are the following three levels, which complement each other and promote the development of the Internet celebrity economic industry.

3.1 Upstream of the Industrial Chain

The upstream of the Internet celebrity economic industry chain is the content production side. The rapid rise and growth of the Internet celebrity economy is inseparable from the compatibility and demand of the current market. In order to cater to consumer preferences, Internet celebrities need to have the talents and outstanding personalities that can withstand the test of the market. And the first way to achieve this is through the packaging and building of Internet celebrity brokerage companies and MCN companies, that is, "Somebody made it". The company will work on many aspects, such as cultivating the ability of Internet celebrity artists to create video content, buying a press release for them to attract attention, planning various online activities, and creating topics to stabilize the personality of Internet celebrity artists (Foodie, innocence, grumpy blogger, etc.). The support of brokerage companies and professional teams can help Internet celebrities win more traffic and maximize their actual commercial interests. The other way is to "self-build". As the name suggests, individuals create themselves to become Internet celebrities, and packaging and popularity rely on their own efforts to develop. However, compared with the former, this method has obvious disadvantages. Without the assistance and support of the team, it will be a lot of pressure to complete all the work alone, and it may also lead to imperfect and inefficient output in the field they involve. So it is difficult to attract audiences and be recognized and sought after. For example, a self-built internet celebrity who sell goods not only needs to plan live broadcast scripts, event themes, and benefits, but also strive to establish the IP image of the store and individual, and control the rhythm and respond to emergencies during the live broadcast. The need for Internet celebrities to handle it themselves is indeed not something that can be done solely by personal strength.

For the upstream of the industry chain, that is, the content production side, this article believes that the excellent Internet celebrities cultivated must eventually be marketed. As public figures, their words and deeds will inevitably affect their audience. Therefore, it is their responsibility to maintain a positive image in front of the camera, to establish their own personal settings and to lead fans to become better themselves, and it is also a prerequisite for the development of the upstream of the Internet celebrity industry chain. For example, on May 15th, the Chinese Internet celebrity, Wei Ya, jointed the United Nations to help the public welfare live broadcast in Rwanda, Africa, focusing consumers' attention on Rwanda, and she relied on her excellent business capabilities plus the accumulated followers in the domestic market since she started broadcasting, related products are sold out almost in one second. Not only that, so far, Wei Ya has done 59 public service live broadcasts, she have explained what she said: "I hope your guys give this industry time, but also give it the right guidance, e-commerce live broadcast can also become an upward force in China in the future" in a practical action. From this, it can be seen that the point that cannot be ignored in the upstream of the Internet celebrity industry chain is the improvement of the Internet celebrity's own capabilities. The products in Wei Ya's live broadcast room were out of stock in a few seconds. Moreover, she teamed up with the United Nations to carry out the live broadcast, occupying the Center position in the propaganda posters. All of the above demonstrates Wei Ya's professional level in the field of Internet celebrity economy.

In addition, the key to developing the upstream of the Internet celebrity economic industrial chain is innovation. With the great strides of the 21st century, although the Internet celebrity economy industry is still named as an emerging industry, more and more people have set up accounts of social media and entered the ranks of Internet celebrities, further promoting the development of internet celebrity economy and even leading to the saturation of the Internet celebrity market. In the great era of Internet development, Internet celebrities who imitate others will only be replaced by a new generation that survive from the severe competition. Therefore, whether Internet celebrities can produce innovative content is particularly important. Internet celebrities with novel features and unique styles are becoming more precious. They can attract more audiences, help themselves achieve a long journey, and also promote the prosperity of the Internet celebrity economy.

3.2 Midstream of the Industry Chain

The midstream is the content dissemination and distribution channel. With the rapid development of technology, new media such as Douyu and Kuaishou are in the ascendant, and the dissemination channels are no longer limited to offline. Take Douyin, a short video platform that is currently very popular in China as an example. It is a short video software, and many anchors and internet celebrities have developed better on this platform. As of the end of 2019, Li Jiaqi, the Chinese "Lipstick Brother", had 36.05 million fans, 210 million likes, and 284 published works on Douyin. In the past 90 days, each work of his had an average of 310,000 likes and 21,000 comments, ranking 11th on the entire platform. And these data are no less than some of the popular Chinese stars. In this link, various platforms are the main body, and Internet celebrities rely on the platform for content distribution. For platforms, Internet celebrities are an important "resource", attracting audiences to strengthen the development of the platform; for Internet celebrities, the ecology and influence of the platform help them accumulate popularity. In view of this, this article believes that in the middle of the industry chain, the platform plays a vital role in the Internet celebrity industry that relies on new media. It is almost the main way for new Internet celebrities to gain exposure and win audiences. For Internet celebrities, not only need to enhance their own capabilities, but also need a platform with a large user base, a stable audience, and the opportunity to obtain resources and be promoted to participate in activities. For example, a Chinese audio-sharing platform, "Himalaya", which has more than 600 million mobile phone users and more than 50 million overseas users, has been very active in organizing and launching activities, and has also achieved remarkable results in helping to build contracted authors, as well as making clusters of publishing house radio stations and writers' radio stations. For the platform, they also need excellent Internet celebrities to bring the fan economy to achieve transformation of economic benefits, such as through fan appreciation and other ways to realize the flow of cash. But whether it is a platform or an individual, contribution and gaining profit are not unilateral. Helping each other and achieving a win-win situation is a vital link between both parties. As the famous economist, Milton Friedman, once said "The most important single central fact about afree market is that no exchange takes place unlessboth parties benefit."

3.3 Downstream of the Industrial Chain

The downstream is the monetization link. Brokerage companies spend a lot of money to train Internet celebrities and various companies to introduce content production incentive plans to compete for Internet celebrities is not unreasonable. Internet celebrities rely on their outstanding output content and use the platform as a springboard to gather audiences, ultimately to achieve a goal-to obtain revenue. This link is based on the name of responsibility is net red to complete the transformation from their own audiences to income, and this link will generally be achieved through three ways. The first is to sell products, relying on their accumulated popularity, audiences and marketing methods, through some price reductions or benefits, to promote products to consumers as Internet celebrity. For example, Wei Ya, a top-level Taobao anchor, is an Internet celebrity anchor who makes a profit by selling products. She has already been involved in beauty, snacks, skin care products field and so on. The second is to sell services. The anchors cater to consumers' preferences by virtue of their professional level and obtain their affirmation. Most of the manifestations of this behavior are rewards that we are very familiar with. For the Internet celebrity who obtain profits through this method, the platform will be divided into various fields according to their professional abilities, such as good-looking anchors, seiyuu anchors, etc. Audiences will choose different fields and types of anchors according to their different needs. Watching and doing entertainment, then they will give a cash feedback. The third is to sell advertisements. Internet celebrity and anchors develop commercial cooperation with advertisers, and use their professional nature of the large audiences of Internet celebrities to spread the products of advertisements to their consumer audiences. This monetization method is a test of how Internet celebrities can control their fans' perception and acceptance. They need to be cautious in interspersing advertising content so as not to lose their existing audiences. For the downstream of the industry chain, this article believes that regardless of the three methods, fans and followers are a necessary prerequisite for Internet celebrities to realize the conversion of audiences to income. For the first type of monetization method to sell products, Internet celebrities using this method generally focus on selling some goods, and the key to gaining consumer groups, repeat customers and even being fans of Internet celebrities lies in having characteristics, such as outstanding appearance. But more importantly, the Internet celebrity themselves must have excellent sales ability and the product itself is worth buying by consumers, otherwise it is very likely to lose potential fans and audiences in the process of selling the goods, and even implicate them. In addition, wearing traffic does not mean shaking up the pillow and sleep in peace. The scope of business and the operation and maintenance of fans must be carefully considered by the Internet celebrity. For example, the fan base of beauty bloggers is mostly young women. Recommending beauty products to them meets the consumption scene and the needs of the crowd. In this way, the Internet celebrities can firmly grasp the needs and preferences of fans and maintain the activeness of fans. However, if an Internet celebrity sells cars to fans, there are still a few fans who will buy it.

For the second type of monetization method to sell services. At a time when the Internet celebrities of good-looking appearance are everywhere and the Internet celebrity economy is booming, it is not uncommon to see Internet celebrity and anchors selling services, but those who stand out are those with unique charm and value in the professional field. For example, in 2016, China's short video self-media blogger papi was suddenly well-known. In the entertainment-oriented social environment, her video content has novel topics and novel viewpoints, which rarely reflects a collective social sentiment, which are extremely resonated. In the nascent age of the Internet celebrity industry, her style is unique. As of this year, Papi has over 30 million fans on Weibo.

For the third type of monetization method to sell advertisements, it is a more risky way to monetize for Internet celebrities. This method has both risks and benefits. Internet celebrities need to balance the proportion of advertising content with their daily output, use their creativity and characteristics to combine advertising with output, and present a better visual experience to fans, so that they will not lose followers and even stimulate sales of that products.

4 Conclusions and Policy Recommendations

The Internet celebrity economy is a new economic model brought about by the development of new media. It opens up a new path for the development of the Internet economy, while at the same time providing a full range of more convenient services to the social audience, and playing a role in spreading positive energy. Based on the analysis perspective of the industrial chain, this article conducts research on the Internet celebrity economy and its marketing model. Specifically, the following three conclusions and policy recommendations are drawn: (1) The upstream of the industry chain of the Internet celebrity economy is the content production end, which includes two ways of creating and self-creating. The key to development lies in a positive and innovative Internet celebrity content. (2) The middle reaches of the internet celebrity economic industry chain are content dissemination and distribution channels, including various new media platforms. For Internet celebrity, the key to operation is a media platform with a large user base and a stable audience; for platforms, the key to development is excellent output content and a strong fan economy. (3) The downstream of the industrial chain is the monetization link, which includes three ways of selling products, selling services, and selling advertisements. For Internet celebrities who are monetized by selling products, characteristics and sales ability are the key to development; for Internet celebrities who are monetizing through selling services, proprietary characteristics and innovation are the key to development; for For Internet celebrities who are monetizing through selling advertisements, balancing advertising content with everyday output is key.

In addition, this article points out three other suggestions for the development of the Internet celebrity economy: (1) Internet celebrities are an indispensable part of this emerging industry, mostly for the younger generation. The words and deeds of Internet celebrities themselves will have a certain impact on them. The country and platform should relentlessly ban and block disqualified Internet celebrities to help their audiences develop healthily and maintain this new format. (2) The platform should balance the resources given to the contracted Internet celebrities, and not restrict the flow of amateur bloggers to ensure that the platform tends to be a good atmosphere. (3) Virtual products, such as virtual currency used to reward Internet celebrities, should be protected like personal property. In order to avoid unnecessary disputes, the user rules issued by the platform should recognize the ownership of property while rewarding Internet celebrities owned by the platform and the Internet celebrities themselves.

References

1. Shen, X., Wang, G.H., Yong, T.F., Zhong, S.Y.: The development process, characteristics analysis and governance of my country's internet celebrity phenomenon countermeasures. J. Inf. **35**(11), 93–98+65 (2016)
2. Sun, J., Wang, X.X.: Internet celebrities and internet celebrity economy: based on celebrity theory. Foreign Econ. Manage. **41**(04), 18–30 (2019)
3. Cui, M.H.: Internet celebrity economy-live broadcast with goods. Shop. Mall Modern. **12**, 5–7 (2020)
4. Wu, B.J.: Research on the Business Model and Development of Content-Based Internet Celebrity Economy. Zhengzhou University, Zhengzhou (2017)
5. Yu, L.Y.: The innovation of the business model in the form of new media-with the "Internet celebrity economy" The Rise of China as an example. New Media Res. **2**(15), 81–82 (2016)
6. Xiao, Z.J., Kang, L.J.: The business model of the internet celebrity economy. Media Obs. **9**, 15–16 (2016)
7. Liang, X.M.: The marketing strategy and prospects of internet celebrity brands. Int. Public Relat. **3**, 60–65 (2016)

8. Li, Y.Q.: The rise of "Internet celebrity" marketing and the development strategy of advertising industry. House Drama **23**, 214 (2020)
9. Wang, M.: Taobao Internet Celebrities Research on the Influence of Weibo Marketing on Audience's Consumption Attitudes. Hunan Normal University, Changsha (2017)
10. Liu, J.L.: Research on Short Video Marketing Strategies of Mobile Internet Celebrities. Inner Mongolia University, Hohhot (2019)
11. Fang, X.: Research on the "Internet Celebrity" Marketing Model from the Perspective of Subculture. Shandong Normal University, Jinan (2019)
12. Li, P.P., Hu, M., Qin, M., Luo, M.H.: Analysis of internet celebrity marketing model and economic development trend of Internet celebrity in the self-media era. China Bus. Rev. **10**, 13–15 (2019)
13. Yin, H.: Research on Influencer Marketing of Korean Companies on Chinese Social Platforms. Zhejiang University, Hangzhou (2018)

Impact of Stock Split on Stock Return

Geyao Zhang(✉)

University of Bristol, Bristol, UK
er20500@bristol.ac.uk

Abstract. This article examines to what extent do stock splits affect stock returns of Chinese companies. Stock splits may affect investor's judgement of stocks by transmitting some signals or information, and then influence stock return. Based on the event study method, we establish a research model and draw a conclusion by analyzing Abnormal Returns (AR) and Cumulative Abnormal Return (CAR). We analyses a dataset of Tencent and WeTrade Group around their announcement of split. According to the research results, we conclude that stock split has no obvious effect on stock returns. This conclusion is in contrast with some previous research results, which may be due to the influence of China's market characteristics.

Keywords: Stock split · Stock return · China

1 Introduction

Stock split refers to the allocation of additional shares to shareholders according to the existing shareholding ratios. When a stock split occurs, the company "repurchases" its outstanding shares, and then changes the original one into more shares. The split ratio can be 2:1,3:1 or higher, and par value will be reduced, while maintain the total amount of shareholders' equity and increase the liquidity of the stock. There are a number of famous companies made a stock split in their process of growing up, like Berkshire Hathaway and Apple. The research by Guo, Liu and Song (2008) show that acquiring firms' manager use stock splits as a way to raise the share value before announcing stock-exchange mergers and acquisitions [1].

Investors in the stock market are pursuing higher returns, many companies use corporate strategies to affect their stock returns. Stock split, as one of these strategies, always considered to pass on a positive signal about the development prospect of the company, which helps to improve investors' confidence in the company. The signaling theory was shown in the model of Leland and Pyle, which conclude that the management can use financial decision to transfer information that served as a signal about the firm value [2]. Brennan and Copeland show that management use stock split as a costly signal of firm value to transfer private information about company prospects to investors [3]. While some other researches like Ross also set model about this theory and make similar conclusion [4–6].

Therefore, whether the stock split will have a significant impact on stock returns has attracted attention, and many people have carried out research on it. Fama, Fisher,

C. Yuan et al. (Eds.): *Proceedings of the 4th International Conference on Economic Management and Green Development*, AEPS, pp. 367–372, 2021.
https://doi.org/10.1007/978-981-16-5359-9_42

Jensens and Roll research that firm tend to make stock split during “abnormally” good time and stock split can only effect market by its dividend implications [7]. Conroy, Harris and Benet find stock split have a negative effect on shareholder liquidity, and the observed rise in the variability of return is partly considered to be result of increase in bid-ask spread [8].

Although previous studies have analyzed the possible impact of stock splits, there are few related studies in China due to the uncommon of stock splits. Thus, we select two Chinese companies with large difference in scale and stock price, using the method of event study to explore whether the stock split will have a significant effect on stock returns.

2 Data and Methodology

2.1 Model

The aim of the research is to test if the stock split will have a positive influence on the stock price around the implementation of the policy. To solve the problem, we use the event study method and set regression model as following:

$$R_t = \beta_0 + \beta_1 R_{mt} + \mu. \tag{1}$$

Where.

R_t = individual stock return.

R_{mt} = contemporaneous market return.

μ = regression error term.

In the model, we use stock’s daily closing price to calculate stock return, and select S&P500 index on behalf of the market return. The variables is represented as following:

$$R_t = \ln P_{t+1} - \ln P_t. \tag{2}$$

$$R_{mt} = \ln I_{t+1} - \ln I_t. \tag{3}$$

Where.

P_t = stock closing price on day t.

I_t = Index closing price on day t.

In order to test the effect of the strategy, we pick estimation period and event period from each stock, and conduct regression analysis respectively. If we set the implement day as $t = 0$, the event period should consist of $t = -E,\ldots, 0,\ldots, E$, and the estimation period should even add some “quarantine” period before the event period.

We predict the returns that should occur during the event window a following:

$$R_t' = \beta_0 + \beta_1 R_{mt}. \tag{4}$$

and then compare with the actual returns that happened Rt.

To do the comparison, we establish the following two formulas:

$$AR_t = R_t - R_t'(t = -E, \ldots, E) \tag{5}$$

$$\mathrm{CAR_t} = \sum\nolimits_{\mathrm{s}=-\mathrm{E}}^{\mathrm{t}} \mathrm{AR_S}. \tag{6}$$

While AR_t is defined as the so-called Abnormal Returns, and CAR_t is Cumulative Abnormal Return for each t in the event period.

2.2 Viable Selection

According to the main purpose of the research, the data is collected from Yahoo!Finance, about three company–Tencent, WeTrade Group and Yunji, as well as S&P 500 Index. These data are intercepted from some time period during July 2013 to September 2020, about seven years. To ensure the authenticity and rationality of the research, all the data are obtained from this website.

We each variable based on its reliability and the statistical requirement of the model. These variables are talked over then.

The dependent variable, R_t, is stock return calculated by the closing price. To research stock split's influence on both big business and the small one, we intercept the data from two sample, Tencent and WeTrade Group. Tencent had divided its stock by a ratio of 5:1 on May 5th, 2014, while Wetrade Group did a 3:1 one in Aug 30 this year.

The independent variable, R_{mt}, is used to represent the market return at time t. Here we choose S&P 500 index, which is one of the most authoritative stock indexes, for the calculation of return from the market portfolio.

The key periods of calculate R_t, as well as R_{mt}, are defined separately in the two companies. For Tencent, we defined E = 47, defined the event period from March 6 to July 24, and drawn estimation period from June 3, 2013 to Feb 5 2014. But when we come to WeTrade Group, we find this company just went public on July, 23, only one month before the split. Since its history is too short to support our model, we decide to use another company with similar business model and scale to calculate the estimation period model. Therefore we select the data of YunJi from May 6019 to March 9, 2020 to represent the estimation period, and still use the data of the original company from Aug 5 to Sep 28 when calculating R_t'.

In addition, we assumed that other omitted factors will not change the stock price significantly during the test period.

3 Rt

The result from the regression of two companies are shown as Table 1:

The coefficients of R_t show the relationship between market return and stock return, during estimation period, the increase of market return would boost stock return of both companies, and AR_t show the difference between expected stock return and the real one, which is entirely assumed to attributable to the split strategy.

In order to find out whether the stock return fluctuation is in the normal range, we set a null hypothesis that stock split has no effect. If we accept this hypothesis, then AR_t should subject to normal distribution $N\left(0, \sigma_\varepsilon^2\right)$, and CAR_t should also submit to normal distribution $N\left(0, (t+E)\sigma_\varepsilon^2\right)$, where σ is the standard deviation during the estimation

Table 1. Result of regression.

	Tencent			WeTrade Group		
		R_t'	AR_t	R_t	R_t'	AR_t
β_0	0.003**	−0.00009	−0.003	−0.006**	−0.0002	0.005
β_1	0.522***	0.148	0.374	0.746***	1.91	1.165

Note: the value with ***and** represents that its significant level is less than 0.01and 0.05

window. In general, we use t-test to test the authenticity of the original hypothesis at 95% confidence level. However, there are a number of data in our sample, so we decide to use the method of gph to test the significance. The results are shown as Fig. 1:

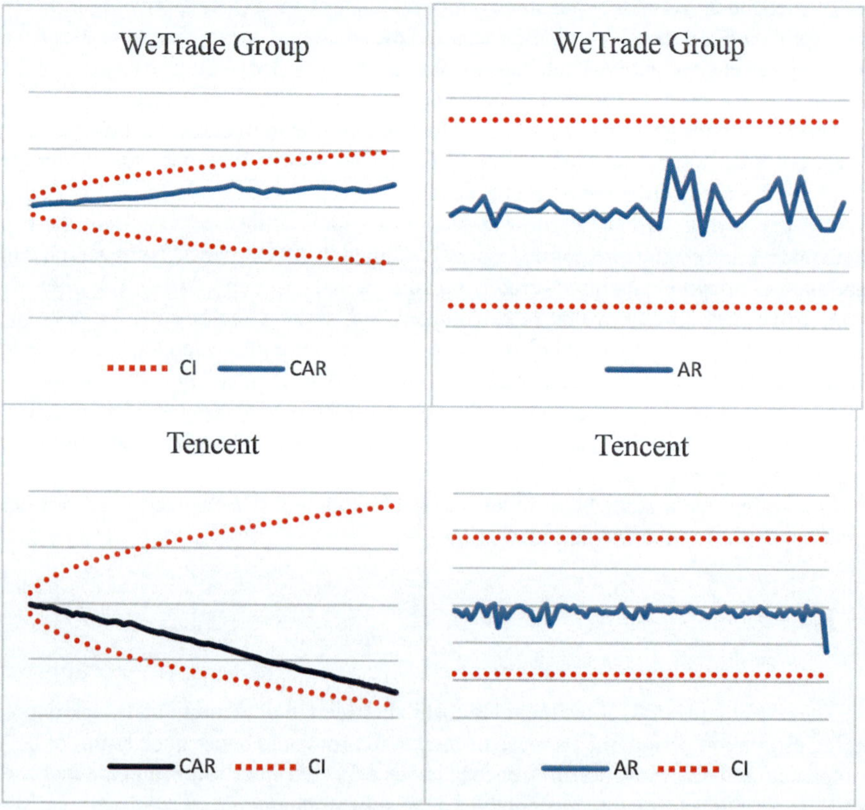

Fig. 1. AR and CAR.

Figure 1 above show the line graphs of the AR, CAR and confident interval during the event window. From the chart, we can see that the lines of AR and CAR are all stay inside the symmetric bands. Which means that the stock return does fluctuate to a certain extent, and the volatility in the days before and after the announcement of stock split and the implementation of the strategy is relatively larger compared with other days but the range of fluctuation does not exceed the normal range. Therefore, we can accept H_0 that stock split has no significant effect on stock return at 95% confidence level.

4 Conclusion

In this paper, we use event study to set a model to research to what extent does the split affect stock returns through the data analysis of the two companies--Tencent and WeTrade Group. The empirical study shows that for the companies, splits were followed by average abnormal return and cumulated abnormal return, which means the market reaction on stock return is not significant at 95% confidence level. This is not consistent with the research results of Grinblatt and other authors [9]. That may be why Chinese major companies are not keen on stock split.

Chinese stock market possesses one of the largest individual investor groups in the world. In 2014, the proportion of retail investors holding shares in China was as high as 50%,while that in the US stock market was only 19% [10]. Margin trading and short selling are limited in China, and these policies have a significant impact on the trading of the less affluent individual investors who are considered to be the net buyer after split announcement [11]. In addition, stock splits are sometimes used as a tool for insiders to sell stocks due to their low average education and lack of sufficient investment knowledge. Therefore, after the Shanghai stock Exchange increased the information disclosure requirements for the stock split announcement in 2015, the investors' reaction to the split was not as positive as before [12].

The companies we chose are both engaged in internet related industries. Due to the popularity of the internet, the transparency of the internet company is relatively high, which is not conducive to the implementation of the signal effect of stock split as investors can make their own judgement according to the information disclosed. Therefore, managers cannot pass information to the market thereby affect the stock returns through stock split.

Although we can make such a conclusion under this situation, but it not completely accurate due to the defects on the number of companies, since there are not that many Chinese company do the split in recent years. According to the increase on Tencent share price in the next few year, we can doubt if the result will remain the same in the test during a longer period. As the research by Da and Warachka show the difference between long term and short term analysis [13]. And for the other aspects and factors that may be affected by stock split, further research is needed in the field of split.

References

1. Guo, S., Liu, M.H., Song, W.: Stock splits as a manipulation tool: evidence from mergers and acquisitions. Financ. Manage. **37**(4), 695–712 (2008)
2. Leland, H.E., Pyle, D.H.: Informational asymmetries, financial structure, and financial intermediation. J. Financ. **32**(2), 371–387 (1977)

3. Brennan, M.J., Copeland, T.E.: Stock splits, stock prices, and transaction costs. J. Financ. Econ. **22**(1), 83–101 (1988). https://doi.org/10.1016/0304-405X(88)90023-2
4. Ross, S.A.: The determination of financial structure: the incentive-signalling approach. The Bell Journal of Economics **8**(1), 23–40 (1977)
5. Spence, M.: Job market signaling. Q. J. Econ. **87**(3), 355–374 (1973)
6. Bhattacharya, S.: Imperfect information, dividend policy, and "the bird in the hand" fallacy. The Bell Journal of Economics **10**(1), 259–270 (1979)
7. Fama, E.F., Fisher, L., Jensen, M.C., Roll, R.: The adjustment of stock prices to new information. Int. Econ. Rev. **10**(1), 1–21 (1969)
8. Conroy, R.M., Harris, R.S., Benet, B.A.: The effects of stock splits on bid-ask spreads. J. Financ. **45**(4), 1285–1295 (1990)
9. Grinblatt, M.S., Masulis, R.W., Titman, S.: The valuation effects of stock splits and stock dividends. J. Financ. Econ. **13**(4), 461–490 (1984). https://doi.org/10.1016/0304-405X(84)90011-4
10. Pujing: How to institutional investors affect corporate governance of their company: Evidence from Chinese companies on the Shanghai Stock exchange and the New York Stock exchange(Master's thesis, Nanjing University) (2017). https://kns.cnki.net/KCMS/detail/detail.aspx?dbname=CMFD201801&filename=1017168568.nh
11. Barber, B.M., Odean, T.: All that glitters: the effect of attention and news on the buying behavior of individual and institutional investors. Rev. Financ. Stud. **21**(2), 785–818 (2008)
12. Wang, G., Wang, Y., Yang, D., Zhang, L., Zhu, Q.: Enhanced disclosure environment and stock dividend/split in china. Appl. Econ. Lett. **1–5**, 1–5 (2020). https://doi.org/10.1080/13504851.2020.1752896
13. Da, Z., Warachka, M.: The disparity between long-term and short-term forecasted earnings growth. J. Financ. Econ. **100**(2), 424–442 (2011). https://doi.org/10.1016/j.jfineco.2010.10.015

The Direction of Technical Changes –The Theoretical Research and the Empirical Research on the Chinese Economy

Junjun Li(✉)

Graduate School of Asia-Pacific Studies, Waseda University, 1-21-1 Nishi-Waseda, Shinjyuku-ku, Tokyo, Japan
lijunjun@asagi.waseda.jp

Abstract. Recent studies show that the long-run trend of a mature economy grows with only labor-augmenting technological change. This study theoretically inspects the direction of technical change through the perspectives of labor productivity growth and profit maximization, and empirically investigates the direction of technical change in China. The theory study suggests that the direction of technical change to ensure stable growth of labor productivity is the proportion of labor-augmenting and capital-augmenting technical change converges to labor output elasticity and capital output elasticity. The empirical study result shows that labor-augmenting technical change is the mainstream of technical change in China as well. However, profit maximization is not the driven force that decides the direction of technical change, while policy-driven growth in fixed capital investment has promoted labor-augmenting technical change in the Chinese economy. Meanwhile, there are more capital-saving technical changes in Japan and other advanced countries in recent years. These findings raise important policy suggestions that both labor-augmenting and capital-augmenting technical change are essential, and the orderly directed technical change is critical to economic growth.

Keywords: Direction of technical change · Labor-augmentation · Capital-augmentation · Labor productivity · Economy of China

1 Introduction

Induced innovation is a hypothesis first proposed by John Hicks (1932). He proposed that a change in the relative prices of production factors is a spur to innovation, a directed innovation that will develop technologies to economize on the costly production factors [1]. It means that technical changes are hardly neutral in their factor-saving potential. He classified technical change to labor-augmenting (labor-saving) and capital-augmenting (capital-saving) technical change according to its effect on the ratio of marginal product of capital to marginal product of labor. Kennedy (1964) argued that induced innovation will push the economy to an equilibrium with a constant relative factor share [2]. Recent theories and empirical studies show that the long-run trend of a mature economy grows

C. Yuan et al. (Eds.): *Proceedings of the 4th International Conference on Economic Management and Green Development*, AEPS, pp. 373–386, 2021.
https://doi.org/10.1007/978-981-16-5359-9_43

with only labor-augmenting technical change. Meanwhile, the shares of national income allocated to labor and capital tend to be constant.

Since the market-oriented institutional and economic reform, China has experienced enormous economic growth. Low labor costs have attracted many investments and make "made in China" spread all over the world. Goodwin (1967) argues that higher labor productivity increases the profitability of investment, generates a tighter labor market, and causes real wages to increase [3]. With the growth of the economy, the level of productivity is constantly improving, accompanied by a substantial increase in average wages and labor costs. Meanwhile, China accelerated its higher education industrialization in the year 2003. The number of the high-level labor force increases rapidly. The total number of university graduates is more than 8.3 million in 2019 [4]. The growth of the economy, the growth of labor productivity and the increase of human capital make China a country with rising labor costs.

On the one hand, the Chinese government is promoting the transformation of the economic structure. On the other hand, in order to reduce labor costs, a large number of labor-intensive firms began to migrate from China to countries with lower labor costs, for example, Vietnam. Figure 1 shows that from 2008 to 2017, average real wages doubled in China, while the wage growth rates are less than 10% in most advanced countries [5].

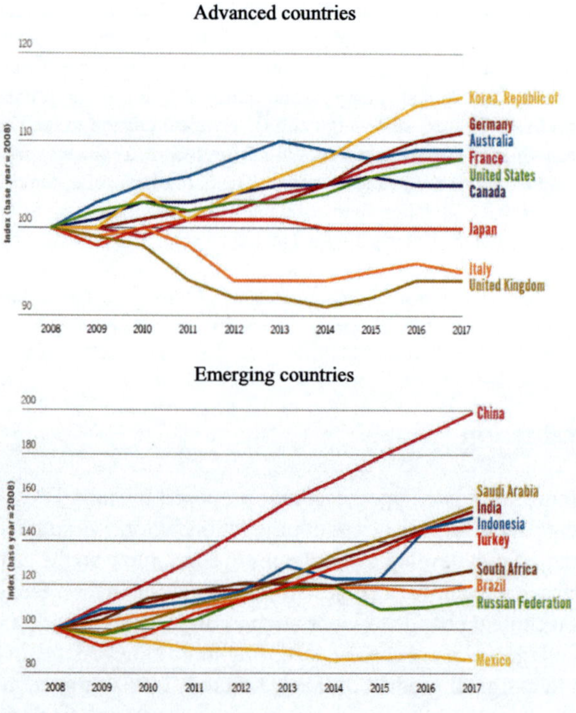

Source: Global wage report 2018/19 (ILO, 2018)

Fig. 1. Average real wage index for advanced and emerging countries, 2008–17

The Kaldor's "stylized" facts for long-term economic growth indicates that capital-output ratio and functional distribution of income are roughly constant as well. However, the factor income shares are no longer constant. Although not universal, the decline in labor income share is a global trend in recent decades [6]. The labor income share in 35 advanced economies fell from around 54% in 1980 to 50.5% in 2014 [7]. Figure 2 is the decline of labor income shares in China, Japan, and the United States. The labor income shares in China have increased in recent years, and the one in Japan has also risen slightly after 2015. However, they are still well below the previous levels. Some studies suggest that the decline of labor income share could be dampened but continue [7]. It means that decreasing labor cost is a long-term global trend. The related literature shows that the decline of labor income share across the world is a result of globalization, the weakening of labor market institutions, the growing pressure from financial markets to shift surpluses generated by large businesses towards investors, and capital substitution and technology [6, 7]. Huang and Xu (2009) argue that the key cause of a declining labor share in China is labor-augmenting technical change [8].

The result of recent studies that the long-run trend of a mature economy grows with only labor-augmenting technical change means labor-augmenting technical change is more critical to an economy in the long term, especially a mature economy. Meanwhile, changes in the relative prices of production factors induce innovation to economize on the costly production factors. As a developing country, China is a labor-surplus economy. However, the labor costs have increased significantly in China, while rental rates of fixed capital have not changed much. The increasing labor cost should induce technical change bias toward labor-augmenting innovations in China, a developing and labor-surplus economy. This research addresses the following questions:

Q1. Can pure labor-augmenting technical change ensure steady economic growth, the growth of labor productivity?

Q2. Does technical change is also directed toward labor-augmenting innovations in the high-tech manufacturing sectors in China, a developing and labor-surplus economy?

2 Related Literatures

The existing theoretical studies and empirical studies are developed on the basis of the pursuit of maximum profit for the firms. Acemoglu studies an aggregate production with two inputs, labor, and Z, which could be capital, land, or other critical inputs [9–11]. The framework of his model of the factor bias of technical change is an analysis from both the demand side and the supply side for new technology (innovation). He pointed that the forces determining the direction of technical change are the price of factors (the price effect), the market size (the market size effect: a larger market for the technology leads to more innovation), and the substitution effect that "the more abundant factor is substituted for the less abundant one, and has a lower marginal product [9]." He finds that the economy grows with the purely labor-augmenting technical change in the long run. Tax policy, labor supply, and savings change factor shares in the short run but have

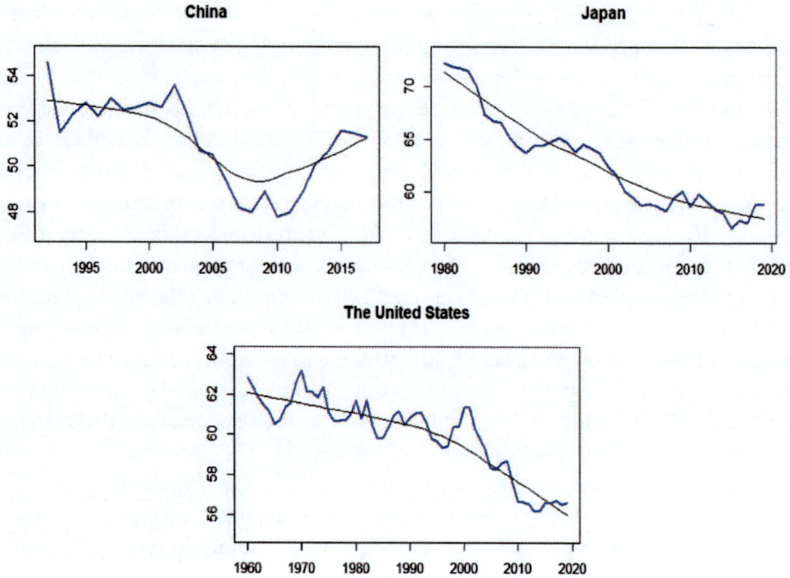

Notes: Unadjusted labor income share for China. Adjusted labor income share for Japan and the United States. The unadjusted labor income share excludes the income from self-employment, and is lower than the true share of labor income.

Source: Author's graphing base on the labor share data from AMECO for the United States and Japan, and from ILO for China.

Fig. 2. The decrease in labor share (whole economy)

no or little effect in the long-run. He expands his model of directed technical change to different inputs for policy suggestions, for example, the clean and dirty inputs with environmental constraints [12].

Irmen extends Uzawa's steady-state growth theorem and uses a similar study framework with Acemoglu. His studies focus on the analysis of the endogenous labor-augmenting technical change. He pointed out that population ageing intensifies the relative scarcity of labor with respect to capital, and leads to more labor-augmenting and less capital-augmenting technical change as well [13]. He argues that only labor-augmenting technical change in the steady-state does not mean to neglect capital-augmenting technical change. He suggests that the steady-state must induce sufficient capital-saving technical progress to offset the depreciation of this stock [14].

Based on the theoretical studies, several empirical studies to the theory of directed technical change are carried out [15–18]. They adopt Kennedy's innovation possibilities frontier and analyzes the demand side and the supply side. Time series data for the United States and panel data for a sample of countries are used for estimations. These studies show that the long-run trend of a mature economy grows with labor-augmenting technical change. Meanwhile, the trend which holds in mature economies also holds in the manufacturing sector of developing countries [18].

This study carries out from the analysis of the relationship between directed technical change and economic growth, growth of labor productivity. It can make up for the shortcomings of only focusing on profit maximization. This study is the first study from the perspective of labor productivity growth. The value of this study is significant. Meanwhile, previous empirical studies verify the situation in the United States and the general transnational situation. There are a few studies on the economy of China. This research conducts an empirical analysis in China.

For the analysis, this paper theoretically inspects the direction of technical change through the perspective of labor productivity growth, which is based on Kennedy's innovation possibilities frontier and the Cobb-Douglas production function. Second, this paper empirically investigates the direction of technical change in China through a comparative study among the economies in China, Japan, and the United States. Here uses the model extended from the definition of labor-saving and capital-saving technical changes from Kennedy. Furthermore, a Granger causality test is conducted to verify the causes of labor-augmenting technical change in China.

3 Direction of Technical Change: The Illustrative Models

3.1 Growth of Labor Productivity and Direction of Technical Change

Kennedy's innovation possibilities frontier indicates that the development and adoption of innovations is a resource-consuming process characterized by a trade-off between the maximum feasible rates of labor-augmentation and capital-augmentation [2]. Here set δ as the proportional rate of labor-augmentation technical change. Then, the proportional rate of capital-augmentation technical change is 1 − δ. Thus, the production function in Cobb-Douglas format is

$$\begin{gathered} Y = (A(1-\delta)K)^{\alpha}(A\delta L)^{\beta}, \\ 0 < \delta < 1,\ 0 < \alpha, \beta < 1,\ \alpha + \beta = 1 \end{gathered} \tag{1}$$

where Y is output, K is capital input, L is labor input, and A is technical change. Then, the labor productivity, the output per capita y is,

$$y = A((1-\delta)k)^{\alpha}\delta^{\beta} \tag{2}$$

where y is output per capita, and k is capital per capita. Take logs and differentiate Eq. (2),

$$\frac{\dot{y}}{y} = \frac{\dot{A}}{A} + \alpha\frac{(1\dot{-}\delta)}{1-\delta} + \alpha\frac{\dot{k}}{k} + \beta\frac{\dot{\delta}}{\delta}$$

Here, technological progress A is exogenous. Its growth rate is constant, $\frac{\dot{A}}{A} = g$. Let g_y, g_k, g_δ be the growth rate of y, k and δ. The growth rate of labor productivity g_y is

$$g_y = g + \alpha g_k + \frac{\beta - \delta}{1-\delta} g_\delta \tag{3}$$

In Eq. (3), when $\delta < \beta$, expanding the proportion of labor-augmentation technical change, $g_\delta > 0$, the growth rate of labor productivity will be higher. When $\delta > \beta$, if expanding the proportion of labor-augmentation technical change, however, the growth rate of labor productivity will be lower. In contrast, reducing the proportional rate of labor-augmentation technical change, $g_\delta < 0$, the growth rate of labor productivity will be higher. Inspection of Eq. (3) reveals that the proportional rate of labor-augmenting technical change converges to labor output elasticity, and the proportional rate of capital-augmenting technical change converges to capital output elasticity. The situation concludes from other studies that only labor-saving technical change can occur and capital-saving technical change will vanish, is not realistic and could not exist.

When $\delta = \beta$, $g_\delta = 0$, the proportional rate of labor-augmenting and capital-augmenting technical change remains constant, respectively equaling to the output elasticity of labor, and the output elasticity of capital. Then, the growth rate of labor productivity is

$$g_y = g + \alpha g_k \tag{4}$$

In the steady-state, y and k will be growing at the same rate, $g_y = g_k$. Output per worker grows at the rate of technical change per labor-augmentation unit

$$g_y = \frac{g}{\beta} \tag{5}$$

3.2 Profit Maximization and Direction of Technical Change

Profit is the difference between value of products and value of input factors. Setting the price of products as 1, and r as the lending interest rate of capital and w as the average wage, the profit π is

$$\pi = (A(1-\delta)K)^\alpha (A\delta L)^\beta - rK - wL \tag{6}$$

For K, profit maximization means $\frac{\partial \pi}{\partial K} = 0$,

$$\alpha(A(1-\delta))^\alpha K^{\alpha-1}(A\delta L)^\beta - r = \frac{\alpha Y}{K} - r = 0.$$

For L, it is the same. Then,

$$\frac{\alpha Y}{K} = r,\ \frac{\beta Y}{L} = w \tag{7}$$

According to Kennedy's definition, labor-saving technical changes will reduce the amount of labor required to produce a unit of product (unit labor saving, p), capital-saving ones will reduce the amount of capital required to produce a unit of product (unit capital saving, q). Following the definition, the unit labor saving p and the unit capital saving q from technical change are

$$p_t = 1 - \frac{L_t}{Y_t}\frac{Y_{t-1}}{L_{t-1}}, p < 1 \tag{8}$$

$$q_t = 1 - \frac{K_t}{Y_t}\frac{Y_{t-1}}{K_{t-1}}, q < 1 \quad (9)$$

where the subscript t refers to the time. An improvement will be labor-saving, neutral or capital-saving according as p is greater than, equal to or less than q [2]. According to Eq. (7), (8) and (9), when profit is maximized, the relationship between p and q is

$$p_t - q_t = \frac{r_{t-1}}{r_t} - \frac{w_{t-1}}{w_t} \quad (10)$$

When $\frac{w_t}{w_{t-1}} > \frac{r_t}{r_{t-1}}$, $p_t > q_t$. It means when the wage w, the price of labor, increases sharply relative to the interest rate r, the price of capital, profit-maximizing firms develop labor-saving technologies. When $\frac{w_t}{w_{t-1}} < \frac{r_t}{r_{t-1}}$, $p_t < q_t$. It means when the interest rate r increases sharply relative to the wage w, profit-maximizing firms develop capital-saving technologies. And, when $\frac{w_t}{w_{t-1}} = \frac{r_t}{r_{t-1}}$, profit-maximizing firms develop neutral technologies, $p_t = q_t$.

Equation (10) and Eq. (3) suggest that both profit maximization and labor productivity growth should be considered when choosing the direction of technical change. Equation (10) shows that profit-maximizing firms will generate innovation to economize on the costly production factors to gain greater profits. However, Eq. (3) suggests that only one direction of technical change will decrease labor productivity. In long-run steady growth, the proportions of labor-augmenting and capital-augmenting technical change should be constant as labor output elasticity and capital output elasticity, respectively.

4 Empirical Estimate, Result and Interpretation

4.1 Models and Data for Empirical Studies

The empirical studies exams the unit labor savings p and the unit capital savings q from technical change in China, Japan, and the United States to analyze the situation in China. Equation (8) and (9) will be used to estimate. Equation (10) will be used in Granger causality test to identify the causes which decide the direction of technical change in China.

Data used in the empirical analysis are from NBSC database (the National Bureau of Statistics of China), AMECO database (the annual macro-economic database of the European Commission's Directorate General for Economic and Financial Affairs), ILO data (the International Labor Organization), and World Bank data.

Capital stocks for the whole economy of China are the estimation from Herd (2020)[1]. Capital stocks for Japan and the United States are available directly from AMECO. All data are at constant prices to eliminate the impact of inflation. Table 1 lists the information of all variables and data sources.

[1] Herd, Richard. 2020. Estimating Capital Formation and Capital Stock by Economic Sector in China: The Implications for Productivity Growth (English). Policy Research Working Paper No. WPS 9317. Washington, D.C.: World Bank Group. http://documents.worldbank.org/curated/en/846601594661216544/Estimating-Capital-Formation-and-Capital-Stock-by-Economic-Sector-in-China-The-Implications-for-Productivity-Growth.

Table. 1 Variables, data and sources

	Variable	Data	Sources
China (1980~2017)	Output (Y)	GDP	Herd, Richard. 2020
	Capital stock (K)	Capital stock	Herd, Richard. 2020
	Labor (L)	Total employment	NBSC
	Capital price (Rent)	Lending interest rate	WDI, World Bank
	Labor price (Wages)	Average wages	ILO
Japan, the United States, and G7 countries (The time span varies by country, the longest is 1960~2019)	Output (Y)	GDP	AMECO
	Capital stock (K)	Capital Stock	AMECO
	Labor (L)	Total employment	AMECO
	Capital price (Rent)	Lending interest rate	WDI, World Bank
	Labor price (Wages)	Real unit labor costs	AMECO

4.2 Unit Labor Savings and Unit Capital Savings in China, Japan, and the United States

Table 2 reports the statistical distribution of the savings of labor to produce a unit of product p and the savings of capital to produce a unit of product q which come from the labor-augmenting and capital-augmenting technical change in China, Japan, and the United States. Figure 3 diagrammatically shows the changes of unit labor savings p and unit capital savings q in these three countries. According to Kennedy (1964), an improvement will be labor-saving, neutral or capital-saving accordingly as p is more significant than, equal to or less significant than q. The first quartile, the third quartile, and the mean of unit labor savings p are greater than unit capital savings q in China, Japan, and the United States. Furthermore, in most years, unit labor savings p is more significant than unit capital savings q in all three countries. It means that the mainstreams of technical change in these three countries are labor-augmenting technical change.

Figure 3 shows that unit labor savings in China is relatively stable with a slight decline in recent years, while unit capital savings have been declining in the past 40 years. In the 1960s and 1970s, the Japanese economy experienced a rapid decline in both unit labor savings and capital savings. However, its unit capital savings began to increase slowly in the early 1980s, and the decline in unit labor savings began to slow down. Both unit labor saving and capital saving are relatively stable, with a slight downward trend in the United States. In China, the gap between unit labor savings and capital savings is widening. In Japan, the situation is the opposite. The gap is gradually narrowing after the early 1980s, and unit capital savings have even surpassed unit labor savings in recent years. Most of the time, the gap between unit labor savings and capital savings was relatively small in the United States. The gap was relatively large from 2001 to 2009, the period of the real estate bubble.

Not only in Japan, in Group of Seven, unit capital savings in Germany, Italy, and the United Kingdom are also greater than unit labor savings in recent years. It means that there are more capital-saving technical changes in these countries, although the gap between the two savings is not big. The real data shows that capital-saving technical

Table. 2 Statistical distribution of unit labor savings p and unit capital savings q in China, Japan, and the United states.

Statistic	Periods	N	Mean	St. Dev.	Min	Pctl(25)	Pctl(75)	Max
p in China	1980~2018	38	7.534	2.322	1.803	6.202	8.721	12.056
q in China	1980~2018	39	- 0.413	3.898	- 9.371	- 2.606	2.981	7.665
p in Japan	1960~ 2019	59	2.899	3.080	- 4.486	0.774	4.153	9.998
q in Japan	1960~2019	59	0.060	2.670	- 8.976	- 1.214	1.360	6.526
p in US	1960~2019	59	1.600	1.232	- 2.409	0.849	2.476	4.307
q in US	1960~2019	59	0.446	1.730	- 3.735	- 0.164	1.499	4.095

Notes: GDP and capital stock are constant at 2010 price for China, 2015 price for Japan and the United States. The statistic method for total employment in China changed after 1990. In consequence, *P* in China for 1990 is not available.
Source: made by the author

change will not vanish and it is unrealistic that only labor-saving technical change can occur in the mature economy.

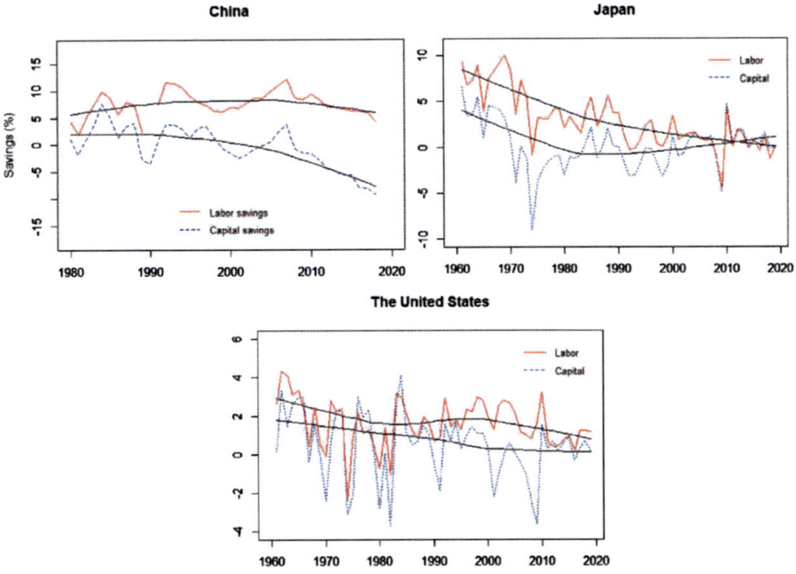

Notes: GDPand capital stock are constant 2010 price for China, 2015 price for Japan and the United States.The statistic method for employment of China changed after 1990. P in China for 1990 is not available.

Source: made by theauthor

Fig. 3. Changes of unit labor savings p and unit capital savings q in China, Japan, and the United states

Analyzing it together with the GDP growth and the capital stock growth data (Fig. 4), we can see that the growths of capital stocks in these three countries are different. In Japan, the growth rate of fixed capital investment declined since 1970s. After experienced four consecutive years of negative growth from 2009 to 2012, it has increased slightly in recent years. The growth of fixed capital investment in the United States is relatively stable. However, it had a drop during the global financial crisis and has not yet returned to the level before the crisis.

The growth of capital stocks in China is increasing since the year of 2000. After experiencing a growth phase dominated by labor-intensive production, the Chinese government began to strengthen fixed capital investment and take fixed capital investment as an essential driving force for economic growth. In the first 20 years of China's reform and opening up, investment mainly concentrated in the eastern region. In 1999, the Chinese government launched the Western Development Plan. In 2003, the strategy of revitalizing the industrial base in the Northeast was proposed. Then, the plan of promoting the growth of the central region was carried out in 2004. Driven by the policies, the growth of capital investment in these three regions has accelerated significantly.

Meanwhile, the government began to encourage and support the development of the non-public economy actively. The legitimate rights and interests of the non-public economy were written into the Constitution in 2004. In consequence, private investment has become a vital force in China's economy. In 1980, investment in the state-owned and collective firms accounted for 86.9% of total investment. However, it accounted only for 29.6% in 2017. Investment in the shareholding firms, private firms, and foreign-invested firms accounted separately for 28.9%, 33.8%, and 3.9% [19]. China has also increased investment in infrastructures such as high-speed railway, highway, telecommunication networks, ecology and urban infrastructure. In China, the rapid growth of fixed capital investment, which has exceeded the GDP growth in the past ten years, mainly comes from the impact of fiscal policy [20]. The fall of unit capital savings are a consequence of the rapid growth of fixed capital investment.

In addition, the unit labor savings in China and Japan are more correlated with their GDP growth, comparing to that in the United States. It means that employment in China and Japan is relatively stable, while employment in the United States is fluctuating. The economy in the United States is more market oriented. It is common to cut the workforce to achieve satisfactory quarterly or annual performance. On the contrary, in China and Japan, firms will avoid layoffs due to short-term performance and seek stable growth. Table 3 lists the statistical distribution of the growth of GDP and capital stock in these three countries at constant prices.

4.3 Granger Causality Test Results

The difference of unit labor savings and unit capital savings, $p - q$, represents labor-augmenting, neutral or capital-augmenting technical change according to it is greater than, equal to or less than 0. Equation (10) suggests that $p - q$ will equal to $\frac{r_{t-1}}{r_t} - \frac{w_{t-1}}{w_t}$ when profit is maximal. However, Pearson correlation test shows that $\frac{r_{t-1}}{r_t} - \frac{w_{t-1}}{w_t}$ and $p - q$ are negative weak correlations (−0.3478) in the economy of China (Table 4). It means that profit maximization is not the driven force that decides the direction of technical

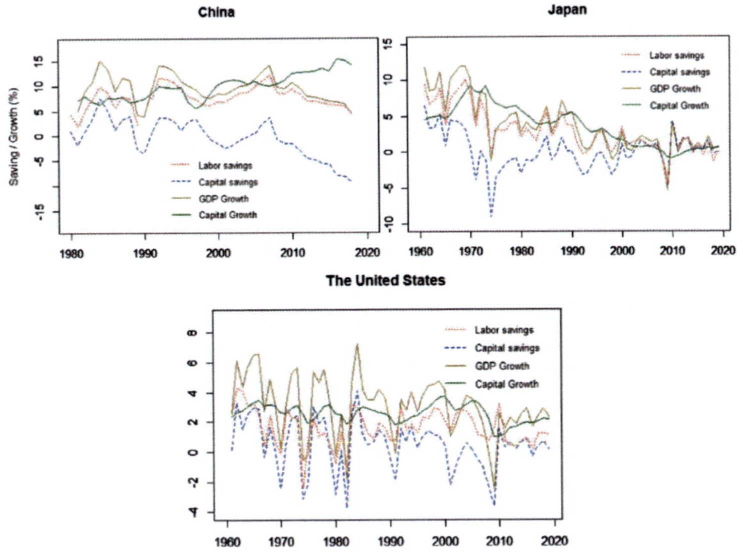

Notes: GDP and capital stock are constant at 2010 price for China, 2015 price for Japan and the United States. The statistic method for employment of China changed after 1990. P in China for 1990 is not available.

Source: made by the author

Fig. 4. GDP and capital stock growth vs. unit labor and unit capital savings in China, Japan, and the United states

change in China. However, average wage (0.9050) has significant positive correlations with $p - q$.

Figure 4 shows that the unit labor savings is positively correlated with the GDP growth, and the unit capital growth is negatively correlated with the capital growth in China. However, the GDP growth and $p - q$ are negative weak correlations (−0.3635), while the unit capital growth (0.9982) has significant positive correlations with $p - q$, or labor-augmenting technical change.

Table 5 summarizes the Granger causality test results. Columns (1) to (3) list estimates of the causality coefficient from capital growth to $p - q$ with the number of lagged differences ranging from one to three. Columns (4) to (6) list the coefficient from average wage to $p - q$. It shows that the growth of capital stock and average wage drive labor-augmenting technical change for long-run. Despite the growth of wages cause labor-augmentation, the relative change of factors' prices has not correlated with technical change direction in China.

5 Conclusions

Recent studies show that the economy goes with a purely labor-augmenting technical change in the long run as firms pursue profit maximization in the short run or long run process. From the perspective of labor productivity growth, however, the only

Table. 3 Statistical distribution of GDP and capital stock growth in China, Japan, the United States.

Statistic	Periods	N	Mean	St.Dev.	Min	Pctl(25)	Pctl(75)	Max
GDP G_China	1980~2018	39	9.444	2.826	3.918	7.793	11.087	15.192
Capital G_China	1980~2018	39	9.811	2.620	5.512	7.500	11.164	15.297
GDP G_Japan	1960~2019	59	3.754	3.834	- 5.416	0.970	5.528	11.953
Capital G_Japan	1960~2019	59	3.625	2.832	- 0.814	0.709	5.391	9.274
GDP G_US	1960~2019	59	3.059	2.038	- 2.537	2.068	4.400	7.237
Capital G_US	1960~2019	59	2.566	0.629	0.987	2.143	3.063	3.714

Notes: GDP and capital stock are constant 2010 price for China, 2015 price for Japan and the United States.
Source: made by the author

Table. 4 Pearson correlation test results in China

	$p - q$	$w - r$	GDP Growth	Capital Growth	Ave wage	Lending IR
$p - q$	1					
$w - r$	-0.3478	1				
GDP Growth	-0.3635	-0.4119	1			
Capital Growth	0.9982	-0.3857	-0.3372	1		
Ave wage	0.9050	-0.1974	-0.4075	0.8835	1	
Lending IR	-0.3647	0.4207	-0.4390	-0.3805	-0.2753	1

Notes: $w - r$ represents $\frac{r_{t-1}}{r_t} - \frac{w_{t-1}}{w_t}$
Source: made by the author

Table. 5 Granger causality test results in China, 1980–2018

$p - q$	Capital Growth			Average wage		
	(1)	(2)	(3)	(4)	(5)	(6)
Pearson's cor	0.9982	0.9982	0.9982	0.9050	0.9050	0.9050
Lags	-1	-2	-3	-1	-2	-3
F-test	24.024***	7.723**	4.569**	5.8482*	6.5267**	6.5655**
p-value	<0.001	0.001833	0.009706	0.02526	0.007883	0.005353
Obs.	37	36	35	22	21	20

Note: · p<0.1; *p<0.05; **p<0.01; ***p<0.001
Source: made by the author

labor-augmenting technical change will damage the growth of the economy. Equation (3), shows that to achieve labor productivity growth, the proportional rate of labor-augmenting technical change converges to labor output elasticity, and the proportional rate of capital-augmenting technical change converges to capital output elasticity. Both labor-augmenting and capital-augmenting technical change are essential to economic growth. When the proportion of labor-augmenting technical change is low, promoting labor-augmenting technical change will enhance economic growth. When the proportion of labor-augmenting technical change is too high, higher than labor output elasticity, promoting capital-augmenting technical change will enhance economic growth. It is essential for economic growth to properly guide the technical change.

From the perspective of the gap between unit labor savings and unit capital savings, the mainstreams of technical change are labor-augmenting technical change in China even it is a developing and labor-surplus economy. The growth rates of wages are high in China, while the lending interest rate is relatively stable. However, profit maximization is not the driven force that decides the direction of technical change in China. Policy-driven growth in fixed capital investment has promoted labor-augmenting technical change in China. Meanwhile, in recent years there are more capital-saving technical changes in Japan, as well as in Germany, Italy, and the United Kingdom. It means capital-augmenting technical change is going with labor-augmenting technical change in advanced economies.

An extensive literature shows that labor income shares in a majority of countries across the world is falling. It means that average wage growth has lagged behind average labor productivity growth. In recent decades, the financial market oppresses businesses to take care more to stockholders and ignore stakeholders, for example, employees. However, low wage growth will dampen household consumption and thus reduce aggregate demand. It is very critical that both capital and labor share the results of labor productivity improvement. The benefit of stockholders and the welfare of stakeholders should be balanced. A further study on the influence of the financial market in the direction of technological change is essential to economic growth.

References

1. Hicks, J.: The Theory of Wages. Macmillan, London (1932)
2. Kennedy, C.: Induced bias in innovation and the theory of distribution. Econ. J. **74**(295), 541–547 (1964). https://doi.org/10.2307/2228295
3. Goodwin, R.: A Growth Cycle. Socialism, capitalism and economic growth (1967)
4. Xinhua: China sees steady employment in January-July. China.org.cn, 14 August 2019 (2019). http://www.china.org.cn/business/2019-08/14/content_75098851.htm
5. ILO: Global Wage Report 2018/19: What lies behind gender pay gaps. Geneva: International Labour Office (2018). https://www.ilo.org/global/research/global-reports/global-wage-report/2018/lang--en/index.htm
6. ILO: Global Wage Report 2016/17: Wage inequality in the workplace. Geneva: International Labour Office (2016). https://www.ilo.org/global/research/global-reports/global-wage-report/2016/lang--en/index.htm
7. Manyika, J. et al.: A new look at the declining labor share of income in the United States. McKinsey Global Institute (2019). https://www.mckinsey.com/featured-insights/employment-and-growth/a-new-look-at-the-declining-labor-share-of-income-in-the-united-states
8. Huang, X., Xu, S.: Reasons for the decline of labor share—from the angle of labor-saving technical progress. Econ. Res. J. **7**, 34–44 (2009)

9. Acemoglu, D.: Directed Technical Change. NBER Working Paper Series (2001). https://doi.org/10.3386/w8287
10. Acemoglu, D.: Labor- and capital-augmenting technical change. J. Eur. Econ. Assoc. **1**(1), 1–37 (2003). https://doi.org/10.1162/154247603322256756
11. Acemoglu, D.: Introduction to Modern Economic Growth/Daron Acemoglu. Princeton University Press, Princeton (2009)
12. Acemoglu, D., Aghion, P., Bursztyn, L., Hemous, D.: The environment and directed technical change. Am. Econ. Rev. **102**(1), 131–166 (2012). https://doi.org/10.1257/aer.102.1.131
13. Irmen, A.: Capital- and labor-saving technical change in an aging economy. Int. Econ. Rev. **58**(1), 261–285 (2017). https://doi.org/10.1111/iere.12216
14. Irmen, A.: A generalized steady-state growth theorem. **22**(4), 779–804 (2018). https://doi.org/10.1017/S1365100516000407
15. Marquetti, A.: Do rising real wages increase the rate of labor-saving technical change? Some econometric evidence. Metroeconomica **55**(4), 432–441 (2004). https://doi.org/10.1111/j.1467-999X.2004.00201.x
16. Sasaki, H.: Classical biased technical change approach and its relevance to reality. Int. Rev. Appl. Econ. **22**(1), 77–91 (2008). https://doi.org/10.1080/02692170701745903
17. Basu, D.: Marx-biased technical change and the neoclassical view of income distribution. Metroeconomica **61**(4), 593–620 (2010). https://doi.org/10.1111/j.1467-999X.2009.04079.x
18. de Souza, J.P., A.: Real wages and labor-saving technical change: evidence from a panel of manufacturing industries in mature and labor-surplus economies. Int. Rev. Appl. Econ. **31**(2), 151–172 (2017). https://doi.org/10.1080/02692171.2016.1225017
19. NBSC: The rapid growth of investment in fixed assets strengthens the foundation for economic and social development: the eighth series of reports on economic and social development achievements in the 40 years of reform and opening up (2018). http://www.stats.gov.cn/ztjc/ztfx/ggkf40n/201809/t20180906_1621360.html
20. Xu, X., Wang, B., Xiong, F.: How China's fiscal policy contributes to investment and economic growth: experience and lessons. China Econ. (06), 4–17 (2013)

Mechanism of the Effect of Financialization on Economic Growth from the Perspective of Talent Flow

Keyan Li(✉)

95 Morgan St, Apt 10A, Stamford, CT, USA

Abstract. Since the 1980s, financial sectors began to expand rapidly and have a growing proportion in GDP which is referred as "financialization". Thus a lot of scholars had studied the relationship between financial development and economic growth. The study first discusses the effect of financial development and financialization on economic growth. And then, the paper discusses the effect of talent inflow in financial industry on economic growth. Finance industry attracts a lot of talents when financial has developed to a certain degree. Meanwhile, financial sectors will occupy relatively excessive talent so that the lack of talent in production industry will hinder the economic growth. The third section first showed the relationship between the talent inflow and financialization. They positively correlated at the beginning and then level off gradually with the increasing of talent inflow. This section also provides an inverted "U" relationship between the increment of talent in financial sectors and economic growth. The last section provides a summary and the shortcoming of the inverted "U" relationship. Moreover, this paper suggests that the government should strengthen financial regulation and limit financial innovation to prevent financial industry from occupying excessive talent.

Keywords: Finance · Financial development · Economic growth · Talent inflow

1 Background

1.1 Introduction of Financialization

In the context of economic globalization, the financial development becomes more and more important to the development of the real economy. By allocating capital to maximize the value of resources, the financial sectors can promote economic growth. On the other hand, the rapid development of the financial industry is not always conducive to economic development. It may lead to inefficient and unreasonable allocation of resources, thus hindering economic development.

Since the wave of financial deregulation in the 1980s, the financial sector has grown rapidly. Finance and economy gradually infiltrate into each other and act together as a whole. As the proportion of the financial sector in the national economy gradually

C. Yuan et al. (Eds.): *Proceedings of the 4th International Conference on Economic Management and Green Development*, AEPS, pp. 387–393, 2021.
https://doi.org/10.1007/978-981-16-5359-9_44

increased, the scale of financial assets expanded, and the profit of the financial sector soared. Thus the financial sectors has attracted a large number of highly educated and high-quality talents. Some scholars began to worry whether the expansion of the financial sector will hinder economic growth.

Due to the increasing financialization of economic relations, that is, the economic relations in the society are more and more manifested as bond, debt, equity, dividend, risk, insurance and so on; the financialization of social assets meaning that the proportion of total financial assets in GDP increasing gradually and the economic financialization becomes more and more obvious. Economic financialization refers to the increment of the ratio of the total value of monetary and non-monetary financial instruments to the total value of national wealth in a country's national economy. It is noteworthy that the rapid development of financial economy still has two sides.

This article discussed the promotion and obstruction of the economic financialization on the economic development from the perspective of talent structure. According to Foster, Magdoff and Sweezy (1966) came up with that the increasing role of finance which was a new form of capitalism was termed as "financialization" [1]. Krippner regarded the growing weight of finance in the American Economy as "financialization" [2]. He put forward that the financialization of the economy caused extreme wealth and income polarization in the US which was eroding the social bases of American democracy [2]. We can see financialization as an economic phenomenon of the increasing importance of finance, financial markets, and financial institutions to the working of the economy.

1.2 Effect of Financialization on Economic Growth

Most academics had a point that finance can improve the efficiency of resource allocation and lower the cost of finance so that it can support the development of economy. For instance, financial repression constrained the economic development in developing countries [3]. There was a significant and strong positive correlation between the development of financial intermediaries and economic growth in China—a developing country, which means that the development of financial intermediaries were likely to promote economic growth [4].

A large number of empirical researches using the increasing amount of data of related indexes with various econometric methods indicated that financialization were positively associated with economic growth. For instance, Bencivenga and Smith who developed an endogenous growth model with multiple assets found the financial intermediaries shifting the composition of savings toward capital and reducing socially unnecessary capital liquidation tended to promote growth [5]. Aghion et al. found the financial development result in the improvement of level of material capital accumulation and resource allocation efficiency [6].

However, there are also a few economists were worried about the disproportionate growth of finance like James Tobin referred to 'the casino aspect of our financial markets' [1]. Moderate financial development is necessary for the optimal economic growth while financial repression or financial excess will hinder the economic growth [7]. And until 2007, a growing number of economists began to pay close attention to 'financialization'.

With the rapid expansion of finance industry, a number of economists revisited the effect of financial development on economic growth especially after the subprime crisis and their understanding had changed a lot. One of the major factors to the subprime crisis was the soaring defaults in subprime housing mortgages and credit crunch caused by the collapse of a housing bubble. It was exactly a time that financial deregulation prevailed. Lack of financial regulation resulted in the rapid expansion of financial sector and financial innovation which significantly improved the level of risk taking so that systematic risk accumulated and financial crisis appeared [8]. More and more economist began to pay attention to the potential downsides of financial development.

By reexamining the empirical relationship between financial development and economic growth, Favara found there was no significant indication that financial development spurs economic growth and the relationship was weak even negative for some specifications [9]. There is also evidence indicates that at current levels of household and business credit further expansion slows growth instead of promoting [10]. Cecchetti and Kharroubi investigated how financial development affects aggregate productivity growth and focusing on advanced economies, they found that that a fast-growing financial sector is detrimental to aggregate productivity growth [11].

According to the findings "opposite" to previous cognition that finance facilitate economic growth, what can be said for sure is that financial development or "financialization" definitely wouldn't spur economic growth all the time. There is a threshold effect in the finance-growth relationship which means finance exerts different effects on economic growth on different level and more finance is not always better [12].

In addition, the enterprise-level empirical research showed the financialization of economy had significant negative impact on the investment rate of non-financial enterprises which reduced material capital investment [13]. As a result, it lowered the added value by production department.

We can't insist the point that financial development will be favourable to economic growth anymore in the present era. According to recent researchers, there must be a complicate relationship between finance and growth.

2 The Effect of Talent Inflow in Financial Industry on Economic Growth

The expansion of financial sector and high-paying jobs will certainly influence the structure and efficiency of human resource allocation. Because of the scarce human resource, the increase in financiers means high-quality talents flowed to financial sector must result in the decrease in entrepreneurs [14]. According to Goldin, from 1970s to 1990s, there was an obvious increase in the proportion of people working in financial sector who graduated from Harvard University [15]. In developed countries, financial deregulation leads to the increasing proportion of telent in financial sector while that decrease in production department and productivity become lower.

However, there are also some economists argue that the increasing number of people pursuing career in financial sector will facilitate financialization of economy and be a stimulus to the economic growth [16]. If financial development is beneficial to economic growth regardless of the level of financialization, we can conclude that the increasing share of finance in employment will be beneficial to economic growth as well because more workforce in financial sectors will improve efficiency of the industry. But this is just an ideal result before most economists reexamine the relationship between financialization and economic growth so that we can't say a wealth of talent flowing to financial sectors can help economic growth all the time.

Finance accounts for a higher share of GDP than of employment before 1940s and after 1980s and they have roughly the same growth tend [14]. Talent is the critical factor for all sectors. More talent in one industry means more efficiency and output of per unit capital. According to career choices of MIT graduates, Shu finds that the financial industries attract high-quality talent. In consequence, the increasing amount of talent brings about more effectively capital allocation for financial sectors [17]. In other words, financial sectors need talent while talent is a scarce resource and the problem is that the increase talent of one industry will cause the decrease in another industry. Due to the two sides of financialization, merely promoting or preventing the talent flow to financial industry doesn't work for current economy situation.

According to the diminishing marginal effect, assuming that all else equal and workforce input increases by the same amount continuously, the increment of the product provided will decrease after increasing to a certain value, then the marginal output of the workforce will decrease. Conversely, after the output increasing to the certain value, the same increment of output will need increasing talent input on the basis of the law of marginal cost. However, according to Philippon, instead of increasing returns to scale, financial sectors have constant returns to scale and the cost doesn't reduce with the expansion of scale [18]. Just by looking at the development of financial sectors without considering other factors we can find that there must be a most appropriate portion of talent input or else too little or excessive talent will lower the efficiency of finance industry.

The early development of finance can truly help economic growth by improving financing efficiency, for example, before subprime crisis, most studies showed that financial development was positively correlated with growth. In fact, subprime crisis was a warning sign of too much finance. Too much finance can even hinder economic development due to holding excessive economic resources which causes crowding out effect to real economy [18]. In the light of crowding out effect, the economic growth caused by increment of talent in financial sectors may be counteracted by the decrease of talent in production industry or even impede growth. Attracted by high-paid jobs and rapid expansion in financial industry, high-quality talent tends to work in finance so that financial development contributes to economic growth by improving efficiency of resources allocation. But at the same time, the production industry will lose part of the scarce resource—talent resource which are originally expected to flow to. With the continuously expansion, the gradually significant reduction of talent can even cause reduction in GDP.

3 An Inverted "U" Model of the Effect of Talent Inflow to Financial Sectors on Economic Growth

There may be an inverted 'U' relationship between talent inflow in financial sectors and economic growth. At the beginning when the level financiallization of economy have not reached a point after which financialization will be negatively correlated with economic growth, talent inflow can increase the output of financial sectors. But when financialization has already began to occupy excessive resources, it will hinder economic growth.

For the relationship between talent inflow and financialization which to some extent can represent the level of financial development, as Fig. 1 shows, it will be positively correlated at the beginning and then level off gradually with the increasing of talent inflow.

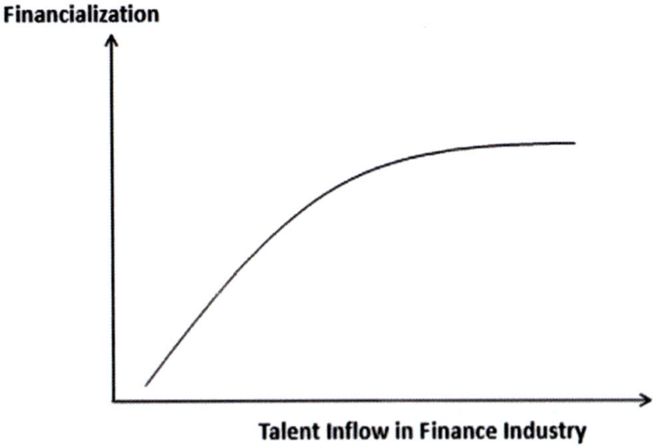

Fig. 1. The relationship between talent inflow in finance industry and financialization.

Given the relationship between financialization and economic growth, the increase of high-quality talent in financial sectors may have an irregular inverted "U" relationship with economic growth. As Fig. 2 shows, the increment of talent resource will definitely improve the efficiency of financial sectors, but the improvement is relatively less obvious than other industries due to the constant returns to scale. With the rising level of financialization, the economic growth gradually slow down after it has reached the top of the "inverted U". As mentioned earlier, the unit auxo-action from the unit of increasing financialization for growth will decrease. But during this period, financial sectors keep expanding which will attract a great number of high-quality talents and even occupy the talent resource for other industries. Later excessive financialization will hinder economic development. Thus, after the slower growth, financialization begins to have negative effect to the growth. At the same time, because talent resource is a kind of scarce resource, the lack of talent in production sectors will hinder economic growth. Therefore, a certain proportion of the number of high-quality talent, for instance, people with bachelor degree or above work in financial sectors need to be found to maximize the positive effect of finance on economic growth.

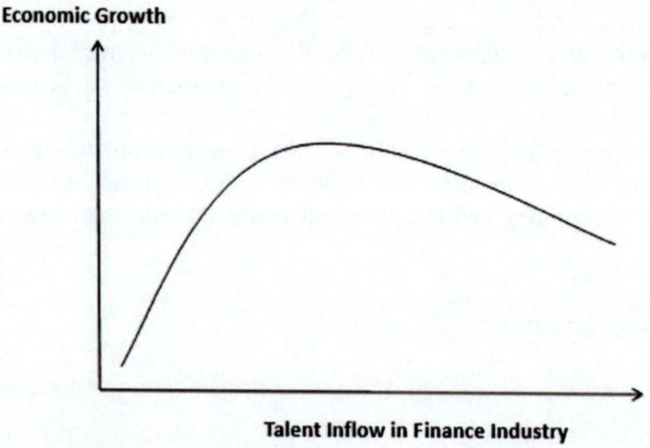

Fig. 2. The relationship between talent in flow in finance industry and economic growth.

4 Conclusions

This paper discussed the relationship between talent inflow in financial sectors and economic growth and indicate an inverted "U" relationship between them. Through discussing how the level of financialization react with talent inflow to indicate the relationship between the inflow and economic growth. More talent in production sector means higher efficiency of output and more talent in financial industry means higher efficiency of capital allocation. But excessive talent inflow in financial sectors will make talent resource in production industry which can hinder economic growth. In the future, exploring a proportion of talent in financial sectors which can maximize the positive effect of financialization on economic growth is necessary. This is also the shortcoming of this paper because lack of empirical data to determine a best proportion. Scholars have some discussion about it, but there is no direct proof of the "inverted U" relationship. In addition, some scholars also showed that when the financial sectors boom, financial institutions draw high-quality talents away from the regulator sectors and equilibrium misbehavior by financial workers increases [19]. This will hinder financial development at first which may indirectly hinder economic growth. Thus, excessive talent inflow to financial sectors should be evaded.

The appropriate proportion of talent inflow to financial sectors is a topic of considerable importance. Alvarez concluded the potential relationship between dependence on finance and wage share which shows increased dependence on financial profits is likely to decrease wage share in non-financial corporations [20]. In other word, the decreased wage share is also likely to hinder the talent inflow to production sectors. In this sense, the government should tighten financial regulation and limit excessive financial innovation lacking real value. At the same time, it is necessary to encourage credit to flow to real economy to sustain long-term economic development. Moreover, financial regulation should be attached more importance. The high profits of financial sectors attracted excessive talent, the managers tend to engage in speculation. In order to ensure appropriate proportion of talent in financial sectors, improving regulation system to increase transparency and limit exceptionally high profit are necessary to prevent the financial sectors from occupying superfluous talent resource.

References

1. Foster, J.B.: The financialization of capitalism. Mon. Rev. **58**(11), 1–12 (2007)
2. Krippner, G.R.: The financialization of the American economy. Soc. Econ. Rev. **3**(2), 173–208 (2005)
3. Galor, O., Zeira, J.: Income distribution and macroeconomics. Rev. Econ. Stud. **60**(1), 35–52 (1993)
4. Tan, R.: Empirical research on the relationship between financial development and economic growth in China. Econ. Res. **10**(15), 337–363 (1999)
5. Bencivenga, V.R., Smith, B.D.: Financial intermediation and endogenous growth. Rev. Econ. Stud. **58**(2), 195–209 (1991)
6. Aghion, P., Howitt, P., Mayer-Foulkes, D.: The effect of financial development on convergence: theory and evidence. Q. J. Econ. **120**(1), 173–222 (2005)
7. Mi, J., Li, J.: Theoretical thinking and empirical analysis on the relationship between financial development and economic growth in China. Manage. World **4**, 23–30 (2002)
8. You, H.: Macroeconomic impact of financial development: from the perspective of economic financialization. Central University of Finance and Economics, Diss (2017)
9. Favara, G.: An empirical reassessment of the relationship between finance and growth (2003)
10. Cournède, B., Oliver, D., Peter, H.: Finance and inclusive growth (2015)
11. Stephen, G.C., Enisse K.: Reassessing the impact of finance on growth (2012)
12. Law, S.K., Singh. N.: Does too much finance harm economic growth? J. Banking Financ. **41**, 36–44 (2014)
13. Tomaskovic-Devey, D., Lin, K.-H.: Income dynamics, economic rents, and the financialization of the US economy. Am. Sociol. Rev. **76**(4), 538–559 (2011)
14. Shakhnov, K.: The allocation of talent: finance versus entrepreneurship. Available at SSRN 3109910 (2017)
15. Goldin, C., Katz, L.F.: Transitions: career and family life cycles of the educational elite. Am. Econ. Rev. **98**(2), 363–369 (2008)
16. Ren, Y.: An Empirical Study on the Influence mechanism of Scientific and Technological Innovation on The Financialization of China's economy [A]. Institute of Emerging Economies, China Center for International Cultural Exchange, Guangdong University of Technology. Proceedings of the 2018 Annual Conference of the Emerging Economies Institute and the 6th Emerging Economies Forum on the Community with a Shared Future for Mankind (Ii) [C].Institute of Emerging Economies, China Center for International Cultural Exchanges, Guangdong University of Technology: Guangdong Institute of Emerging Economies, vol. 13 (2018)
17. Shu, P.: Innovating in science and engineering or'Cashing In'on Wall Street? Evidence on Elite STEM Talent. Harvard Business School Technology & Operations Mgt. Unit Working Paper 16–067 (2016)
18. Philippon, T.: Has the US finance industry become less efficient? (2011)
19. Bond, P., Glode, V.: The labor market for bankers and regulators. Rev. Financ. Stud. **27**(9), 2539–2579 (2014)
20. Alvarez, I.: Financialization, non-financial corporations and income inequality: the case of France. Soc. Econ. Rev. **13**(3), 449–475 (2015)

Research on the Transmission Mechanism of Land Mortgage Scale to the Rise of Commercial Housing Price—Take the Data of Prefecture Level Cities in China as an Example

Ziyan Li(✉)

School of Humanities and Social Sciences, Beijing Institute of Technology, Liang Xiang Campus, Fang Shan District, Beijing, China

Abstract. The economic development of the real estate industry in recent years has been a good economic performance. As the transmission signal of real estate economic development trend, house price ushered in a development peak in 2015. The rise of house prices has always been a hot research topic. Previous studies mostly focus on the relationship between supply and demand, macro tax and other aspects, but less from the perspective of land mortgage. This paper selects the land mortgage and commercial housing price data of 297 prefecture level cities to establish the fixed effect model, and finds that the expansion of land mortgage scale will lead to the rise of commercial housing price. As the core force of macro-control, the government should strengthen the supervision of the land mortgage market, strictly control and improve the access threshold; The government should release and develop the land mortgaged by the government and the land not opened by the developers as soon as possible, so as to alleviate the soaring house prices from the supply side. At the same time, in the process of regulation and control, we should also pay close attention to the housing demand, guide the housing price to return to rationality, try to meet the housing demand of the regression due to high housing price, and improve people's livelihood.

Keywords: Land mortgage · House price · Panel data

1 Introduction

As an important driving industry of China's economic development, real estate has been widely concerned by all walks of life. Although the rapid development of real estate has been normalized and the growth rate is slowing down, and the government has always implemented the policy concept of "the house is for living, not for speculation", but in the past few years, the development of real estate is still at a high level. Since 2015, the real estate market ushered in a small peak: in 2015, the national real estate development investment was 9597.9 billion yuan, a nominal increase of 1.0% over the previous year.

C. Yuan et al. (Eds.): *Proceedings of the 4th International Conference on Economic Management and Green Development*, AEPS, pp. 394–401, 2021.
https://doi.org/10.1007/978-981-16-5359-9_45

However, by 2016, the national real estate development investment reached 10258.1 billion yuan, a nominal increase of 6.9% over the previous year, and has maintained a high growth rate since then. Until 2019, national real estate development investment is 13219.4 billion yuan, an increase of 9.9% over the previous year. Real estate investment has been expanding, on the other hand, the demand for real estate is not weak at the initial stage: in 2016, the sales area of commercial housing reached 157349 million square meters, an increase of 22.5% over the previous year. With the double support of investment and demand, house prices continued to rise, and the growth rate of sales area in the following years slowed down, and the sales area of commercial housing in 2019 decreased by 0.1% compared with the previous year.

House prices are also rising at a faster pace. The average price of commercial housing sales in China was 6324 yuan in 2014 and 6793 yuan in 2015. Only one year later, the average price of national commercial housing sales jumped to 7476 yuan in 2016 and 8726 yuan in 2018.

Another element that accompanies the house is land. The value of land itself is directly related to the house price. Due to the public ownership system of land with Chinese characteristics, land mortgage has become the main way of land appreciation. As early as 2013, there were 403900 hectares of land in 84 key cities nationwide, with a total mortgage loan of 7.76 trillion yuan. In 2013, 367000 hectares of state-owned construction land were transferred nationwide, with the contract price of 4.20 trillion yuan, increasing by 13.7% and 56.3% respectively. Thus, the scale of land mortgage is gradually expanding.

Real estate, as an industry which is widely concerned by the public, is also an important pillar industry of China's economic development. As the core element of the industry, house price plays an important role in the whole economy. Therefore, from the perspective of land mortgage scale, it is very necessary to study the path and transmission mechanism of real estate price rising.

2 Literature Review

For the problem of high real estate prices, many scholars have summarized it from the perspectives of school district housing, geographical location, infrastructure construction and so on. Zhang et al. (2020) believed that the consistent education policy would still lead to the rise of housing prices, that is, the more educational resources attached to the house, the higher the price; Zhang et al. (2018) proposed that the high housing price in school districts actually covered the cost of choosing high-quality schools; Gu and Zhou (2015) took Wuhan as an example to explore that the spatial distribution of housing prices in Wuhan is the result of the comprehensive effect of urban planning, geographical location, environmental landscape and housing attributes [1]. Most of these studies are inseparable from the basic analysis framework of the economic principle that supply and demand affect prices. There are also scholars from the perspective of real estate tax analysis, from the macro perspective of the government to explore real estate prices. Sun and Cui (2018) found that personal housing property tax will reduce the house price, but also reduce the housing demand of families [2]. Among them, there are many researches on the real estate price from the perspective of land, According to the analysis of land market by Zhang et al. (2019), China's local economic development and large-scale urban construction rely heavily on the land market, which land acts as a lever.

However, there is little research on the perspective of land mortgage, and to study land mortgage, it is necessary to analyze the land system. China's land system has a strong socialist color. The socialist system with Chinese characteristics determines the implementation of public ownership of land and the ownership of land by the state. Fu Hongcai et al. (2018) divided the three rights of land into ownership, contracting right and management right. The state holds the ownership of the land, which provides a fundamental possibility for the government to mortgage land and obtain financing. In essence, the financing channel of local finance not only has the support of land system, that is, a large number of land resource reserves in the hands of the government, but also has the guiding factors on administrative guidance. When exploring the interaction mechanism between land transfer and land price, Zheng et al. (2014) pointed out that in the early stage of economic development and market transformation, the GDP oriented local assessment mechanism prompted local governments to find out the breakthrough point of economic development by expanding urban construction in the face of pressure, thus turning a large number of land resources inventory into financing tools. Similarly, not only the government, but also with the continuous development and improvement of the financial market and the continuous relaxation of policies, the influence of land mortgage in other areas is also increasing. For example, under the development goal of Inclusive Finance, China's rural land contractual management right mortgage loan reform has brought timely financial support to rural small farmers, and farmers themselves have high enthusiasm for the financial product model (Li and Luo 2015).

3 Research Methods and Data

3.1 Data Source and Model Establishment

The data used in this paper comes from the database "city annual database of Guoxin real estate information network". In the process of data collection, the statistical yearbooks and Statistical Bulletins of 297 prefecture level cities are selected in the period of 1999–2015. On the basis of collecting widely used macroeconomic indicators, great emphasis is also placed on the data of land mortgage and real estate prices. Finally, in terms of land mortgage, three indicators are collected: land mortgage loan, land mortgage area and land mortgage valuation; and the most representative commodity housing price is selected for real estate price.

The following equation is constructed to investigate whether the land mortgage has a transmission mechanism to the price rise of commercial housing in prefecture level cities.

$$\ln poch_{it} = \alpha_0 + \alpha_1 \ln lml_{it} + \alpha_2 \ln grp_{it} + \alpha_3 \ln nrrp_{it} + \beta_0 findper_{it} + \beta_1 sindper_{it} + \mu_{it} + \varepsilon_{it} \tag{1}$$

$$\ln poch_{it} = \alpha_0 + \alpha_1 \ln lma_{it} + \alpha_2 \ln grp_{it} + \alpha_3 \ln nrrp_{it} + \beta_0 findper_{it} + \beta_1 sindper_{it} + \mu_{it} + \varepsilon_{it} \tag{2}$$

$$\ln poch_{it} = \alpha_0 + \alpha_1 \ln lmv_{it} + \alpha_2 \ln grp_{it} + \alpha_3 \ln nrrp_{it} + \beta_0 findper_{it} + \beta_1 sindper_{it} + \mu_{it} + \varepsilon_{it} \tag{3}$$

Among them, *poch* represents the price of commodity housing in prefecture level cities, *lml* indicates the local land mortgage loan, *lma* indicates the land mortgage area, *lmv* expresses the value of the land mortgage, *grp* indicates the gross domestic product, *nrrp* indicates the number of registered residence population, *findper* represents the proportion of the primary industry, *sindper* represents the proportion of the second industry, μ indicates the individual fixed effect of the various cities, and ε indicates random disturbance. i and t in each variable indicate the corresponding prefecture level city and year respectively. $\alpha_1 \sim \alpha_3, \beta_0$ and β_1 are regression coefficients.α_0 is a constant term.

In the process of data preprocessing, it is found that the data of commercial housing price, land mortgage loan, land mortgage area and land mortgage valuation are skewness. Therefore, logarithmic processing is carried out. At the same time, in the process of analyzing data, the proportion of gross domestic product and household registration population is fixed, and the proportion between primary industry and the second industry is added to improve the actual economic relevance of the study.

4 Result Analysis

4.1 Descriptive Analysis of Scatter Plot

Firstly, the logarithm variables are described and analyzed by scatter plot. Taking $\ln lml, \ln lma$ *and* $\ln lmv$ as independent variables and $\ln poch$ as dependent variables, three scatter plots were generated by SPSS software (Figs. 1, 2 and 3).

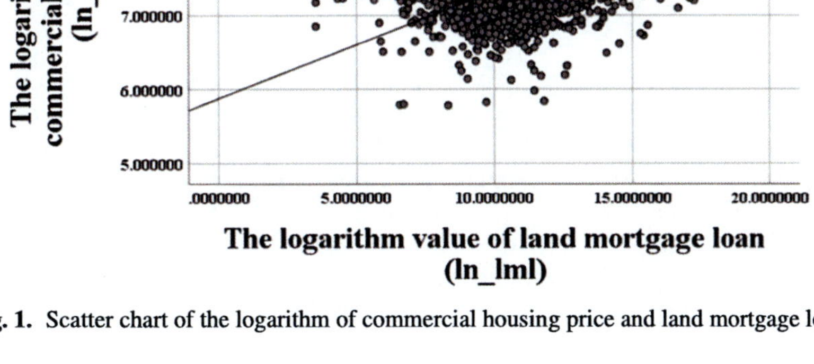

Fig. 1. Scatter chart of the logarithm of commercial housing price and land mortgage loan.

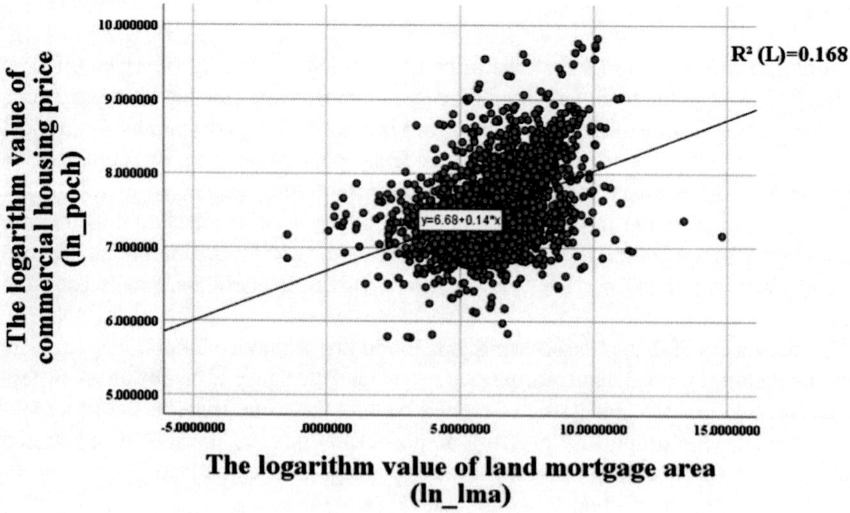

Fig. 2. Scatter chart of the logarithm of commercial housing price and land mortgage area.

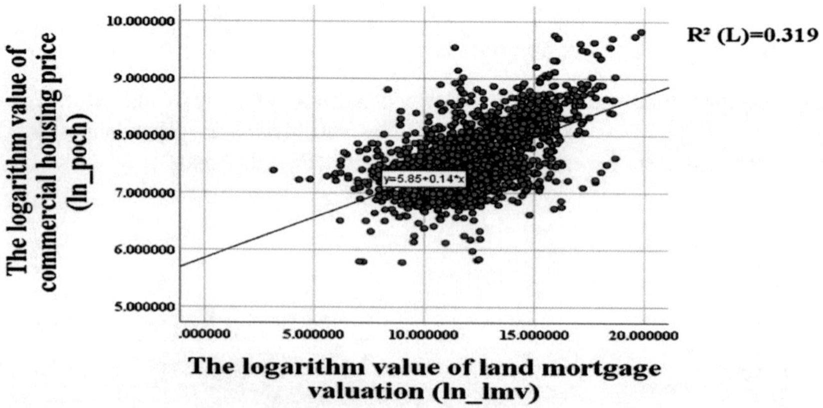

Fig. 3. Scatter chart of the logarithm of commercial housing price and land mortgage valuation.

The abscissa of scatter chart is the logarithm value of land mortgage loan, land mortgage area and land mortgage valuation respectively. The ordinate is the logarithmic value of commercial housing price. The slope of regression trend line is about 0.15. Through the analysis of scatter diagram, it is obvious that there is a certain correlation between land mortgage and the price of commercial housing in prefecture level cities.

Then, from the logic theory level, we can roughly infer the transmission mechanism behind it: land mortgage brings capital inflow to investors, and then drives the construction of land infrastructure. After the completion of construction and supporting measures, the added value of land will increase. As an important guide of house price, the improvement of land value will inevitably lead to the increase of house price. Of course, to get such a transmission mechanism, we need further scientific test.

4.2 Regression Analysis Based on Fixed Effect Model

Although the scatter plot shows a relevant trend, the simple regression analysis involving only two variables lacks rigor, and the factors influencing the real estate price are far more than that, and the goodness of fit R2 is not ideal on the whole. Therefore, in order to explore the transmission mechanism more scientifically, this paper adopts a fixed effect model to fix the two variables that may affect the price of real estate and the household registration population. After that, the above three equations are analyzed and the results are as follows (Tables 1, 2 and 3).

Table 1. Regression results of commercial housing price with land mortgage loan as core variable.

ln_poch	Coef.	Std. Err.	t	P>|t|	[95% Conf.	Interval]
ln_lml	.0205285	.0046027	4.46	0.000	.0115014	.0295557
ln_grp	.7019874	.0167596	41.89	0.000	.669117	.7348578
ln_nrrp	-.1065873	.0920309	-1.16	0.247	-.2870862	.0739116
findper	.0032395	.0026356	1.23	0.219	-.0019298	.0084088
sindper	-.0072092	.0015262	-4.72	0.000	-.0102025	-.0042158
_cons	3.864203	.5146899	7.51	0.000	2.854749	4.873658
sigma_u	.4185447					
sigma_e	.19591193					
rho	.82027871	(fraction of variance due to u_i)				

F test that all u_i=0: F(282, 1795) = 9.48 Prob > F = 0.0000

Table 2. Regression results of commercial housing price with land mortgage area as core variable.

ln_poch	Coef.	Std. Err.	t	P>|t|	[95% Conf.	Interval]
ln_lma	.0072729	.0046278	1.57	0.116	-.0018036	.0163494
ln_grp	.7270044	.0149409	48.66	0.000	.6977012	.7563076
ln_nrrp	-.1595029	.0837489	-1.90	0.057	-.3237572	.0047514
findper	.0042454	.002651	1.60	0.109	-.000954	.0094447
sindper	-.0066215	.0015215	-4.35	0.000	-.0096056	-.0036374
_cons	4.164273	.4640704	8.97	0.000	3.254105	5.074441
sigma_u	.40145429					
sigma_e	.19755308					
rho	.8050518	(fraction of variance due to u_i)				

F test that all u_i=0: F(282, 1816) = 9.84 Prob > F = 0.0000

The Coef. in the table represents the regression coefficient of the variable. Std.err represents the standard error, and the probability values of t and P jointly measure whether the regression coefficient is significant. From the above table, we can draw a conclusion

Table 3. Regression results of commercial housing price with land mortgage valuation as core variable.

ln_poch	Coef.	Std. Err.	t	P>|t|	[95% Conf.	Interval]
ln_lmv	.0097342	.0044617	2.18	0.029	.0009828	.0184856
ln_grp	.7387849	.0182733	40.43	0.000	.7029424	.7746274
ln_nrrp	-.2111825	.0860698	-2.45	0.014	-.380006	-.042359
findper	.009516	.0028495	3.34	0.001	.0039267	.0151052
sindper	-.0038793	.0015811	-2.45	0.014	-.0069805	-.0007781
_cons	4.104493	.4775116	8.60	0.000	3.167867	5.041119
sigma_u	.38278405					
sigma_e	.18986705					
rho	.80254764	(fraction of variance due to u_i)				

F test that all u_i=0: F(282, 1574) = 9.14 Prob > F = 0.0000

that after the two variables of fixed local gross domestic product and household registered population, the core explanatory variables of land mortgage have basically significant impact on the price of local commercial housing. With the regression coefficient is 0.0205, the tail probability of land mortgage loan is 0, showing a strong significance and a positive impact on the price of commercial housing; The tail probability of land mortgage valuation is 0.029, which also has the significance. The influence to the commercial housing is positive, and the regression coefficient is 0.01. In spite of the influence of land mortgage area on the price of commercial housing is not significant, the basic conduction direction is consistent with the direction described in the scatter diagram. Therefore, we can basically conclude that the expansion of land mortgage scale has indeed caused the rise of commercial housing prices to a certain extent.

In addition, it can be seen from the table that the proportion of the secondary industry has a significant negative correlation with the housing price, and the regression coefficients are generally between 0.004 and 0.007. That is to say, the greater the proportion of the secondary industry in the local total output value, to a certain extent, the higher the degree of local industrialization, the less demand for labor force, and the relative reduction of population. Therefore, from the macro perspective, the overall demand for commercial housing is low, which leads to the decrease of local commercial housing prices.

5 Conclusions and Policy Recommendations

It can be seen that land mortgage plays an important role in the promotion of housing prices. There are many logical transmission mechanisms behind it. From the perspective of government entities, in the market environment of pursuing rapid economic development in the early years, the GDP oriented performance evaluation standard pushed the government to the road of exchanging land resource reserve for local economic development. A large area of land was mortgaged for development, so the land added value increased, and the price of commercial housing rose naturally.

In order to stabilize the price of commercial housing and regulate the real estate market, the government should pay enough attention to the land mortgage market. The expansion of the scale of land mortgage undoubtedly intensifies the risk of the land market to a certain extent, and the investment return cycle of the real estate market is long. If the mortgage chain breaks in the process, it will bring huge losses to the mortgagor; At the same time, with the rising prices of commercial housing, although housing demand is rigid demand, for most people, commercial housing is as consumer goods rather than investment goods, that is, normal goods rather than giffen goods. With the rising prices, the substitution effect of commercial housing is greater than the income effect, and people are more likely to transfer to other markets (such as the rental housing market). As is shown in the above data in 2019, the sales area of commercial housing has decreased slightly. The government should strengthen the supervision of the land mortgage market, strictly control and improve the access threshold; For the land mortgaged by the government itself through the financing platform and the undeveloped land in the hands of developers should be released and developed as soon as possible, so as to alleviate the soaring house prices from the supply side. In the process of regulation and control, government should also pay close attention to the housing demand, implement more preferential tax policies or relief policies, and try to meet the housing demand of the regression due to high housing prices, so as to truly achieve the purpose of improving people's livelihood.

References

1. Xing, G., Liqing, Z.: Spatial differentiation of housing prices and its influencing factors in Wuhan Based on geographically weighted regression. Land Nat. Resour. Res. **03**, 63–68 (2015)
2. Shaoqin, S., Jun, C.: Economic effect of personal housing property tax: theoretical discussion and did empirical test – based on the perspective of local economic entities. J. Central Univ. Financ. Econ. **09**, 12–21 (2018)
3. Bo, Z., Yiding, W., Jiangtao, L.: Will the nine-year admission policy increase housing prices in school districts? – Based on double difference analysis. Educ. Econ. **36**(04), 40–49 (2020)
4. Zhang, K., Zhang, L., Zhu, D.: The impact of urban basic education resources on housing prices: a case study of Haidian District, Beijing . Educ. Econ. 2018(01), 27–34 + 96 (2018)
5. Zhang, L., Wei, H., Ou, D.: Land financing, local debt and leverage: an analysis of land mortgage of local financing platform. Financ. Res. **2019**(03), 92–110 (2019)
6. Fu, H., Gan, Y., Li, R., Fu, M.: Analysis on the reform of Land Limited and limited differential management system in China. Econ. Res. Guide **2018**(29), 24–27 + 32 (2018)
7. Zheng, S., Sun, W., Wu, J., Wu, F.: Research on investment and financing mode of urban construction with Chinese characteristics. Econ. Res. **49**(08), 14–27 (2014)
8. Tao, L., Jianchao, L.: Behavior response of farmers' mortgage loan of land contractual management right: a micro empirical study based on Poisson hurdle model. Manage. World **07**, 54–70 (2015)
9. Yuanchun, L., Jinzhi, C.: Land system, financing mode and industrialization with Chinese characteristics. China Ind. Econ. **03**, 5–23 (2020)
10. Zhou, L.: Easy and healthy. rising house prices, household debt and urban residents' Consumption: a loan to value ratio perspective. Chinese Manage. Sci. 1–12 (2020)
11. Tang, A.: The impact of house price fluctuation on systemic financial risk. Shanghai Academy of Social Sciences (2018)
12. Ran, T., Fei, Y., Guangzhong, C.: Regional competition, land transfer and local fiscal effect: an analysis based on the panel data of Chinese prefecture level cities from 1999 to 2003. World Econ. **10**, 15–27 (2007)

Ability to Absorb Knowledge and Endogenous Economic Growth: Expansion of Romer Economic Model

Zhiyuan Zhu(✉)

Department of Math and Statistics, American University, 4400 Massachusetts Ave NW, Washington, DC, USA

Abstract. Above the solid base founded by Alfred Marshall, Robert Solow, Paul Romer, and other famous economists around the world, this article focuses on exploring new aspects of economic growth based on the Romer Economic Growth Model. By adding two new variables, the heterogeneity and the ability to absorb knowledge, into the model, which are two main conditions that allows the spillover effect of knowledge, this article realizes that such an effect is, in fact, a mutual communication between industries, in which they absorb knowledge from each other. In order to set up a new model based on the Romer Economic Growth Model, this article imports the two variables into the model, and finally achieve a new endogenous economic growth model. Theoretically, this will be a more inclusive economic growth model: Romer Economic Growth Model, Solow Growth Model, and AK Model are all special situations of this model. Mathematically, since this model emphasizes the impact of ability to absorb knowledge in the economic growth, comparing to previous economic growth models, this one will hold a stronger point in explaining endogenous economic growth.

Keywords: Romer economic model · Endogenous economic growth

1 Introduction

Seeking the origin of economic growth is one of the most archaic and crucial problems in Economics in the history of the globe. In the 17th century, William Petty had proposed that "Labour is the Father and active principle of wealth, as lands are the Mother." Afterward, Adam Smith labeled "capital" as a new factor of production. With the growth of the modern economy, knowledge and innovation started to be noticed as essential roles in economic growth, which indicated new standards of theories in explaining how this growth happened continuously. A large number of famous economists around the world, such as Kenneth Arrow, Nicholas Kaldor, Robert Solow, and Hirofumi Uzawa, had emphasized that knowledge, technique, and innovation represent the major elements in pushing modern economic growth [1–4]. However, what they built in their models, normally, illustrates that the improvement of technique and the acquisition of new knowledge are exogenous, which means they happen naturally. That actually limited the most

C. Yuan et al. (Eds.): *Proceedings of the 4th International Conference on Economic Management and Green Development*, AEPS, pp. 402–407, 2021.
https://doi.org/10.1007/978-981-16-5359-9_46

tremendous momentum of economic growth into a box. Whereas, Paul Romer (1986), the winner of the Nobel Memorial Prize in Economics, thought differently [5]. In his model, the production of new knowledge and the activity of innovation are nominated as variables, explained how market mechanism and economic policies spur the forward research and innovation in enterprises, which, in the deepest degree, gave the problem a reasonable explanation in the modern world. Romer's theory is still called "the Endogenous growth theory" (1998) [6].

The dimensions of society, such as culture, politics, laws, etc., can be varied from country to country, it seems impossible for Romer and all successors to create a single model to explain the world. Therefore, Romer pinpointed knowledge as his variable to have further research. The best part of Romer's Endogenous growth model is to endogenesis the knowledge and technique innovation, which excluded the intervention of government policies. While Romer believes that since new knowledge cannot be rigged by a single company, which means when new knowledge has been produced, all other enterprises can learn the knowledge totally and use it for their own good to create profit. However, is gaining new knowledge as easy as it tells us? Therefore, this article will dig more into knowledge, based on Romer's Endogenous growth model.

Experiences of years of studying tell the story that learning consumes time and wealth. Riches can always acquire better education sooner and faster than the poor. Such a theorem can be applied in the economic growth that abilities to learn, absorb, and transform knowledge of different companies are different. For example, the management philosophy of large enterprises and small enterprises are different. While a large company gives birth to a new management philosophy, which could be regarded as new knowledge in the economic system, the ability of small enterprises to learn is tiny. Whatsmore, Romer's Endogenous growth model suggests that the economy of large countries will always develop faster than small countries, which is challenged by the fact that large countries, such as India, are not necessarily growing faster, now. The purpose of this article is to import two new variables about knowledge based on Romer's Endogenous growth model to make the model more convincing and explanatory.

However, limitations of the Romer Economic Growth Model is amplified due to the progress of modern economic system, which can be described as two major points: (1) it had not detailed the necessary preconditions of the spillover effect of knowledge; (2) though it was not obvious yet in 1986, now we can find a dilemma that, according to Romer, large countries will always be able to develop faster than small countries; however, it can still not be able to explain why large countries like India are not necessarily develop faster than small countries [7]. In this article, two new variables (preconditions) will be imported to improve the Romer Economic Growth Model. After all, the Endogenous Growth Model will be more convincible and explanatory.

2 The Romer Economic Growth Model

The Romer Economic Growth Model hypothesizes that the production function of the identical industry can be contributed by the private knowledge k, the total knowledge that a society obtains K, and other inputs x. First of all, Romer supposed that no other inputs other than knowledge growth, which can be refer to $x = \bar{x}$. Therefore, the production function of the industry can be represented by:

$$y = f(k, K) = F(k, K, \overline{x}) \tag{1}$$

In the function, F indicates that for k and x, the returns to scale remains unchanged, and for k and K, it's a progressive increase model for return. In the Romer Economic Growth Model, the total knowledge of society K measures the externalities of knowledge, or technique, to the contribution of output. Therefore, when private industries are deciding about new investments, they do not have to consider K as a changing variable, which means K(t), as the level of the total knowledge, is fixed. However, for social planners, such as politicians, will need to consider the change of K and how its internality and externality will affect the output when deciding on new investment. For these social planners, the total knowledge of society K is the sum total of knowledge possessed by each individual industry in the given society, which can be figured by $K = sk$. There are some constraint conditions for the increasing rate of knowledge of the private industry, which can be referred to as the relationship between the total investment for producing new knowledge l and knowledge production $\dot{k}$. The rate of increase in private industry's knowledge possession is the ratio between the private industry's investments and the private industry's knowledge possession, which can be expressed by:

$$\frac{\dot{k}}{k} = g\left(\frac{l}{k}\right) \tag{2}$$

In this function, g is an increasing function which has an upper limit γ, $g(0) = 0$, $g^{'}(x) > 0$, $g^{''}(x) < 0$. We can have the conclusion that the profits of producing new knowledge is decreasing.

Under this circumstance, the artificial planning $P_{\infty}(K)$ and social planning PS_{∞} can be expressed by:

$$P_{\infty}(K){:} \quad max \int_0^{\infty} u(c(t))e^{-\delta t}dt$$

$$s.t. \frac{\dot{k}}{k} = g\left(\frac{l}{k}\right) = g\left(\frac{[f(k, sk) - c]}{k}\right) \tag{3}$$

$$PS_{\infty}{:} \quad max \int_0^{\infty} u(c(t))e^{-\delta t}dt$$

$$s.t. \frac{\dot{k}}{k} = g\left(\frac{l}{k}\right) = g(\frac{[f(k, K) - c]}{k}) \tag{4}$$

Romer explained the beingness of these two questions and gave an assumption that competitive equilibrium and social optimum are inconsistent. Competitive equilibrium is a condition of social suboptimum, since all the competitive industries in the market regard K(t) and price of set values and the private marginal product of knowledge is $\frac{\partial}{\partial k} f(k, K)$. However, for social planners, the social marginal product of knowledge is $\frac{\partial}{\partial k} f(k, K) + s * \frac{\partial}{\partial K} f(k, K)$. Since the private marginal product of knowledge is smaller than the social marginal product of knowledge, the producers will choose a knowledge production that is lower than the social optimum level, which will eventually lead to the increasing rate of competitive equilibrium being smaller than the increasing rate of social optimum [8]. Therefore, Romer stood steady on the side of government intervention.

3 Improved Model

Because of the heterogeneity of k, the total knowledge that society obtains K cannot be simply equal to sk; instead, it should be related to the heterogeneity degree of the information of knowledge each industry possesses. Moreover, the spillover effect of the total social knowledge should be related to the ability that each industry absorbs different kinds of knowledge, such as management and techniques. Depend on all discussed above, we can set a more realistic expression about the total social knowledge K:

$$K = (sk)^{\alpha\beta} \tag{5}$$

In the expression, $0 \leq \alpha \leq 1$, $0 \leq \beta \leq 1$, α represents the ability that an industry absorbs specific new knowledge, which is largely related to the level of education, and β represents the diversity of the new knowledge (in other words, how different is the new knowledge compare to the existing ones). If one of α and β, or both, equals 0, which means, if the knowledge obtains by different industries remains consistent and/or the industry's ability to absorb new knowledge is none, the knowledge will not experience spillover effect. On the other hand, if and only if $\alpha = 1$ and $\beta = 1$, we can conclude that knowledge has a total spillover effect. When $0 < \alpha < 1$ and/or $0 < \beta < 1$, the knowledge has a partial spillover effect.

After the modification from $K = sk$ to $K = (sk)^{\alpha\beta}$, the new artificial planning $P_\infty(K)$ and social planning PS_∞ can be expressed by:

$$P_\infty(K): \quad max\int_0^\infty u(c(t))e^{-\delta t}dt$$

$$s.t.\frac{\dot{k}}{k} = g\left(\frac{l}{k}\right) = g(\frac{[f\left(k, (sk)^{\alpha\beta}\right) - c]}{k}) \tag{6}$$

$$PS_\infty: \quad max\int_0^\infty u(c(t))e^{-\delta t}dt$$

$$s.t.\frac{\dot{k}}{k} = g\left(\frac{l}{k}\right) = g\left(\frac{[f(k, K) - c]}{k}\right) \tag{7}$$

Theorem 1: The existence of solutions in social planning situation. We suppose that U, f, and g are real continuous functions, and U and g are concave functions. If $f\left(k, (sk)^{\alpha\beta}\right) \leq \mu + k^\rho$, $\rho > 0$, $0 \leq g(x) \leq \gamma$, and μ,ρ,γ are real numbers. If $\gamma\rho$ is smaller than the discount rate δ, and the value of $\alpha\beta$ is equal to the critical value, which leads to $k \to \infty$, the marginal profit of $f\left(k, (sk)^{\alpha\beta}\right)$ about k is a constant $m > \delta$, or when the value of $\alpha\beta$ is larger than the critical value, PS_∞ has the equilibrium solution of the increasing rate over 0. When the value of $\alpha\beta$ is equal to 0 or the value of $\alpha\beta$ cannot exceed the critical value, which makes the marginal profit of $f\left(k, (sk)^{\alpha\beta}\right)$ about k decreases, the equilibrium increasing rate of the social optimum is going to be 0.

Because of $f\left(k, (sk)^{\alpha\beta}\right) \leq \mu + k^\rho$ and the upper limit of $g(x)$ is γ, the upper limit of the increasing rate of spending c is $\gamma\rho$, $\gamma\rho$ smaller than the discount rate δ will guarantee the existence of finite solutions.

4 Results

One thing that worth to notice is that after modified artificial planning reaches equilibrium, it is going to be $K = (sk)^{\alpha\beta}$ instead of $K = sk$ in the Romer Economic Growth Model. When $\alpha = 1$ and $\beta = 1$, which, as above, means the knowledge has a total spillover effect, the modified model will be in the same shape as the Romer Economic Growth Model. When one of α and β, or both, equals to 0, or the value of $\alpha\beta$ cannot exceed the critical value, which makes the marginal profit of $f\left(k, (sk)^{\alpha\beta}\right)$ about k decreases, the modified model will be regarded as the Solow Growth Model. If it happens, the competitive optimum and the social optimum will be the same. When the value of $\alpha\beta$ is equal to the critical value, which leads to $k \to \infty$, the marginal profit of $f\left(k, (sk)^{\alpha\beta}\right)$ about k is a constant m, the modified model will be the same as the AK Model (Khaled Hussein & Thirlwall, A. 2000) [9].

In the modified model, the new private marginal product of knowledge is $\frac{\partial}{\partial k} f\left(k, (sk)^{\alpha\beta}\right)$; on the counterpart, for social planners, the new social marginal product of knowledge is $\frac{\partial}{\partial k} f\left(k, (sk)^{\alpha\beta}\right) + s(sk)^{\alpha\beta-1} * \frac{\partial}{\partial K} f\left(k, (sk)^{\alpha\beta}\right)$. Under this circumstance, in order to reach the Pareto Optimality (Yew-Kwang Ng, 1973) [10], Romer's political advice will also need to be utilized as well. Since the private marginal product of knowledge and the social marginal product of knowledge are both increasing functions related to $\alpha\beta$, the larger $\alpha\beta$ is, the larger private and social marginal product of knowledge will be. The consequence is: not only both the socially optimal knowledge stock and equilibrium knowledge stock will be larger, but both the socially optimal rate of economic growth and the equilibrium optimal rate of economic growth will be larger.

5 Conclusion

By importing new variables about heterogeneity and the ability to absorb knowledge, there are several further conclusions can be made depend on the results above.

The spillover effect of knowledge is related to the diversity of the new knowledge β. Therefore, duplicate construction will be tremendously harmful to one country's economic power, since duplicate construction will result in lowering the index of the diversity of new knowledge, the marginal product of knowledge, socially knowledge stock (both the social optimal and the equilibrium optimal), and increasing rate of economic growth (both the social optimal and the equilibrium optimal).

From the modified model, we can conclude that international trade will have a positive effect on economic growth. During the international trade, the number of different kinds of knowledge that countries can learn will increase substantially; on the other side, it can also open a whole new market for a country, which as a result increasing the value of s, total industries in the function, and the marginal product of knowledge. Eventually, it will reflect to increasing the rate of economic growth.

The ability to absorb new knowledge α is highly related to the citizen's overall educational level. Therefore, reinforcing the investment in educational programs and stocking more and more high-valued human capitals will accelerate a country's economic growth.

References

1. Arrow, K.J.: The economic implications of learning by doing. In: Readings in the Theory of Growth, pp. 131–149. Palgrave Macmillan, London (1971)
2. Kaldor, N.: A model of economic growth. Econ. J. **67**(268), 591–624 (1957)
3. Solow, R.M.: A contribution to the theory of economic growth. Q. J. Econ. **70**(1), 65–94 (1956)
4. Uzawa, H.: Optimum technical change in an aggregative model of economic growth. Int. Econ. Rev. **6**(1), 18–31 (1965)
5. Romer, P.M.: Increasing returns and long-run growth. J. Polit. Econ. **94**(5), 1002–1037 (1986)
6. Aghion, P., Howitt, P., Howitt, P.W., Brant-Collett, M., García-Peñalosa, C.: Endogenous Growth Theory. MIT Press, Cambridge (1998)
7. Barro, R.J.: Determinants of economic growth: a cross-country empirical study, No. w5698. National Bureau of Economic Research (1996)
8. Cass, D.: Optimum growth in an aggregative model of capital accumulation. Rev. Econ. Stud. **32**(3), 233–240 (1965)
9. Hussein, K., Thirlwall, A.P.: The AK model of "new" growth theory is the Harrod-Domar growth equation: investment and growth revisited. J. Post Keynes. Econ. **22**(3), 427–435 (2000)
10. Ng, Y.K.: The economic theory of clubs: Pareto optimality conditions. Economica **40**(159), 291–298 (1973)

Confirmation Bias and Gambler's Fallacy Effect with Bayesian Method

Qijun Zhu(✉)

Whiting School of Engineering, Johns Hopkins University, Baltimore, USA
qzhu21@jhu.edu

Abstract. This paper mainly focuses on the influence of confirmation bias and gambler's fallacy effect on people's behaviors in the short-term markets. The model sets the gambler's fallacy effect as an endogenous variable of the representative heuristic, and uses the information to explain confirmation bias. After using the Bayesian method to assign different prior distributions according to the different market conditions, I find these two biases both have effects on the stock price and they will be changed by the different market types. Besides, results also show that in 2019–2020, confirmation bias plays a more important role in the tech-stock market. When the market increased more in the last week, the gambler's fallacy effect is more significant.

Keywords: Confirmation bias · Gambler's fallacy effect · Bayesian method

1 Introduction

Today, people always want to invest in the financial markets, so, how to make a reasonable evaluation and prediction for the market is important. The assumptions of the Black-Sholes model indicate that the fluctuation of the stock return can be described by the Brownian Motion. In other words, it suggests that every step of the price change is random and cannot be predicted. However, if we relax the assumptions, as the Farmer, Nourry, Vendittiwe say, we would notice that the markets are not efficient because of the complex environmental factors and people's behaviors, and that is the foundation for our analysis, where we aim to use the past information to predict the future activities [1].

Poteshman and Serbin point out the existence of irrational behaviors in financial markets through the evidence of the stock options [2]. This paper will focus mainly on two biases of people's behaviors. One is the gambler's fallacy effect; the other one is the confirmation bias. The gambler's fallacy effect is derived from the law of the small numbers, which means that people always use the small sample to evaluate the whole situation. Rabin tries to model this bias from the probabilistic perspective with the Bayesian method and puts forward two useful views. The one is the explanation that the people always consider the situation without replacement. The other one is the assumption that every draw in the experiment is identical and independent. These two

C. Yuan et al. (Eds.): *Proceedings of the 4th International Conference on Economic Management and Green Development*, AEPS, pp. 408–414, 2021.
https://doi.org/10.1007/978-981-16-5359-9_47

things provide the direction for future research [3]. Besides, there are many practical researches. In the lottery play, Clotfelter and Philip J. Cook show that in the "three digit games", the winning numbers will not be chosen in a while [4]. Suetens and Tyran focus on whether the gender is the factor of the bias and show on average that men are about 1% less likely to bet on numbers drawn in the previous week than on numbers not drawn [5]. Besides gender, Fong, Rob Law and Lam also examine the other three factors: the latest outcome, streak length and outcome frequency [6]. In the financial world, K. Loh, Warachka and Kudryavtsev, Cohen, and Hon-Snir also show that people have this bias in the stock markets [7, 8]. It means if people see stock returns have the same sign in the past several days, they may predict that the sign will reverse in the next following days. However, confirmation bias means that people will insist more on their belief so that when they observe the news or events which are consistent with the belief, they will get more information and deepen their confidence. Research always needs to focus on using some ways to simulate the prior belief, in other words, confidence in something. In political beliefs, Taber, Lodge and Westerwick, Mothes, Polavin both use the experiment to show that when people choose the news they want to read, they have the confirmation bias [9, 10]. Kassin, E.Dror b, Kukucka show that in criminal events, the prior belief such as confession and fingerprint always influence the judgment [11]. Moreover, as Santamaria proposes, it also influences the investment strategy [12]. It means that if investors believe the company and industry are great and the stock will increase, when they see the same-sign sequence of returns, they will believe the momentum can continue.

Then, it is obvious that these two biases have some similarities. They both consider people's prior belief in certain thing and influence their prediction for the future. However, they may lead to two opposite directions. The one is a reversion; the other one is momentum. Therefore, this paper will focus on using the Bayesian method to combine these two biases to model people's behaviors in the financial market since the method is always a suitable way to deal with problems under different prior conditions. First, for the gambler's fallacy effect, Tversky and Kahneman, who raise the bias firstly, explain it by the representative heuristic, which means the people's thoughts are in a certain pattern [13]. Therefore, the paper treats the effect in an endogenous way and assumes it is only related to the prior belief and time. For the confirmation bias, many researchers, such as Taber, Lodge and Cipriano and Gruca, are more willing to set a stimulation mechanism in dealing with political and financial topics [14]. After these procedures of handling the prior belief, people are more likely to show confirmation bias. Thus, I consider people's different acceptance ratio based on the different tendencies of the information. Finally, we combine these two models to analyze people's choices in different types of the stock market with different prior settings. By Metropolis-Hastings method, we have verified that these two biases have different effects on the stock price and with changes in the type of the markets, gambler's fallacy effect will have different influences.

2 Model

This paper will mainly focus on the short-term issue, where people tend to show gambler's fallacy effect and confirmation bias. To model the people's behaviors in the market, we consider a time interval: $[T_0, T]$. Based on the characteristics of the biases, we can divide the interval into two parts: $[T_0, 0]$ and $[0, T]$. The first part is the time to decide people's prior belief by observing some sequences or some customs and habits, and it will influence the intensity of the following process. We can use the Bayesian method to model them. The second part is the time that the biases have happened and influenced people's behaviors and choices. Based on the different situations between T_0 and T, people will show their behaviors in different ways.

2.1 Gamble's Fallacy Effect

We should notice that people use the representative heuristic to guide their future activities, which is related to considering the small sample. For example, if we flip a coin and get the head many times, people may think the next toss is very likely to be tail, but the actual probability is just 0.5. Therefore, we can model this logic through a constant representative process, and it is not influenced by the outside environment. If we set w_t as the gambler's fallacy effect at time t, we can get the following model:

$$w_0 = d_0$$

$$w_t = w_{t-1} + q(T - t)^2 \tag{1}$$

In this model, d_0 is influenced by the prior belief of the representative heuristic from time T_0 to 0. The larger the d_0, the scarcer the prior situation and the gambler's fallacy effect will be stronger. For each w_t, it is only influenced by the time factor q.

2.2 Confirmation Bias

As for the confirmation bias, people will be influenced by the different kinds of information. If the type of news is the same as their prior confidence, they will pay more attention to it. However, if the news is in the opposite direction, it will have less effect on people. We define the x_{ti} as the i th information people receive at time t:

$$x_{ti} = \begin{cases} 1 & news\ is\ positive \\ 0 & news\ is\ neutral \\ -1 & news\ is\ negative \end{cases} \tag{2}$$

Then, we can define the confirmation model as follows:

$$k_0 = c_0$$

$$k_t = \sum_i (aD_n + bD_m) * x_{ti} \tag{3}$$

c_0 is the prior confidence, which is also decided between time T_0 and T. When the c_0 is larger, people will insist more on their belief and have more reaction to the information which is consistent with their mind. D_n and D_m are the dummy variables of the information which represent the same and opposite direction with c_0. Therefore, if x_{ti} is the good news compared with the prior confidence, $D_n = 1,\ and\ D_m = 0$, and if x_{ti} is the bad news compared with the prior confidence, $D_n = 0,\ and\ D_m = 1$. a and b are two parameters to control the extent that people receive different information.

2.3 Logistic Regression with Bayesian Method

The Bayesian method considers prior factors, so it is a suitable method to deal with these two biases based on the different prior conditions. Ignoring other biases in the financial markets, if we categorize the response variable into binary format, the model can be a logistic regression. Setting $x_1 = k_t, x_2 = w_t, P(Y_i = 1|x, \gamma, \beta) = p$, we can get the following model:

$$\log\left(\frac{p}{1-p}\right) = \beta_0 + \beta_1\gamma_1x_1 + \beta_2\gamma_2x_2 \tag{4}$$

β_0 means the the whole prior environment. I assume the d_0 and c_0 are subject to two independent normal random variables, which are $N(\mu_1, \sigma)$ and $N(\mu_2, \sigma)$. Because of the opposite directions of the two biases, the prior distribution of β_0 is subject to $N(\mu_1 - \mu_2, \sigma)$. β_1 and β_2 represent the power of the confirmation bias and gambler's fallacy effect respectively. γ_j is binary variable, which evaluate whether the bias has the significance influence on the reponse variable. We also should set the appropriate prior distribution for them based on different conditions.

3 Practical Research

I have used the real financial data to realize the model and compare the influences of the two biases in the different market types.

3.1 Data Selection

Concentrating on the stock markets, I first collect the close price of the Apple Inc(AAPL) from 12-06-2019 to 12-04-2020. Then I change them to the categorized response variable.

$$y_i = \begin{cases} 1 \ Price_{i+1} > Price_i \\ 0 \ Price_{i+1} \le Price_i \end{cases} \tag{5}$$

Then for finding the agency variable of the information, I use other three related close prices in technology fields (MSFT, TSM, NVDA). People can always see the whole market changing at the same time, so if they see three or two of them is increasing, they may have positive confidence in the market. Otherwise, they may lose confidence in investment. At this time, we do not consider the negative effect and different influences of news.

$$x_{it} = \begin{cases} 1 \ \textit{Two or three related stocks are increasing} \\ 0 \ \textit{Otherwise} \end{cases} \tag{6}$$

For endogenous variable of gambler's fallacy effect, I choose to evaluate it by weeks. Therefore, as $T = 5$, it is changed from $t = 1$ to 5 in a weekly cycle.

In addition, I choose to divide the market into several types using the different numbers of increasing or decreasing of the AAPL in the last five weekdays, which means T_0 is 5. For example, the type 2 market means AAPL increases two days in the last weeks.

3.2 Bayesian Processing and Results

Because of the limitation of the samples in each market types, I only put the type 2, 3, 4 market in practice, which the sample numbers are 160, 135, 90. The prior distribution of β_0 is different in each market types. In the type 4 or type 3 market, the prior mean of the gambler's fallacy effect is larger than the type 2 market. It is shown in Table 1.

Table 1. Prior of distribution of β_0

Market type	Prior of distribution of β_0
Type 2	N(1,4)
Type 3	N(0,4)
Type 4	N(−1,4)

Table 2. Mean of Parameter Value

Market type	Only confirmation bias	Add gambler's fallacy effect (mean)
Type 2	(β_0) −0.845; (β_1) 2.007	(β_0) −0.900; ($\beta_1 * \gamma_1$) 2.002; ($\beta_2 * \gamma_2$) 0.095
Type 3	(β_0) −1.182; (β_1) 2.498	(β_0) −1.285; ($\beta_1 * \gamma_1$) 2.553; ($\beta_2 * \gamma_2$) −0.006
Type 4	(β_0) −0.506; (β_1) 1.331	(β_0) −0.616; ($\beta_1 * \gamma_1$) 1.423; ($\beta_2 * \gamma_2$) 0.115

For the parameter for confirmation bias (β_1), I set the prior distribution is N(1, 4) since in 2019–2020, investors always have high expectations in the tech-related stocks. In addition, I set the parameter for gambler's fallacy effect (β_2) is N(0, 4) because of the endogenous feature.

Using Metropolis-Hastings method, the results are shown in Table 2 and Table 3. We can notice that confirmation bias always has the decisive effect on the all three market types and cases, but only in about half of the cases, gambler's fallacy effect can have its influence. It means in the 2019–2020, people believe more on momentum for the tech-stock market. Besides, only in the type 3 market, the gamblers fallacy effect will have the mean negative relationship with the confirmation bias.

Table 3. Relationship

Market type	Probability of $\gamma_j = 1$	Negative effect
Type 2	(p_1) 0.9996; (p_2) 0.3790	0.0784
Type 3	(p_1) 0.9997; (p_2) 0.4443	0.5488
Type 4	(p_1) 0.9982; (p_2) 0.5862	0.2073

However, when we research on the details about the two biases, in the type 4 markets, the gambler's fallacy effect has the highest probability of $\gamma_2 = 1$, and in the type 3 and

4 markets, it has a larger negative effect. Therefore, it is shown that when the marketing is more consistent in increasing in the last week, the negative feature of gambler's fallacy effect is always more obvious. In other words, people will think more about the representative heuristic for reverse in the stock price.

4 Conclusions

In this paper, I analyze the confirmation bias and gambler's fallacy effect, which both have some prior conditions, but finally go to the different directions. Combining with Bayesian Method, I model and realize them by using some financial stock prices. I have found that in the tech stock markets of 2019–2020, confirmation bias will have a more important impact. Besides, with the increasing of the market types, the gambler fallacy effect will be more distinct.

However, there are actually other biases in the financial markets, so getting rid of the influences of them is another essential topic for the following research. Moreover, there are some developments I hope to find a way to deal with such as changing the extension of the confirmation bias or trying different time for prior distributions and prediction period since these two bias may produce some opposite effects in the long-term based on the research of Barberis, Andrei Shleifer, and Robert Vishny and Wilkinson [15, 16]. I hope that we can use the model in the real markets according to its usefulness. In several different markets' situations such as volatile markets and stable markets, different types of stocks, and different information, I also hope that the model can give investors and governments some perspective to decide whether and when investing, find some suitable entering time points, and monitor the risk and volatility in the markets to make some appropriate policies in time.

References

1. Farmer, R.E., Nourry, C., Venditti, A.: The inefficient markets hypothesis: why financial markets do not work well in the real world (No. w18647). National Bureau of Economic Research (2012)
2. Poteshman, A.M., Serbin, V.: Clearly irrational financial market behavior: evidence from the early exercise of exchange traded stock options. J. Finance **58**(1), 37–70 (2003)
3. Rabin, M.: Inference by believers in the law of small numbers. Q. J. Econ. **117**(3), 775–816 (2002)
4. Clotfelter, C.T., Cook, P.J.: The "gambler's fallacy" in lottery play. Manage. Sci. **39**(12), 1521–1525 (1993)
5. Suetens, S., Tyran, J.R.: The gambler's fallacy and gender. J. Econ. Behav. Organ. **83**(1), 118–124 (2012)
6. Fong, L.H.N., Law, R., Lam, D.: An examination of factors driving Chinese gamblers' fallacy bias. J. Gambl. Stud. **30**(3), 757–770 (2014)
7. Loh, R.K., Warachka, M.: Streaks in earnings surprises and the cross-section of stock returns. Manage. Sci. **58**(7), 1305–1321 (2012)
8. Kudryavtsev, A., Cohen, G., Hon-Snir, S.: 'Rational' or 'Intuitive': are behavioral biases correlated across stock market investors? Contemp. Econ. **7**(2), 31–53 (2013)
9. Taber, C.S., Lodge, M.: Motivated skepticism in the evaluation of political beliefs. Am. J. Polit. Sci. **50**(3), 755–769 (2006)

10. Knobloch-Westerwick, S., Mothes, C., Polavin, N.: Confirmation bias, ingroup bias, and negativity bias in selective exposure to political information. Commun. Res. **47**(1), 104–124 (2020)
11. Kassin, S.M., Dror, I.E., Kukucka, J.: The forensic confirmation bias: Problems, perspectives, and proposed solutions. J. Appl. Res. Mem. Cogn. **2**(1), 42–52 (2013)
12. Duong, C., Pescetto, G., Santamaria, D.: How value–glamour investors use financial information: UK evidence of investors' confirmation bias. Eur. J. Finance **20**(6), 524–549 (2014)
13. Tversky, A., Kahneman, D.: Belief in the law of small numbers. Psychol. Bull. **76**(2), 105 (1971)
14. Cipriano, M., Gruca, T.S.: the power of priors: how confirmation bias impacts market prices. J. Predict. Mark. **8**(3) (2014)
15. Barberis, N., Shleifer, A., Vishny, R.: A model of investor sentiment. J. Financ. Econ. **49**(3), 307–343 (1998)
16. Wilkinson, N.: An Introduction to Behavioral Economics. Macmillan Education, UK

Social Housing in Guadalajara: A Viable Strategy to Mitigate the Negative Externalities of Metropolitan Urban Sprawl?

Eugenio Arriaga Cordero(✉) and Paola Romero Gutiérrez

ESARQ - Escuela Superior de Arquitectura y Artes, Libertad #1745, Col Americana, Guadalajara, Jalisco, Mexico
{earriaga,promero}@esarq.edu.mx

Abstract. The urban population in Mexico doubled between 1980 and 2010, while urban sprawl increased tenfold on average [1]. This unsustainable growth is exemplified in the Guadalajara Metropolitan Area (GMA). Among the negative effects of urban sprawl are environmental, social, and economic problems, many of which stem from increasing dependence on the private car in a dispersed, fragmented and low-accessibility urban environment. One of the driving forces behind this urban sprawl is an increase in the cost of housing in Guadalajara: 11% between 2018 and 2020 [2] combined with a decrease in the average income of its inhabitants [3]. In addition, an increase in the supply of social housing in the municipality of Tlajomulco de Zúñiga, at the fringe of the metropolitan area, has encouraged urban sprawl. This article explores a series of social housing policies for low-income groups to access affordable housing in the city urban core, in order to limit or prevent further urban sprawl. Among the housing policy alternatives explored here are rental housing, inclusive housing and land banks.

Keywords: Sustainability · Social housing · Urban sprawl

1 Introduction

At present, there is legitimate concern about major environmental and social problems associated with the phenomenon of climate change, in large part aggravated by overdependence on cars, which are mandatory in a context of urban sprawl. The urgent need to reduce greenhouse gas emissions has been placed on the public agenda, and for this it would be necessary to reduce the use of cars.

However, this presents a serious challenge given that this mode of transport is inextricably linked to the design of our cities. For this reason, the urban form of our cities has become critically important. The GMA has followed a model based on urban expansion that produces scattered, fragmented cities with low accessibility. This urban expansion model also implies the conversion of agricultural land, forests and ecologically fragile zones to urban uses such as housing; this negatively affects the quality of water, soil and biodiversity [4].

C. Yuan et al. (Eds.): *Proceedings of the 4th International Conference on Economic Management and Green Development*, AEPS, pp. 415–423, 2021.
https://doi.org/10.1007/978-981-16-5359-9_48

Additionally, the sprawl-type urban expansion model has exacerbated residential segregation of the most vulnerable population groups, driving them to the metropolitan periphery of Guadalajara.

This is the result of what CEPAL [5] refers as "territorial heterogeneity", what Soja [6] calls "distributional inequality," and what Harvey [7] defines as "territorial injustice" . These concepts essentially describe how residents of consolidated municipalities, compared to municipalities on the periphery, enjoy more and better public services and opportunities for material well-being. The municipalities in the periphery are characterized by high levels of poverty, low quality and fewer public services; greater public insecurity and environmental injustice; and longer travel distances to access services, employment and entertainment activities, which produces high levels of social exclusion [8].

Urban sprawl in Mexican cities have worsened in the last two decades. The urban population in Mexico doubled between 1980 and 2010, while territorial expansion increased tenfold on average over the same period [9]. This trend was mirrored in Guadalajara, where the metropolitan urban expansion tripled between 1980 and 2015, while the population only doubled (see Table 1). IMEPLAN [9] further estimates a 57% expansion in urban land for the period 2015–2045, which will undoubtedly increase the magnitude of social and environmental problems and negative externalities.

According to Ewing et al. [10] urban expansion is characterized by dispersed, disconnected and low-density residential developments and a rigid separation of land uses, where residences are isolated from places of consumption, employment and entertainment. Anderson et al. [11] add that such urban expansion implies a growth of the metropolitan limits towards the outside. This implies a reduction in the intensity of all forms of land use, measured in terms of population and business density. This "urban expansion" model is thus the antithesis of a "compact" city model.

The adoption of the sprawl-based urban expansion model in Mexico has multiple causes. Among them are (1) the construction of road infrastructure to facilitate motorized mobility, and (2) the wide access to credit for the purchase of new social housing in remote places of the metropolitan periphery, where residential developments are built on cheap land regardless of the insuficient provission of public services, such as education and health. In the metropolitan periphery it is difficult to find neccesary day to day services close enough to access by foot or bicycle, and public transit is limited. As a consecuense, car use becomes essential to function efficiently [12]. Therefore, it would be very difficult to reduce car use and the negative impacts it has on the environment and society.

2 Environmental Externalities of Automobile Dependence

The increase in traffic in Guadalajara has contributed to a drop in average speed (23.5 km/h) and an increase in the average time per trip [13]. Urban sprawl with its associated low-density urban development model has contributed to this increase in the average distance of trips. The transport and construction sectors in Mexico jointly produce about 35% of greenhouse gas emissions [14].

In Guadalajara, the massive construction of subdivisions (both social housing and market rate) in increasingly remote areas of the metropolitan periphery, together with the longer trip distances to access these destinations, have important environmental consequences. These include negative impacts on the recharge of water tables, contributing to floods and rising temperatures (paved urban areas generate "heat islands") [15].

Table 1. Population growth and urbanized territory in the GMA, 1980 and 2015.

Variable	Guadalajara metropolitan area	
	1980	2015
Territory sprawl	22,329 ha	69,250 ha
Population	2′371,278	4′865,122

Source: IMEPLAN 2016.

3 Social Externalities

It is estimated that 92% of the world's population breathes highly polluted air and every year 3 million people die from exposure to pollutants in the air [16]. Infants and children are at especially high risk: 570,000 children under the age of five die from respiratory infections due to exposure to air pollution. Air pollution also increases the risk of suffering from heart disease, stroke, and cancer [17].

In Mexico there are 20 thousand deaths per year due to air pollution, mainly among those over 65 years [18]. One of the problems in Mexico is the high number of older vehicles on the road, which pollute at greater rates than newer vehicles. During 2011 in the GMA, 39% of the vehicles were more than 10 years old, which polluted 400% more than the most recent models [19].

Injuries resulting from road incidents are the leading cause of death among adolescents in the world: In 2015 they caused the death of 115,000, most of them pedestrians, cyclists and motorcyclists [20]. In Mexico, traffic incidents cause 24,000 deaths, 40,000 disabled, and 750,000 injured per year [21]. Jalisco is the second state in Mexico with the highest number of deaths in road incidents.

Urban form affects the proclivity to exercise. Various studies have found that people who live in compact communities (dense, connected, and mixed-use) tend to be more physically active than those who live in suburban areas [22, 23]. This is important because physical inactivity contributes to an increased risk of chronic diseases and conditions such as obesity, hypertension, diabetes, cancer, and heart disease. All of this adds up to increased costs for the public health sector. The sedentary lifestyle, partly a product of the high rate of motorization and longer trips in part due to urban sprawl, has contributed to an obesity rate of 33% in Mexican cities (and a 39% overweight rate).

As mentioned above these conditions lead to chronic diseases such as diabetes, which are very expensive to treat [24]. In Guadalajara, noise pollution mostly from cars also causes severe consequences, such as the loss or decrease of hearing capacity, and physiological and psychological diseases that affect balance, the nervous system, sleep, and work performance [25].

4 Economic Externalities

The average annual cost for the diagnosis and treatment of diabetes mellitus in 2008 in Mexico was 77 thousand pesos, and 70 thousand pesos for cardiovascular diseases [26]. In the metropolitan area of Guadalajara during 2009, air pollution cost more than 4,000 million pesos; traffic incidents, 4.970 million; road congestion, 10,635 million; noise, 1,615 million; diseases [27, 28]. These data confirm the perniciousness of the negative externalities associated with car use in a urban sprawl context.

5 Discussion: Population Loss in Guadalajara and Intra-municipal Migration

The case of the municipality of Guadalajara does not fit the typical pattern of urban expansion in Mexico. Guadalajara has lost nearly 200,000 inhabitants in recent years, dropping from 1,650,000 inhabitants in 1990 to 1,460,000 in 2015 [29]. In contrast, the metropolitan periphery increased its population in the same period of time. Intra-municipal migration from the city of Guadalajara to the peripherical metropolitan municipalities is explained in part by (i) the increase in the cost of land and housing in Guadalajara [30], (ii) the low incomes of the population in the GMA: 69% of the employed population earns around 7,300 pesos per month [31, 32], and (iii) the increase in the supply of social housing in the metropolitan periphery.

This is the context in which urban sprawl in the metropolitan area of Guadalajara occurs, with the negative externalities mentioned above resulting from car use in a urban sprawl context. In the next section, a series of housing policies which can contribute to provide access to affordable housing to low-income groups in the inner city of Guadalajara at the time that may contribute to reduce urban sprawl are discussed. The policy alternatives are rental housing, inclusive housing, land trusts/land banks, and the "Tanteo y retracto".

6 Social Housing Policies

6.1 Rental Housing

Rental housing reaches very high levels in countries like Germany and Austria: 60% and 50%, respectively. The case of the city of Berlin is paradigmatic: 89% [33]. Rental housing in Mexico has been growing recently, nearly doubling between 2000 and 2010, from 2.8 to 4 million units. Guadalajara is one of the cities where it has grown the most: currently 33% of housing is for rent [34, 35].

Rental housing programs in the consolidated city would enable low-income groups to access housing at a lower rent than what the market offers. The rental housing programs work through (i) subsidies for suppliers, such as tax incentives and preferential financing for private developers for the construction of rental housing; and (ii) support for consumers, particularly the socially disadvantaged who require help to pay rent.

As Ponce posits [36] the increase in the cost of housing in cities, as well as social and demographic changes have contributed to the growth of rental housing, both in absolute numbers and as a proportion of all housing. Among the advantages of rental housing in the urban core are its contribution to the renovation of urban centers; access to homes for socially disadvantaged groups; the flexibility it offers to seasonal workers; the increase in employment and services (which reduces the social exclusion to which low-income groups displaced to the metropolitan periphery are exposed); and an increase in the mixing of socio-economic strata. The rational is that through access to rental housing in the urban core, socially disadvantage groups currently living in the metropolitan periphery would be able to move to the consolidadated city.

6.2 Land Banks and Community Land Trusts

Land Banks emerged in the 1960s to make it easier for governments to acquire surplus properties, in part, for the purpose of increasing social housing units [37]. This policy flourishes in a context in which the real estate market does not function efficiently, which keeps land prices very high (ibid).

Community land trusts acquire the land, build housing, and then rent units often for 99 years to cooperatives that provide housing for low-income families. The families acquire all the rights of a homeowner except the right to sell the unit at a profit. Its sale price is established through a formula approved by the trust, which generally allows the recovery of the purchase price, plus home improvements, and some cost-of-living adjustment, but excludes the value of the land, which remains in the trust rent [38].

This approach can contribute to increasing the supply of affordable housing for low-income families. The trust establishes ground rules to ensure that housing is permanently available to families in need, at the most affordable rental prices possible.

According to Madden and Marcuse [39] among the advantages of these trusts, it stands out that they contribute to controlling and keeping the rent price low, when the land acquired by the government is not sold, but rented. They enable the creation of permanently affordable housing for low-income households, including future residents. They reduce the intensity of inequality induced by a private market that considers housing for its exchange value and not for its use value. They also help develop communities and stabilize neighborhoods [38].

On the other hand, they can also generate conflicts if the compensation offered by the government to the owners of the land is considered low, which does not ensure their stability. In any case, these approaches eliminate the variable of the high cost of land, facilitating the possibility of offering affordable housing for low-income groups in the consolidated city, and avoiding their displacement to the metropolitan periphery. This is possible, in part, because in both cases the possibility of speculating on an abnormal increase in the value of land and housing is waived.

6.3 Inclusionary Zoning

This policy seeks to combine low- and high-income families in the same development [40, 41]. It regularly works through an agreement with private developers to provide a certain number or proportion of low-income housing units in a development in exchange for permission to build more units. Inclusive housing programs are local policies that harness the economic benefits of rising real estate values to create affordable housing.

These programs often require developers to sell or rent between 10 and 30% of new residential units to low-income individuals [40].

Inclusionary housing policies may offer different terms, type of support, and percentages of low income units. In some cities it is a mandatory policy while in others it is voluntary [42]. The central idea is that local authorities are given additional powers to ensure that new housing developments include a proportion of affordable housing to meet a growing demand for housing for socially disadvantaged groups [43]. This policy allows for denser and taller private housing developments in established areas. There are different variants of this policy in terms of type of support.

Essentially, the units reserved for low-income groups are paid for with the profits generated by the authorized surplus housing, at market prices. The key questions of this policy, according to Maden and Marcuse [40] are: who is included in the program and in what proportion? The lower the income of the groups served and the higher the required proportion of these to be included, the better.

This is a program designed to act against discrimination exercised by the housing market, by ensuring that socially disadvantaged people have access to decent housing in central areas of the city. Therefore, this program acts to reduce residential segregation, a form of discrimination that restricts opportunities for access to quality employment and services to low-income groups, and inhibits social diversity.

In New York City, this program is known as the "80–20." That is, 80% of households in an inclusionary housing project are high income, while 20% are middle and low income [44].

In San Francisco and Washington, D.C., state laws require cities to give developers "density bonuses" to build affordable housing [42]. Those in favor of this policy argue that it promotes social housing without having to spend public resources via subsidies, while those who are against claim that this policy raises house prices by reduces the supply of market rate units [42]. In California from 1988 to 2005, the policy increased multi-family housing units and reduced single-family housing units [45].

6.4 Right of Urban Preference "Tanteo y retracto"

An alternative that local governments have to control the cost of land and housing and real estate speculation is that used in the cities of Barcelona and Paris: the right of preference "Tanteo y retracto" [46]. This instrument empowers cities to offer to buy land and housing on the private market in previously selected areas, at a regulated price, and convert it to social housing, which increases the number of social housing units.

Usually, the acquisition of a home occurs through the integration of a public fund for that purpose. Ampudia [47] points out that the Barcelona City Council has acquired land and housing by this means in central neighborhoods, at a price below the market, to develop public housing or provide spaces for economic activities of social interest, protecting the businesses that are being expelled by the phenomenon of gentrification. The Barcelona City Council in some cases uses a model to transfer control over housing for 75 years to cooperatives, to manage social housing.

The Paris City Council has the "right of urban preference" through which it can acquire land and real estate for sale in the private market in "areas of action" defined for it. Like Barcelona, it does so at prices below the market. This has been a useful strategy to temper gentrification.

The purchase process works as follows: the City Council has the right to be the first to make a purchase offer (with a regulated price) on any property that goes on sale in the private market [48]. The notary is obliged to notify the City Council of the intention to sell a property, so that within a period of less than two months the city government can determine whether the offer is of interest. If the property is of interest to the municipality the price will be determined, which may vary downwards from what is established by the owner. If the new price is agreed upon, the purchase proceeds. Otherwise, the owner can either ask a judge to determine a new price or abandon the transaction [47].

7 Conclusions

The GMA's urban sprawl has led to a harmful urban model of development that is disproportionately dependent on private car use. The environmental, social and economic cost of the negative externalities of urban sprawl closely associated with car use are very high. One of the main causes of urban expansion and, therefore, of the intra-municipal migration of the poorest inhabitants from Guadalajara city to the metropolitan periphery, is (i) the high cost of land and housing in the consolidated city of Guadalajara, (ii) the low wages of its inhabitants, and (iii) excessive growth in the supply of social housing in the remote urban periphery.

The social housing policies discussed in this article can help to remove the high cost of land in Guadalajara from the intra-urban social housing equation, and therefore offer affordable housing for socially disadvantaged populations, avoiding their involuntary displacement towards the metropolitan periphery, and thereby decreasing pressures to continue expanding the metropolitan territory.

It is imperative to design public policies that ensure efficient coordination between the policy areas of transport, land use and housing to produce more sustainable and compact cities, where the problem of urban sprawl can be tempered. In addition, it is still necessary to increase accessibility within the municipality of Guadalajara by means of more and better mass public transport; increase the population density of downtown, and the mix of land uses [49, 50]; while increasing the supply of affordable housing, a pending issue. Without these conditions, we will inevitably continue to suffer the negative externalities of the current urban sprawl model of urban development. The design and implementation of a social policy around housing that offers opportunities for the poorest groups to live in the "compact" city and benefit from accessibility to public services, urban facilities, and job opportunities, should be one of the highest priorities of local governments not only in the city of Guadalajara but in Mexico.

References

1. Instituto Metropolitano de Planeación del Área Metropolitana de Guadalajara (IMEPLAN): PotMet. Plan de Ordenamiento Territorial Metropolitano del GMA (2016)
2. Sociedad Hipotecaria Federal: Índice SHF de Tercer Trimestre de Precios de la Vivienda en México 2018, rescatado el 3 de mayo de 2019, de (2018). https://www.gob.mx/shf/documentos/indice-shf-de-precios-de-la-vivienda-en-mexico-2018

3. Observatorio de Salarios, Ibero Puebla: Informe Anual del Observatorio de Salarios 2016. Los salarios y la desigualdad en México, recuperado el 25 de octubre de 2019 (2016).http://redsalarios.org/informes/leer/
4. Anderson, W.P., Kanaroglou, P.S., Miller, E.J.: Urban form, energy and the environment: a review of issues, evidence and policy. Urb. Stud. **33**(1), 7–35 (1996)
5. Comisión Económica para América Latina y el Caribe (CEPAL): La matríz de la desigualdad social en America Larina (2016)
6. Soja, E.: Seeking spatial justice.University of Minnesota Press. UN-Hábitat. 2003. Rental Housing An essential option for the urban poor in developing countries. UN-Hábitat, Nairobi (2010)
7. Harvey, D.: The Enigma of Capital and the Crisis this Time. Profile, Londres (2010)
8. Arriaga, E.: No somos iguales: exclusión social y movilidad en el Área Metropolitana de Guadalajara. Retos vigentes de la gestión metropolitana. Navarrete, C. (coord.) *Modelos y políticas públicas de la gobernanza metropolitana.* El Colegio de Jalisco, pp. 135–154 (2019)
9. IMEPLAN (2016)
10. Ewing, R., Schmid, T., Killingsworth, R., Zlot, A., Raudenbush, S.: Relationship between urban sprawl and physical activity, obesity, and morbidity. Am. J. Health Promot. **18**(1), 47–57 (2003)
11. Anderson, W.P., Kanaroglou, P.S., Miller, E.J.: (1996)
12. Giuliano, G.: Low income, public transit, and mobility. Transp. Res. Rec. **1927**(1), 63–70 (2005)
13. Jalisco Cómo Vamos: Así vamos en Jalisco. Reporte de indicadores sobre calidad de vida 2012 (2012)
14. Instituto Nacional de Ecología y Cambio Climático (INECC): Estudio de emisiones y actividad vehicular en la zona metropolitana de Guadalajara, Jalisco. 2018. Presentación del Inventario de Emisiones de Gases y Compuestos de Efecto Invernadero 1990–2015 (2011)
15. Sánchez, A.: Las Villas Panamericanas son un crimen ecológico anunciado. Rescatado de (2014). http://cucea.udg.mx/es/ noticia/03-dec-2014
16. Word Health Organization (WHO): Mapa de calidad de aire en tiempo real (2019). https://aqicn.org/map/world/es/
17. WHO: La herencia de un mundo sostenible? Atlas sobre salud infantil y medio ambiente. 2019. Mapa de calidad de aire en tiempo real. Rescatado el 11 de julio de 2019 de (2018). https://aqicn.org/map/world/es/
18. Encuesta Nacional de Salud y Nutrición de Medio Camino (ENSANUT): Informe final de resultados. Instituto Nacional de Salud Pública (2016)
19. INECC (2011)
20. Word Health Organization (WHO): No contamines mi futuro! El impacto de los factores medioambientales en la salud infantile (2017)
21. SEDATU: Estrategia Nacional de Movilidad Sustentable, Subsecretaría de Desarrollo Urbano y Vivienda, Gobierno de México (2014)
22. Frank, L., Engelke, P.: The built environment and human activity patterns: exploring the impacts of urban form on public health. J. Plan. Lit. **16**(2), 202–218 (2001)
23. Frank, L., Kerr, J., Chapman, J., Sallis, J.: Urban form relationships with walk trip frequency and distance among youth. Am. J. Health Promot. **21**(4_Suppl.), 305–311 (2007). https://doi.org/10.4278/0890-1171-21.4s.305
24. OECD: Panorama de la Salud 2017 (2017). OECD. Governing the city. OECD Publishing, Paris (2015)
25. ITESO, Ceit y SVyT: Movilidad. Una visión estratégica en la Zona Metropolitana de Guadalajara, México (2002)

26. Gutiérrez, C., Guajardo, V., Álvarez, F.: Costo de la obesidad: las fallas del mercado y las políticas públicas de prevención y control de la obesidad en México. Obesidad en México: recomendaciones para una política de Estado, pp. 348–359. UNAM, México, (2013)
27. SEDATU (2014)
28. ITDP: Transformando la movilidad urbana en México, Instituto de Políticas para el Transporte y Desarrollo México, Embajada Británica en México (2012)
29. Instituto Nacional de Estadística y Geografía (INEGI): Encuesta Intercensal 2015 (2015)
30. Sociedad Hipotecaria Federal (SHF): Índice SHF de Precios de la Vivienda en México: Tasas de apreciación (+) / depreciación (-), 2015.I - 2020.I. del primer trimestre de 2018, 2019 y 2020. Recuperado de (2020). www.gob.mx/shf/artículos
31. Observatorio de Salarios, Ibero Puebla: informe_2016_observatorio_de_salarios/56 (2016)
32. Instituto Nacional de Estadística y Geografía (INEGI): Encuesta Nacional de Ocupación y Empleo (ENOE). INEGI (2017)
33. UN-Hábitat: Rental Housing An essential option for the urban poor in developing countries. UN-Hábitat, Nairobi (2003)
34. INEGI: XII Censo General de Población y Vivienda 2000 (2000)
35. INEGI: Censo de Población y Vivienda. Censo de Población y Vivienda (2010)
36. Ponce: Posibilidad de una nueva política habitacional: La vivienda en renta en México. Instituto Belisario Domínguez. Senado de la República (2014)
37. Alexander, F.: Land Banking as Metropolitan Policy. Blueprint for American Prosperity, Metropolitan Policy Program at Brookings (2008)
38. Alexander, F.: (2008)
39. Madden, D., Marcuse, P.: In defense of housing. The politics of crisis (2016)
40. Madden, D., Marcuse, P.: (2016)
41. Schuetz, J., Meltzer, R., Been, V.: 31 Flavors of inclusionary zoning: comparing policies From San Francisco, Washington, DC, and Suburban Boston. J. Am. Plan. Assoc. **75**(4), 441–456 (2009)
42. Schuetz, J., Meltzer, R., Been, V.: (2009)
43. Varady, D., Walker, C.: Housing vouchers and residential mobility. J. Plan. Lit. **18**(1), 17–30 (2003)
44. Schwartz, A., Tajbakhsh, K.: Mixed-income housing: unanswered questions. Cityscape **3**(2), 71–92 (1997). US Department of Housing and Urban Development. Stable. http://www.jstor.org/stable/41486511 Accessed 11 Dec 2015. 17:13 UTC
45. Bento, A., Lowe, S., Knaap, G., Chakraborty, A.: Housing market effects of inclusionary zoning. Cityscape **11**(2), 7–26 (2009)
46. Ampudia Farias, A.: Políticas, instrumentos de gestión y ampliación del parque de vivienda social: del contexto europeo a estudios de caso Barcelona, España y Guadalajara, México (2017)
47. Ampudia Farias, A: (2017)
48. Burón Cuadrado, J.: Las reservas de suelo para la vivienda protegida: Lecciones del caso de Victoria-Gasteiz. Archit. City Environ. **1**(2), 85–103 (2006)
49. Boarnet, M., Crane, R.: Travel by Design: The Influence of urban Form on Travel. Oxford University Press, New York (2001)
50. Handy, S., Cao, X., Mokhtarian, P.: Correlation or causality between the built environment and travel behavior? Evidence from Northern California. Transp. Res. Part D **10**, 427–444 (2005)

Research on International Cross Border National Culture Communication Strategy——Take Ewenki as an Example

Anqi Hu[1](✉) and Zhaohui Huang[2]

[1] Ningbotech University, No. 1 Xuefu Road, Yinzhou District, Ningbo, Zhejiang, China
[2] Network and New Media Major of Ningbotech University, Ningbo, China

Abstract. Based on the relatively stable development of China's border areas and the high willingness of neighboring countries to develop friendly cooperation with China, in September and October 2013, Chinese President Xi Jinping put forward the cooperation initiative to build the "Belt and Road", and actively carried out cooperation with countries along the route Cooperation to create a community of interests. In the process of developing mutually beneficial cooperation, the cross-border ethnic groups on the border have great advantages and become an indispensable part of the country's external communication. The value and future development expectations of the new era explain the importance of ethnic cultural exchanges in the context of cross-border ethnic cultural integration in the new era, and propose some strategies for cultural communication.

Keywords: Cross-border ethnic groups · Cultural communication · Strategic research · Ethnic minorities

1 Introduction

Nation refers to a group of people who are objectively distinguished in culture, language, history and other groups. It is a concept formed by studying human evolutionary history and race in modern times. Due to historical reasons, a country can have different ethnic groups, and a nation can live in different countries.

In the context of modern nation-states, cross-border ethnic groups are a special existence. Different scholars have different perspectives and understandings of cross-border ethnic groups. Mr. Ma Rong proposed in "How to Understand "Cross-Border Ethnic Groups"" that the concept of "cross-border ethnic groups" means that members of a certain "nation" live in an internationally recognized On both sides of the national border, their place of residence "crosses" the national border. An important prerequisite for recognizing the concept of "cross-border ethnicity" is to recognize that members of the group living on both sides of the border belong to the same "nation".

Therefore, due to the artificial division of national borders, an ancestor's blood has a certain origin, and the same or similar groups in language, culture, religious beliefs, etc. are divided into different political entities, this group can be called cross-border nation.

C. Yuan et al. (Eds.): *Proceedings of the 4th International Conference on Economic Management and Green Development*, AEPS, pp. 424–432, 2021.
https://doi.org/10.1007/978-981-16-5359-9_49

The Ewenki nationality is an ethnic group in Northeast Asia. It mainly lives in Siberia, Russia, and China's Inner Mongolia and Heilongjiang provinces. It also has a small distribution in Mongolia, while the Ewenki nationality is called the Ewenki in Russia. Ewenki means "people living in the mountains and forests". They are divided into three parts: Sauron, Tunguska, and Shilu Ewenki. In history, they were called Sauron, Tungusic, or Yakut by other peoples. The Manchus called them Sauron, the Yakuts called the neighboring Evenkis Tunguska, and the Russians called the part of the Evenkis mixed with Yakuts called Yakuts. The Ewenki language belongs to the Tungusic branch of the Man-Tungus language family of the Altaic language family. There are three dialects of Buteha, Tungusic and Aoluguya, and there is no written language. This article takes the Ewenki ethnic group located on the Sino-Russian border as an example, analyzes the value of cross-border ethnic cultural research and future development expectations through related documentation, and elaborates the importance of ethnic cultural exchanges in the context of cross-border ethnic cultural integration in the new era, and it is cultural Spread some strategies.

2 The Value of Cross-Border Ethnic Cultural Research

2.1 Social Value

The Need to Resolve Cultural Conflicts and Maintain National Cultural Security. Cross-border ethnic groups can become the link of cultural exchanges between neighboring countries. Benedict Anderson believes that the common culture is the decisive factor of national identity, which is fundamentally cultural identity. The conflicts of nation-states are mainly caused by conflicts of ethnic cultures. Therefore, respecting the differences and diversity of ethnic cultures, and respecting the religious beliefs and customs of cross-border ethnic regions are important prerequisites for winning their support and recognition. In the process of globalization, the exchanges between various regions and ethnic groups in the world are becoming more frequent. Cross-border ethnic cultural exchanges are conducive to mutual understanding and friendship, and thus resolve cultural conflicts.

At the same time, due to the diversity of ethnic cultures and strong cultural heterogeneity, for the two multi-ethnic countries of China and Russia, strengthening the study of cross-border ethnic groups will help deepen the reform of ethnic regions and promote their economic, cultural and social development. Comprehensive development and consolidation of the country's unity and stability.

The Need for International Cooperation. Culture is the link of exchanges and cooperation between countries, and culture first will promote the development and progress of the cultural field. The cross-border ethnic groups on both sides of the national border have similar cultural accumulations and ideological foundations, so they have more advantages in international exchanges and cooperation, and it is easier to form cultural identity during cultural exchanges. At the same time, the country's "One Belt One Road" strategy is also encouraging our nation Culture "going out", facing the huge cultural market demand and the need to enhance the country's cultural soft power, cross-border ethnic culture can provide the foundation and meet the needs.

2.2 Cultural Value

Meet the Needs of National Cultural Diversity and Commonality. The "World Declaration on Cultural Diversity" adopted at the 31st session of UNESCO stated: "Cultural diversity is the source of communication, innovation and creativity. Indispensable. In this sense, cultural diversity is the common heritage of mankind, and it should be recognized and affirmed in consideration of the interests of contemporary people and future generations".

Cross-border ethnic cultures are characterized by cultural diversity. The Ewenki ethnic group located on the Sino-Russian border is associated with geographical location, environment, and properties, so the culture is more distinctive. For example: birch bark culture, fur culture, hunting culture, music and dance culture, horse culture, dog culture, etc. These cultures are completely different from the cultures of neighboring nations.

The Need for National Cultural Protection and Development. Most cross-border ethnic cultures are ethnic minority cultures. As a multi-ethnic country composed of 56 ethnic groups, more than half of them are cross-border ethnic groups. Ethnic cultures are diverse, but in the process of modernization, the unique culture of cross-border ethnic groups it is also suffering from shocks. For example, the languages of the indigenous peoples on the Sino-Russian border are disappearing. Therefore, studying cross-border ethnic culture helps to propose effective ways to protect and inherit ethnic minority cultures to maintain cultural diversity.

3 Status Quo of Chinese and Russian Cultural Tendencies

According to data from the Russian Public Opinion Survey Foundation (a survey conducted from June 30 to July 1, 2018), overall, 49% of Russians prefer to live in Europe, and only 28% prefer to choose China. This tendency is stronger among 18–30-year-olds. In the survey of these groups, 61% of the respondents tend to Europe, and 27% tend to China.

This shows that the Russian people have a certain willingness to contact Chinese culture, but compared to European culture, Chinese culture is still not attractive enough. Therefore, there is still a lot of room for improvement in my country's national image and Sino-Russian cultural diplomacy.

4 An Analysis of the Status Quo of Cross-Border Ethnic Groups (Ewenki) Between China and Russia

4.1 Historical Evolution

The Ewenki nationality is a nationality scattered in Northeast Asia. Before the 17th century, part of the Ewenki nationality lived within the territory of China, and some lived outside the territory. Heilongjiang was originally an inland river in China. Tsarist Russia first invaded the Heilongjiang basin in the 17th century and translated Heilongjiang as

"Amur". Since then, Heilongjiang has become the boundary river between China and Russia.

On September 7, 1689, China and Russia signed the Nerchinsk Treaty. The treaty stipulates that the large areas of territory east of Lake Baikal, south of the Lena River, north of the Ergun River, and west of the Gerbizi River are Russian territory. As a result, the Hezhe, Ewenki, Oroqen and other ethnic groups who live here all year round are classified as ethnic minorities in the Russian Far East. The Ewenki nationality has become a cross-border ethnic group, which is called the Ewenki nationality in China and the Ewenki nationality in Russia.

4.2 Main Distribution

Population and Distribution in China. According to the 2010 China National Census, there are 30,875 Ewenki people in China, of which 26,201 are located in the Inner Mongolia Autonomous Region, accounting for about 87% of the national Ewenki population. They mainly live in the Daxinganling and Hulunbeier grasslands in the northeast. The ethnic groups, Han, Daur, Oroqen and other ethnic groups live in a scattered manner, forming the distribution characteristics of large dispersion and small settlement.

Population and Distribution in the World. The Ewenki nationality is most distributed in Russia, and a few are distributed in Mongolia. In Russia, the Ewenki people are called the Evenki. According to the 2010 Russian census, there are 37,843 Evenki people. In Mongolia, there are also a small number of Ewenki people with a population of about 3,000.

5 Prospects for the Future Development of Cross-Border Ethnic Cultural Communication

5.1 Facing New Opportunities

China-Russia Relations are Developing Rapidly. On June 5, 2019, China and Russia signed the "Joint Statement of the People's Republic of China and the Russian Federation on the Development of a Comprehensive Strategic Partnership for a New Era", indicating that Sino-Russian relations have entered a new era. The statement proposed in the statement to jointly promote China's initiative The "Belt and Road" initiative and the "Eurasian Economic Union" initiated by Russia will help China and Russia to carry out friendly cooperation in culture, education, and economy. After the Ukrainian crisis, the Far Eastern society's favor with China has further increased. 2013 ~ In 2015, a social survey conducted by the Russian Levada Center in Russia showed that the proportion of people who think "China is an ally of Russia" has doubled, from the previous 20% to 43%. However, in recent years, the social xenophobia in the Russian Far East has increased compared with the past, and Russia's awareness of China lags behind the level of political cooperation between the two countries. Therefore, under the background of national strategic support, deepen cultural exchanges between the two countries and draw closer the ethnic groups distance, cross-border ethnic cultural communication has good development prospects.

Technological Innovation. Now entering the era of "Internet +", the development of new media has changed the carrier and method of cultural communication. Realistic and substituting webcasting provides a broad space for the spread and innovation of national culture and accelerates culture. The speed and breadth of communication will help further expand the influence of Chinese cultural confidence. The popularity of Tibetan boy Ding Zhen in Japan also reflects the widespread nature of Internet communication.

Economic and technological development has also enhanced transportation advantages. The railway, highway, water transportation, aviation and other transportation infrastructures in the eastern border area of Northeast China are relatively complete. Dandong Port, located at the northernmost coastline, is the most convenient sea-rail logistics channel connecting Russia, Mongolia, North Korea, South Korea, and Japan. It can provide transportation between China and Russia. Materials provide favorable conditions.

5.2 Challenges Faced by Cross-Border Ethnic Cultures

Differences in Cultural Identity. Due to differences in countries, in different cultural places and cultural spaces, and affected by different social systems, history and cultural values, the differences in the thinking and behavior of cross-border ethnic groups have formed, and the understanding of national lifestyles and historical destiny It is also different from concern. Every nation makes a judgment based on its own cultural identity, which will inevitably lead to contradictions or conflicts, which will restrict the communication between the two nations. Russian culture has a certain degree of duality. For example, it not only hopes to develop the Far East, but is also wary of the rapidly rising China. The difference in cultural identity constitutes an invisible obstacle to the development of cultural and educational interaction between China and Russia.

The original Oroqen language of the Oroqen people is currently only used by a small number of elderly people in their 70s and 80s. The Oroqen people in China speak Chinese and use Chinese characters on a daily basis. The Ewenki people in Russia speak Russian and use Russian. In terms of eating habits, Oroqen the ethnic group has been influenced by the national culture and has fully adapted to the local food customs.

Lack of Support for Cultural Communication in Border Areas. Border areas are often areas where the country's economic development strength is relatively weak, and local governments cannot afford the cost of cultural construction. Therefore, cultural activities are often not deep and regular. At the same time, the lack of funds has also led to the lack of resources for the government to train talents specializing in cultural communication, which has brought certain difficulties to cross-border cultural communication.

Relatively Weak Cultural Status. In the context of the strengthening of globalization, the squeeze of the weak culture by the strong culture has become more and more serious. During the communication process, they have been forced to accept the huge impact from the strong culture. The emergence of new media has expanded the cultural communication influence of ethnic minorities. However, due to the influence of the mainstream society and the subtle influence of the media, the trend of ethnic minorities' sinicization is also very obvious. Even many ethnic groups have the characteristics of losing their own national language. This is especially obvious among the Ewenki people and in the Soviet Union. The compulsory implementation of Russian in the period has made Russian the language of inter-ethnic communication among the Evanki people. Only the elderly over 60 years old continue to use the Evanki language because they have not received formal education. The protection of weak culture is also a current dilemma one.

The Concept of Cultural Inheritance is Relatively Backward. In terms of school education in the two countries, under the bilingual teaching in our country and the mother tongue teaching advocated by Russia, graduates have gradually integrated with the local society, and boarding education has also caused problems in the way parents and children teach traditional national culture. Although the Ewenki nationality's ability to accept mainstream culture and integration into modern life has been enhanced, many ethnic members of the Ewenki nationality have shown a lack of interest in their own culture, unwilling to learn their own language and lifestyle, and inherit and spread their own traditional culture. There was conflict.

6 Cross-Border National Cultural Communication Strategies

6.1 Use Media Resources to Achieve Effective Communication

It is a necessary means of cultural communication in the Internet age to spread national culture through new media channels. At home, intangible cultural heritage culture can be spread to young groups through live network, and at abroad, national culture can be transmitted to foreign people through the same means. Therefore, it is a way to show the living conditions of Oroqen to foreign audiences by making Oroqen nationality documentaries or related programs. China has also paid attention to Oroqen nationality documentaries, but they have a long time ago, and most of the audiences are domestic people. At present, there are no documentaries from the perspective of foreign audiences.

Dissemination is only one link. How to make the national culture truly recognized and accepted by foreign cultures is still the key. Therefore, dissemination must not only be broad, but also effective. At the same time, we must also pay attention to not only let the national culture "go out", but also let the excellent foreign culture "in" in order to achieve the purpose of inclusiveness and mutual integration.

6.2 Show National Characteristics and Build Cultural Brand

The Ewenki nationality in China is mainly distributed in Heilongjiang and Inner Mongolia. The diverse ethnic minority cultures can eventually rise to the image of the city and the country, and the Ewenki people in the neighboring country must have closer contact with the neighboring cities, so based on the further extraction of local ethnic cultural characteristics in regional cities is also the need for external cultural communication. There is an introduction method in "Hulunbuir City Image and Ethnic Culture External

Communication", that is, by summarizing Hulunbuir's long grassland culture and refining the spiritual connotation of the city, to enable foreign people to fully understand that grassland culture is an important part of Chinese culture.

At the same time, it is necessary to actively export cultural products and build national cultural brands. The cultural products and services in Heilongjiang and Inner Mongolia are relatively weak. It is recommended to cooperate with domestic and foreign film and television companies, cultural institutions and other relevant departments through various channels. Set out to create cultural products with distinctive characteristics.

6.3 Highlight Ethnic Origin and Strengthen Cultural Identity

The cross-border ethnic groups living on the Sino-Russian boundary river-Heilongjiang have the same roots in ethnic and historical development, and share the values of respecting the old and loving the young, bearing hardships and standing hard work. The clan homology of cross-border ethnic groups is a great advantage. In the celebration of religion and festivals, the Oroqen people and the Ewenji people have retained the Shamanism beliefs and some common ancient customs. For example, the Oroqen people of the two countries respect the god of fire. The tradition of the Bonfire Festival, so on New Year's Eve, bonfires will be lit in front of every house, praying that the god of fire will give the tribe more prey next year.

Cultural dissemination should start from these common points. Different food and art performances of the two countries can be incorporated into the bonfire festival, and cultural elements of the two countries' literature, entertainment, fashion, clothing and apparel can be incorporated into the design of cultural tourism products; cross-border tourism product design It can be integrated into the Sino-Russian bridge, transnational affection, immigrants and other elements to show the characteristics of Sino-Russian border culture.

6.4 Strengthen the Construction of Digital Economy and Support National Cultural Innovation

Support the development of the Internet industry, use 5G, virtual reality technology, big data and other intelligent interactive technologies to apply in cultural consumption, education and health, transportation and travel, and encourage national digital creative industries, such as digital museums, digital cultural centers, digital art galleries, etc. At the same time, it supports the original research and development of digital content, and uses creative design as the driving force to focus on the development of film and television animation, software games, and digital publishing. It is similar and different in the traditional characteristics of national culture.

As the Ewenki traditional culture relies on precepts and deeds, oral inheritance and other characteristics, VR virtual reality, artificial intelligence and other technologies are restored to the daily life of nomads, and its characteristic costume culture and food culture are preserved to a great extent, with national characteristics and creativity. The design attracts young people, and builds an intra-ethnic communication platform through the unique customs and habits of the Ewenki ethnic group, so that the Ewenki ethnic group is no longer a single ethnic group, and communicates and exchanges in a modern way to a greater extent, strengthening the Ewenki culture at home and abroad Agree.

6.5 Build a Self-media Communication Mechanism

In the era of new media, ethnic minorities have more diverse choices for disseminating their national culture. Establishing a diversified and multi-level cultural dissemination mechanism has become one of the important ways to promote minority culture. The main body of dissemination has gradually transitioned from traditional media reporters to Ewenki people. Due to the inaccurate reports of traditional media, the Ewenki people have resisted. Therefore, each Ewenki people propaganda through the media, which is more conducive to the outside world to understand the real Ewenki At the same time, the filming of the Ewenki nationality documentary has formed a positive influence, so that more people have a strong interest in the Ewenki nationality, and it is also conducive to the self-promotion of the ethnic group by means of live broadcast and short video, in the protection and development of ethnic minorities. At the same time of culture, it also helps it maintain its authenticity.

7 Conclusions

In China with as many as 32 cross-border ethnic groups, these cross-border ethnic groups are not only a member of the Chinese nation, but also an important part of the soft power of Chinese culture. Under the advocacy of China's "One Belt, One Road" initiative, more and more countries and ethnic groups have participated in it. Each cross-border ethnic group on the border is a postcard displayed by the country. In the wave of modernization and modernization, foreign culture cannot be allowed to influence their destiny and future, which becomes a difficult problem in the protection and dissemination of national culture. Today under the development of new media, for cross-border ethnic groups, it is both an opportunity and a challenge. How to develop cross-border ethnic culture has also become an issue that cannot be ignored among the nations. As an important part of the "One Belt, One Road" initiative, cross-border ethnic groups are also of great significance to every country.

References

1. Rong, M.: How to understand "cross border ethnic groups". Open Era (06), 199–211+11 (2016)
2. Chen, P.: One belt, one road, the value implication of Guangxi's cross border ethnic culture heritage. J. Beibu Gulf Univ. **35**(03), 46–51 (2020)
3. Shao, Y.: Cultural diversity, cross-border interaction and cultural inheritance in the border areas between China and Laos -- based on the investigation of cross-border ethnic groups in Mengla County, Xishuangbanna. J. Chuxiong Norm. Univ. **30**(01), 77–82+92 (2015)
4. One belt, one road, Song, L., Jiang, J.: Comparative analysis and development strategy of cross border ethnic cultural education between Heilongjiang and China. Ethn. Minor. Educ. Res. **30**(05), 73–81 (2019)

5. Xu, H.: Cultural mutual learning between China and Russia: how to effectively cultivate the intimacy between the two peoples. Extern. Commun. (09), 30–33+1 (2018)
6. Zheng, Y.: Cross border communication strategy of frontier minority culture – taking the same root culture of Guangxi Zhuang and Vietnam Danong as an example. J. Guangxi Norm. Univ. Natl. **34**(03), 95–99 (2017)
7. Ni, J., Liu, Q.: Hulunbuir city image and external communication of minority culture. Pract. (Ideol. Theor. Ed. (06), 48–50 (2015)
8. Zang, J.: Comparative study on social life and customs of Oroqen people, a cross-border ethnic group between China and Russia. J. Heihe Univ. **9**(08), 205–206 (2018)
9. Yu, X.: Public opinion analysis of Russian Far East cooperation with China since the Ukraine crisis. Rus. Eastern Eur. Cent. Asia Stud. (06), 58–76+149 (2019)
10. Xia, T.: Exploration of ethnic minority culture communication path in the second half of the Internet. New Media Res. **5**(04), 95–96 (2019)
11. Li, D.: The dilemma and strategy of minority culture communication in the new media era. J. Hubei Univ. Natl. (Philos. Soc. Sci. Ed. **33**(02), 113–117 (2015)
12. Zhang, F.: On the choice of cultural modernization of ethnic groups with small population. Central University for Nationalities (2013)
13. Deng, L.: Investigation on the Media Contact of Shilu Ewenki People. Inner Mongolia Normal University (2019)
14. Wu Chuk, S.: Ewenki: An ancient ethnic group living in Northeast Asia. China Natl. Dly. (008) (2016)

Prospects of the Global Precision Medicine Market

Muge Yang[1] and Bin Li[1,2(✉)]

[1] Washington Institute for Health Sciences, 4601 N Fairfax Drive, Arlington, USA
[2] Georgetown University Medical Center, 3900 Reservoir Road, N.W., Washington D.C., USA
bl444@georgetown.edu

Abstract. With the rapid development of next-generation gene sequencing technology and information technology, the medical model is changing from the traditional "one-size-fits-all" model to the precision medical model, which is based on the specificity of individual genetics, environment and lifestyle to develop individualized treatment and prevention plans. In this study, through the analysis of publicly published precision medicine market reports, we summarized the driving factors and obstacles to the development of precision medicine, and estimated the global market value for the current and next decade. The factors that promote the development of precision medicine include technological progress, market demand, changes in medical models, and public attention and government action. The factors hindering the development of precision medicine include high price, technical limitations, scarcity of professionals, and lack of appropriate policies and regulations. The global precision medicine market size is estimated at USD 53.7–87.2 billion in 2019 and it is expected to reach USD 146.8–278.6 billion by 2030 and the compound annual growth rate (CAGR) from 2020 to 2030 is 11.13%. The precision medicine market can be segmented by ecosystem, end user, treatment and geography. By ecosystem and end user, diagnostic-related segments occupy the main part of the precision medicine market. By therapeutics, the oncology segment dominates the precision medicine market with an estimated more than 30% market dominance. North America is the largest and most potential regional precision medicine market, in 2019, its market was 31.4–54.6 billion and contributed approximately 35.0–46.8% of the global market values. Europe is the second-largest market with a market size of 15.1–34.3 billion in 2019. The Asia-Pacific region is the fastest growing precision medicine market with a CAGR of 14.4–16.6% in next few years. It is foreseeable that precision medicine is one of the fastest-growing industries in the world in the next ten years.

Keywords: Precision medicine · Global market value · Influencing factors

1 Introduction

Health and disease are the result of complex interactions between genetics and the environment. Due to the limited information collection and processing capabilities in the past, traditional medical models have to adopt a "one-size-fits-all" approach. Most

C. Yuan et al. (Eds.): *Proceedings of the 4th International Conference on Economic Management and Green Development*, AEPS, pp. 433–438, 2021.
https://doi.org/10.1007/978-981-16-5359-9_50

drug treatments are designed for "average patients". As a result, clinical practice based on the traditional medical model may be successful for some patients, but ineffective or have side effects for others.

With the rapid development of next-generation gene sequencing technology and information technology (IT), the genetic information of each patient can be obtained quickly and accurately. In addition, the widespread use of electronic medical records and the artificial intelligence processing of medical big data has promoted the transformation from traditional medical models to precise medical models.

Precision medicine is defined as "an emerging approach for disease treatment and prevention that takes into account individual variability in genes, environment, and lifestyle for each person". Compared with the traditional medical model, the advantages of precision medicine are obvious. Therefore, incorporating precision medicine into routine clinical practice has become an agenda for global health care development. In this study, based on the analysis of publicly published precision medicine market reports, we summarized the driving and hindering factors of the development of precision medicine, and estimated the global precise medicine market size for the current and next decade.

2 Method

We used "Precision Medicine Market" as the keyword to search the literature on ProQuest Central and downloaded the 200 most relevant articles. Based on the data in the literature, we estimate the value of the current global precision medicine market, including ecosystems, end users, therapies, and geographic regions, as well as the global market value in the next 10 years.

3 Results

3.1 The Factors Affecting the Development of Precision Medicine

The main factors that promote the development of precision medicine include technological progress, market demand, changes in medical models, and public attention and government action. Specifically, technological progress includes advances in low-cost gene sequencing technology, precision imaging technology, next-generation monoclonal antibody-based platforms, proper storage of genome data, cloud-based solutions, big data analytics, artificial intelligence, integration of wireless technologies with portable healthcare devices, etc. Market demand includes ageing population, rising prevalence of cancer and chronic diseases, etc. The changes in medical models include the shift from treatment to prevention, personalized treatment, targeted therapies, gene therapy, companion diagnostics, digitization of healthcare, reduce adverse drug reactions through pharmacogenomics testing, etc. The public attention and government action include patient's involvement in personal healthcare, high public awareness, potential to reduce the total cost of healthcare, payers willing to reimburse the cost of high-value drugs, government investment in research and development, amendments to health policies and regulations to make precision medicine more adaptable, etc.

The main factors hindering the development of precision medicine include high price, technical limitations, scarcity of professionals, and lack of appropriate policies and regulations. Specifically, the high price includes high diagnostic cost, high cost of deployment, lack of a systematic approach to fund high-cost personalized drugs, lack of robust reimbursement landscape, etc. Technical limitations include lack of research and evidence on molecular mechanisms/interactions, limited knowledge of precision medicine, lack of a unified data integration framework due to the multiple data systems used in patient data collection, lack of data sharing mechanism, challenge of secure storage of large volumes of sequenced data, potential risk of hardware or software failure, etc. Scarcity of professionals refers to lack of well-trained professionals with knowledge of both clinical medicine and IT. Factors in policies and regulations include security and privacy of sensitive patient information, lack of appropriate health policies and regulations, and lack of awareness about the possible applications of precision medicine, etc.

3.2 The Global Precise Medicine Market Size for the Current and Next Decade

The global precision medicine market size is estimated at USD 53.7–87.2 billion (Bn) in 2019 and it is expected to reach USD 146.8–278.6 Bn by 2030. The global precision medicine market is expected to exhibit a CAGR of 11.13% between 2020 and 2030 (Table 1 and Fig. 1).

Table 1. Global precision medicine market size.

Year	Market size (USD billion)
2015	38.9–39.1
2016	40.0–47.0
2017	45.7–54.0
2018	52.0–78.9
2019	53.7–87.2
2020	54.1–70.0
2021	59.8–78.8
2022	66.1–88.6
2023	87.7–98.3
2024	80.7–96.6
2025	85.5–119.9
2026	84.0–141.7
2028	120.3–216.8
2030	146.8–278.6

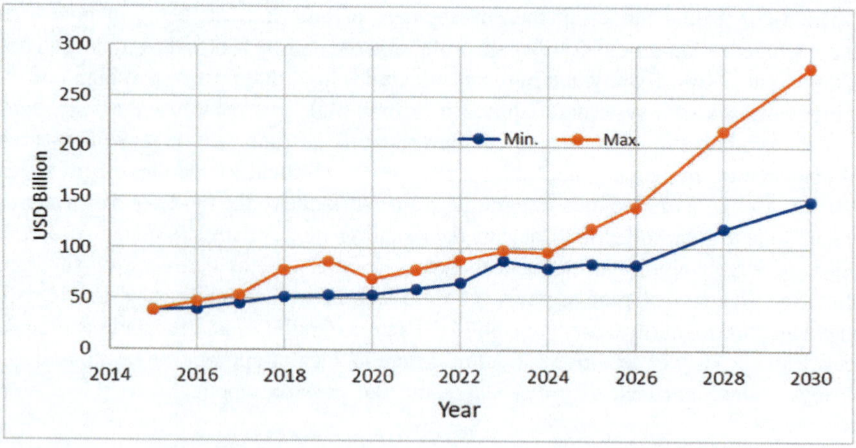

Fig. 1. Global precision medicine market size (USD billion).

3.3 Precision Medicines Market Size and CAGR in Different Segments in 2019

Precision medicines market has been segmented based on ecosystem, end user, therapeutics and geography. By ecosystem and end user, diagnostic-related segments occupied the main part of the precision medicine market. It is estimated that the market of global diagnostic tool companies is 19.9–24.3 Bn in 2019, and the CAGR in future several years is 11.5%; the diagnostic companies is 11.0 Bn, and the CAGR is 11.5%; the market of companion diagnostics is 10.6–15.4 Bn, and the CAGR is 11.4%. The drug discovery segment also occupies a considerable market. In 2019, its market was 12.4 Bn and CAGR in future several years is 8.3%. Precision medicine software market has a high growth rate, and its market is 1.1–1.2 Bn in 2019, the CAGR is 11.4–13.6%.

By therapeutics, the oncology segment dominates the precision medicine market with an estimated more than 30% market dominance due to the growing cases of cancer. Its market is 33.1–36.6 Bn in 2019, the CAGR is 11.9–13.9%.

By geography, North America is the largest and most potential regional precision medicine market. In 2019, its market was 31.4–54.6 Bn and CAGR is 10.8% and it contributed approximately 35.0–46.8% of the global market values. Europe is the second-largest market for global precision medicine with a market size of 15.1–34.3 Bn in 2019 and CAGR of 10.3–13.4%. Germany and U.K are expected to experience significant market growth. The Asia-Pacific region is the fastest growing precision medicine market in the world with a market size of 11.3–22.0 Bn in 2019 and CAGR of 14.4–16.6%. Due to the aging population and the rising incidence of chronic diseases, China has a dominant position in the Asia Pacific precision medicine market. India and Japan are expected to experience significant market growth. The precision medicine market in Latin America will be driven by the development of health infrastructure and population growth. Due to the poor socio-economic status, growth in the Middle East and Africa is slow (Table 2).

Table 2. Precision medicine market size and CAGR in different segments in 2019.

	2019 (Bn)	CAGR
Diagnostic tool companies	19.9–24.3	11.5%
Diagnostic companies	11.0	–
Companion diagnostics	10.6–15.4	11.4%
Drug discovery	12.4	8.3%
Precision medicine software market	1.1–1.2	11.4–13.6%
Oncology segment	33.1–36.6	11.9–13.5%
North America	31.4–54.6	10.8%
Europe	15.1–34.3	10.3–13.4%
Asia-Pacific	11.3–22.0	14.4–16.6%

4 Conclusions

With the rapid development of IT technology and DNA sequencing technology, precision medicine will replace the traditional medical model in the future. Especially in recent years, significant progress has been made in the diagnosis and treatment of cancer and the development of new drugs. Developed countries in North America, Europe and the Asia-Pacific region have invested a lot of research funds and formulated corresponding health policies and regulations to promote the development of precision medicine. It is foreseeable that precision medicine is one of the fastest-growing industries in the world in the next ten years, and the growth rate will remain over 10%.

References

1. Stern, A.D., Alexander, B.M., Chandra, A.: How economics can shape precision medicines. Science **355**, 1131–1133 (2017)
2. Gavan, S.P., Thompson, A.J., Payne, K.: The economic case for precision medicine. Expert Rev. Precis. Med. Drug Dev. **3**, 1–9 (2018)
3. Ginsburg, G.S., Phillips, K.A.: Precision medicine: from science to value. Health Aff. **37**, 694–701 (2018)
4. Dugger, S.A., Platt, A., Goldstein, D.B.: Drug development in the era of precision medicine. Nat. Rev. Drug Discov. **17**, 183–196 (2018)
5. Krzyszczyk, P., et al.: The growing role of precision and personalized medicine for cancer treatment. Technol. (Singap. World Sci.) **6**, 79–100 (2018)
6. Reportlinker: Global Precision Medicine Market: Focus on Ecosystem, Technology, Application, Country Data (21 Countries), and Competitive Landscape - Analysis and Forecast, 2020–2030 (2020). https://search-proquest-com.proxy.library.georgetown.edu/docview/2441014502/2BC441E11FF149DFPQ/1?accountid=11091
7. BIS Research: BIS Research Report Highlights the Global Precision Medicine Market to Reach $278.61 Billion by 2030 (2020). https://www-proquest-com.proxy.library.georgetown.edu/docview/2443366481?accountid=11091

8. Global Market Insights Inc.: Precision Medicine Market size revenue to exceed $87.7bn by 2023, to grow at 10.5% CAGR during the forecast period: Global Market Insights Inc. (2016). https://search-proquest-com.proxy.library.georgetown.edu/docview/1815476368/9C988DEA2819424EPQ/1?accountid=11091.
9. Zion Market Research: Europe Precision Medicine Market Will Reach USD 72,800.0 Million by 2022: Zion Market Research: According to the report, Europe precision medicine market was valued at around USD 33,200.0 million in the year 2014 and it is expected to reach approximately USD 72,800.0 Million by 2022. The Europe precision medicine market is expected to exhibit a CAGR of more than 10.2% between 2017 and 2022 (2018). https://search-proquest-com.proxy.library.georgetown.edu/docview/1994326826/DCCF5DC3CDCC4AD0PQ/1?accountid=11091
10. Clinical Trials Week: Research and Markets; Asia-Pacific Precision Medicine Market 2018–2023: A $20.9 Billion Opportunity-ResearchAndMarkets.com (2019). https://search-proquest-com.proxy.library.georgetown.edu/docview/2198094905/F4B4771D02934F82PQ/1?accountid=11091
11. Market Research Future: Precision Medicine Market to Demonstrate Progress at a ~12.48% of CAGR by 2022 (2018). https://search-proquest-com.proxy.library.georgetown.edu/docview/1988938188/3C59BD7EB2674150PQ/1?accountid=11091
12. Acumen Research and Consulting: Precision Medicine Market Size to Worth US$ 84 Bn by 2026: Acumen Research and Consulting (2019). https://search-proquest-com.proxy.library.georgetown.edu/docview/2302400655/BD5CDD20BB184725PQ/1?accountid=11091
13. Market Research Future: Precision Medicine Market USD 88.64 Billion Revenue Witnesses High Demand due to Prevalence of Genetic Diseases Forecasts by 2022 (2018). https://search-proquest-com.proxy.library.georgetown.edu/docview/2099048703/830A4E9E6247490DPQ/1?accountid=11091
14. Global Market Insights: Precision Medicine Market to register over 11% CAGR through 2026: Global Market Insights, Inc. (2020) https://search-proquest-com.proxy.library.georgetown.edu/docview/2369753246/9002D8F3DFBC47A9PQ/1?accountid=11091

Did Cultural Finance Policies Improve Financing Efficiency of Cultural Corporates in China? Based on the Empirical Analysis of Listed Companies in 2006–2018

Liu Yijun[1], Jin Xuetao[2(✉)], and Zhang Tianchang[3]

[1] School of Economics and Management, Communication University of China, Beijing 100024, P. R. China
[2] Faculty of International Media, Communication University of China, Beijing 100024, P. R. China
jinxuetao-cuc@cuc.edu.cn
[3] Department of Economic History, The London School of Economics and Political Science, London WC22AE, UK

Abstract. Financial capital plays a very important role in the cultural industry which has high sunk costs and volatile returns. Therefore, studying the impacts of cultural finance policy on the financing efficiency of cultural corporates has practical significance. This paper used a sample of listed non-financial firms from 2006 to 2018 and studied the association between cultural finance policy and the financing efficiency of cultural corporates based on the information asymmetry theory and principal-agent theory. The empirical research results show that cultural finance policy significantly increases cultural corporates' financing efficiency. These results enlarged the research content of cultural corporate and provide theoretical guidance for supporting the development of cultural corporate.

Keywords: Public finance · Cultural industry in mainland China · Financing efficiency

1 Introduction

Cultural enterprises building upon creativity and innovation and relying on human resources as the main assets often are confronted with financing difficulties due to the comparatively large initial investment and the unpredictability of future return. In 2010, nine major government departments, collectively issued the Guiding opinions on Financial Support for Invigorating and Developing Cultural Industry Prosperity, which is China's first special policy document on cultural finance at the national level. Since then, competent cultural authorities and local governments have issued measures to support the development of cultural finance successively, targeting the establishment of a multi-level capital market and the enlargement of cultural enterprises' direct financing

C. Yuan et al. (Eds.): *Proceedings of the 4th International Conference on Economic Management and Green Development*, AEPS, pp. 439–446, 2021.
https://doi.org/10.1007/978-981-16-5359-9_51

scale; indeed, investigation of the impact of cultural financial policy on the financing efficiency of cultural enterprises of certain burden practical implications.

Although scholars reviewed the financing efficiency and contributed substantial research findings, the comparatively longer period samples and impact of policy environment changes on the financing efficiency of cultural enterprises are less investigated. Building upon information asymmetry and principal-agent theory, this paper takes all Chinese non-financial listed enterprises from 2006 to 2018 as samples to study the impact of cultural finance policy on the financing efficiency of cultural enterprises.

2 Theoretical Foundation

2.1 Financing Cost

Cultural enterprises have labour and cultural resources as the core input, and capital as an important input factor, namely, cultural industry is categorized as knowledge-intensive and capital-intensive industries. The production of cultural enterprises depends on labour's intellectual work to excavate and select cultural resources, to combine cultural resources with symbolism, and to produce cultural products and services as a consequence. This complexity confers the ultimate value of cultural products and services high uncertainty. The financial capital is crucial to Cultural enterprises with high sunk costs and fluctuating returns. As for Chinese scholars, Zeng firstly proposed the concept of 'financing efficiency' in mainland China, claiming that financing efficiency and financing cost would influence financing decisions [1]. Financing efficiency has been broken down into fund-raising efficiency and allocation efficiency, in which fund-raising efficiency refers to the enterprise's ability to obtain capital at the lowest cost, while allocation efficiency refers to the enterprise's ability to employ the capital for the most efficient use [2]. The fund-raising efficiency is determined by financing cost, while the latter is up to return of capital allocation. The financing cost of cultural enterprises falls into external financing costs and internal financing costs. The external financing cost stems from the following three aspects.

The first aspect is the information asymmetry. The ultimate value of the products and services of cultural enterprises is highly uncertain; the operation risk of enterprises is large, and the profit sustainability is difficult to estimate.

Secondly, financial products and services are lacking pertinence. Traditional enterprises rely on tangible collateral for debt financing from financial institutions such as banks or non-financial institutions, while cultural enterprises mainly rely on intellectual property rights and human capital which fail to be scientifically and effectively assessed under the Chinese current evaluation system. The majority of financial institutions cannot satisfy the financing requirements of cultural enterprises and exhibit the shrunken motivation to develop and tailor products and tools suitable for cultural enterprises [3].

The last aspect is the absence of the effective mechanism in capital market of cultural sector. Now, in mainland China, there are public cultural organizations and pure cultural enterprises, and they play different roles for the development of culture. However, it should be noted that the development of capital market of cultural sectors is lagging behind other sectors such as manufacture, Hi-tech, and transportation, etc. In reality, the lack of a systematic culture finance market construction regime and corresponding preferential policy issues induced to the low financing efficiency.

The preceding three phenomena aggrandized the external financing cost of cultural enterprises. Since external financing could not acquire sufficient funds, must funds be supported by internal financing, which also increases the opportunity cost of internal financing. Therefore, the financing cost of cultural enterprises is far exceeding that of other types of enterprises. The particularity of cultural enterprises further aggravates the conflicts of interest among shareholders, management, and creditors, due to high uncertainty of the ultimate value of cultural enterprises' products, services, and relatively great risk of target investment projects. Besides, as cultural enterprises are mainly large state-owned cultural enterprises and small and medium-sized private cultural enterprises, their governance capabilities are generally low, which would intensify the contradiction between shareholders, management, and creditors. In the face of these issues, the return of capital allocation of cultural enterprises is lower than that of other companies.

2.2 Culture Finance Policy

After the release of the Plan on Reinvigoration of the Cultural Industry in 2009 and the Guiding opinions on financial support for invigorating and developing cultural industry prosperity in 2010 (hereafter referred to as Guiding opinions), the cultural industry has been promoted to national strategic industry, and the financial environment of the cultural industry was greatly improved. We proposed that the promulgation of cultural finance policy improves the financing efficiency of listed cultural companies at least from the following two aspects.

Reduce the Financing Costs. Firstly, the implementation of cultural finance policy could reduce the financing costs of listed cultural companies and improve the fund-raising efficiency in financing efficiency. As mentioned above, the high financing cost of cultural enterprises mainly attributes to information asymmetry, deficient targeted financial products and services, and insufficient institutions. For information asymmetry, the Guiding Opinions required financial institutions to establish a scientific credit rating system and business evaluation system and establish a multi-level loan risk-sharing and compensation mechanism for the cultural industry. To tackle the lack of pertinency of financial products and services, the Guiding Opinions required the formulation and the perfection of the measures for the evaluation, pledge, registration, custody, circulation, and realization of intangible assets such as patent rights and copyrights. The Guiding Opinions also demanded the financial institutions to develop credit products suitable for cultural enterprises. For the lack of institutions, the Guiding Opinions supported the central and local finances to set up investment funds for the cultural industry, on that basis, to develop the multi-level capital market vigorously and increase the scale of direct financing of cultural enterprises.

Enhance the Return of Capital Allocation. Secondly, the promulgation of the cultural financial policy could enhance the return of capital allocation of listed cultural companies and promote the allocation efficiency in financing efficiency. The return of capital allocation of cultural enterprises is lower than that of other enterprises, largely ascribing to biggish conflicts of interest among shareholders, management, and creditors. The Guiding Opinions required the cultural enterprises to establish a modern corporation system and consummate the company's financial management system and governance structure. The aforementioned measures were capable of improving the governance level of cultural enterprises, moderating the conflicts of interest among shareholders, management and creditors, raising the return on the capital allocation of cultural enterprises, and then improve the allocation efficiency in the financing efficiency of cultural enterprises.

The preceding theoretical paved the firm foundation to garner further insights into the following hypothesis relating to financing efficiencies.

H1: The promulgation of the cultural financial policy was conducive to the financing efficiency of cultural enterprises.

3 Sampling and Data Description

This paper took all A-share listed companies from 2006 to 2018 as the initial data, since the Ministry of Culture has promulgated the Outline for the Culture Development Planning during the National 11th Five-Year Plan Period in 2006, which stressed the development of cultural and creative industries. To guarantee the robustness and pertinence of the research results, the process of data screening was deliberately kept up to the following rules: (1) all financial companies are eliminated due to differences in accounting standards; (2) all samples with missing financial data were removed. The sample dataset was composed of 28,204 samples extracted from China Stock Market & Accounting Research Database (CSMAR). This article employed Stata15.0 for initial data processing. To control the influence of some special values, a 1% winsorization level has been applied to the sample dataset (Table 1).

Table 1. Descriptive statistics result.

Variable	Sample size	Mean	Median	Standard deviation	Maximum	Minimum
DEA	28204	0.753	0.749	0.059	1.000	0.457
Culture	28204	0.092	0.000	0.289	1.000	0.000
Nature	28204	0.436	0.000	0.496	1.000	0.000
Size	28204	21.960	21.810	1.309	25.910	19.010
CF	28204	0.042	0.042	0.076	0.255	−0.200
Relate	28204	0.497	0.000	0.500	1.000	0.000
GDP	28204	21.330	24.000	8.263	31.000	1.000

The descriptive statistical results of all variables provide an overview of the dataset. The evaluation object (Culture) has an average at 0.092, indicating that 9.2% of enterprises belong to cultural enterprises and the amount of cultural enterprises is relatively modest (SD = 0.289). For control variables, the average value of Nature is 0.436, indicating that 43.6% of enterprises have state-owned property rights. The maximum and minimum of the scale of enterprises (Size) is 25.910 and 19.010, and there's a significant difference in the scales of sample enterprises (SD = 1.309). The maximum and minimum of cash flow (CF) is 0.255 and −0.200, while the difference in cash flow is narrow among sample enterprises (SD = 0.076). The average level of the macroeconomic situation (GDP) is 21.330, with a median of 24.000, illustrating that the sample enterprises are mainly registered in economically developed regions.

4 Regression Analysis

The Tobit model combined with the difference-in-difference model (DID model) was applied to verify hypothesis 1. The model (1) has been set up as follows:

$$DEAit = \alpha 0 + \alpha 1 Cultureit + \alpha 2 Afterit + \alpha 3 Cultureit \times Afterit + \sum k\alpha k\ Controlit + \phi it \quad (1)$$

Dependent variable is DEA in the model (1), referring to the variable of financing efficiency. The majority of existing studies relied on fuzzy comprehensive evaluation method, the ratio of return on investment (ROI) to cost of capital (ROI), multiple regression linear model, and DEA model for evaluation. DEA model, as a linear programming model, eliminates the influence of subjective factors and has good objectivity. By referring to the previous research, we derived the financing efficiency of cultural enterprises and other enterprises from DEA's constant return to scale model (CCR) and variable return to scale model (BBC) [4,5]. The comprehensive technical efficiency calculated by the two models is the financing efficiency of enterprises. The input index of the DEA model conveys the financing scale and financing cost, and the output index reflects the return of capital allocation (Table 2).

Table 2. Input and output indicator definition.

Indicator category	Indicator name	Measurement method
Input indicator	Aggregate liability	Total indebtedness
	Aggregate stock rights	Paid-in capital
	Aggregate internal financing	Sum of surplus reserves and undistributed profit
	Financial expense	Total financial expense
Output Indicator	Business growth	Increase rate of operating income
	Return on total asset	Net profit/total assets at the beginning of the year
	Operating income	Sum of operating income

Since the DEA model requires input-output data to be non-negative, we nondimensionalized the data of input and output indicators to ensure all data to be positive without any impact exerted on the final research results. The specific treatment method is shown in the model (2).

$$y_{ij} = 0.1 + 0.9 \times \frac{x_{ij} - min(x_{ij})}{max(x_{ij}) - min(x_{ij})} \tag{2}$$

The application of DID requires a corresponding experimental group and control group. In model (1), the cultural enterprise and other enterprises are the experimental groups and the control group respectively, and Culture is the variable to distinguish the treatment group and the control group. If the enterprise belongs to the cultural enterprise, it is 1; otherwise, it is 0. The method to judge a cultural enterprise is whether the description of the main business in its financial report conforms to the criteria of Cultural and Related Industries Classification (2018). After represents the dummy variable to distinguish the period before and after the release of the Guiding Opinions. The release time of the Guidance is 2010, however, there was a time lag for the effect taking place; variable After 2010 is 1, and the variable before 2010 is 0. Culture × After regression coefficient is the observed object, indicating differences in the financing efficiency between cultural enterprises and other enterprises before and after the issue of the Guiding Opinions.

In model (1), Control indicates the Control variable. The selection of control variables, using the nature of property rights, the scale of enterprises, cash flow, governance level, and macroeconomic situation were derived from previous research [6,7]. The detailed implications and measurement methods are in Table 3.

Table 3. Variable definitions and measurement methods.

Variable	Implication	Measurement methods
DEA	Financing efficiency	The comprehensive technical efficiency is calculated according to the DEA and BBC models
Culture	Appraisal object	Dummy variable. Cultural enterprise (experimental group) is 1, other enterprises (control group) is 0
After	The period of cultural finance policy implementation	The dummy variable of the release of Guiding opinions. After 2010, the variable is 1; otherwise, it is 0
Nature	Nature of property right	When the enterprise is state-owned stock, it is 1; otherwise, it is 0
Size	Scale of enterprise	The natural log of the total assets of listed companies at the end of the year
CF	Cash flow	The ratio of net cash flow from operations to total assets
Relate	Governance level	If there is no correlation between the top ten shareholders, it is 1; otherwise, it is 0
GDP	Macroeconomic situation	The GDP ranking of the company's registered location in the corresponding year. The higher the ranking, the higher the GDP

Table 4 is the regression results of the model (1). In column (3), the coefficient of Culture is −0.016, and the coefficient of Culture × After is 0.015, which is significant at the 1% level as well. The results validate hypothesis 1. The Guiding Opinions reversed the relationship and enabled the financing efficiency of cultural enterprises to be superior to that of others. The control variables, nature of property rights, the scale of enterprise and governance level are negatively correlated with financing efficiency, while cash flow and macroeconomic situation are positively related to financing efficiency, with all of five correlation significant at 1% level.

Table 4. Cultural financial policies and financing efficiency.

Variables	(1)		(2)		(3)	
	DEA		DEA		DEA	
	Coefficient	T-value	Coefficient	T-value	Coefficient	T-value
Culture	−0.004***	−2.66	−0.013***	−5.00	−0.016***	–
Culture × After			0.013***	4.45	0.015***	5.38
After			−0.008***	−9.99	−0.006***	−7.15
Nature					−0.011***	−14.50
Size					−0.005***	−18.19
CF					0.192***	42.29
Relate					−0.002***	−3.37
GDP					0.001***	5.77
Constant	0.741***	180.88	0.746***	181.10	0.859***	116.69
Ind	Control		Control		Control	
Observations	28204		28204		28204	
P > chi2	0.000		0.000		0.000	
LR chi2	1519.430		1621.200		49.686	
LL	40212.07		40262.96		41460.82	

* $p < 0.10$, ** $p < 0.05$, *** $p < 0.01$

5 Conclusions

The financing efficiency of cultural enterprises was generally lower than that of other enterprises before the release of the Guiding opinions, and the financing efficiency of cultural enterprises has been significantly promoted by the policy. By facilitating the financial innovation of the cultural industry, and ameliorating the issues such as information asymmetry between the supply and demand side of the capital in the industry, the financing efficiency is improved.

Although the computing method of financing efficiency in this paper is drawn on the results of existing scholarship, it may not accurately reflect the financing efficiency of cultural enterprises and may reduce the accuracy of the conclusion in this paper. Nevertheless, the research content of this paper is still of certain theoretical and practical implication, which enriches the research on cultural enterprises and financing efficiency. We aim to shed light on a comprehensive understanding of the impact of cultural finance policy on cultural enterprises and provide theoretical guidance to better facilitate the development of cultural enterprises.

References

1. Zeng, K.: How to view direct and indirect financing. Southwest Finan. **11**, 30–32 (1993). (in Chinese)
2. Zhao, D., Li, G., Zhou, H.: The reformation of Chinese financing system. Economic Press China (1999). (in Chinese)
3. Chang, Y.: Research on financial support for the development of cultural Industry. Econ. Res. Guid. **000**(012), 84–85 (2009). (in Chinese)
4. Luo, C., Zhang, P., Li, X., Liang, S.: Research on the financing efficiency of cultural finance industry based on DEA method. Stat. Decis. **000**(023), 107–109 (2016). (in Chinese)
5. Xiong, Z., Ding, L., Wan, J.: Equity financing efficiency measurement and promotion strategy of listed companies of cultural industry: In the perspective of doubling plan. Econ. Manag. **36**(8), 109–116 (2014). (in Chinese)
6. Cui, J., Hu, H., Zhang, D.: Research on factors influencing financing efficiency of unlisted small and medium-sized enterprises—Evidence from manufacturing unlisted SMES. Soft Sci. **12**, 88–92 (2014). (in Chinese)
7. Zhou, C., Xun, C.: Evaluation and research on financing efficiency and influencing factors of sea-related industrial enterprises—empirical data based on DEA-random effects models. J. Ocean Univ. China **2**, 19–26 (2015). (in Chinese)

Analysis of Stock Investment Behavior: Case Study on College Students from Tianjin

Rong Chen(✉)

Department of Management, Tianjin University Renai College, Tianjin, China

Abstract. In recent years, more and more college students have joined the stock market due to a bull market in China. However, it is controversial whether they should invest in the stock market during the period of university. In this paper, through the questionnaire survey of college students in Tianjin, in order to analyze their stock investment behavior, Logistic model is used to research on college students' willingness to invest in stocks, and factor analysis is used to investigate the reasons of investing in stocks for college students. According to the results, it is suggested how to guide the stock investment behavior of college students in the future.

Keywords: Stock investment behavior · College students · Logistic model · Factor analysis

1 Introduction

As an important part of future investors, college students get lots of attention and their status and habits of investment would affect their future investment tendency. The phenomenon of investing in stocks among the group of college students heated increasingly. How to treat the phenomenon of investing in stocks of college students correctly is always a subject to be thought and studied deeply by college educators. Financial investment can enable contemporary college students to have a better concept of wealth and financial awareness, but college students lack of financial investment and management knowledge [1]. The financial management consciousness of college students is still in a stage of start. It should put forward opinions on college students' proper guidance in financial investment from the perspective of university, family and society to make students establish good financial management concepts, to create a good environment of finance and investment for college students [2]. Majority of scholars hold agree opinion and think that it could establish a healthy investment concept for college students by positive education and leading from the perspective of school, parents and society.

Then what kinds of college students are more likely to enter into the stock market? What is the motivation for college students to enter the stock market? Before education is being conducted, it should urgently make a thorough research on the stock investment behavior of current college students. The most existing literatures mainly focus on the research of investment behavior of residents and institutions, even though there are

C. Yuan et al. (Eds.): *Proceedings of the 4th International Conference on Economic Management and Green Development*, AEPS, pp. 447–457, 2021.
https://doi.org/10.1007/978-981-16-5359-9_52

some studies to research financial investment behavior of the university students as investors, concentrating on college students born after 90's in south of China. As times are changing, learning and living environment are different, it is dying to know how it is going in north of China. Therefore, college students in these five years are taken as respondents in Tianjin, the financial center of northern China, to conduct questionnaires investigation and empirical analysis of intention of college students in stock market investment in-depth to answer the two questions mentioned above.

2 Literature Review and Questionnaire Design

The research on college students' stock investment behavior is mainly carried out from two aspects mentioned above. The first problem can be solved by analyzing the factors influencing college students' stock investment intention. Most of the students who have entered the stock market are motivated by the desire to learn financial knowledge, accumulate experience and investment [3]. Young people, especially students in school, have the characteristics of group life and are easily affected by the surrounding crowd [4]. Second, on the basis of the previous research questions, this paper further explores the reasons of college students' stock investment, so as to solve the second question what the motivation for college students to enter the stock market is. Zheng, Zhang and Fang [5] finds that more than 80% of college students are motivated to enter the market because of earning the price difference. Liu [6] concludes that the number of college students involved in investment and financial management is more and more and their willing to invest in stock become stronger, but there are also some phenomena of following trend blindly as students.

Based on the research achievements of predecessors, this paper draws on existing questionnaires combining with the characteristics of the present college students. The questionnaire was distributed randomly to several universities in Tianjin for researching from November to December in 2019. There are 235 questionnaires being collected and 195 questionnaires are valid in total. The effective of collecting is 83%. The respondents are involving undergraduates, graduates and PHD students. The proportion of undergraduates is 82.06%. There are 52 students investing in stocks and the ratio is 26.67%. The other 143 students don't have any experience of investing stocks and the ratio is 73.33%. The proportion of these two groups is 1:2.75. From the gender perspective, there are 84 males (43.08%) and 111 females (56.92%). The ratio of these two groups is approaching 1:1. From the major perspective, there are 147 students (75.38%) from majors in economic and management and 48 students (24.62%) from non-economic and management majors. The ratio is nearly 3:1. Thus, the samples are highly representative.

3 Analysis of Stocks' Investing Intention of College Students

On the basis of the results of the existing research [7], and combining with the basic situation of the questionnaire, the willingness of college students to invest in stocks can be expressed as whether they have ever invested in stocks. Therefore, this part will use the data obtained from the first part of the questionnaire and try to establish a Logistic model to analyze the influencing factors of college students' willingness to invest in stocks. The willingness to invest in stocks can be expressed if any experience in trading stocks, which is taken as dependent variable (y_1). There are 13 independent variables, which are Gender (x_1), Major (x_2), Grade (x_3), Household income (x_4), The number of relatives and friends investing in stocks (x_5), Familiarity with financial media (x_6), Average monthly living expenses (x_7), Knowledge of stocks (x_8), Expected returns of stocks (x_9), Tolerance for volatility of market prices (x_{10}), Maximum loss tolerance of assets (x_{11}), Maximum tolerance for assets volatility (x_{12}), Degree of risk appetite (x_{13}).

Before doing Logistic model, the independent sample t-test and test of variance are used to observe if any significant differences in the variance and mean of each variables between the two groups, one is investing in stocks and the other is not. From the results, the grade, household income, familiarity of financial media, average of monthly living expenses, knowledge of stocks, tolerance for market prices, maximum loss tolerance of assets and maximum tolerance for assets volatility, the variances of these 8 variables are not equal significantly, and the mean of them is significant different at different significance levels (10%, 5% and 1%). It may tell that these 8 variables may be significantly relative to willingness to invest in stocks. The variance of other variables can be considered as being equal between the two groups. While the mean of gender, the number of investing stocks of relatives and friends and degree of risk appetite is all significant different at 5% significance level. The mean of major is significant different at 10% significance level. The mean of expected returns of stocks is not significantly different between the two groups, which means that this variable may be not relative to the willingness to invest in stocks.

It can be found from the correlation coefficient matrix between variables that it is a significant correlation between each variables. The KMO test value is 0.661 which is more than 0.6. Bartlett spherical test value is 532.814 and the probability is 0.000 which is less than the significance level 0.05. These results tell that variables are highly correlated. In order to avoid collinearity, principal components of variables other than dummy variables are extracted before Logistic regression is used. In addition, all data samples are randomly divided into the training group and the test group, with the training group sample size accounting for 2/3 of total and the test group sample size accounting for 1/3. The training group sample is used for regression analysis, and the test group sample is used to test the accuracy of the established model.

Four principal components have been extracted using the principal component analysis program in SPSS19.0. The cumulative variance contribution rate is 65.067%, basically preserving the original variable information. From the results of coefficient matrix of principle components (shown in Table 1), the absolute values of zx_{12},zx_{11} are the maximum relatively from the component F_1, which represents the risk tolerance of college students. The absolute values of zx_7,zx_3,zx_4 are maximum relatively from the component F_2, which represents the economic capability of college students. The absolute value of zx_8 is the maximum relatively from the component F_3, which represents the knowledge of stocks. The absolute value of zx_9 is the maximum from the component F_4, which represents the expected return of stocks.

It takes the extracted four principal components and two dummy variables, which are gender and major, as independent variables, and whether trading in stocks, which is a binary variable for Logistic regression, is taken as the dependent variable. The Forward method (Wald) was selected as the filtering method for variables to the model. The results show in Table 2 and only F_1 and F_2 pass the Wald test. The Logistic model can be established as shown in Eq. (1). The Logistic model can be expressed as standardized variables shown in Eq. (2).

Table 1. Coefficient matrix of principle components

Variables	Coefficients of principle components			
	F_1	F_2	F_3	F_4
zx_3	0.167	0.392	−0.425	0.260
zx_4	0.317	0.327	0.477	0.085
zx_5	0.368	0.255	−0.015	−0.072
zx_6	0.313	0.088	−0.318	−0.232
zx_7	0.241	0.475	0.422	−0.117
zx_8	0.295	0.123	−0.499	0.226
zx_9	0.124	−0.243	0.215	0.738
zx_{10}	0.281	−0.330	0.127	−0.125
zx_{11}	0.375	−0.354	−0.032	−0.320
zx_{12}	0.415	−0.300	0.031	−0.118
zx_{13}	0.292	−0.208	0.022	0.354

Note: zx_i represents the standardized x_i, F_i represents the extracted principle components.

Table 2. Logistic model abstract

Variables	Coefficient	S.E.	Wald	Sig.
F_1	−0.356	0.126	7.926	0.005
F_2	−0.394	0.164	5.794	0.016
Constant	1.247	0.225	30.616	0.000

$$ln\frac{p}{1-p} = 1.247 - 0.356F_1 - 0.394F_2 \quad (1)$$

$$ln\frac{p}{1-p} = 1.247 - 0.214zx_3 - 0.242zx_4 - 0.232zx_5 - 0.146zx_6 - 0.273zx_7 - 0.153zx_8 + 0.052zx_9 + 0.030zx_{10} + 0.006zx_{11} - 0.030zx_{12} - 0.022zx_{13} \quad (2)$$

From Eq. (2), there are 3 independent variables are positive relative with the willingness to invest in stocks, which are the expected returns of stocks, the tolerance for volatility of market prices and the maximum loss tolerance of assets. It means that the willingness to invest in stocks will increase when the expected returns of stocks, the tolerance for volatility of market prices and maximum loss tolerance of assets increase respectively. Other variables are all negative with the willingness to invest in stocks. The higher values of the grade, the household income, the number of relatives and friends investing stocks, the familiarity with financial media, the average monthly expenses, the knowledge of stocks, the maximum volatility of assets and the degree of risk appetite, the weaker the willingness of investing stocks for college students. The possible reason is that college students do not establish an awareness of stock investment and financial management. They don't pay much attention on finance or stock news in daily life. Most of them just hear about something related investment in stocks from surrounding people without researching by themselves. The coefficients of average monthly expenses, the household income, the number of relatives and friends investing in stocks, and the grade are relatively maximal, which means these 4 variables may influence the willingness to invest in stocks much more. It can be stated that the college students' willingness to invest in stocks is mainly influenced by their economic capability, relatives and friends as well as their ages. The average monthly living expenses has the lowest correlation with college students' willingness to invest in stocks. The possible reason is that they don't have extra money to invest in stock market, and also they may not have enough financial awareness. Due to the loss of information in both the principal component extraction and Logistic progress, the regression results are partially different from those obtained by the independent sample t test.

It can be used to predict if college students will make stock investment by the Logistic model obtained by regression. Generally, 0.5 is used as the probability limit. That is, when the probability is larger than 0.5, college students will be judged to make stock investment. While when the probability is less than 0.5, college students will not make stock investment. In this paper, the accuracy of the prediction of Eq. (2) is tested with samples from the test group. The prediction accuracy is as high as 76.92%.

4 Reason Analysis of College Students' Investing in Stocks

This part focuses on the reasons of college students to make stock investment. The objective of research is only for those college students who have experience of stocks investment, so the data collected from the second part of the questionnaire is used with an effective sample size of 52. There are 18 questions being proposed with certain correlations among the questions. Therefore, this part intends to use factor analysis method to extract representative influence factors and to analyze the importance of each influence factor.

According to Wu and Yao [7] as well, we put forward the possible reasons from four aspects which are the self-fulfillment factors, the psychological factors, the market factors and the economic factors. The Likert's five-point scale is used. The values of 1–5 are representing the "Strongly Disagree" to "Strongly Agree" respectively. All questions are settings from the angle of subjective perception of college students. The design and descriptive statistics of each indicator are shown in Table 3.

SPSS19.0 is also used for factor analysis. The reliability and validity of sample data are tested before factor analysis. Cronbach's Alpha coefficient is used to measure the reliability test. The Cronbach's Alpha of the 18 test items is 0.929, reaching above 0.8 indicating that the data shows internal consistency and high reliability. The KMO test and Bartlett test are used for the validity test. The KMO test value is 0.826, and the Bartlett spherical test statistic Sig < 0.01, indicating that there is a significant difference between variables correlation, which is suitable for factor analysis.

The principal component analysis method is used to extract common factors, and the common factors are extracted according to the condition that the characteristic value is greater than 1. The results show that a total of 4 common factors can be extracted. The factor loading matrix is rotated according to the maximum equilibrium method. The rotation results show that the variance contribution rates of the four common factors are 19.062%, 18.380%, 18.242%, and 17.292%, respectively, with a total contribution rate of 72.974%, indicating that these four factors could better explain the reasons for college students to invest in stocks. The extraction rate of the variables by common factors is above 60%, so it can be seen that common factors can reflect most of the information of each variable.

Table 3. Indicators and descriptive statistics

Influence factor	Indicator content	Mean	S.E.
Self-fulfillment factor	Knowledge of stocks	2.962	1.154
	Achieve financial independent	2.885	1.114
	Reveal intelligence	3.058	1.178
	Feel happy to make money	3.019	1.111
	Improve financial management skills	3.308	1.147
	Exercise psychology quality	3.327	1.024
Psychological factor	Envy to make money in the stock market	2.885	1.182
	People around to talk about stocks a lot	2.750	1.169
	Consider the stock market as legal casinos	3.115	1.149
	People around to invest stocks	2.846	1.092
	Feel out of style not investing stocks	2.481	1.196
	Feel interesting to invest stocks	3.250	1.186
Market factor	Strong policy support	3.077	0.987
	Stock market getting better	2.846	0.937
Economic factor	Not enough living costs	2.962	1.400
	Make money fast	2.827	1.150
	Rising prices	3.115	1.182
	High yield investment	3.115	1.182

From the results shown in Table 4, it is high correlation between factor 1 with the four following statements: envy to make money in the stock market, not enough living costs, consider the stock market as legal casinos and reveal intelligence. It can be named as economic benefit factor. It is high correlation between factor 2 with exercise psychology quality, improve financial management skills, high yield investment, feel interesting to invest stocks and strong policy support. It can be named as self-fulfillment factor. The factor 3 is high correlated with knowledge of stocks, feel out of style not investing stocks, stock market getting better, achieve financial independent and feel happy to make money. It can be named as financial management consciousness factor. The factor 4 is correlated with rising prices, people around to talk about stocks a lot, make money fast and people around to invest stocks. It can be named as environment around factor.

According to the ratio of variance contribution rate of each factor to the total variance contribution rate, the influence weight of each factor on college students' stock investment behavior can be calculated, as shown in Table 5. Among them, the economic benefit factor has the largest influence, the self-fulfillment factor is the second, and the financial management consciousness factor is the third one. The environment around factor has the least influence. As we can see, there is a few difference between the influence weights of each factors. It is indicating that the factors affecting college students in Tianjin to invest in stocks are variety and each factors seem important to influence

Table 4. Rotated component matrix

Variables	Component			
	1	2	3	4
Envy to make money in the stock market	0.758	0.197	0.207	0.232
Not enough living costs	0.752	−0.066	0.027	0.036
Consider the stock market as legal casinos	0.732	0.209	0.097	0.201
Reveal intelligence	0.679	0.339	0.115	0.352
Exercise psychology quality	0.263	0.834	0.084	0.282
Improve financial management skills	0.228	0.748	0.182	0.359
High yield investment	0.107	0.664	0.206	0.113
Feel interesting to invest stocks	0.030	0.551	0.462	0.354
Strong policy support	0.375	0.467	0.445	0.295
Knowledge of stocks	−0.118	0.411	0.747	0.075
Feel out of style if not investing stocks	0.315	−0.136	0.732	0.314
Stock market getting better	0.546	0.306	0.670	0.062
Achieve financial independent	0.532	0.359	0.568	0.072
Feel happy to make money	−0.047	0.496	0.555	0.349
Rising prices	0.106	0.389	0.018	0.819
People around to talk about stocks a lot	−0.025	0.140	0.546	0.760
Make money fast	0.462	0.219	0.034	0.724
People around to invest stocks	0.308	0.118	0.500	0.673

college students' investing stocks. But they don't have a certain understanding of the volatility of the stock market and the impact of nations' policy. With the progress of the times and the diversification of information, college students can understand the knowledge of investment and financial management from a variety of channels. However, it is still worrying that the speculative psychology of college students who want to make fast money by speculating in the stock market still accounts for most of the reasons, and there is a certain phenomenon of blind following and a lack of self-discrimination ability.

Table 5. Name and weighting of each factor

Name of factors	Dependent variables	Weighting
Economic benefit factor	Envy to make money in the stock market, Not enough living costs, Consider the stock market as legal casinos, Reveal intelligence	26.12%
Self-fulfillment factor	Exercise psychology quality, Improve financial management skills, High yield investment, Feel interesting to invest stocks, Strong policy support	25.19%
Financial management consciousness factor	Knowledge of stocks, Feel out of style not investing stocks, Stock market getting better, Achieve financial independent, Feel happy to make money	25.00%
Environment around factor	Rising prices, People around to talk about stocks a lot, Make money fast, People around to invest stocks	23.70%

5 Conclusion and Suggestion

This paper takes college students in Tianjin as the respondents of investigation. Logistic model and factor analysis are used to make an empirical study on college students' willingness and reasons of stock investment. From the perspective of investment intention, firstly, the monthly average living expense is an important factor to identify whether college students invest in the stock market. The grade, the household income and the number of relatives and friends investing in stocks also have influence on whether investing in stocks of college students, and they are all negative impact. Secondly, the expected returns of stocks, the tolerance for volatility of market prices and the maximum loss tolerance of assets have positive impact on whether college students' stock investing. It is indicating that the college students may seek for risk so that they don't have enough risk awareness and they are lack of financial quotient from the other side. From the perspective of investment reasons, the economic benefit is the primary reason for college students to invest in the stock market, the self-fulfillment factor and the financial management consciousness factor are the second and third reasons respectively. It can be seen that college students have more diversified motivations to invest in the stock market, and they tend not only to use the stock market to make money, but also to take it as a platform to exercise their mind and improve their financial management ability. However, the phenomenon of speculation and following the trend is still prevalent among college students.

According to the above conclusions, it puts forward the following suggestions on how to guide college students to invest in stocks rationally. First, from the perspective of colleges, it should conduct necessary investment and financial education for all students not only for those majored in finance and management. It should be conducted some general courses like public compulsory courses to develop financial concept so as to cultivate students' financial intelligence. Second, colleges should hold some stocks simulation trading competition among college students collaborating with the security companies. These kinds of competitions can be set up as one of practice courses for college students to select, it not only can complete several practice credits, but also can exercise professional knowledge and psychology during these competitions. Third, from the perspective of students, college students need to set up a rational investment view. Before entering the stock market, it should make their own motives and assess their risk tolerance clearly. It shouldn't follow the crowd blindly, shouldn't have gambler psychology. In the courses of investing, students should learn to allocate time reasonably, and shouldn't spend excessive energy to go up in stock market quotation so as to affect normal study and life.

Appendix

Questionnaire on Stock Investment Behavior of College Students.

Dear students,

Thank you very much for your participation in our survey. This questionnaire aims to understand the stock investment situation of college students and is only used for academic research. Your careful answer will contribute to the successful completion of our investigation. Thank you again for your support for this survey!

Part 1 Basic Information and Stock Knowledge Survey

1. Your gender
2. Your major
3. Your grade
4. Your family's annual income (including salary, bonus, securities investment income, deposit interest, etc.)
5. How many of your relatives and friends are in the stock market?
6. Which of the following financial channels/software/websites are you familiar with? (multiple choices)
7. In addition to tuition fees, your average monthly living expenses are about
8. How much do you know about stocks and other related knowledge?
9. What can ordinary foreign investors invest in domestic stock exchanges?

10. At present, the implementation of the delivery system of A shares in stock markets of Shanghai and Shenzhen is
11. The daily price limit of ordinary listed A shares is
12. Investors called the stock of a large company with an important position in the industry, good performance, and good dividend
13. At present, the capital account of Chinese investors is generally opened at
14. When the return rate reaches what, you will enter the market (consider the opportunity cost of stock market risk, capital, time, etc.)
15. When the market price fluctuation degree is less than how much, you will choose to enter the stock market
16. The maximum loss you can accept on your assets is
17. The maximum volatility you can accept on your assets is
18. A has an 80% probability of losing 4000 yuan (20% probability of no loss), and B has a 100% probability of losing 3000 yuan. You are inclined to choose
19. Assuming the following four investment possibilities, you think the most likely investment portfolio you will choose is
20. Do you have any experience of stock investment?

Part 2 Investigation on Stock Investment Motivation

The following questions are used to investigate the motives of college students investing in stocks. There is no right or wrong answer. Please tick the number you choose according to your true opinion:1. Strongly disagree 2. Almost disagree 3. Not sure 4. Almost agree 5. Strongly agree.

	1	2	3	4	5
21. I think stocks are a high-yielding investment					
22. I know a lot about the stock market					
23. I envy my friends and relatives who make money by investing in stocks					
24. In my opinion, the national policy has given great support to the development of stock market					
25. I think the stock market will continue to do well within a year					
26. I think investing in stocks will help me achieve financial independence					
27. I think the stock market is a legal casino					
28. I think investing in stocks can show my intelligence					
29. I often don't have enough money to live on					
30. It makes me happy to make money by investing in stocks					
31. It helps me to improve my financial skills					
32. Many people around me often talk about stocks					
33. I think it can make quick money by investing in stocks					
34. The stock market is good for training my psychological quality					
35. I would feel behind if I didn't invest in stocks					

(continued)

(*continued*)

	1	2	3	4	5
36. I find it very interesting to invest in stocks					
37. The rise in prices (inflation) has raised my awareness of financial management					
38. Many people around me are investing in the stock market					

References

1. Li, D.H., Shi, C.R., Liu, Y.S.: Analysis and discussion on the current situation of financial investment of college students. Chin. Market 30, 40+55 (2019)
2. Shou, Z.K., Zheng, Z.L., Bai, W.R.: Research on financial behavior in the era of big data finance-based on college student. Educ. Teach. Forum **6**, 100–101 (2020)
3. Rong, J.: Investigation on undergraduates security investment in application-oriented Colleges. Educ. Teach. Forum **2**, 128–129 (2020)
4. Feng, M.: Analysis on the behavior characteristics of young people investing in the stock market. Chin. Youth Soc. Sci. **34**(06), 47–48 (2015)
5. Zheng, Y., Zhang, L., Fang, F.: Investigation and analysis on influencing factors of securities investment behavior of college students in Shenyang Institute of Technology. Farmers Consult. **19**, 186 (2019)
6. Liu, N.: University students investment and financial management statues and analysis of influencing factors. Master's Thesis. Hebei University (2018)
7. Wu, Y.T., Yao, H.X.: Research on the stock investment behaviors of college students—a case study of college students in Shanghai. Commun. Finan. Account. **8**, 3-6+129 (2017)

Factor Models: Theory and Development

Jiacheng Yang(✉)

International College Beijing, China Agricultural University, Qinghua East Road, Beijing, China

Abstract. Factor models play an important role in the domain of asset pricing. In the recent years of research, factor models are usually developed and analyzed separately and topics are scattered. Thanks to the efforts of economists, a variety of findings have been discovered and presented. Due to the complexity of factor models, a comprehensive understanding is very essential to better understand them. To fill this gap, we examine different kinds of factor models, including the CAPM model, the q-factor model, and Fama-French's three- and five-factor models. Based on a systematic review of the previous research, we find that all of these models are created by data mining. In this case, a theory-based model is needed for further development.

Keywords: Factor models · Factor theory · Development

1 Introduction

In the modern world, the stock market is a very important means of pursuing return. However, due to its uncertainty, investments in the stock market often do not reap the expected returns. In fact, economists have long been studying the laws of the stock market to find the factors that can influence stock market returns. William Sharpe established the Capital Asset Pricing Model (CAPM) in 1964, which argues that systematic risk is the main factor that affects stock returns when asset portfolios are fully diversified. This model is a milestone in the field of capital asset pricing. However, subsequent researchers find that stock returns are poorly correlated with market risk and there are plenty of studies that have rejected the validity of the CAPM. The interpretation of the CAPM's error can be mainly divided into the two categories of the risk-based and nonrisk-based alternatives. The former which consists of multi-factor asset pricing models develops under the premise that investors are rational and capital markets are sound, which is difficult to detect statistically, while the latter includes deviations due to data snooping, investor irrationality and market frictions, transaction costs, and liquidity effects, which is easily detectable [1].

Furthermore, Gibbons, Ross and Shanken also question the testing of the CAPM in their paper, arguing that some of the model's testing premises have not been validated. They use a multivariate statistical approach to test the questionable central point of the CAPM: all market portfolios are mean-variance efficient, and the result explains why the null hypothesis is rejected. When we use GRS test to check the stock returns with this model [2], the value of interception is still pretty large (pricing error is not jointly

C. Yuan et al. (Eds.): *Proceedings of the 4th International Conference on Economic Management and Green Development*, AEPS, pp. 458–464, 2021.
https://doi.org/10.1007/978-981-16-5359-9_53

equal to zero), which means a large number of factors which may affect the result still have not been considered [3].

Based on this result, a body of literature has shown that certain neglected indicators have better explanatory power for stock returns. Basu and Bhandari find that the P/E and leverage ratios of firms respectively have a surplus effect on stock returns [4, 5]. Thus, the search for efficient pricing factors has become a major research component of capital asset pricing. Building on their predecessors, Fama and French further state that the CAPM cannot explain the variation in returns on different stocks, and that besides beta risk, firm size and book-to-market ratio are also significant elements in capital asset pricing which give rise to the improvement of the factor models [6].

In this paper, we will show the development process and the impact of different factor models by introducing the concept of model factors and analyzing their mathematical formulas. In addition, this paper contains the critique of the factor models' explanatory scope and the model test methods, reflecting the characteristics of current factor models and the existing problems from a more comprehensive perspective.

This paper consists of five separate sections. Section 1 describes the background of the asset pricing field and the beginning of factor model development. Section 2 gives the introduction of different factor models and their concepts. Section 3 offers the comments on and development of factor models, including their limitations and test methods. Concluding remarks are offered in Sect. 4.

2 Factor Model Theory

2.1 Introduction of Factor Model Theory

The first breakthrough is the three-factor model proposed by Fama and French, which, as the name implies, is constructed from three different factors: market factor, size factor and value factor. They state that it has stronger explanatory power than the CAPM and can explain about 60% to 70% of the return anomalies in the market. This caused a dramatic stir in the financial field at that time and people began to discover and value the stability and certainty that factor models could bring.

However, with further research on factor models, some scholars find that the three-factor model is still invalid to explain some of the anomalies in stock return pricing. Thanks to successive related studies, more and more stock return anomalies have been discovered. In order to better explain the stock return, researchers have made continuous revisions to the three-factor model in terms of these anomalies. Hou, Xue and Zhang point out that there are still some asset pricing anomalies are not covered by the Fama-French three-factor model. In this case, they come up with a four-factor model (dubbed the q-factor model) which can capture more anomalies of stock market. In their related paper, the q-factor model adds the profitability and investment dimensions to the market and size factors, consisting of four factors. They provide a detailed list of the current market anomalies. While we using the q-factor model, most of the market anomalies including momentum, earnings, intangibles and trading frictions are well resolved [7].

In the meantime, the development of the factor model keeps continuing. In 2015, using their three-factor model as the base, Fama and French added two more factors (i.e., profitability and investment effect) to the model to come up with a new five-factor model.

Compared with the three-factor model, the five-factor model is more powerful for the anomalies that about 80% to 90% of the anomalies can be explained by the five-factor model, which proves the success of the addition of profitable factor and investment factor [8].

2.2 Factor Model Building and Regressions

This section focuses on the brief details of each factor model mentioned below. In this section, we will introduce their respective regression formulas and explain the specific definition and generation of each factor. Finally, we will test their ability to explain expected returns.

Three-Factor Model

$$R_{it} - R_{ft} = a_i + b_i\big(R_{mt} - R_{ft}\big) + s_i(S - B)_t + h_i(H - L)_t + e_{it} \quad (1)$$

Equation (1) presents the three-factor model. In this equation, R_{it} is the return of the portfolio over period t; R_{ft} is the risk-free return (which in the stock market can be expressed as the return on shares of banks or government bonds); R_{mt} is the return of the entire market, and the size factor is formed by small minus big, which represents the difference in returns between small- and large-size companies (dubbed as $S - M$). In fact, in the stock market, since the growth space is more significant, small-size companies are more likely to have higher expected returns. While the value factor, on the other hand, is formed by high minus low, which represents the difference between the returns of high B/M companies and the returns of low B/M companies (dubbed as $H - L$). Where B/M is the book-to-market value ratio, which refers to the ratio of a company's book value to its stock market value. This factor shows that in the stock market, low B/M companies tend to be more profitable than high B/M companies. At last, the final e_{it} is the residual term, which includes all the factors that fail to be covered by the above factors but can still affect the stock market return outcome (e.g., rational human thinking, etc.).

Fama and French propose three grouping methods to construct risk factors. The first is the 2 × 3 method which firstly divides all stocks into two groups of small- and large-market capitalization in light of the median market capitalization of the stocks, and then divides all stocks into three groups according to the 30% and 70% quantile of book-to-market ratio; secondly, with the combination of the two parameters of market capitalization and book-to-market ratio, all stocks can be separated into 6 portfolios. Again, Fama and French replace the book-to-market ratio with operating profit margin and investment style respectively and repeat the above steps to divide the stocks into 12 portfolios. Next, they calculate the market capitalization-weighted average return for each of the above portfolios for each period. Finally, they construct four factors on the basis of the difference between the returns of the different portfolios. The second is the 2 × 2 method which substitutes the 30% and 70% quantile with the 50% quantile. The third is the 2 × 2 × 2 × 2 × 2 method in which the four indicators are crossed simultaneously and the whole stock is separated into 16 portfolios.

However, the test result suggests that there are other factors that still have a significant impact on asset returns. Thus, the three factors of market, size and value are insufficient to fully interpret the expected returns of different portfolios, and other valuable factors still need to be explored to improve the level of explanation [6].

Four-Factor Model

$$E\left[r^i - r^f\right] = \beta^i_{mkt}E[r_{MKT}] + \beta^i_{ME}E[r_{ME}] + \beta^i_{i/A}E\left[r_{I/A}\right] + \beta^i_{ROE}E[r_{ROE}] \quad (2)$$

Among these four factors, MKT expresses the market excess return, while r_{ME} represents the result of subtracting the return on a portfolio of big-size stocks from the return on a portfolio of small -size stocks, which is the same as the "$R_{Mt} - R_{Ft}$" and "*S-B*" in the Fama-French three-factor model. However, the other two factors are different from Fama-French's. Instead of focusing on B/M, they add two factors (i.e., investment and profitability) to explain the return, in which $r_{I/A}$ means the result of subtracting the return on a portfolio of low investment stocks from the return on a portfolio of high investment stocks and r_{ROE} refers to the result of subtracting the return on a portfolio of high profitability stocks from the return on a portfolio of low profitability stocks [7].

When constructing the factors, Hou, Xue and Zhang use a $2 \times 3 \times 3$ independent triple ranking of market capitalization, single-quarter ROE, and total asset change I/A to reflect the relationship between the expected returns in different conditions. The size value is divided at the median, while ROE and total asset change are divided at 30% and 70% quartile. In this case, the independent triple ranking yields a total of 18 portfolios, with the stocks in each portfolio weighted by market capitalization.

In order to show the efficiency of their work, they list the 447 anomalies put forth by academics, covering six sorts: momentum, value/growth, investment, earnings, intangibles and trading frictions. However, by using the four-factor model as the pricing model, they find that 436 (98%) of these anomalies are no longer significant, which lays a solid foundation for the subsequent Fama-French five-factor model [7].

Five-Factor Model

$$R_{it} - R_{Ft} = a_i + b_i(R_{Mt} - R_{Ft}) + s_i(S - B)_t + h_i(H - L)_t + r_i(R - W)_t + h_i(C - A)_t + e_{it} \quad (3)$$

The Fama-French five-factor model is constructed on the three-factor model, which inherits the market, size and value factors from the prior one, and the additional two factors are constructed by the difference of return between stocks (firms) with low and high profitability *(R-W)*, and the difference between firms with high and low investment *(C-A)*. When constructing the portfolio model on their original three-factor model, Fama and French still divide the market capitalization into two groups through the median, then divide the return on equity (ROE) and total asset change into three groups through the 30% and 70% deciles, thus perform a 2×3 arrangement to obtain six different portfolios, and test the relevant variables through the weighting method.

The result shows that the five-factor model has more explanatory power than the three-factor model, but it still remains some deficiency that fails to explain some market-specific anomalies. For example, when the profitability is low, investing in more small-capitalization stocks will have the problem of below-average returns [8].

GRS Test. The GRS method is employed to test the explanatory power of all the factor models to the expected returns. If the factor models can fully explain the return of different scenarios in the market, there should be no abnormal return, which means that the interception of the portfolio regression in the GRS test is zero. Conversely, if the value of the intercept term of the GRS test is significantly large than zero, it means that the explanatory power of the tested model is weak and there are other unincluded factors that influence the regression results [2].

As for the three-factor model, the result shows that the GRS statistic (p-value) is greatly different from the table value, but the q-factor model performs better and more robustly than the Fama-French three-factor model in the test, with a smaller intercept term and the ability to capture the anomalies of momentum. In addition, the value of the GRS statistic for the five-factor model is smaller than that of the three-factor model, and the absolute value of the intercept term of the regression (which represents abnormal returns) is also smaller [2].

3 Development of Factor Models

As for the development of factor models, a number of questions have been raised by different scholars. For example, the test method used to check the accuracy of the models is somewhat problematic. According to Kan, Robotti and Shanken, although the two-pass cross-sectional regression (CSR) has been a very popular way to test the factor models, it still remains critical issues because it has an implicit premise that expected returns are completely linear in the valuation of beta [9]. But in fact, when non-trading factors are involved, many models do not satisfy the linear relationship well, and thus CSR test does not explain the expected returns in those cases well. Roll and Ross question the linear relationship between expected returns and the value of β. Subsequently, other professors have demonstrated the limitations of OLS and R^2, replacing non-trading factors with simulated portfolios and using multivariate statistical methods for regression modeling. Through examining the performance of the CAPM model at sample cross-sectional R^2, they find that many of these models are rejected at conventional statistical levels. Furthermore, when they test whether model irregularities affect the standard errors of the β_0 and risk premium estimates. The results show that standard errors vary significantly when the underlying factors do not trade-off (e.g., macroeconomic factors). Although the paucity and noise of statistics often prevent the conclusions from being sufficiently convincing, the advantage of the Fama-French three-factor model over the CAPM is still statistically significant [10].

Fama-French's factor model is also not invulnerable. Over the years of 1972 to 2015, the alphas for *H-L* when regressed on *Mkt* or *Mkt* and *S-B* by Barillas and Shanken are highly significant, whereas the alphas for *S-B* when regressed on *Mkt* or *Mkt* and *H-L* are modest. Through aggregating all of the evidence, they show the result which arrives at posterior probabilities with the benchmark prior of 60% for the two-factor model (*Mkt* and *H-L*) and only 39% for the Fama-French three-factor model, with the remaining 1% split between CAPM and *Mkt* and *S-B*. In this case, it is necessary to have simple statistical tools that can interpret the various models jointly in a model-comparison framework. By grouping factors of the same type and comparing the CAPM, the three-factor and five-factor models, the four-factor model, Barillas and Shanken conclude that the six-factor model has the highest posterior probability, with six factors, including *Mkt, I/A, ROE, S-B, H-L* and *U-D*, and that the value and momentum factors are not redundant any more when more timely *H-L* is taken into account. The validity of the six-factor model is also demonstrated after we test 25 portfolios and allow for smaller departures. On this basis, the other top models are closely related to the six-factor model in terms of composition. All this evidence supports that the explanatory power of the six-factor model is better than that of the Fama-French three-factor model [11].

In addition, due to the fact that the factors in Fama-French's models are supposed to span the space of the unknown state factors, the possibility of specification errors in the model has been raised. Modified Hausman artificial regression is a valuable way to test for measurement errors. Through the generalized method of moments (GMM) based panel data approach, Racicot and Rentz conclude that the only consistently significant factor is the market factor [12].

4 Conclusions

By reviewing factor models and their development in the previous literature, we find that the explanatory power of the factor models for different market anomalies rises with recent research. Among all the mainstream models mentioned in the paper, the market factor and size factor have high generalizability as the basics. These two factors also pass model comparison tests. Although the model factors have been changing and developing all over time, all models have certain limitations and cannot provide reasonable explanations for certain market anomalies. There are also some corresponding limitations in the test methods, which may lead people to wrong results and insights about the true market risks.

Based on the above analysis, we find that the theoretical base of the factor models is still weak since most of the new model factors are discovered by data construction and mining, while the portfolio of different factors is coming up through data mining. In fact, there is no completely perfect factor model that can solve all the market anomalies. There is still a long way to build a parsimonious model with high explanatory power. Therefore we suggest more research should be done in the factor model field since the limitations of data mining techniques are about to be reached and thus a clear theoretical base is crucial for further development.

References

1. MacKinlay, A.C.: Multifactor models do not explain deviations from the CAPM. J. Financ. Econ. **38**(1), 3–28 (1995)
2. Gibbons, M., Ross, S., Shanken, J.: A test of the efficiency of a given portfolio. Econometrica **57**(5), 1121–1152 (1989)
3. Sharpe, W.F.: Capital asset prices: a theory of market equilibrium under conditions of risk. J. Financ. **19**(3), 425–442 (1964)
4. Basu, S.: The relationship between earnings' yield, market value and return for NYSE common stocks: further evidence. J. Financ. Econ. **12**(1), 129–156 (1983)

5. Bhandari, L.C.: Debt/equity ratio and expected common stock returns: empirical evidence. J. Financ. **43**(2), 507–528 (1988)
6. Fama, E.F., French, K.R.: Common Risk Factors In The Returns On Stocks And Bonds. J. Financ. Econ. **33**(1), 3–56 (1993)
7. Hou, K., Xue, C., Zhang, L.: Digesting anomalies: an investment approach. Rev. Financ. Stud. **28**(3), 650–705 (2015)
8. Fama, E.F., French, K.R.: A five-factor asset pricing model. J. Financ. Econ. **116**(1), 1–22 (2015)
9. Kan, R., Robotti, C., Shanken, J.: Pricing model performance and the two-pass cross-sectional regression methodology. J. Financ. **68**(6), 2617–2649 (2013)
10. Roll, R., Ross, S.A.: On the cross-sectional relation between expected returns and betas. J. Financ. **49**(1), 101–121 (1994)
11. Barillas, F., Shanken, J.: Comparing asset pricing models. J. Financ. **73**(2), 715–754 (2018)
12. Racicot, F.-E., Rentz, W.F.: Testing Fama-French's new five-factor asset pricing model: evidence from robust instruments. Appl. Econ. Lett. **23**(6), 1–5 (2015)

International Portfolio Management: A Volatility-Based Method

Yijin Chi(✉)

University of California, Davis. 1 Shields Avenue, Davis, CA 95616, USA

Abstract. Volatility-based portfolios behave differently to returns in different countries, for different factors, and over different time periods. In this paper, an adjusted management portfolio is constructed and distributed to both developed countries and the Asia-Pacific region (excluding Japan) to observe how the management strategy alters and influences the return of the five factors Mkt, SMINUSB, HMINUSL, RMINUS, and CMINUSA over the three-time period. My strategy is to use the reciprocal of the variance realized last month to approximate the conditional risk-return trade-off. I find that the returns of the five factors in developed and developing countries react differently to the new portfolios. This paper is not only a typical risk-based portfolio interpretation but also a study of the structure of expected returns over time.

Keywords: Factor models · Volatility-based method · Developed and developing countries

1 Introduction

The adjusted-managed portfolio is constructed to produce a greater excess return, increase Sharpe ratios, and generate greater utility gains for investors. From other studies [1], it is obvious that a risk-adjusted portfolio strengthens and produces significant risk average return by managing and diversifying the allocation. However, the effect of managed strategies on the rate of return of various factors in different countries and time periods remains to be studied and verified.

I distribute the adjusted-managed portfolio to both developed and Asia-Pacific (ex-Japan) region to see if there is any conspicuous increase or decrease in return of the five factors: the size of the company (SMINUSB), book-to-market values (HMINUSL), excess market returns (Mkt), profitability (RMINUSW), and investment (CMINUSA). I calculated the risk-benefit return by the reciprocal of the variance recorded from last month. Then, I observe the results of average and excess returns for developed and developing countries over three time periods, i.e. 2000–2007, 2008–2019, and 2000–2019. The results show that the five factors in developed countries all have higher excess returns in the adjusted portfolio throughout the three-time period, while only the average returns of Mkt factor in Asia are better during 2000 and 2007.

C. Yuan et al. (Eds.): *Proceedings of the 4th International Conference on Economic Management and Green Development*, AEPS, pp. 465–472, 2021.
https://doi.org/10.1007/978-981-16-5359-9_54

This paper can be divided into the following sections. Section 2 is the literature review as the introduction of economic models, factors, and facts. Section 3 presents how the data is formed. Section 4 discusses my empirical strategy and results of the return based on the volatility management portfolio. Section 5 is the conclusion and discussion.

2 Literature Review

It is essential to understand the relationship between expected return and systemic risk by considering the equilibrium condition of the capital market and observing the capital asset prices model (CAPM). "Individual preferences" and "natural relations" interact are considered to determine the equilibrium net interest rate and the market risk premium [2]. Investors should manage and adjust their portfolios according to changes in risk to obtain a well-performing portfolio. Usually, the risks inherent in assets can be avoided through diversification, so the total risk of assets is not the relevant influencing factor of their prices. Moreover, to obtain equilibrium in the capital market, researchers usually assume all investors can borrow or lend equally at a common net interest rate and investors' expectations are homogeneous. On the other hand, Lintner [3] discusses that there are increasing effects of investment uncertainty and risk on asset prices and rational decision-making of individual investment. Investors should select the optimal security portfolio based on risk and be risk-averse for lower investment volatility and stable return. Also, investors should be aware of the different expected rates of return in each portfolio so that standard deviation, variance, and covariance are necessary because they provide direct evidence for figuring out the most appropriate functional relationship between the required rate of return and relative risk. Furthermore, the general equilibrium model of capital asset pricing has received considerable attention in recent years. The stocks with high systematic risk and volatility mean the stocks are underperforming, where that high-beta stocks have negative alpha (return), and low-beta stocks have positive alpha [4].

The Fama-French (FF) three-factor, four-factor, and the five-factor model are proposed to capture risks and explain anomalies [5]. First, the FF three-factor model includes the size factor, value factor, and market factor. It is designed to describe stock returns in the asset pricing and portfolio management. The idea behind the model is that the companies with small-cap and high value usually have regularly better performance in the market [6]. Following, the FF four-factor model adds a fourth factor, momentum (UMD), to a market size model that improves portfolio returns at a given level of risk. The four-factor model increases the "explanatory power of the model to 95%" [7], compared to 90% in the three-factor model. The FF five-factor asset pricing model capturing two more factors, profitability and investment, for an average stock return performance, and it can explain some anomalies and cross-sectional variations in expected average returns that are not targeted by the other two models. Also, it captures the uncertainty of the impact of state variables on expected returns because many of the changes in average rates of return are related to profitability and investment because they are made by natural selection.

Further factors are introduced to explain the risk and return. Frazzini and Lasse [8] propose a model with leverage and margin constraints that vary with investors and time and find out that the factor betting against beta (BAB) is important in explaining the impact of capital friction on asset pricing. High-risk beta assets require lower risk-adjusted returns than low beta assets so that leverage is required. Also, there is a result that implies the investors who are more constrained should increase their holdings of high beta assets in their portfolios. On the other hand, downward losses are the financial risk associated with losses and the risk of the actual return that is below the expected return [9]. If the downward trend of an asset in the down market is greater than the upward trend in the upmarket, its return is often very low. Also, the downside beta value in the past is a good predictor of future covariance and market decline. Moreover, there are different measurements, such as valuation ratios, that can be used to predict return. The valuation ratios were positively correlated with subsequent returns and the implied predictability of returns was considerable in the long run [10].

Some tests show that a volatility-managed strategy is better than other strategies since it is less vulnerable to volatility shocks. The volatility-based portfolio adjusts the monthly rate of return based on the inverse of the variance realized last month by reducing the risk exposure when there is high variance, and vice versa [1]. Investors should reduce risk in the event of a volatile shock and gradually increase the risk exposure as the volatility shock subsides. When volatility is high, a less risky managed portfolio generates a larger return and significant utility gains for investors. Usually, after recessions, financial crises, and market crashes, volatility tends to be high, so investors should be risk-averse because the risk is higher when the stock market has high volatility. Overall, the volatility management portfolio and strategy can be used to price a wider range of dynamic strategies and greatly reduce pricing errors.

3 Data

In order to study how the portfolio should be constructed and managed to get relatively less risk and achieve the best performance in return, I analyze the daily return data from the Fama and French (FF) five-factor model [5] of both developed and Asia Pacific ex Japan countries. The five factors are composed of the size of the company (SMINUSB), book-to-market values (HMINUSL), excess market returns (Mkt), profitability (RMINUSW), and investment (CMINUSA) in average stock return performance.

The developed districts include Australia, Austria, Belgium, Canada, Switzerland, Germany, Denmark, Spain, Finland, France, Great Britain, Greece, Hong Kong, Ireland, Italy, Japan, Netherlands, Norway, New Zealand, Portugal, Sweden, Singapore, and United States, while the Asia Pacific ex-Japan countries include Australia, Hong Kong, New Zealand, and Singapore.

All the data come from the Kenneth R. French Library website, and all the returns for the five factors in the portfolio are calculated in United States dollars on a daily basis by the FF library in 2000–2019, including dividends and capital gains, rather than consecutive compound interest. In order to construct the Mkt, SMINUSB, HMINUSL, RMINUSW, and CMINUSA factors, the regional stock is split into two market values with big and small stocks and three groups with B/M, profitability (OP), and investment (INV) at the end of June each year. First of all, the size of the company is measured by SMINUSB (small minus big), as the average return of small stock portfolios minus the big stock portfolios. The second factor BM is measured by HMINUSL (high minus low), as the average return of the two value portfolios minus the average return on the two growth portfolios. Thirdly, the excess market return (Mkt) is measured by Rm–Rf, which is the return of the portfolio minus the risk-free return. Forth factor profitability is measured by RMINUSW (robust minus weak) as the average return on the two robust operating profitability portfolios minus the average return on the two weak operating profitability portfolios. Finally, the investment factor is measured by CMINUSA (conservative minus aggressive) as the average return difference between two conservative portfolios and two aggressive portfolios. For each of the 5 factors, there are 5217 daily returns being observed and recorded from 2000 to 2019.

4 Empirical Strategy and Results

4.1 Empirical Strategy

The return of the portfolio usually varies with the investment method, time, size, and special risk. In order to compare the result effectively, I first divide two datasets, i.e. developed and Asia Pacific ex Japan region, into 3 time periods, as from 2000 to 2007, 2008 to 2019, and 2000 to 2019. Driven by empirical evidence that volatility is highly variable, persistent, and unpredictable [1], I approximate the conditional risk-benefit trade-off by the reciprocal of the conditional variance. Also, the way I build a portfolio with volatility management is to measure excess returns by the reciprocal of conditional variances.

$$f_{t+1}^{m} = \frac{c}{\hat{\sigma}_t^2(f)} f_{t+1} \tag{1}$$

$$\hat{\sigma}_t^2 = \frac{1}{22}\sum_{d=0}^{21}\left(f_{t-d} - \frac{1}{22}\sum_{d=0}^{21} f_{t-d}\right) \tag{2}$$

The f_{t+1} measures excess return of the "buy and hold portfolio". The variance ($\hat{\sigma}_t^2$, Eq. (1, 2)) realized last month is used as a proxy for conditional variances for building the portfolio. Then, the first 22 days are used to predict the variance in day 23rd (Eq. (2)), in order to make portfolio construction simpler. The constant c is used to control the investment strategy's average exposure.

Then, based on the measure of conditional variance, the investment strategy will either increase or decrease the exposure to the portfolio each month. Moreover, we can figure out the performance of the adjusted investment strategy over time and whether it brings any effect on the five factors by observing the calculated alpha value, as the intercept. The classical alpha value of the volatility management portfolio represents the excess return from various factors that people can receive after controlling for the risk in the market. In this paper, it explains the effectiveness of the newly formed and unexplained portfolio.

Finally, I construct the time series regression on the original factors for the portfolio managed by volatility (Eq. (3)). The significantly positive alpha means that the timing of the fluctuations increases the Sharpe ratio relative to the original factor. The Sharpe ratio is a measure of the return on a risk-adjusted financial portfolio, and a portfolio with a higher Sharpe ratio is considered superior to a similar portfolio because it represents the portfolio brings a higher excess return for the specific factor. On the other hand, the alpha measures whether the volatility management strategy boosts the return, while beta is the measure of volatility and systematic risk.

$$f^{m}_{t+1} = \alpha + \beta f_{t+1} + \varepsilon_{t+1}. \quad (3)$$

4.2 Results

Table 1 is the statistical summary that shows the corresponding average returns for investment strategies for the five factors over three time periods. For the performance of the developed country, by comparing the mean, as the average return, of the five factors Mkt, SMINUSB, HMINUSL, RMINUSW, and CMINUSA throughout the three-time period: 2000–2007, 2008–2019, and 2000–2019, the results indicate that there are positive average returns for all of the 5 factors from 2000–2019 for developed countries. Also, the portfolio brings significantly high return for factors HMIUNSL and Mkt during 2000 and 2007 in Asia region. However, by comparing the mean of Asia-Pacific ex-Japan countries within the three-time period, the average return decreases for the 5 factors over time. Also, there is a negative average return for the SMINUSB factor over the three periods.The investment strategy performes well overall, but it brings different effects over different time periods and factors.

On the other hand, Table 2 shows whether the adjusted management portfolio has yielded excess returns to the five factors in the developed countries by observing the alpha value. The standard errors are showed in parentheses, and *** represents $p < 0.01$, ** represents $p < 0.05$, * represents $p < 0.1$. The rate of return for the Mkt factor doubled between 2008 and 2019 compared to 2000–2007, and the Mkt factor has the largest significant average excess return. Also, the return of the factors SMINUSL and HMINUSL increased significantly between 2000 and 2007. Although this new investment strategy has shown a less rewarding performance on the CMINUSA factor than before, it contributed significantly to the overall return on all five factors.

On the other hand, Table 3 shows whether the managed portfolio brings excess returns to the five factors in the Asia-Pacific ex-Japan countries by observing the alpha value. The modified investment strategy is significantly effective for Mkt and RMINUSW factors according to the increasing alpha value because it brings a significant and higher excess return between 2008 and 2019. The larger alpha indicates that investors are able to benefit from volatility time. For example, an alpha increase from 0.147 to 0.168 from 2000–2007 to 2008–2019 for Mkt and almost double the excess return for factor RMINUSW. However, there is negative alpha for factor SMINUSB and HMINUSL, which indicates that the specific factor is underperforming under the risk adjusted portfolio.

Finally, by comparing the mean between developed countries and the Asia-pacific region, the average rates of return for the five factors in developed countries have better overall performance between 2000 and 2019 than those in Asia region. Also, developed countries have lower standard error for the five factors, which indicates that the market is more complete. However, the Asia-pacific region has higher return for all five factors during 2000–2007. For example, Mkt factor has the highest return of 0.043 in Asia pacific. In addition, by comparing the alpha value of the two regions between 2000 and 2019, we can tell that the adjusted investment portfolio has a better performance for the five factors in the developed countries, which that there is positive and higher excess return, despite there is a negative return for factor CMINUSA. A positive alpha indicates that volatility management strategy extends the mean-variance boundary and increase the Sharpe ratio [1] compare to the original portfolio, which indicates that the adjusted portfolio brings excess return for investors.

Overall, a portfolio managed by volatility produces a significant risk-adjusted return on factors Mkt and RMINUSW since there is no negative return throughout the years. Also, volatility-based strategy is less vulnerable to volatility shocks and takes less risk during bad times. When volatility is high, a less risky managed portfolio generates larger alphas, increases the Sharpe ratio, and produces significant utility gains for mean variance investors.

Table 1. Summary statistics.

	Mkt	SMINUSB	HMINUSL	RMINUSW	CMINUSA
2000–2007 (Developed)					
Mean	0.009	0.019	0.055	0.020	0.036
Standard Error	0.018	0.010	0.010	0.007	0.009
2000–2007 (Asia)					
Mean	0.043	−0.004	0.042	0.020	0.014
Standard Error	0.020	0.011	0.013	0.012	0.009
2008–2019 (Developed)					
Mean	0.025	−0.002	−0.006	0.016	0.003
Standard Error	0.018	0.007	0.006	0.004	0.004
2008–2019 (Asia)					
Mean	0.016	−0.015	0.014	0.017	0.017
Standard Error	0.020	0.010	0.009	0.008	0.008
2000–2019 (Developed)					
Mean	0.019	0.007	0.018	0.017	0.016
Standard Error	0.013	0.006	0.005	0.004	0.004
2000–2019 (Asia)					
Mean	0.027	−0.011	0.025	0.018	0.016
Standard Error	0.015	0.007	0.007	0.007	0.007

Table 2. Regression results for developed region (Alpha).

	Mkt^m	$SMINUSB^m$	$HMINUSL^m$	$RMINUSW^m$	$CMINUSA^m$
2000–2019	0.225*** (0.019)	0.087*** (0.038)	0.087*** (0.052)	0.103*** (0.054)	−0.239*** (0.071)
2000–2007	0.167** (0.023)	0.184** (0.061)	0.461*** (0.110)	0.177 (0.098)	−0.141 (0.124)
2008–2019	0.264*** (0.027)	0.039 (0.047)	−0.100 (0.044)	0.026 (0.054)	−0.232 (0.077)

Table 3. Regression results for Asia-pacific ex-Japan region (Alpha).

	Mkt^m	$SMINUSB^m$	$HMINUSL^m$	$RMINUSW^m$	$CMINUSA^m$
2000–2019	0.164*** (0.016)	−0.017*** (0.027)	−0.005*** (0.020)	0.083*** (0.024)	0.008*** (0.038)
2000–2007	0.147*** (0.019)	−0.0006 (0.035)	−0.0003 (0.023)	0.022 (0.028)	−0.062 (0.053)
2008–2019	0.168*** (0.022)	−0.030 (0.038)	−0.001 (0.029)	0.121*** (0.034)	0.054 (0.053)

5 Conclusion and Discussion

In this paper, volatility-based portfolio models are established for developed regions and the Asia-Pacific region (excluding Japan) in three time periods to capture heterogeneous markets features. The investment strategies are used in the 5 factors Mkt, SMINUSB, HMINUSL, RMINUSW, and CMINUSA to figure out the effective of the models on the returns. This paper shows that the modified management portfolio and investment strategies have different performances for different factors in different time periods and countries. Compared to developing countries, developed countries have significantly higher return. There are fundamental differences between developing and developed countries since developed countries have better regulations and complete markets. Overall, a portfolio managed by volatility tends to offer huge risk-adjusted returns for investors. Investment portfolios should be adapted from time to time to suit the specific needs of specific countries.

Investors should notice that the same portfolio strategy approach does not apply to all countries and different countries' investment strategies should be discussed separately. The adjusted investment portfolio delivers higher overall excess returns for the five factors in the developed countries than the developing countries, and the Mkt factor has the best performance throughout the time. In this study, the former 22 trading periods' return data are used to predict the variance on the day of 23 and then the variance is used to adjust portfolios. However, the time length could be changed to analyze whether a better performance of the portfolio can be found. This study could provide investors a better understanding of volatility-based investment strategies.

References

1. Moreira, A., Muir, T.: Volatility-managed portfolios. J. Financ. **72**(4), 1611–1644 (2017)
2. Sharpe, W.F.: Capital asset prices: a theory of market equilibrium under conditions of risk. J. Financ. **19**(3), 425–442 (1964)
3. Lintner, J.: The valuation of risk assets and the selection of risky investments in stock portfolios and capital budgets. In: Stochastic Optimization Models in Finance, pp. 131–155. Academic Press (1975)
4. Jensen, M.C., Black, F., Scholes, M.S.: The capital asset pricing model: some empirical tests. In: Jensen, M.C. (ed.) Studies in the Theory of Capital Markets. Praeger, New York (1972)
5. Fama, E.F., French, K.R.: A five-factor asset pricing model. J. Financ. Econ. **116**(1), 1–22 (2015)
6. Hou, K., Xue, C., Zhang, L.: Digesting anomalies: an investment approach. Rev. Financ. Stud. **28**(3), 650–705 (2015)
7. Kyle: The Four Multi-Factor Models You Should Know (3, 4, and 5 Factor Models), Incremental Returns (2019)
8. Frazzini, A., Pedersen, L.H.: Betting against beta. J. Financ. Econ. **111**(1), 1–25 (2014)
9. Ang, A., Chen, J., Xing, Y.: Downside risk. Rev. Financ. Stud. **19**(4), 1191–1239 (2006)
10. Campbell, J.Y., Thompson, S.B.: Predicting excess stock returns out of sample: can anything beat the historical average? Rev. Financ. Stud. **21**(4), 1509–1531 (2008)

Scene Marketing Strategy in E-commerce Era

Liao Sihan(✉)

Jiangxi Normal University, Nanchang, China

Abstract. The development of mobile Internet technology has greatly promoted the expansion of scenes in the e-commerce field, and users are increasingly inclined to feel products and brands from marketing scenes. Based on this, this paper studies a series of challenges faced by scene marketing in the e-commerce era: difficulty in obtaining traffic, excessive experience consumption and small supervision of enterprises. Starting from the marketing strategy, this paper explores how to realize the effective use of scenes in the e-commerce era, providing new ideas and weights for the long-term competitive development of enterprises.

Keywords: E-commerce · Scene marketing strategy · Experience · Date

1 Literature Review

Professor Bernd Schmitt of Columbia University proposed "experiential marketing" for the first time in his book *Experiential Marketing*, which mentioned that scene marketing is to provide different experiences according to different customers, thus arousing consumers' sympathy and is a new stage in the development of experiential marketing. Peng Lan pointed out in *Scenes: New Elements in the Mobile Media Era* that the arrival of the e-commerce era has resulted in fragmentation of information and time, and scenes have become a new link connecting information flow, relationship flow and service flow. Cai Xueyong mentioned in his article "*Finding Your Commercial Value in the Scene Era*" that through the application of intelligent devices and advanced means, the future scene era will change and affect various industries and their subdivided vertical fields, and people's scene fields will undergo great changes. Wu Dingming and Shi Yibin in *Reproduction of Meaning and Desire: Research on Innovation of Online Advertising under Scene View Valve* expounds the advantages of application scenes in the Internet era. Consumers enter a "real-time" scene under the catalysis of consumption concept, service experience, aesthetic taste, advertisement penetration, etc., thus making consumers immersed in the situation to generate consumption desire and commodity significance. According to the existing research, application scenarios have become an important means of marketing in the e-commerce era. Scenario marketing conforms to the general trend of modern marketing and has been widely used now. However, there are still many hidden dangers and challenges behind the prosperity of scene marketing, which is worth pondering. This paper will put forward future marketing strategies and expectations for the field of e-commerce scene design based on its shortcomings.

C. Yuan et al. (Eds.): *Proceedings of the 4th International Conference on Economic Management and Green Development*, AEPS, pp. 473–479, 2021.
https://doi.org/10.1007/978-981-16-5359-9_55

The advent of the e-commerce era has changed the information flow mode of consumers and formed a new consumption scene. As a medium, scene is loaded with the functions of transmitting information and creating experience environment, which is influenced by space environment, social environment and psychological environment. Now, the bonus period for traffic as an entrance has passed. Consumers not only pay attention to simple commodity trading, but also value the process value between buying and selling, so as to meet their own psychological needs such as identity and sense of belonging. The scene will reshape the products, marketing, channels and connection methods of e-commerce, become a new tool for brand communication needs in the era of big data, and provide new ideas and weights for the long-term competitive development of enterprises.

2 Application of Scene Marketing in E-commerce

In the era of mobile Internet, almost every life scene is facing the possibility of being defined. The wide application of intelligent devices has fragmented people's time and information, making it possible for any life scene to be converted into actual consumption. The e-commerce market has begun to change from traditional price orientation to scene orientation.

Robert Scober and Sher Israel introduced the concept of "scene" into the application level in their book *The Coming Age of Scenes*. Based on this, the "five scene forces" are proposed: mobile devices, social media, big data, sensors and positioning systems, which play the roles of experience carriers, information adaptation, logical tools, information acquisition and physical positioning respectively. Users are more and more inclined to feel products and brands from marketing scenes. Carefully built scenes are easier to stimulate customers' interactive experience and consumption behavior, thus forming an effective cycle of communication between scenes and users and realizing efficient links between enterprises and consumers, which has become an indispensable and important tool in the e-commerce era.

In the application field, the ultimate goal of scene marketing is to provide adaptation information and services under specific scenes. Different from the traditional marketing method, which takes economies of scale as the leading factor, scenario marketing takes "people" as the core, that is, customers, customers as the center, starting from customer needs. Based on the scene requirements of virtual space, social applications such as WeChat, Weibo and Little Red Riding Book solve the needs of users for instant messaging and realize interaction and association on the Internet. The application of short videos and live broadcasts, such as fast hands and trembles, promotes the high integration of technology, content and space, and uses scenes to construct emotional mobilization mechanisms to promote viewers' consumption. E-commerce platforms such as Taobao and Jingdong seamlessly connect online and offline through dynamic "links" formed by mobile Internet technology and intelligent terminals, and carry out accurate marketing through big data to enhance consumers' information experience. With the application of scenes in the e-commerce field, the single physical environment marketing communication has changed to the compound scene marketing communication. The spatial dimension, the time dimension and the relationship dimension further highlight the dominant position of consumers, meet the needs of higher levels in Maslow's demand theory, and satisfy the psychological value of individual consumers, thus realizing the improvement of brand image transformation into sales effect.

3 Challenges Faced by E-Commerce Scene Marketing

3.1 It is Becoming More and More Difficult to Obtain Traffic

Traffic usually refers to the number of visits and views of e-commerce, reflecting the public's preference for different platforms. Judging from the current market, the entry threshold for e-commerce is not high. As long as you have a We Media account, you can apply for an e-commerce platform. Moreover, the platform has no requirements on the content and quality of content editing. For example, common graphic forms, sound, short video forms and live broadcast forms have all become the media for its platform to publicize. The creation of new scenes, new links and new experiences emerge one after another, but it is very difficult to really attract market traffic. It is also because of the low marketing threshold in We Media that market competition has become very fierce. The effective realization of flow is accidental and uncertain. To realize the goal of profit transformation, we must need the support of large We Media platforms, in addition to flow support.

3.2 Excessive Experiential Consumption Leads to Disorderly Consumption

The concept of scene marketing has been widely recognized by the market since China's accession to the WTO, and various enterprises have even used all their efforts to struggle in the fierce market competition environment. For example, the upsurge of live broadcast in recent years has appeared on major e-commerce platforms in a short period of time. While draining, it also shows the characteristics of similarity and compulsion. It is very difficult for the new anchor to make further breakthroughs, and the audience will also have visual and auditory fatigue, resulting in a decrease in the retention rate. In order to stimulate consumers' unnecessary consumption, some cosmetics anchors filled the studio with commodities of major brands to stimulate the audience's senses. There are also some anchors who use marketing skills to increase the added value of commodities, giving the audience the illusion that the value of products is increasing, thus falling into the pseudo desire of consumption. Although the pulling effect of live broadcast on the economy cannot be ignored objectively, it is also easy to cause some audiences to fall into the whirlpool of irrational consumption and impulsive consumption such as entertainment and ostentatious display of wealth, leading to disorderly consumption.

3.3 Inadequate Supervision of Enterprises, Creating the Illusion of "False Prosperity"

Behind the barbaric growth of e-commerce, there are problems of safeguarding rights and difficult platform supervision. The first is the issue of user privacy. When predicting consumer demand, scene data will be widely collected. Nowadays, everyone will carry their mobile phones with them. When mobile phone users enter a certain scene, they can be identified and complete information collection. This work will involve the privacy of consumers. If the privacy issue is not handled properly, it will lead to the disclosure and even abuse of user information. In addition, the supervision of the national network security policy is not perfect, and it is difficult for enterprises to supervise in all directions. Once the information is disclosed, it will seriously market the reputation and reputation of e-commerce platforms.

In addition, some merchants and platforms increase commodity sales through false data streams. For example, consumers are misled by hidden rules in the industry such as bill-brushing and data fraud, and consumers' explicit needs are met by gimmicks, concessions and other means, resulting in "false prosperity" of e-commerce and affecting the industry atmosphere.

4 Scene Marketing Strategy in E-Commerce Era

Marketing strategy type	Specific marketing strategy	Specific representative measures
Product strategy	Pay attention to user experience and accurate push to meet the personalized needs of consumers	Through links, the fragmented information of consumers is connected, the user behavior trend is grasped, and accurate marketing is carried out
	The scene is regarded as a part of the product, and a certain demand and desire of consumers are increased through intangible products to meet the potential demand of customers	Use graphic posts, short videos and live broadcasts to create immersive scene experience and enhance users' telepresence
Promotion strategy	Make proper use of public relations, give consideration to social interests while developing oneself, assume social responsibilities, maintain a harmonious social and political environment, and establish a good image of the enterprise	Make proper use of big data, abide by relevant laws and regulations, and establish a good corporate image and public opinion propaganda

4.1 Regard Links and Experiences as the Core of E-Commerce Scenarios

What is currently referred to as scenario marketing, It is based on the needs of consumers in information, products and other aspects as the starting point, Around the marketing of a series of scenes such as consumers' life and work, the scenes are reconstructed based on the dynamic "link" formed by mobile Internet technology and intelligent terminals, so that the scenes can form a multi-dimensional fragmentation, thus improving consumers' information experience and carrying out accurate marketing communication [1].

O2O has created full-channel links, and consumers have realized seamless connection of contacts based on online and offline. Marketing logic is no longer to attract attention with traffic, but to take people as the center and scene evolution as the lifestyle. Scene labeling, multi-screen linkage and full-channel method are all scene linking methods. The combination of social media and mobile devices, big data, sensors and positioning systems can produce extremely rich and personalized content and user data, thus accurately making user portraits, mastering the behavior trends of the Internet as a whole and individual users, and providing diversified and personalized consumption experiences for them.

The scene brings experience by reconstructing the link between the enterprise and the consumer. Experience determines the user's consumption will, consumption motivation and even the core of the final conversion link. It endows the user with a wider range of psychological feelings and social significance when making individual actions and purchases, and makes the information about brands, products and other aspects of the enterprise have a more positive impact on consumers. It can also encourage users to form active content creation, and continuously derive new scenes and new links through sharing, thus providing more communication opportunities for enterprises.

4.2 Establishing Immersive Marketing Experience to Enhance Telepresence and Commodity Trust

The establishment of the scene is a kind of social and commercial activities to attract users and bring immersive experience to users. Real scenes can increase users' telepresence and commodity trust. Telepresence is the perception of the disseminator when interacting with others, which is enough to enable both parties to understand each other's characteristics and the meaning behind communication in the communication process. Therefore, increasing telepresence will help individuals to obtain more information and experience in the communication process [2]. In the content field guided by KOL, graphic posts, short videos and live broadcasts are the three magic weapons that fully embody the immersion effect. The connection between KOL and fans enables consumers to immerse themselves in the scene consumption.

Take Li Ziqi, a popular online blogger, as an example. She showed her living environment, countryside, streams, rice and other rural scenes in the video, and filmed the production process, materials and procedures of the product. Using real rural scenes, daily life reproduces and shows the original appearance of the product, making users feel as if they were in their surroundings, as if the purchased product was made by Li Ziqi, enhancing the telepresence and trust of the product, thus increasing the purchase intention.

In addition, immersion marketing is also widely used in e-commerce live broadcast, such as the process of picking fruits from fruit trees to packing boxes, which enables users to directly understand the planting environment and product quality, and also provides the opportunity for live broadcast fruit selection. When selling food, try it out in the live broadcast room to show the real color and quality of the food, and give voice description and visual stimulation, which greatly stimulates users' purchasing desire.

The sensory upgrade brought by live broadcast enables us to feel the product more truly, understand the product in a more three-dimensional way, generate a brand-new marketing scene, quickly realize the focus of users and sales transformation, make brands closely connected with consumers, effectively eliminate the blindness of traditional marketing, and greatly improve the overall marketing efficiency. At the same time, the live broadcast has created a platform for enterprises to communicate with consumers and provided opportunities for direct communication. It is an indispensable marketing method for enterprises to understand consumers' feedback more accurately and quickly and establish immersive scene experience.

4.3 Make Proper Use of Data and Information to Protect the Rights and Interests of Consumers

To do a good job of scene analysis and user portrait, we need to rely on the support of various information data. This information includes the consumer's location, preferred brands, daily browsing records, actual purchasing behavior, etc. Enterprises will draw pictures of users according to big data, and consumers' preferences and behaviors will become clear and predictable. Using AR, VR and other identification and sensing systems, focus on consumer scene interaction and emotional experience. Through face recognition, consumers are accurately recognized and their consumption preferences and characteristics are recorded. In the next consumption, face recognition is used to provide more effective recommendations and targeted services for its personalized information, thus enhancing consumers' respect for needs and consumption experience.

After obtaining this information, the work organization must do a good job in the confidentiality of the data information, not only to supervise the collection and use of the data, but also to ensure the effective use of the data. Nowadays, consumers have a sense of privacy protection, but China is still far from the legalization of the Internet. In the data analysis of user scenarios, enterprises should consciously and strictly follow the network security management regulations and information confidentiality standards, protect the rights and interests of consumers, and establish sincere cooperation, equality and mutual benefit public relations with the public. At the same time, establish a good corporate image, let the public know the relevant information of the enterprise in a timely manner, consciously assume the publicity function, actively publicize the corporate culture, business philosophy, employee cultivation, and establish a good image in the hearts of the target consumers and the public.

5 Conclusion

To sum up, in the e-commerce era, seizing the scene is the best time to understand users. The scene realized the logical change from the traditional "product-centered" to "consumer-centered". Through data and links, the behavior trend of individuals is predicted, and a premium is paid for the scene through experience, bringing the experience of "integration of time and space". Each enterprise, brand is a group of independent perception and experience, Only by creating marketing with temperature and emotion, integrating and evolving commodities into a way of life, realizing the effective use of the

We Media platform, and making brands the only choice for consumers in this scenario, can we increase users' support in the long run and provide core advantages for the growth of enterprise value.

References

1. Xi, L.Y.: Link and experience: scene marketing communication in mobile internet era. News Knowledge, Issue 10 (2019)
2. Yu, T.T., Dou, G.H.: The Application of Social Telepresence in the Study of Online Purchasing Behavior, International Press, No. 5, pp. 133–146 (2014)

Capacity Reserve Decision for Emergency Supplies with Government Subsidy Policy

Yongning Shen, Hui Yang(✉), and Fei Sun

School of Economics and Management, Nanjing University of Science and Technology, Nanjing, Jiangsu, China
{yanghui,fsun}@njust.edu.cn

Abstract. The purpose of this paper is to investigate how a government sets precise subsidy policy to stimulate a manufacturer to reserve capacity for emergency supplies. Firstly, we develop a benchmark model where a manufacturer should determine its capacity strategy for future emergency supplies without government subsidy. We identify the thresholds for the manufacturer on reserving capactiy or not. Secondly, we put forward and compare four types of subsidy polices, including no subsidy, fixed subsidy, marginal subsidy, and hybrid subsidy. Thirdly, we study the government's precise subsidy policies. The results show that the fixed subsidy policy is optimal if the fixed investment for capacity reserve is high but the marginal cost is low. When the fixed investment is low but the marginal cost is high, marginal subsidy policy is the optimal policy. The government should choose the hybrid subsidy policy if the fixed investment and marginal investment are both high. Numerical examples are included to illustrate the major results.

Keywords: Capacity reserve · Emergency supplies · Subsidy policy

1 Introduction

The capacity reserve for emergency supplies plays an important role in disaster relief and mitigation. When a disaster strikes, the manufacturers should quickly produce the emergency supplies. Unavailability of supplies leads to ineffective emergency response. To enhance emergency service level, government should provide subsidy policies for the manufacturers when they have no incentives to invest in capacity reserve for uncertain emergency demand.

There is a growing literature that studies capacity reserve and the impact of government subsidy. Whybark (2007) studied the emergency supplies capacity reserve and argued that capacity reserve should be considered in the emergency supplies inventory management. Goyal and Netessine (2011) showed that volume flexibility can mitigate aggregate demand uncertainty for both complementary and substitutable. Chen et al. (2016) pointed out that capacity reserve is an effective measure to reduce physical reserve cost and improve distribution efficiency. Li and Zhao (2016) showed that capacity reserve can effectively alleviate the demand for emergency supplies when supply interruption is

C. Yuan et al. (Eds.): *Proceedings of the 4th International Conference on Economic Management and Green Development*, AEPS, pp. 480–486, 2021.
https://doi.org/10.1007/978-981-16-5359-9_56

caused by emergency. Serel et al. (2001) found that capacity reserve contracts for manufacturers and long-term suppliers can reduce the manufacturers' purchasing cost, and the risk of supply interruption. Although capacity reserve has many advantages, manufacturers might have no incentives to invest in it as it incurs both fixed cost and marginal cost. Some literature studies government subsidy policies which support manufacturers to reserve capacity. Zhang and Li (2013) studied the optimal investment decision of manufacturers on capacity reserve and supplies reserve under government subsidy. Yao et al. (2018) showed that the government's outsourcing of capacity reserve can increase both government and companies' benefits. Chen et al. (2017) studied a risk-sharing plan for joint storage of emergency supplies by non-profit organizations under government's subsidy.

This paper proposes precise government subsidy policy for emergency supplies, which differs from the aforementioned studies. We seek to answer two research questions: (i) When does a manufacturer reserve capacity for emergency supplies? How does it decide its optimal reserve capacity? (ii) How does a government set an optimal subsidy policy if it is needed? To address these questions, we develop and compare the capacity decision models without and with subsidy policy. We show how the government induces the manufacturer to reserve capacity to reach a certain service level through optimal subsidy policies.

This paper is structured as follows. Section 2 proposes a benchmark model. Section 3 extends it to the model with government subsidy. The precise subsidy policy for government is placed in Sect. 4. Section 5 concludes the paper.

2 Capacity Decision Model Without Government Subsidy

We first consider a benchmark model where a manufacturer should make capacity decision without government subsidy. Consider the manufacturer supplies a kind of product to the market with capacity Q_0 in a period. The length of the period is assumed to be one unit. If there is no emergency during the period, the demand of the product is equal to Q_0. If an emergency occurs during the period, there will be another stochastic demand x for the product from a distribution function $F(x)$ with a density of $f(x)$, besides the certain demand Q_0. We assume the probability of emergency is k and the emergency occurs at the time t, where t is a random variable and is uniformly distributed from 0 to 1 with density of $g(t)$. Faced with the uncertainty of emergency, the manufacturer has two choices for capacity decision: to reserve capacity Q ($Q \geq 0$), or not to reserve capacity. If the manufacturer decides to reserve capacity, the total capacity for the product is $Q_0 + Q$. Otherwise, the total capacity is Q_0. We assume that the fixed investment in reserve capacity is I, and the marginal cost of unit reserve capacity is ρ.

If the manufacturer decides not to reserve capacity, its profit is:

$$\Pi_N = mQ_0 \tag{1}$$

If the manufacturer decides to reserve capacity, its expected profit is:

$$\begin{aligned}\max_{Q_1} E\Pi_F = k\textstyle\int_0^1 \{t(mQ_0 - \rho Q) + (1-t)[mQ_0 + \int_0^Q (mx - \rho(Q-x))f(x)dx \\ + \textstyle\int_Q^U mQf(x)dx]\}f(t)dt + (1-k)(mQ_0 - \rho Q) - I \\ s.t. \qquad Q \geq 0\end{aligned} \tag{2}$$

The Kuhn-Tucker condition is applied to find the optimal solution. We can get Lemma 1.

Lemma 1. If the manufacturer decides to reserve capacity, its optimal capacity (Q^*) is:

$$Q^* = \begin{cases} 0, & \rho > \overline{\rho} \\ F^{-1}(1 - \frac{2\rho}{k(\rho+m)}), & 0 \leq \rho \leq \overline{\rho} \end{cases}, \text{ and } \overline{\rho} = \frac{mk}{2-k}.$$

Define $\overline{I} = \frac{k}{2}(m+\rho)(Q^* - \int_0^{Q^*} F(x)dx) - \rho Q^*$. Comparing $E\Pi_F^*$ with Π_F, we can get the decision rule for the manufacturer, as shown in Proposition 1.

Proposition 1. *When $\rho \leq \overline{\rho}$ and $I \leq \overline{I}$, the manufacturer decides to reserve capacity Q^*; Otherwise, the manufacturer decides not to reserve capacity.*

Proposition 1 shows when both types of costs are below a certain threshold, the manufacturer's optimal decision is to reserve capacity. One can see that when the manufacturer applies optimal production volume, the service level is $F(Q) = 1 - \frac{2\rho}{k(\rho+m)}$, as shown in Lemma 1.

3 Subsidy Policy Type

In this section, we consider the government's subsidy policies. Let α denotes the target service level of the emergency supplies. If the manufacturer does not reserve capacity or the service level ($F(Q)$) is lower than the target service level (α), the government needs to provide subsidy to stimulate the manufacturer to improve its capacity reserve. Policies that stimulate the manufacturer to satisfy the target service level α are called effective policies. The subsidy policies can be divided into four types: no subsidy (N), fixed subsidy (φ), marginal subsidy (δ), hybrid subsidy ($\delta + \varphi$). Assume the government's subsidy investment is C, the government should minimize subsidy investment. The effective policy that incurs minimum cost is called an optimal policy.

3.1 No Subsidy

If the manufacturer has motivation to reserve capacity and the capacity level can satisfy the target service level α without any subsidy, the government applies the no subsidy policy (N-policy). From Proposition 1, we can obtain the feasible region for the policy N is $\{0 < \rho \leq \rho_\alpha\} \cap \{0 < I \leq \overline{I}\}$ and $\rho_\alpha = \frac{km(1-\alpha)}{2-k(1-\alpha)}$. In this case, the manufacturer will voluntarily reserve capacity and the service level ($F(Q)$) is not lower than α. The manufacturer's maximum expected profit is $E\Pi_F^*$ and the government's investment is $C_N^* = 0$.

3.2 Fixed Subsidy

Fixed subsidy policy (φ-policy) denotes the government invests a fixed amount of φ to push manufacturers to reserve capacity. Under this policy, the manufacturer's maximum expected profit and the government's minimum investment are:

$$\begin{aligned} \underset{Q}{max} E\Pi_\varphi = k\int_0^1 \{t(mQ_0 - \rho Q) + (1-t)[mQ_0 + \int_0^Q (mx - \rho(Q-x))f(x)dx \\ + \int_Q^U mQf(x)dx]\}g(t)dt + (1-k)(mQ_0 - \rho Q) + \varphi - I \\ s.t. \qquad Q \geq 0 \end{aligned} \tag{3}$$

$$\begin{aligned} &\min_{\varphi} C_\varphi = \varphi \\ &s.t. \qquad E\Pi_\varphi^* \geq \Pi_N,\ F_\varphi(Q^*) \geq \alpha,\ \varphi > 0 \end{aligned} \tag{4}$$

We now analyze the decision-making behavior of the manufacturer and government. With Eq. (3), we can obtain the service level $F_\varphi(Q^*) = 1 - \frac{2\rho}{k(\rho+m)}$, which is independent of government's fixed subsidy (φ). Therefore, when $\rho > \rho_\alpha$, the fixed subsidy policy is not an efficient policy. If $\varphi = 0$, Eq. (3) is equivalent to Eq. (2). We can obtain the necessary conditions for effective fixed subsidy policy is $I > \overline{I}$, and the minimum government investment is $\varphi^* = I - \overline{I}$. In summary, the feasible region for the fixed subsidy policy is $\{0 < \rho \leq \rho_\alpha\} \cap \{I > \overline{I}\}$, and the government's investment is $C_\varphi^* = \varphi^* = I - \frac{k}{2}(m+\rho)(Q^* - \int_0^{Q^*} F(x)dx) + \rho Q^*$.

3.3 Marginal Subsidy

Marginal subsidy policy (δ-policy) denotes the government provides a marginal subsidy of δ for each emergency product produced by the manufacturer. Under this policy, the manufacturer's maximum expected profit and the government's minimum investment are:

$$\begin{aligned} \max_Q E\Pi_\delta = k\int_0^1 \{t(mQ_0 - \rho Q) + (1-t)[(m+\delta)Q_0 + \int_0^Q [(m+\delta)x - \rho(Q-x)]f(x)dx \\ + \int_Q^U mQf(x)dx]\}g(t)dt + (1-k)(mQ_0 - \rho Q) - I \\ s.t. \qquad Q \geq 0 \end{aligned} \tag{5}$$

$$\begin{aligned} &\min_\delta C_\delta = k\int_0^1 (1-t)[\delta Q_0 + \int_0^Q \delta x f(x)dx + \int_Q^U \delta Q f(x)dx]g(t)dt \\ &s.t. \qquad E\Pi_\delta^* \geq \Pi_N,\ F_\delta(Q^*) \geq \alpha,\ \delta > 0 \end{aligned} \tag{6}$$

From Eq. (5), we can obtain the service level $F_\delta(Q^*) = 1 - \frac{2\rho}{k(m+\rho+\delta)}$, which is related to government's marginal subsidy (δ). The manufacturer's profit and service level increases as the marginal subsidy (δ) increases. When $\rho > \rho_\alpha$, the government can use δ to change the manufacturer's optimal capacity. We define $\overline{I}_\alpha = \frac{\rho}{1-\alpha}[\alpha Q^* + Q_0 - \int_0^{Q^*} F(x)dx] - \frac{k(m+\rho)Q_0}{2}$. In this case, if $0 < I \leq \overline{I}_\alpha$, the marginal subsidy

$\delta^* = \delta_\alpha = \frac{2\rho}{k(1-\alpha)} - (m + \rho)$; If $I > \overline{I}_\alpha$, we obtain $\delta^* = \delta_0 = \{\delta | E\Pi^*_\delta - \Pi_N = 0\}$. When $\rho \le \rho_\alpha$ and $I > \overline{I}$, we obtain $\delta^* = \delta_0 = \{\delta | E\Pi^*_\delta - \Pi_N = 0\}$. In summary, the feasible region for the marginal subsidy policy is $\{0 < \rho \le \rho_\alpha\} \cap \{0 < I \le \overline{I}\}$. When $\delta^* = \delta_\alpha$, the government's subsidy investment $C^*_\delta = \frac{k\delta_\alpha}{2}[Q^* + Q_0 - \int_0^{Q^*} F(x)dx]$; When $\delta^* = \delta_0$, $C^*_\delta = \frac{k\delta_0}{2}[Q^* + Q_0 - \int_0^{Q^*} F(x)dx]$.

3.4 Hybrid Subsidy

Although the feasible range for the marginal subsidy policy is large, in some cases it is not the optimal policy. In addition, the service level is independent of government's fixed subsidy. Hybrid concept can effectively overcome the two disadvantages. Through above analysis, there is an overlap area ($\{0 < \rho \le \rho_\alpha\} \cap \{I > \overline{I}\}$) between the fixed subsidy policy and the marginal subsidy policy. By comparing the investment of the two subsidy policies, we can get Proposition 2.

Proposition 2. The fixed subsidy is always the optimal policy when the fixed subsidy policy and the marginal subsidy are both effective policies.

We use numerical examples to show Proposition 2 intuitively. The parameter values are set as $m = 10$, $Q = 100$, $k = 0.1$, $\alpha = 0.8$, $I = 100$ (Fig. 1).

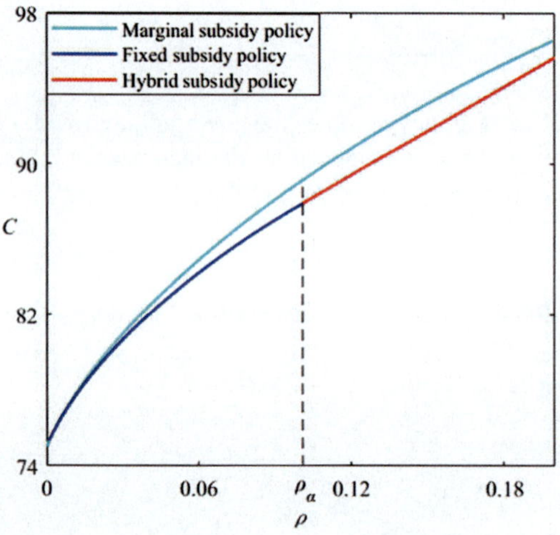

Fig. 1. Comparison of different policies.

Inspired by Proposition 2, we can provide hybrid subsidy policy: the government uses the marginal subsidy to satisfy the target emergency service level, and uses the fixed subsidy to satisfy the manufacturer's profit level. The marginal subsidy policy and the fixed subsidy policy are both special cases of the hybrid subsidy policy. The feasible region of the hybrid subsidy policy is the largest. The government should use the hybrid subsidy policy when the fixed policy and the marginal policy are both ineffective or non-optimal policies. When $\rho > \rho_\alpha$ and $I > \overline{I}_\alpha$, the analysis is the same as in Sect. 3.3, the service level $F_{\delta+\varphi}(Q^*) = F_\delta(Q^*)$, and $\delta^* = \delta_\alpha$. To satisfy the profit level, let $\varphi^* = I - \overline{I}_\alpha$. The feasible region for the optimal hybrid subsidy policy is $\{\rho > \rho_\alpha\} \cap \{I > \overline{I}_\alpha\}$, and $C^*_{\delta+\varphi} = \varphi^* + \frac{k\delta_\alpha}{2}[Q^* + Q_0 - \int_0^{Q^*} F(x)dx]$. We can get Proposition 3.

Proposition 3. When $\{\rho > \rho_\alpha\} \cap \{I > \overline{I}_\alpha\}$, the government's optimal policy is the hybrid subsidy policy.

4 Precise Subsidy Policy

Comparing the four subsidy policies, we can get the government's optimal strategies in different cases. The government should classify manufacturers according to the marginal investment and fixed investment, and then decides precise subsidy policy according to different types of manufacturers. With the precise policies, the government can stimulate the manufacturers to reach the target emergency service level with the minimum investment (Table 1).

Table 1. The government's precise subsidy policy.

Precise subsidy policy	Marginal investment cost (ρ)	Fixed investment cost (I)
No subsidy (N-policy)	$0 < \rho \leq \rho_\alpha$	$0 < I \leq \overline{I}$
Fixed subsidy (φ-policy)	$0 < \rho \leq \rho_\alpha$	$I > \overline{I}$
Marginal subsidy (δ-policy)	$\rho > \rho_\alpha$	$0 < I \leq \overline{I}_\alpha$
Hybrid subsidy ($\delta + \varphi$-policy)	$\rho > \rho_\alpha$	$I > \overline{I}_\alpha$

5 Conclusion

In this paper, we study how a government induces a manufacturer to reserve capacity to reach a certain service level through precise subsidy policy. We firstly study a benchmark model where a manufacturer should determine its capacity strategy for future emergency supplies without government subsidy. The results show that when the marginal investment and the fixed investment are both lower than the thresholds, the manufacturer will choose to reserve capacity. Secondly, we propose four types of subsidy policies, including no subsidy, fixed subsidy, marginal subsidy, and hybrid subsidy. The results show that when the manufacturer has the motivation to reserve capacity and satisfy the target emergency service level, the government applies the no subsidy policy; When the fixed investment for capacity reserve is high but the marginal cost is low, the fixed subsidy is the optimalpolicy; When the fixed investment is low but the marginal cost is high, the marginal subsidy is the optimal policy; When the fixed investment and marginal investment are both high, the hybrid subsidy is the optimal policy. Our reseach provides management insights for the government to implement precise subsidy policy for emergency supplies.

Acknowledgment. The authors gratefully acknowledge financial support from the National Social Science Foundation of China (Grant No. 20VYJ074), the Fundamental Research Funds for the Central Universities (Grant No. 30920010018), and the Postgraduate Research & Practice Innovation Program of Jiangsu Province (Grant No. KYCX20_0394).

References

Chen, J., Liang, L., Yao, D.-Q.: Pre-positioning of relief inventories for non-profit organizations: a newsvendor approach. Ann. Oper. Res. **259**(1–2), 35–63 (2017). https://doi.org/10.1007/s10479-017-2521-4

Chen, L., Wang, F., Wu, Y.: Research on selecting emergency material production enterprises based on productivity reserve. In: International Conference on Intelligent Networking and Collaborative Systems (INCoS), pp. 423–427 (2016)

Goyal, M., Netessine, S.: Volume flexibility, product flexibility, or both: the role of demand correlation and product substitution. Manuf. Serv. Oper. Manag. **13**(2), 180–193 (2011)

Li, J., Zhao, Q.: Research on the allocation decision of emergency material productivity reserve under the cooperation between the government and the enterprise. In: International Conference on Logistics, Informatics and Service Sciences (LISS), pp. 1–6 (2016)

Serel, D.A., Dada, M., Moskowitz, H.: Sourcing decisions with capacity reservation contracts. Eur. J. Oper. Res. **131**(3), 635–648 (2001)

Whybark, D.C.: Issues in managing disaster relief inventories. Int. J. Prod. Econ. **108**, 228–235 (2007)

Yao, X., Huang, R., Song, M., Mishra, N.: Pre-positioning inventory and service outsourcing of relief material supply chain. Int. J. Prod. Res. **56**(21), 6859–6871 (2018)

Zhang, Z., Li, X.: The optimal manufacturer's reserve investment and government's subsidy policy in emergency preparedness. J. Inequalities Appl. **62** (2013)

Theoretical Research on the Mechanism of Improving Digital Literacy for Optimizing Doing-Digital-Business Environment

Zhen Wang[1], Xiaolong Li[2], Jiyin Li[1], and Chunhui Yuan[1](✉)

[1] School of Economics and Management, Beijing University of Posts and Telecommunications, Xitucheng Road, Haidian District, Beijing, China

[2] School of Modern Post, Beijing University of Posts and Telecommunications, Xitucheng Road, Haidian District, Beijing, China

xiaolongli@bupt.edu.cn

Abstract. Referring to the theory of digital literacy and digital Doing-business environment from the theoretical point of view, this article analyzes the fundamental and driving role of digital literacy in optimizing doing-digital-business environment from the perspective of human capital theory and digital dividend theory. This article summarizes 'efficiency improving', 'innovation encouraging' and 'inclusiveness promoting' as three major mechanisms of improving digital literacy in optimizing the digital production environment, digital innovation environment and education environment, that is, the doing-digital-business environment, which benefit the forming of a fair, balanced, healthy and sustainable development environment for the digital economy.

Keywords: Digital literacy · Doing-digital-business environment · Human capital · Digital dividend

1 Introduction

After the Second World War, along with the active innovation in the field of information technology, human society has entered an important stage of transformation. Based on the development of digital information technology, information civilization and industrial civilization blend with each other, constantly producing new modes of production and living conditions. In this process, scholars from various countries have put forward many new concepts, such as "information revolution", "information industry", "information economy", "information society", "information age", "information technology", etc. In China, the word "informatization" is widely used to describe the process of the revolution, and the reform direction is to rely on "information industry", develop "information economy" and move towards "information society". In the information society, data has become more important than land, material and energy resources. In order to develop and utilize data and information resources, digital economic activities are expanding rapidly, gradually replacing industrial production activities and becoming the main content of

C. Yuan et al. (Eds.): *Proceedings of the 4th International Conference on Economic Management and Green Development*, AEPS, pp. 487–498, 2021.
https://doi.org/10.1007/978-981-16-5359-9_57

national economic activities. The status of digital economy in the national economy is highlighted.

Since entering the information society, the extensive use of digital technology has changed the economic activities and economic environment, booming the digital economy. In 2019, the scale of digital economy of 47 economies in the world will reach 31.8 trillion US dollars, an increase of 1.6 trillion US dollars over the previous year, and the GDP of the same period will increase by 1.2 trillion US dollars. The growth value of digital economy has exceeded the value of GDP growth. Among them, the added value of China's digital economy in 2019 has reached 35.8 trillion yuan, accounting for 36.2% of GDP, contributing 67.7% to GDP growth.

The digital economy brings huge dividends to countries, but it also puts forward some new requirements for the business environment. In recent years, international organizations have incorporated the relevant indicators of business environment of digital economy into the national evaluation system. In many evaluation systems, most of them measure the level of digital literacy, such as "Internet user skills" and "advanced skills and development" in digital economy and society index (DESI) 2020, and ICT skills (average years of education, total enrollment rate, and total enrollment rate) in ICT development index released by the International Telecommunication Union, "The proportion of people with ICT skills" and "human capital index" issued by the United Nations Department of economic and social affairs. International organizations obviously agree on the important role of digital literacy in the construction of digital doing business environment. However, by combing the relevant literature on digital literacy and digital doing business environment in recent years, we found that there is no research on the relationship and mechanism between the two. Therefore, we will analyze the important position of digital literacy in digital doing business environment from a theoretical perspective, provide a theoretical basis for the follow-up research, and explore the path through which the improvement of citizen digital literacy will affect the digital business environment.

2 Theoretical Background

2.1 The Concept of Digital Business Environment

The emergence of the word "business environment" can be traced back to the "doing business" project which launched by the world bank in 2002. Through the establishment of a set of global comparable business environment assessment system, the survey provides reference for different countries to improve the business environment. The world bank defines the business environment as the sum of the time and cost that an enterprise in an economy needs to spend in opening an account, financial credit, protecting investors, paying taxes and other important fields covering the whole life cycle of an enterprise [2]. It has designed a series of business environment evaluation index systems, and has continuously conducted annual service implementation on business laws and regulations of various countries since 2003. This report includes 11 indicators and 190 economies in the year of 2020.

The concept of a "digital doing business environment" first came from the United Nations Trade and Development Council (UNCTAD) report "Developing Digital Industries" in 2017, and the World Bank proposed the "Digital Doing Business Indicators" in 2018, arguing that the digital business environment should be built to promote healthy competition among countries and enhance the vitality of digital economic development. The World Economic Forum (WEF) and the Harvard Business Review report "Digital Doing Business Environment Convenience 2019", argues that the "Digital Doing Business Environment" refers to the level of convenience that digital platforms enjoy in entering, operating, thriving and exiting markets. Alibaba Group released the Digital Economy Business Environment Report in 2019, arguing that the "digital economy business environment" should focus on the platform to create a good environment for resident enterprises, referring to the technologies promote the formation of a fair, transparent and healthy platform ecology.

The Digital Doing business environment should not only is to use digital technology to enhance the traditional market environment of "digital + business environment", more should be "digital business + environment", a new business environment which adapts to the digital economy and innovation development needs (as shown in Fig. 1). In the digital economy era, the government, enterprise, platforms, interconnected between consumers and colleges, optimizing the production relations to create value, (as shown in Fig. 1) platform use technical tools to integrate the resources to achieve precise matching between supply and demand, the enterprise's products or services through the platform to reach consumers' hands, Consumers use the platform to pay for the product or service, meanwhile left an important data elements on the platform, platform using data elements on technology development to provide better service experience, on the other hand, enterprises can get some technical support from platforms for product development or internal management, which will improve enterprises' operating efficiency and accelerate product iterative innovation speed. In the process of value creating, the enterprises, platforms, consumers the government plays an important role . Ideally, Government lax regulation will create a positive and relaxed production environment, the talents with high digital literacy, high and new technology and strict standards provided by the research institutes of universities and colleges will provide rich and innovative soil for the value creation of the whole digital economy, here we mention two environments: the production environment and innovation environment, there are also the current two important sub environment of digital doing business environment.

From the analysis of the value activities, we believe that the production environment and the innovation environment are two important aspects constituting the digital doing business environment. In combination with the indicators related to the business environment of digital economy in the evaluation system of several international organizations, the production environment is relatively broad, mainly including the digital infrastructure environment, data security environment, competition and consumer protection environment, and regulatory and service environment. The innovation environment is relatively narrow, which mainly includes citizen digital literacy, attitude towards entrepreneurial risks, creative output and other aspects [3]. We can see that digital literacy indicators are included in the digital innovation environment. There is no doubt that digital literacy plays an important role in the digital innovation environment, but there is no relevant

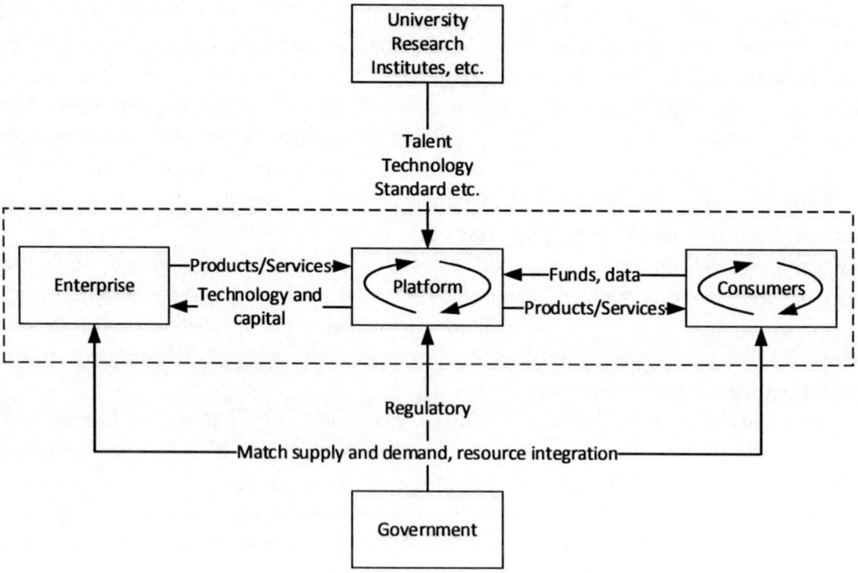

Fig. 1. Digital economy subjects & value activities

literature to study the functional relationship between digital literacy and the digital business environment. It is of great practical significance to understand the interaction between them to help us find a way to optimize the digital business environment and promote the development of digital economy.

2.2 Digital Literacy Concept and its Framework

With the advent of the digital economy era, digital ability has become an essential ability for individuals to adapt to their future work and life. International organizations such as OECD, World Economic Forum, World Bank and United Nations have identified digital ability as the basis for us to face the changing world in the future [4]. Digital ability refers to the integrated use of digital information technology to achieve returns (including economic returns, capacity growth, social participation, promotion and development opportunities, etc.). The ability means that people will be able to use enough skills and ability to solve social task efficiently, but do not cover the cognitive abilities of creative ideas and ethics of social development, cultural attitudes and other advanced capabilities. Digital ability is generally considered as a part of digital literacy, and "literacy" is considered as a holistic concept, a concept that constantly changes and keeps pace with The Times, as well as a high-level capability element system containing digital ability [5].

The word "Digital Literacy" first appeared in 1994. Israeli scholar Y. Eshet- Alkalai sums it up as "the ability to understand and use various digital resources and information displayed on the computer" [6]. Then Paul Gilster in his book "Digital Literacy" formally put forward the concept of "Digital Literacy" [7]. He defines digital literacy as the ability to access, understand and integrate digital information. Specifically including web search, hypertext reading, digital information criticism and integration skills, he describes the Digital Literacy for the use of Digital information and understanding, and stressed the importance of digital technology as a "basic skills", effectively distinguish Digital Literacy and traditional printing, speaking, reading and writing.

Obviously, digital literacy is not a single skill, but a complex and interdisciplinary open literacy, which can promote individuals to acquire a variety of survival skills including digital ability, improve the creative cognitive ability of social groups, and become an important asset in the development of the information society. There is a consensus among academics on the importance of digital literacy in the 21st century, but for the framework and definition of digital literacy has been chaotic, related research has shown that the model and the framework of digital literacy is more than 100 kinds of [8], but there is no universal model, the framework of most frame makers are considering their political, economic and social environment, but digital literacy framework varies from place to place, often across national interagency characteristics [9]. National and international organizations continue to try to develop standardized digital literacy frameworks.

In the world, the influential research on digital literacy mainly included 21st Century Skills of the US and Key Competences of the European Union, among which the digital literacy framework developed by the European Union had a very heavy impact on the subsequent research on digital literacy. As one of the eight core literacy stressed by the European Union, the digital literacy has experienced from the initial "information technology" (emphasis on the technology itself) to the "information technology skills and use", "information technology skills" (emphasis on skills), and "digital literacy" (emphasis on literacy) [10]. And the European Union defines "digital literacy" as "The ability to use information technology confidently, critically and innovatively in work, employment, study, leisure and social engagement", emphasizes people's network interaction and cooperation on the basis of obtaining and using information. The EU digital literacy project, launched in 2011, sets three goals:

i. To identify the core elements that will influence digital literacy;
ii. Building on previous research, improve the EU digital literacy framework/indicators.
iii. Propose an applicable digital literacy framework to all citizens.

The digital literacy framework proposed by the project includes five literacy domains: information, communication, content creation, security awareness, and problem solving, as shown in Table 1. Every competency area was divided into multiple "specific Competences" indicators, which were divided into A (basic), B (medium) and C (advanced) three "competence Levels". Then it is described in three dimensions of knowledge, skill and attitude.

Table 1. EU digital literacy framework

Literacy domain	Overview
Information domain	The ability to identify, locate, retrieve, store, organize, and analyze digital information and determine its relevance
Communication domain	Share resources, communicate and cooperate with others through online digital tools, participate in online community interaction, and have cross-cultural awareness, in short, the ability to communicate in a digital environment
Content creation domain	The ability to create and edit new content, from word processing to graphics, images, videos, etc., to reintegrate previous knowledge and content, to generate creative expression of information and media output and programming, and to apply intellectual property properly
Safety consciousness domain	Personal protection, data maintenance, digital identity protection, security measures, sustainable utilization capability
Problem solving domain	Identify digital information needs, select the most appropriate digital tools according to needs, solve problems through digital means, use technology innovatively, and solve technical problems

3 The Status of Citizens' Digital Literacy

3.1 Enhance the Human Capital of Digital Doing Business

From the perspective of human capital theory, people with digital literacy provide high-quality human capital for digital doing business. The new economic theory holds that human capital is an important factor to explain economic growth by endogenous growth model and emphasizes that the production of human capital is more important than that of material capital. Many scholars have three main understandings about the mechanism of human capital acting on economic growth: One is the factors of human capital is regarded as inputs to the production of final products directly affect economic growth. The second is that human capital is a key input in technology production and indirectly affects economic growth through innovation and technology diffusion. Third, human capital is regarded as both the input factor of final product and the key input product of technological production, which is a joint action mechanism [11]. For the digital economy, we also tend to think of digitally literate citizens as a joint mechanism for fostering growth models in the digital economy.

On the one hand, the citizens with digital literacy ACTS directly on the digital economy as the input factor for the production of final products. The digital economy represented by the new generation of productivity such as block chain, cloud computing, big data, artificial intelligence and digital twins needs people with the above digital skills to promote digital production and create economic value.

On the other hand, the citizens with digital literacy, as a key input in technological production, indirectly affects the digital economy through the diffusion of innovative knowledge and technology, and affects the development of the digital economy by influencing technological progress. The improvement of total factor productivity in digital economy mainly depends on technological innovation and the diffusion and absorption of cutting-edge technologies. The stock of human capital greatly influences these two aspects. The larger the number of citizens with high quality digital literacy, the faster the pace of knowledge innovation and technology diffusion. Accordingly, human capital promotes the growth of productivity of digital economy more obviously.

Obviously, the starting point and end point of digital business environment construction is to protect the development of digital economy. Through the above analysis of the action mechanism of citizens' digital literacy on the growth of digital economy from the perspective of human capital theory, we can see that the population with digital literacy is an important human capital for the development of digital economy and plays a very important role in the development of digital economy.

3.2 Help Unleash the Digital Dividend

The digital dividend refers to the widespread benefits of using digital technologies (Internet, mobile phones and all other tools for collecting, storing, analyzing and sharing information digitally), mainly includes three aspects [12]:

i. Grow economy
ii. Expand employment
iii. Improve services

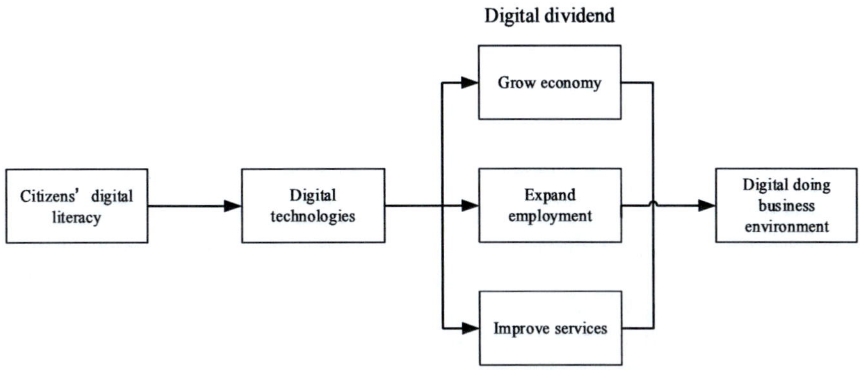

Fig. 2. The path to release digital dividend of digital literacy

As shown in Fig. 2, citizens' digital literacy affects the release of digital dividends by influencing the absorption and diffusion of digital technologies, thus exerting a role in the digital business environment.

Firstly, digital literacy has a positive effect on the absorption of digital technology knowledge. A digital technology that achieves diffusion and commercial application is premised on being used, but different groups of people have different preferences for absorbing knowledge. It is inefficient for people who have never been exposed to computer knowledge to absorb knowledge of digital technology, and the people with certain digital literacy for digital technology knowledge absorption is efficient. They can combine their own digital knowledge reserve to summarize and sort out the absorbed knowledge, and are likely to research and produce new technologies or new knowledge, so as to complete the transformation and upgrading of the existing digital technology knowledge system.

Second, digital literacy affects the diffusion of digital technology. The diffusion model of digital technology basically follows the diffusion of ICT technology and The innovation diffusion model of Rogers, and generally presents an "S"-shaped curve (Fig. 3).

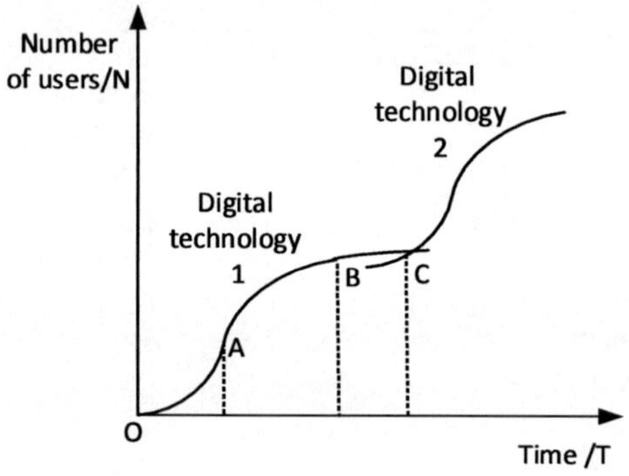

Fig. 3. Multi-digital technology diffusion model

In the very early "O-A" phase of digital technology 1, A necessary condition was that someone would use the technology first and the skills would be spread, and then the digital technology diffused slowly as the number of users (digital skill owners) increased. As the number of people with digital technical 1 skills continues to increase, digital technical 1 enters a period of rapid development, known as the "A-B" period in the figure. When the number of digital technology 1 users reaches a certain proportion in the total population, that is, the number of people with the digital skills reaches saturation, the development of digital technology 1 enters a period of stagnation and slowdown.

In the period of stagnation, the number of users no longer provides the impetus for the development of digital technology, technology can only be innovated with people who have this digital skill. They can improve their literacy within digital literacy domains such as information, knowledge exchange, content creation, and problem solving, and develop new technologies through iterative innovation (digital technology 2), then we move on to the next diffusion phase of the technology. It can be seen that the citizens' digital literacy is the breakthrough of the transformation from old technology to new technology, and plays a key driving role in the iterative innovation of digital technology.

4 Mechanisms for Improving the Digital Business Environment

Through the above theoretical analysis of digital literacy and digital business environment, summarize three mechanisms that can help improve digital doing business environment by improving citizens' digital literacy (Fig. 4):

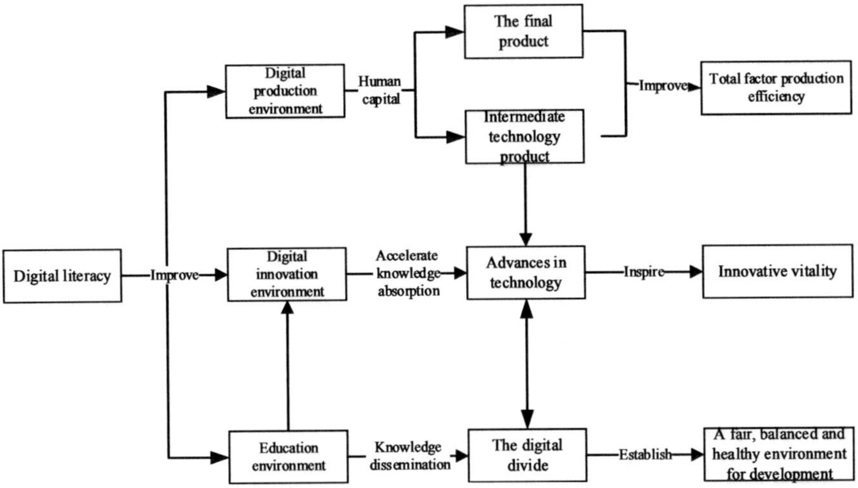

Fig. 4. Mechanism of action to improve the citizens' digital literacy for optimizing the digital business environment

i. Improve efficiency to optimize the digital production environment including digital infrastructure environment, data security environment, competition and consumer protection environment, and regulatory and service environment.

Improving the citizens' digital literacy can significantly improve the human capital in the digital doing business environment, improve the total factor production efficiency in the digital production environment which including the digital infrastructure environment, data security environment, competition and consumer protection environment, as well as the regulatory and service environment.

ii. Promote innovation to stimulate the innovation vitality of digital innovation environment.

Information technology revolution based digital information and communication technology (ICT) is the driving force of the development of social informatization. It is also the key to stimulate technological innovation, product and service innovation, organizational innovation and system innovation in all fields of the whole society, so as to optimize resource allocation and improve economic operation efficiency. The wide application of digital technology in production, finance, scientific research and education, medical care, enterprise and government management, and family and individual social life has exerted a huge and profound impact on economic and social development, fundamentally changing people's lifestyle, behavior and values. A dynamic and innovative business environment is conducive to the accumulation of knowledge and technological innovation.

iii. Promote inclusiveness to create a fair, balanced, healthy and sustainable development environment.

The vigorous development of the digital economy, push the modern social progress, but the construction of the information society is not achieved overnight, the spread of digital technology penetration degree also has never been uniform and indiscriminate. The "digital divide" is widespread across countries, regions and groups of different ages, occupations and genders, and with the further spread of technology. And as technology spreads further, the digital divide deepens.

In the early days, researchers and policy makers generally focused on the differences in ICT infrastructure access, that is, the differences between those who had ACCESS and those who did not. However, with the gradual popularization of Internet access in developed countries, especially the continuous increase of the penetration rate of broadband access, the differences in network use gradually become prominent. Due to different people's different degrees of mastering network application skills, their ability to effectively obtain and use information through the network is different, which is specifically manifested in different aspects such as the duration, intensity and behavior pattern of network use, and the efficiency of information search.

In recent years, Internet access has become highly pervasive in countries leading in ICT development, and the new digital divide is becoming increasingly significant. It has been observed that even when users have autonomous and unrestricted access to ICT infrastructure, there are significant differences in their proficiency in using the network to accomplish specific tasks and achieve specific goals. Even if users have the same level of skills and reach the same level of intensity, the returns (including economic returns, capacity growth, increased social participation, development opportunities, etc.) they can get from Internet use are often different [13–16]. That means The difference in digital literacy has created a new digital divide.

Improving the citizens' digital literacy can make ordinary people more sensitive and receptive to digital knowledge, contribute to the dissemination of knowledge, bridge the digital gap to a certain extent, and contribute to the optimization of the educational environment. The relationship between the educational environment and the citizens' digital literacy is inseparable, which is a two-way correlation. Education can improve the citizens' digital literacy, and the improvement of the citizens' digital literacy can in turn strengthen the education level through technological means.

In recent years, developed countries have increased their citizens' digital literacy education, such as the comprehensive citizen digital literacy cultivation model of the United States [17], the UK has integrated personal digital literacy into the vocational development education model [18], and China is also actively exploring the education model for improving digital literacy.

5 Conclusion

With the popularization of information technology, digital economy plays a prominent role in national economy, and the business environment is constantly changing. Countries have gradually realized the importance of the digital doing business environment, and the continuous improvement of the digital business environment has become a priority for governments around the world. According to the analysis of the relationship between digital literacy and digital doing business environment in this paper, we think that in the process of optimizing digital doing business environment, we should consider the following ways to improve the citizens' digital literacy:

i. We should establish a multi-dimensional digital literacy cultivation model covering vocational education in primary and secondary schools and higher education, popularize computer-related courses, improve citizens' digital literacy from the perspective of education, and provide high-quality human resources for the digital doing business environment.
ii. Attaches great importance to the talent management. Technological progress is an important factor to promote economic development and reform, we should attach greater importance to innovation in the whole society. Give slack and proper management to innovative talents, and establish the coordinated mechanism of rewards and punishments to inspire talent innovation vigor, improve the stock for the optimization of digital production and innovation environment are helpful.
iii. Attach importance to the construction of scientific and technological innovation system. To construct a closed loop of knowledge creation from digital literacy of talents to high-tech breakthrough can fully realize the innovation vitality stimulated in the innovation environment and develop the technology growth economy in practice.

References

1. Basol, O., Yalin, E.C.: How does the digital economy and society index (DESI) affect labor market indicators in EU countries. Hum. Syst. Manage. **39**(1) 2020
2. World Bank Group. Doing business 2020: The World Bank, 2019. https://openknowledge.worldbank.org/bitstream/handle/10986/32436/9781464814402.pdf
3. Development research centre of the state council. Doing business in digital economy: International Indicator Framework and Policy Direction. No. 164.
4. Wang, X.: Future-oriented framework of global digital literacy and competency standards – analysis based on DQ global standards report 2019. Libr. Constr, 1–14, 2020
5. Wang, Y., Yang, X., Wei, H., Wang, J.: From digital literacy to digital competence: conceptual evolution, components and integration models. J. Dist. Educ. **31**(03), 24–29 (2013)
6. Wang, J., Huang, Y., Shen, Y.: Research status and enlightenment of digital literacy. Digit. Educ. **4**(01), 15–21 (2008)
7. Gilster, P.: Digital Literacy. Wiley, New York (1997)
8. Brown, M., Xiao, J.: Challenges of digital literacy: the Leap from limited skills to critical thinking. China Distance Educ. (04), 42–53+79–80 (2018)

9. Mountains. II digital literacy in higher education: the new media alliance horizon project strategy bulletin research. Libr. Constr. (7), 42–47, 53 (2018)
10. Ren, Y., Suo, X., Liu, X.: European union digital literacy framework research. Mod. Distance Educ. Res. **05**, 3–12 (2014)
11. Wei, D., Yang, Z., Xia, G.: Research on the action mechanism of human capital promoting economic growth . China Soft Sci. **08**, 173–183 (2014)
12. World bank group. World bank development report: digital dividends. The World Bank (2016)
13. OECD. Understanding the digital divide (2001)
14. Brandtzæg, P.B., Heim, J., Karahasanović, A.: Understanding the new digital divide—A typology of Internet users in Europe. Int. J. Hum. Comput. Stud. **69**, 123–138 (2011)
15. van Dijk, J.A.G.M.: The evolution of the digital divide: the digital divide turns to inequality of skills and usage. Digital Enlightenment Yearbook (2012)
16. van Deursen, A.J.A.M., Helsper, E.J.: The third-level digital divide: who benefits most from being online. Stud. Media. Commun. **10**, 29–53 (2015)
17. Zhang, J.: Status quo and enlightenment of digital literacy education in the United States. . Books Intell. Work **6**, 135–142 (2018)
18. Li, C.: The practice and enlightenment of digital literacy education in British university libraries. Libr. Constr. **8**, 78–82 (2017)

Doing Business Environment Assessment-A Review Study

Songliang Guo[1], Xiaolong Li[2], Jiyin Li[1], and Chunhui Yuan[1](✉)

[1] School of Economics and Management, Beijing University of Posts and Telecommunications, Xitucheng Road, Haidian District, Beijing, China

[2] School of Modern Post, Beijing University of Posts and Telecommunications, Xitucheng Road, Haidian District, Beijing, China

xiaolongli@bupt.edu.cn

Abstract. The article focuses on the doing-business environment evaluation issues in the context of the digital economy. It uses bibliometrics, knowledge graphs, summaries to sort out relevant domestic and foreign literature on doing-business environment evaluation, extracts and analyzes related research hotspots in this field. The key focus areas the connotation of the doing-business environment and the evaluation index system; how the connotation of the doing-business environment evolves with time; single dimension and comprehensive dimensions for the doing-business environment evaluation index system according to different research perspectives.

Keywords: Bibliometrics · Doing-business environment · Evaluation system · Knowledge graph · Visualization

1 Introduction

To encourage healthy competition between countries and build a more efficient system of digital economy policy framework, many international organizations in the countries to promote the digital economy best practices research in recent years, they actively carry out all kinds of a digital national economy evaluation, the digital economy-related indicators such as digital economy business environment, digital economy, global innovation and competitiveness index into the evaluation system, which for the assessment of digital economy business environment is becoming a national new focus.

The digital economy brings huge dividends and also puts new requirements on the business environment. China attaches great importance to improving the business environment, vigorously promotes the reform of "delegating powers and services" and keeps improving the business environment. At the opening ceremony of the first China International Import Expo in November 2018, General Secretary Xi Jinping pointed out the need to "create a world-class business environment." [1]. On October 22, 2019, the State Council formulated and issued the Regulations on Improving the Business Environment. On October 31 [2], 2019, the fourth Plenary Session of the 19th CPC Central Committee clearly stated that "improving the business environment and stimulating the vitality

C. Yuan et al. (Eds.): *Proceedings of the 4th International Conference on Economic Management and Green Development*, AEPS, pp. 499–515, 2021.
https://doi.org/10.1007/978-981-16-5359-9_58

of various market entities". In November 2019, General Secretary Xi Jinping pointed out at the opening ceremony of the second China International Import Expo that "we will continue to optimize the business environment, which is the soil for the survival and development of enterprises, and constantly improve the market-oriented, legalized and internationalized business environment." [3]. In November 2020, In the third China International Import and Export Expo, General Secretary Xi Jinping once again stressed the importance of "continuously optimizing the business environment" [4]. It can be seen that China's vigorous efforts to improve the business environment and the positive results achieved are the results of the great importance and continuous efforts of the State Council of the CPC Central Committee.

In recent years, China's digital doing Business environment has been significantly optimized. The world Bank's "Doing Business 2020" shows that China's ranking of doing Business environment has risen to 31st, 15 places higher than before [5]. International business environment evaluation system includes digital infrastructure, digital economy innovation environment, competition, consumer protection, data and security environment, supervision and service environment, etc. , with countries can obtain common indicator is given priority to, focused on the development status of evaluation, the evaluation emphasis of different groups and different, it will appear the evaluation results can't reflect each country to do business on the dilemma of the development of the real, like xi general secretary said that all countries should strive to improve their business environment, solve their own problems, can't blame others, always whitewash himself, Don't shine on others and not yourself like a flashlight.

In view of this, this paper will focus on the digital business environment evaluation research overview. Firstly, through literature measurement and knowledge map, the domestic and foreign literature in the field are sorted out to understand their research situation. Then, comparative analysis and induction are made for each research topic, and literature is sorted out and described in detail. Based on the above research, the future research trend of this field is proposed.

2 Overview of Doing Business Environment Research Based on Bibliometrics

2.1 Analysis of Literature Volume Trend

Taking November 5, 2020, as the retrieval time point, data collection was conducted on the literature related to business environment research in CNKI database and WOS database respectively. In CNKI database, the choice of the old advanced search, and the search keyword "doing business", to ensure the quality of research, eliminate class meeting, newspapers and other news texts, in the journal child list check the SCI source journals, EI, core journals, CSSCI source journals and journal of CSCD as a sample, through sorting out again the author unit is empty, the author called this newspaper editorial articles, received 429 articles (including digital business environment with a total of 6) relevant to the subject matter. In the WOS database, articles of economic, management, business types were set for retrieval with "title" as the retrieval scope, "Business environment" as the retrieval term, and the literature type was defined as

"Article". The literature were exported into plain text format, and repeated literature were cleaned in the CiteSpace software. Finally, 1105 related articles were obtained. In the past 23 years, the publication volume of relevant literatures at home and abroad is shown in Fig. 1, and the paper growth and change curve is drawn in Excel. It can be seen in the domestic business environment as the theme of the related literature were identified nearly five years, growth, trend regression formula for y = 6 × 10-6e-0.7294x, determination coefficient R2 value is 0.9; foreign to the study of evaluation about the business environment can be traced back to 1997, the overall number showed a trend of steady rise, trend regression line equation for y = 1.1943x + 31.112, equation determination coefficient R2 is 0.5387, the more as you can see, both at home and abroad for doing business abroad study earlier than domestic, and the research trend is stable, the relevant domestic business environment research started late, but the momentum of rapid development, each year the number of post-show exponential growth way, this may be related to digital development under the background of economic globalization to the attention of the business environment.

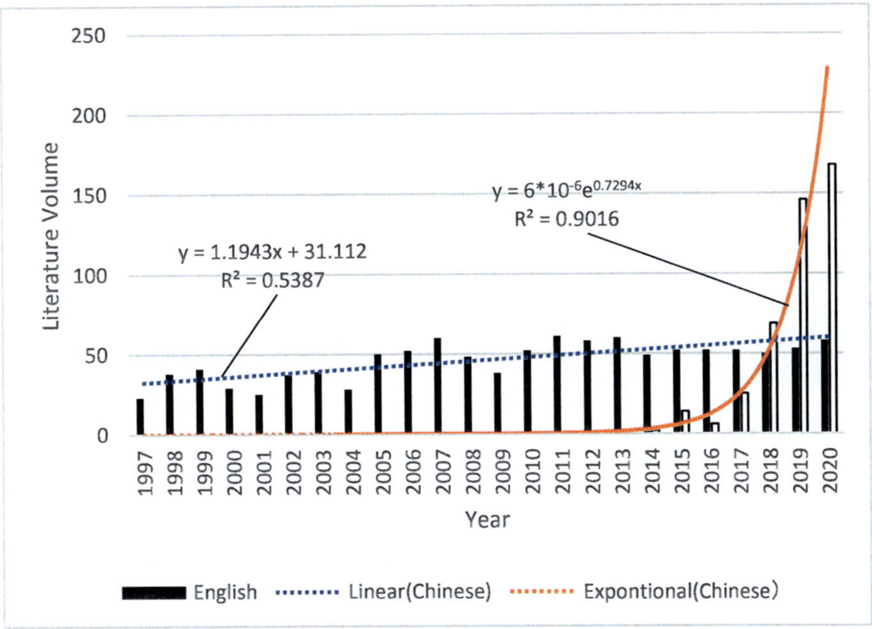

Fig. 1. Statistics of published literatures from 1997 to 2020.

2.2 Literature Metrological Analysis Research Situation

In order to further clarify the research situation in the field of the business environment, Citespace software was used to process the exported literature respectively and construct the network relationship diagram between keywords. The keyword distribution network of Chinese studies is shown in Fig. 2.

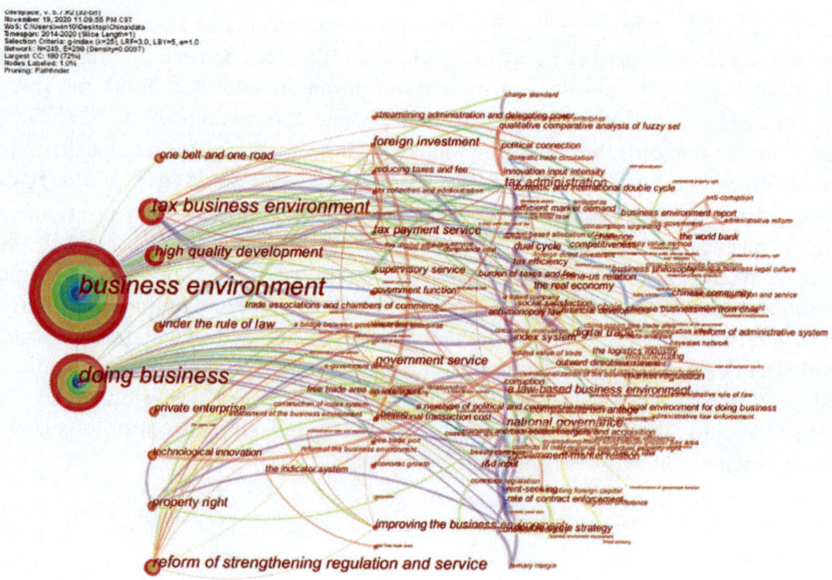

Fig. 2. Keywords distribution network in Chinese research.

The centrality of keywords is reflected by the size of nodes. The larger the node is, the stronger the centrality of keywords is. It can be seen that the business environment is the core node, among which the centrality is 0.62. "Doing business", tax business environment, and reform of strengthening regulation and service are intermediate nodes, among which the core values are 0.33, 0.22, and 0.21 respectively, indicating that scholars focus on these aspects of a business environment. It can be seen from the figure that there are relatively many edge nodes, which, on the one hand, indicates the diversity of relevant research contents in the business environment, and on the other hand, indicates that the research is relatively scattered. Further research on network nodes reveals that although the centrality of edge nodes, World Bank, doing Business report and indicator system is relatively low, their degree is high, indicating that they are hot research issues in the field of doing business.

Considering that the research samples of domestic and foreign scholars are from different sources and the amount of data is quite different, also besides, the research in foreign countries started earlier than that in China. If the data of the two are combined for analysis, only the keywords of English research are in the visualized figure, and the nodes of Chinese research will be marginalized. Taking the above considerations into account, relevant studies on business environment English are also analyzed independently, as shown in Fig. 3.

Different from Chinese studies, foreign studies have four central nodes, namely business, strategy, model, and adoption. Their centrality is 0.12, 0.11, 0.1, and 0.1 respectively, indicating that scholars pay high attention to business and strategy. Environment, Competitive Advantage, Performance, Impact, Innovation, and Organization are intermediate nodes, among which the centrality is 0.09, 0.08, 0.08, 0.07, 0.07, 0.07, and 0.07

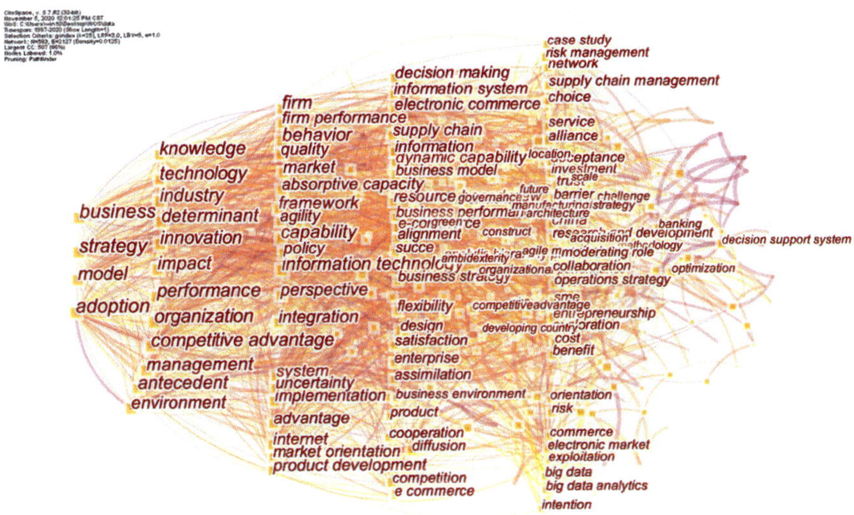

Fig. 3. Distribution network diagram of keywords in English research.

respectively, indicating that scholars focus on these aspects of the business environment. It can be seen from the figure that there are more nodes between different levels of foreign studies than Chinese studies, indicating that the scope of foreign studies is more extensive. It is worth noting that the centrality of the business environment is relatively low, but its degree is relatively high, indicating that there are more studies on the combination of the business environment and other topics.

Given this, this paper will summarize the connotation of business environment and the business environment evaluation research, and propose the development direction of future research in the field of business environment based on this.

3 Research on the Definition of Doing Business Environment

Materialist dialectics thought, all things have been accompanied by the development of advance and twists and turns, the understanding of doing business is based on the practice of reflection, the cognitive process is constantly changing, so the connotation of the business environment has certain timeliness, uncertainty, and fuzziness, the connotation of the current understanding of the business environment has not yet formed, in this article, through combing the literature in the field of the domestic and foreign business environment, to tease out business environment connotation by the timeline route of key research, as shown in Fig. 4.

As early as in 1968, the American scholar in accordance with the west in law, and Peter bunting in international business arrangements the conceptual framework of the article puts forward the evaluation of investment environment of "cold heat analysis method", this method from the standpoint of the investor and the investment main body, with political stability, unified market opportunities, growth status, cultural level, fiat dysfunction, substantial barriers to investment environment, geographical and cultural

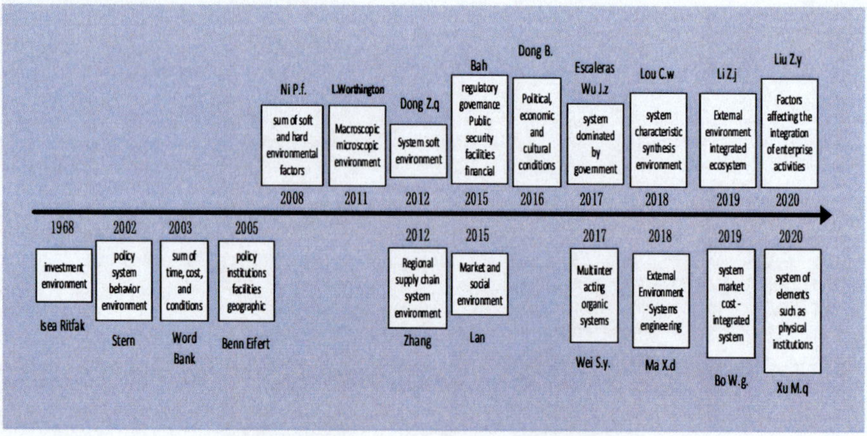

Fig. 4. Connotation of business environment.

differences in the seven factors, evaluate the investor, and the "hot" (excellent investment environment) to "cold" investment environment (poor) one by one sorting [6].

The concept of Doing Business to enter the public eye in recent years, with the released since 2003, the world bank's annual "Doing Business report, the appearance of the word" Doing Business "can be traced back to 2002, the world bank's international finance corporation in order to under different life cycle from multiple aspects of the enterprise Business environment faced by analysis evaluation, launched a" Doing Business "survey, the survey by making a comparable global Business environment assessment system, The World Bank defines the business environment as the sum of the time and cost required for an enterprise to comply with policies and regulations in terms of application for an opening, production and operation, trade activities, tax payment, closure and execution of contracts [5]. Stern regards the business environment as the sum of current and expected policy, institutional and behavioral environments, usually involving macroeconomic policies, government management systems and infrastructure environment, and believes that they can affect the returns and risks of enterprise investment [7]. Benn Eifert et al. believed that the business environment is the policies, institutions, infrastructure, human resources and geographical environment that affect the operating efficiency of different industries and enterprises [8]. Bah and Fang attributes the business environment to five aspects: environmental regulation, corruption governance effect, public security environment, infrastructure improvement and financial development [9]. Worthington and Britten attributed the business environment to the macro environment (such as politics, economy, law, society, etc.) and the microenvironment (such as suppliers, competitors, labor market, financial institutions, etc.) that affect the production and operation of enterprises [10]. Lan pointed out that business environment refers to the requirements of relevant policies, systems, codes of conduct and the market and social environment faced by market players when they engage in production and business activities or trade activities [11]. Escaleras and Chiang takes the business environment as institutional quality, including government efficiency, regulation quality and corruption level, etc. Escaleras and Chiang closely links the business environment with the government, believes that the business environment is an inevitable outcome of the government-led economy and even considers government intervention in the market as a routine part of the business environment [12].

The concept of "business environment" has a relatively short history in China. Guangdong province is the first province to conduct systematic theoretical research and practice on "business environment". Guangdong has organized relevant institutions to carry out research on the topic of "Building a legalized international Business environment", and commissioned Sun Yat-sen University and Guangdong University of Foreign Studies to carry out thematic research on "Building a legalized international business environment by adhering to the socialist market-oriented reform direction" [13]. Before doing business the concept is the investment environment, it is the same as the development of foreign business environment, is the early nineties of the 20th century China's reform and opening up the focus of attention, mainly from the perspective of foreign investment was observed the development of a regional environment, so there is investment business environment, strictly speaking, doing business is the development environment, including software, hardware, hardware is the physical environment, soft environment is the system cultural environment, they constitute the development of a country or a region as a whole environment [14].

Ni Pengfei summarized the business environment as a combination of hard environment, such as market and infrastructure, and soft environment, such as government policies and services, as well as judicial, administrative and tax systems, etc., and summarized the business environment as the sum of various factors in the production and operation activities of enterprises [15]. Some scholars only regard the business environment as a kind of institutional soft environment. The government can effectively shape the business environment through market regulation, tax collection, infrastructure provision, property rights protection and other policy tools [16]. Based on the perspective of globalization, Zhang et al. extended the definition of business environment and pointed out that a systematic environment composed of business environment factors should include regional factors such as labor cost and transportation cost, factors such as exchange rate, tax and tariff, and supply chain factors such as products and trade [17]. From the perspective of social environment, Dong Biao and Li Renyu think business environment refers to the business subject to engage in commercial organizations or business conduct various circumstances and conditions, including affect business subject behavior of the political factor, economic factor, cultural factor, etc., is a country or a region effective exchanges and cooperation and competition, embodies the economy of the country or region soft power [13]. Wu Jingzhou pointed out that the business environment is essentially the regulatory environment of the government. The offside, absence and dislocation of the government will affect the business environment [18]. Wei Shuyan and Sun Feng believed that the investment and business environment is an organic dynamic system, including political system, economy, market, social culture, policies and regulations and other factors that interact and influence each other [19].

Lou Chengwu believed that the business environment is the development environment faced by market entities in a region, including the government environment, market environment, rule of law environment, infrastructure environment, factor environment, etc., and it is the external manifestation of various formal systems provided by the government and various existing informal systems [14]. Ma Xiangdong and Wang Yuesheng believed that business environment generally refers to the sum of various external environments such as political environment, economic environment, legal environment and international environment in the process of establishment, operation and cancellation of an enterprise. It is a systematic project involving many fields of economic and social reform and opening up [20]. Li Zhijun thinks doing business is the enterprise engaged in the business, such as innovation, financing and investment activities such as the external environment faced by a comprehensive ecological system, including enterprise resource conditions facing in the entire life cycle and environmental conditions, such as capital, talent, technology, policy, including the government service efficiency, public services, financial credit services, human resources, innovation environment and market environment six big [21]. Bo Wenguang et al. believed that the business environment is usually a comprehensive system that a series of enterprises must deal with in the establishment and operation process, including the institutional soft environment, market environment, business cost environment, infrastructure environment and ecological environment [22]. Liu Zhiyong thinks doing business is contains the influence enterprise activities such as social, economic, political and legal factors, such as many areas involved in social and economic reform and opening to the outside of a systemic project, reflects the various elements such as system, the rule of law, credit, but different from simple system environment, legal environment and credit environment of the complex [23]. Xu Mingqiang et al. believed that the business environment is an organic system containing many elements such as material, system, society and relationship, and is a systematic project involving economic and social reform and opening up to the outside world [24].

It can be seen that scholars have different definitions of the business environment, and each scholar defines the connotation of the business environment based on their own research perspectives. Although there is currently no unified definition of the concept of a business environment, it has become a consensus that the business environment is an important factor affecting the development of enterprises. With the deepening of scholars' understanding of the business environment, the connotation of the business environment is also more extensive, which is embodied in the transformation of the investment environment to the business environment, the integration of macro and micro perspectives, the combination of hard and soft environments, and finally the realization that the business environment is an organic ecosystem.

In the context of digital economy, digital trade promotes the transformation from consumption Internet to industrial Internet and finally realizes the intellectualization of manufacturing industry, which is the expansion and extension of traditional trade in the era of digital economy [25]. Therefore, digital trade Enders a richer connotation of business environment -- digital business environment. Different from the traditional meaning of Doing Business, the World Economic Forum and Harvard Business Review released a report on the Convenience of Doing Digital Business 2019, arguing that "Doing Digital Business" is the convenience enjoyed by index platforms in the links of entering, operating, prospering and exiting the market [26]. Alibaba Group released the "Doing Business in the Digital Economy" report in 2019, believing that the "doing business in the digital economy" should focus on creating a good environment for enterprises to enter the platform, which means enabling technology to promote the formation of a fair, transparent and healthy platform ecology [27]. The digital business environment should not only be a "digital+business environment" to upgrade the traditional market environment through the transformation of digital technologies, but also a "digital+business environment", a new business environment that is oriented to and meets the innovative development needs of market players in the digital economy. In terms of the evaluation of the digital business environment, the main evaluation dimension includes digital infrastructure and security systems, digital market order, digital Government Services, digital Innovation and talent Cultivation, Digital market size.

4 Doing Business Environment Evaluation Study

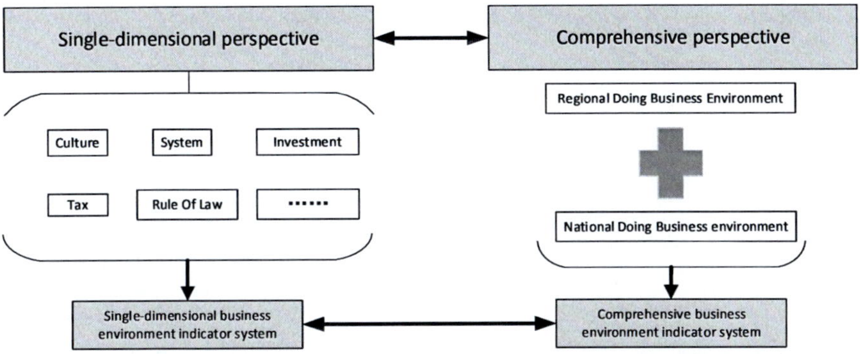

Fig. 5. Construction perspective of doing business environment evaluation index system.

In terms of business environment assessment, was established in 2001, the world bank Doing Business team, responsible for Business environment index system building, and in 2003 published its first "Doing Business report, using the index of 133 economies of 5 sets of small and medium-sized enterprise business environment to conduct a comprehensive assessment, after 10 years of development, has 2020 contains 11 primary indexs (starting a business, property of the construction permits, access to electricity, registration, access to credit, protection of minority shareholders, taxes, cross-border trade, contract enforcement and bankruptcy, hire employees) [6]. The 49 second-level indicators are basically studied by domestic and foreign scholars based on the logic of the World Bank's business environment assessment, which can be divided into two perspectives: the one-dimensional perspective and the comprehensive perspective. The research development process is shown in Fig. 5.

4.1 Single Dimensional Perspective Research

A one-dimensional perspective refers to the evaluation and research on a certain aspect of the business environment, such as taxation, rule of law, culture, etc. In terms of taxation,

Table 1. Evaluation index system of business environment from single-dimensional perspective.

Perspective	Author	Year	Indicators dimension
Tax business environment	Wang Shaole et al. [28]	2014	Rule of law index, efficiency index, cost index, satisfaction index
	Li Cheng et al. [29]	2020	Tax preference, time undertaking, national conditions difference, intermediary market, entrepreneur's own quality
Government and business environment	Peng Xianggang et al. [30]	2018	Demand identification, service function, service ability, service supply
Rule of law business environment	Zheng Fanghui et al. [31]	2019	Legal environment, law enforcement environment, judicial environment, law-abiding environment
	Xie Hongxing et al. [32]	2019	Environment for the formulation of laws and policies, administrative environment according to law, judicial environment, credit environment, social environment
Business environment for insurance industry	Sun Wujun et al. [33]	2020	Policy, rule of law, integrity, investment/operation, talent taxes and fees, cognition, public opinion
Logistics business environment	Huang Yuyi et al. [34]	2019	Logistics demand, development facilities, development environment, operation convenience, operating cost, government support

Wang Shaole established an index system from four dimensions, namely, tax rule of law index, efficiency index, cost index and satisfaction index, and conducted an empirical study with Guangdong Province as the object [28]. Li Cheng et al. evaluated the World Bank's tax and business environment indicator system, and believed that the World Bank did not pay attention to such factors as tax preferences, time cost, national conditions of different countries, and the quality of entrepreneurs [29]. Peng Xianggang et al. [30], from the perspective of government affairs, constructed four evaluation indexes of demand identification, service function, service ability and service supply, which are applicable to the universal evaluation framework. For the content and standard of the rule of law environment evaluation in, Zheng Fang Hui and justice in business law, business law, business, business to obey the law to [31] four dimensions, the system of evaluation index system, Xie Hongxing [32] put forward from the laws and regulations policy environment, administration according to law and judicial environment, credit environment, social environment, the five aspects of the business environment evaluation index under the rule of law.

From the perspective of the industry, other scholars have constructed indicators of the business environment of the industry. Sun Wujun et al. constructed the business environment evaluation system of China's insurance industry from eight perspectives, including policy, rule of law, integrity, investment, talents, taxes and fees, public opinion and cognition, and conducted empirical analysis with Jiangsu Province as the sample [33]. Huang Yuyi et al. proposed to build an evaluation index system of urban logistics business environment from six aspects: urban logistics demand, logistics development facilities, logistics development environment, business convenience, enterprise operating cost and government support [34]. Table 1 summarizes the specific contents of the evaluation index system.

4.2 Comprehensive Perspective Research

A comprehensive perspective refers to the establishment of a multi-dimensional business environment assessment index system based on the overall situation of the business environment. By combing relevant literature, this paper finds that scholars have constructed business environment evaluation index systems from regional and national perspectives.

At the regional level, Taking the Yangtze River Economic Belt as an example, Peng Diyun [35] constructed the regional business environment evaluation index system from four aspects of economic environment, market environment, infrastructure and supportive environment by using the entropy weight method, and found that the business environment was gradually decreasing from east to west. Gu Xueqin [36] selected four first-level indicators of government services, industrial transformation, regional functions and talent environment in the Yangtze River Delta region, and used entropy weighting method to determine index weights and conduct comprehensive scores. The conclusion was that Shanghai's business environment was the best. Zhang Sanbao et al. conducted a quantitative analysis of China's provincial business environment from the four dimensions of market, government affairs, legal policies and humanities, and the results showed that China's provincial business environment has hierarchical characteristics [37]. Yang Tao et al. compared the business environment of Jiangsu, Shandong, Guangdong and Anhui from the perspective of market environment, policy administrative environment and legal environment, and found that there was no significant difference among the four provinces in terms of first-level indicators [38]. Ding Ding et al. [39] constructed the urban business environment evaluation index system from six aspects, namely government environment, public environment, financial environment, human resources, innovation environment and market environment, and selected relevant cities for empirical research. The results show that there is a significant gap between cities in terms of business environment. Radukic and Stankovic [40] used cluster analysis and variance analysis to analyze the business environment of 17 BFC certified cities in Serbia, and put forward the key points of promoting FDI, including regional economic development strategy, efficient construction permit issuing system, financial stability and development of public-private partnership. Tang Leilei [41] designed questionnaires to obtain data from five aspects, including market environment, policy and government environment, socialized service environment, financing environment and legal environment, and analyzed the business environment of small and medium-sized enterprises in Dalian (Table 2).

At the national level, PavelKorner et al. [42] measured the business environment of the Czech Republic, Hungary, Poland and Slovakia from five aspects: corruption perception index, comprehensive governance index, capture index, opacity index and corporate governance risk index. Based on the government and legal environment, economic environment, social and cultural environment and technological environment, Zhang Dahai et al. [43] constructed the business environment evaluation index system, and found that China's business environment improved year by year on the whole. Lou Wenqian et al. [44] used the Doing Business database and indicators of the World Bank to evaluate the Business environment of 62 countries along the Belt and Road, and then graded and classified the Business environment of countries along the Belt and Road. Li Qingchi [45], based on the analysis of the evaluation index system of business environment in Russia, India, Singapore and Hong Kong, proposed that an index system that meets the needs of China's reform practice should be built to allow market players to participate in the evaluation. Hamlova et al. [46], based on the World Bank's 2013 business environment assessment data, constructed an evaluation index system from the aspects of legal regulatory intensity and the complexity and cost of the regulatory process, and analyzed the reasons for the Czech Republic's low ranking in business environment. The following table lists the specific indicator systems for doing business at different levels.

To sum up, business environment evaluation index system of research both unidimensional perspective, and a comprehensive perspective, the research focus is different, but also learn from each other and promote each other, the evaluation index system also contains mostly business environment connotation under the good environment and soft environment dimension, although there are now many business environment evaluation index, the index system is also increasing, but still want to notice the different dimensions of index selection repeatability and redundancy problem, at the same time should be paid attention to in the future will index system with the characteristics of digital economy dimension is analyzed.

5 Conclusion and Future Research

This paper summarizes relevant literature in the field of business environment suing bibliometrics, knowledge mapping, and summary. Study found that the current business environment evaluation study focused on the connotation of the business environment, business environment evaluation index system from two aspects:

i. The connotation of doing business with the passage of time gradually update, scholars' understanding also in deepening, the specific performance in the investment environment to the change of business environment, the integration of macro and

Table 2. Different levels of doing business environment assessment indicators.

Perspective	Author	Year	Indicators dimension
Regional level evaluation	Peng Diyun et al. [35]	2019	Economic environment, market environment, infrastructure, support environment
	Gu Xueqin et al. [36]	2020	Government services, industrial transformation, regional functions, talent environment
	Zhang Sanbao et al. [37]	2020	Market environment, government environment, legal and policy environment, cultural environment
	Yang Tao [38]	2015	Market development environment, policy government environment, science and technology innovation environment
	Ding Ding et al. [39]	2020	Government affairs, public environment, financial services, human resources, innovation environment, market environment
	Radukic S et al. [40]	2015	Economic development strategies, licensing systems, financial stability, development of public-private partnerships
	Tang Leilei [41]	2012	Market environment, policy government environment, socialized service environment, financing environment, legal environment
National level evaluation	Pavel Körner et al. [42]	2002	Corruption perception index, comprehensive governance index, capture index, opacity index, corporate governance risk index

(continued)

micro perspective, the combination of hardware and soft environment, finally realize the business environment is an organic ecosystem, the digital background, based on

Table 2. (*continued*)

Perspective	Author	Year	Indicators dimension
	Zhang Dahai et al. [43]	2019	Government and legal environment, economic environment, social and cultural environment and technological environment
	Lou Wenqian et al. [44]	2020	World Bank Indicator System
	Li Qingchi [45]	2018	Evaluation index research, the actual needs of enterprises, promote reform and innovation, carry out follow-up evaluation, into the performance assessment
	Hamlov et al. [46]	2014	The legal regulation intensity and the complexity of the regulation process

the connotation of the traditional business environment, digital economy will make it has more rich connotation.

ii. The study of business environment evaluation index system has both a one-dimensional perspective and a comprehensive perspective, and the choice of its index dimension is closely related to the connotation of business environment.

Through the systematic review of the field of business environment assessment, it can be found that, on the one hand, the research on business environment assessment has been very rich and diversified, with many studies on business environment assessment from different perspectives, levels and regions. However, on the other hand, the criteria of the doing Business assessment index system have not yet been determined, which is specifically reflected in the different index systems and the different weighting indexes of the same index. Even the most internationally recognized World Bank assessment report on doing Business has some applicability defects. In view of this, future research can be carried out in the following two aspects:

i. To explore accurate business environment evaluation methods. Due to the influence factors of doing business is more, the construction of evaluation index system of the lack of integrity, the existing evaluation index system of draw lessons from the bank's research, mostly USES the questionnaire survey and subjective and objective combination, are greatly influenced by subjective, lead to the lack of scientific nature and accuracy, so the future can be through the identification of internal and external factors affecting business environment, to reveal the mechanism of action of different factors on the business environment, and then build objective and effective evaluation index system, build business environment evaluation model of economics, to achieve accurate assessment of doing business.

ii. Develop an intelligent system for assessing business environment. In the information age, simulation algorithms can be developed to dynamically simulate the formation and evolution of the business environment, reveal the action mechanism of different factors on the business environment, and explore and develop integrated business environment evaluation and simulation system based on emerging technologies such as big data, artificial intelligence and cloud computing.

References

1. XinHua Net: keynote speech by Xi Jinping at the opening ceremony of the first China International Import Expo [EB/OL], 5 November 2018. http://www.xinhuanet.com/politics/leaders/2018-11/05/c_1123664692.htm
2. XinHua Net: (authorized release) decision of the CPC central committee on several major issues concerning upholding and improving the socialist system with Chinese characteristics and promoting the modernization of China's governance system and capacity [EB/OL], 5 November 2019. http://www.xinhuanet.com/politics/2019-11/05/c_1125195786.htm
3. Central People's Government of the People's Republic of China: opening up and cooperation - keynote speech at the opening ceremony of the second China International Import Expo [EB/OL], 5 November 2019. http://www.gov.cn/gongbao/content/2019/content_5453381.htm
4. XinHua Net: keynote speech by Xi Jinping at the opening ceremony of the third China International Import Expo (Full text) [EB/OL], 4 November 2020. http://www.xinhuanet.com/politics/leaders/2020-11/04/c_1126698327.htm#contentTitle
5. World Bank Group Doing Business 2020: The World Bank (2019).https://openknowledge.worldbank.org/bitstream/handle/10986/32436/9781464814402.pdf
6. Hongbing, D.: Investment Environmentology, p. 128. China University of Geosciences Press, Wuhan (2000)
7. Stern, N.: A Strategy for Development. World Bank Publications, Washington (2002)
8. Eifert, B., Gelb, A., Ramachandran, V.: Business Environment and Comparative Advantage in Africa: Evidence from the Investment Climate Data. ICA World Bank Working Paper (2005)
9. Bah, E.-H., Fang, L.: Impact of the business environment on output and productivity in Africa. J. Dev. Econ. **114**, 159–171 (2015). https://doi.org/10.1016/j.jdeveco.2015.01.001
10. Worthington, L., Britton, C.: The Business Environment. 4th edn. Economic Management Press, Beijing (2011). Lei Xu, Xiaoli Hong, Translate
11. Worthington, L., Britton, C.: The Business Environment. 7th revised edn. Pearson Education (2015)
12. Escaleras, M., Chiang, E.P.: Fiscal decentralization and institutional quality on the business environment. Econ. Lett. **159**(8), 161–163 (2017)
13. Biao, D., RenYu, L.: A study on the construction of a legalized international business environment in China-based on the analysis of the doing business report. Bus. Econ. Res. **13**, 141–143 (2016)
14. ChengWu, L., GuoYong, Z.: A framework of doing business assessment based on subjective perceptions of market players - a review of the World Bank's doing business assessment model. Contemp. Econ. Manage. **40**(06), 60–68 (2018)
15. PengFei, N.: What do Chinese cities do to attract investors. Manage. Consult. **1**, 102–105 (2008)

16. ZhiQiang, D., XiaHai, W., CanQing, T.: Institutional environment and economic development-an empirical study of the business environment in 30 major cities. Manage. World **4**, 9–20 (2012)
17. Zhang, A., Huang, G.Q., Liu, X.: Impacts of business environment changes on global manufacturing in the Chinese Greater Pearl River Delta: a supply chain perspective. Appl. Econ. **44**, 4505–4514 (2012)
18. Jingzhou, W.: The revitalization of northeast China should start from optimizing the business environment. Econ. Aspect **1**, 31–35 (2017)
19. Shuyan, W., Feng, S.: Assessment and optimization of investment and business environment in northeast China. Changbai J. **6**, 84–92 (2017)
20. Xiangdong, M., Yuesheng, W.: New strategies for attracting foreign investment in the new era: from preferential investment policies to business environment optimization. J. Party Sch. CPC Cent. Committee **22** (2018)
21. Zhijun, L., Shiguo, Z., Yifei, L.: Assessment and recommendations on the business environment of Chinese cities. Jiangsu Soc. Sci. **257**(2), 30–42 (2019)
22. Wenguang, B., Yanyu, Z., Dingkun, L.: Entrepreneurial talent, business environment and total factor productivity of enterprises - based on micro data analysis of listed companies in China. Bus. Econ. Manage. **8**, 85–97 (2019)
23. Zhiyong, L., Lili, W.: Summary of the research on the construction of business environment in China: development track, main achievements and future direction. Contemp. Econ. Manage. **2**, 22–27 (2020)
24. Mingqiag, X., Dahai, D.: Research on the components and influencing factors of the business environment of China free trade zone. Manage Case Stud. Rev. **13**(4), 460–475 (2020)
25. Zhixin, W.: Statistical measurement of the digital trade and business environment of One Belt and One Road countries. Stat. Decis. Making **36**(19), 47–51 (2020)
26. Global Competitiveness Report 2019: How to end a lost decade of productivity growth. http://www3.weforum.org/docs/WEF_TheGlobalCompetitivenessReport2019.pdf
27. Alibaba: Alibaba Digital Doing Business Report [EB/OL], 15 June 2019. https://max.book118.com/html/2019/0603/6042222141002034.shtm
28. Shaole, W., Zhonghu, L.: A study on the measurement of China's tax business environment. J. Guangdong Univ. Finance Econ. **29**(3), 33–39 (2014)
29. Cheng, L., Wenbo, S.: Study on the World Bank index system of tax paying business environment. J. Xiamen Univ. (Philos. Soc. Sci. Ed.) **5**, 118–130 (2020)
30. Xianggang, P., Ran, M.: Optimization of government and business environment and construction of evaluation index system. Acad. Res. **11**, 55–61 (2018)
31. Fanghui, Z., Zheng, W., Hongzheng, W.: Business rule of law environment index: evaluation system and guangdong empirical study. Guangdong Soc. Sci. **5**, (2019)
32. Hongxing, X.: China's thinking and system of business legal environment evaluation - based on the perspective of rule of law. Hubei Soc. Sci. **3**, (2019)
33. Wujun, S., Le., X.: Construction of business environment assessment system for China's insurance industry - based on application analysis of Jiangsu province. Insur. Res. **5**, 34–49 (2020)
34. Yuyi, H., Han, H.: Research on evaluation theory of urban logistics business environment. Bus. Econ. Res. **13**, 91–93 (2019)
35. Diyun, P., Bo, C., Zhijia, L.: Construction and application of regional business environment evaluation index system - a case study of Yangtze River Economic Belt. Finance Econ. **5**, 49–55 (2019)
36. Xueqin, G., Yudong, L., Hongxing., Y.: Policy measures and effect evaluation of business environment in The Yangtze River Delta region. China Circ. Econ. **34**(6), 86–95 (2020)
37. Sanbao, Z., Bicheng, K., Zhixue, Z.: Evaluation of business environment in Chinese provinces: index system and quantitative analysis. Econ. Manage. **42**(4), 5–19 (2020)

38. Tao, Y.: A study on the establishment of business environment evaluation index system based on the comparative analysis of Shandong, Zhejiang and Guangdong provinces. Bus. Econ. Res. **13**, 28–31 (2015)
39. Ding, D., Qiang, G., Xianxiang, L.: The construction process and evaluation of China's urban business environment-taking 36 provincial capitals, municipalities directly under the central government and cities separately listed in the plan as examples. Macroecon. Manage. **1**, 55–66 (2020)
40. Radukic, S., Stankovic, J.: Evaluation of local business environment in The Republic of Serbia. Procedia Econ. Finance **19**, 353–363 (2015)
41. Leilei, T.: Analysis on the Business Environment of Dalian SMEs. Dongbei University of Finance and Economics (2012)
42. Körner, P., Kudrna, Z., Vychodil, O.: Measuring business environment quality in Central Europe. Finance Uver **12**, 674–697 (2002)
43. Dahai, Z., Zhichuan, Z.: Factor analysis and evaluation of China's business environment under entropy method. Finance Acc. Mon. **18**, 124–130 (2019)
44. Wenqian, L., Lingling, P.: Comprehensive assessment of the business environment of One Belt and One Road countries along the belt and road. Stat. Decis. Making **36**(14), 152–156 (2020)
45. Qingchi, L.: Research on the construction and application of business environment evaluation index. Adm. Reform **9**, 76–81 (2018)
46. Hamlová, E., Provazníková, K.: Assessment of the business environment competitiveness in the Czech Republic and EU. Procedia Soc. Behav. Sci. **109**, 1225–1229 (2014)

Author Index

C. Yuan et al. (Eds.): *Proceedings of the 4th International Conference on Economic Management and Green Development*, AEPS, pp. 517–518, 2021.
https://doi.org/10.1007/978-981-16-5359-9